Ecological Studies, Vol. 160

Analysis and Synthesis

Ecological Studies

Volumes published since 1996 are listed at the end of this book.

Springer
New York
Berlin
Heidelberg
Hong Kong
London
Milan
Paris
Tokyo

Thomas T. Veblen William L. Baker
Gloria Montenegro Thomas W. Swetnam
Editors

Fire and Climatic Change in Temperate Ecosystems of the Western Americas

With 122 Illustrations

Springer

Thomas T. Veblen
Department of Geography
University of Colorado
Boulder, CO 80309-0260
USA
thomas.veblen@colorado.edu

William L. Baker
Department of Geography and Recreation
University of Wyoming
Laramie, WY 82071
USA
bakerwl@uwyo.edu

Gloria Montenegro
Departamento de Ciencias Vegetales
Facultad de Agronomía e
Ingeniería Forestal
Pontificia Universidad Católica de Chile
Casilla 306
Santiago, Chile
gmonten@puc.cl

Thomas W. Swetnam
Laboratory of Tree-Ring Research
University of Arizona
Tucson, AZ 85721
USA
tswetnam@ltrr.arizona.edu

Cover illustration: Photographs courtesy of Laboratory of Tree-Ring Research, University of Arizona, and Thomas T. Veblen.

Library of Congress Cataloging-in-Publication Data
Fire and climatic change in temperate ecosystems of the western Americas
p. cm.—(Ecological studies; v. 160)
Includes bibliographical references (p.).
ISBN 0-387-95455-4 (alk. paper)
1. Fire ecology—West (U.S.) 2. Climatic changes—West (U.S.) 3. Fire ecology—South America. 4. Climatic changes—South America. I. Veblen, Thomas T., 1947–
II. Series. QH104.5.W4 F57 2002
577.2—dc21 2002017655

ISSN 0070-8356
ISBN 0-387-95455-4 Printed on acid-free paper.

Printed in the United States of America.

9 8 7 6 5 4 3 2 1 SPIN 10868329

www.springer-ny.com

Springer-Verlag New York Berlin Heidelberg
A member of BertelsmannSpringer Science+Business Media GmbH

Preface

In the context of global change, there is an increasing urgency for a comprehensive understanding of how climatic variation influences fire regimes across a broad range of spatial and temporal scales. The chapters in this book examine how the spatial and temporal variation of fire occurrence varies in particular ecosystems and broad regions, particularly in relation to climate but also where appropriate in relation to land use. The book also considers the ecological consequences of these variations in fire regimes.

Geographically, we focus on the temperate ecosystems of western North and South America. These regions are broadly similar in climate and vegetation physiognomy but differ in the timing and intensity of human land use. They also strongly contrast in the phylogenetic origins of the biota, which creates the opportunity to test the generality of some climate and fire hypotheses for floras with quite distinct evolutionary histories. Broad similarities in present-day climate and vegetation of these two regions provide the potential for comparative studies of the effects of climate variation and human activities on fire regimes and of the responses of these ecosystems to altered fire regimes.

This volume had its beginnings at two workshops held in Silver Falls, Oregon, in 1996 and in Bariloche, Argentina, in 1997 that were sponsored by the Inter-American Institute and the National Science Foundation. Within the context of fire and global change research, the goals of these workshops were to (1) assess current knowledge of potential influences of global change on fire regimes, (2) define a research agenda on the potential effects of global change on fire regimes, (3) evaluate methodologies for analyzing the influences of climate and land-use changes on fire regimes, and (4) form a network of researchers and research institutions interested in developing an interdisciplinary research agenda that focuses on interhemispheric comparisons of fire regime and global change. The current volume summarizes much of the work achieved at those workshops as well as much research that was conducted subsequently.

Much of the discussion at the 1996 and 1997 workshops was centered on four broad questions: (1) What is the relationship of fire to climate variation across a range of biomes and at a range of temporal scales from seasonal to centennial? (2) How are climate-induced changes in fire regimes linked to broad-scale atmospheric circulation patterns and mechanisms? (3) How have fire regimes been altered by land-use practices by humans including both Native Americans and Euro-American practices? (4) What is the role of landscape heterogeneity in influencing how fire regimes respond to climate variation and human impacts? These four broad questions are strongly reflected in the different chapters of this book.

The book is divided into four sections: (1) methods and models, (2) North American case studies, (3) South American case studies, and (4) practical implications. The initial chapter by Whitlock and Anderson critically evaluates the theoretical and empirical basis for charcoal analysis as a methodology for reconstructing fire history from sedimentary records from lakes and wetlands. This first chapter also presents detailed Holocene fire histories for several study areas in Oregon and in the Sierra Nevada of California. This focus on sedimentary methods is complemented by the discussion of methods of extracting climatic signals from tree-ring-based fire histories in the chapter by Swetnam and Baisan. Several other chapters also apply tree-ring methods to reconstruct fire history.

Modeling perspectives on fire and climate are also considered in Section 1. Simulation approaches are often the only means available to study the interaction of wildland fire, vegetation, fuels, and climate in a spatial domain over long time periods. Keane and Finney use a conceptual simulation model called FESM (fire effects simulation model) as the context for a summary of the important ecosystem processes that need be explicitly simulated to adequately model fire interactions with ecosystems at a landscape scale. Miller uses the simulation model FACET (or FM), developed in the Sierra Nevada of California, to model complex influences of climate on fire and forest dynamics. Simulation results suggest that indirect effects of climatic change on the fire regime can be as significant as the direct effects of climatic change.

The chapters in Section 2 illustrate the richness of the literature and knowledge of fire regimes in western North America. The chapter by Flannigan, Stocks, and Weber on Canadian forests, in particular, western boreal forests, examines current knowledge of fire–climate interactions derived from existing fire–weather/climate analyses, fire history reconstructions, and paleo studies. It applies such knowledge with general circulation models to present possible scenarios of the impact of anticipated climate change on the fire regime and Canadian forests. Growing evidence supports a rapid increase in temperature and increased rates of burning, particularly at higher latitudes. In reviewing fire and climate in the forests of the U.S. Rocky Mountains, Baker stresses the need for greater understanding of how climate, fuels, the landscape, and land-use practices separately and jointly shape fire regimes, thus substantially complicating the task of identifying a climatic signal in historical fire data. For the Rocky Mountains, he contrasts a view that emphasizes how broad-scale patterns of climate and fuels control fire regimes, with a contingent view in which local spatial constraints and historical legacies may limit general trends. Models that represent the broad-scale view tend to stress a rapidly responding, climatically controlled fire regime affecting a passive and independent vegetation in a featureless landscape. In contrast, the contingent view suggests that fire regimes are inherently spatial, are constrained by the physical landscape, and are shaped by climate and vegetation as well as by historical legacies.

In their chapter on the Southwest and the Sierra Nevada, Swetnam and Baisan review time series of fire occurrence derived from extensive networks of tree-ring records. The synchrony of fire across large regions is an effective strategy of separating broad-scale climatic influences from local nonclimatic influences and contingencies of individual sites. An important finding is that annual resolution fire-scar networks can provide an independent indicator of changing temporal patterns of globally important climatic processes, such as the El Niño–Southern Oscillation. ENSO is also shown to be a major driver of fire by Heyerdahl and Alvarado in their tree-ring-based fire history in the pine-oak forests of the Sierra Madre Occidental in north-central Mexico. Changes in land use, rather than climate, however, probably caused the near cessation of fire recorded asynchronously at sites after 1900 to 1950. In their review of past, current, and future fires in California shrublands, Keeley and Fotheringham focus on the issue of human impacts on fire regimes and on vegetation patterns. They critically examine competing models of how fuel cycles and humans constrain fire occurrence in chaparral vegetation.

The chapters in Section 3 on South America illustrate the rapid increase in research on fire regimes in Chile and Argentina since about 1990. For northern Patagonia, Veblen et al. examine the roles of humans in altering fire regimes, and the interaction between landscape patterns and fire behavior. They stress the profound and long-lasting impacts on the landscape of short periods of exceptionally high rates of forest and shrubland burning associated with human activities and severe droughts. Land-use changes, such as grazing by livestock and twentieth-century fire exclusion, have had many of the same ecological effects as in xeric conifer woodlands of western North America. Also for northern Patagonia, Kitzberger and Veblen analyze changes in fire occurrence derived from both tree-ring and documentary records in relation to climatic variation. ENSO is a major driver of the year-to-year variation in fire regimes and also has a detectable influence at longer time scales. They stress the differential responses of fire regimes to interannual climatic variability along the steep vegetation gradient from Andean rain forests to the Patagonian steppe. For the rain forests of southern Chile, Lara et al. document the importance of past fires to the dynamics of these wet forests over periods of many centuries. In this region of intensive deforestation, intentionally set fires during the twentieth century have played a major role in shaping the landscape. Similarly, for relatively xeric forests of *Austrocedrus* in central Chile, Aravena et al. use tree-ring evidence to document the importance of fire, mainly of anthropogenic origin, in stand dynamics. Also for central Chile but at lower elevations, Montenegro et al. review the effects of humans on fire in the region of Mediterranean-type shrublands. They stress the effects of fire on community dynamics, taking into account the relative unimportance of natural fires in the history of this vegetation. For southern Patagonia, Huber and Markgraf use sedimentary records to reconstruct Holocene fire history in the ecotone between Patagonian steppe and *Nothofagus* forests. Peat macrofossil and macroscopic charcoal data suggest that on multimillennial time scales, increased aridity has favored fire occurrence in this region.

In the final chapter, Morgan, Defossé, and Rodríguez focus on the practical, management implications of the fire and climate change research that is reported in the preceding chapters. They describe the strong parallels, as well as important differences, in the vegetation, climate, and history of land use between the temperate zones of North and South America. They consider the varied goals, strategies, and contexts of fire management, and stress the complexity of interactions among fire, climate, and land use.

Thomas T. Veblen
William L. Baker
Gloria Montenegro
Thomas W. Swetnam

Acknowledgments

The editors are grateful to all the contributing authors for their sustained effort in assembling this book and for their patience in seeing to completion this lengthy project.

We wish to thank the many anonymous reviewers who generously helped assure the rigor and accuracy of the individual chapters. In general, each chapter was reviewed by at least two experts in the subject matter of the chapter. We are particularly appreciative of the dedicated editorial assistance provided by Rosanna Ginocchio of the Pontificia Universidad Católica de Chile.

We gratefully acknowledge funding from the Inter-American Institute and the National Science Foundation, which supported the initial workshops from which this volume originated.

Thomas T. Veblen
William L. Baker
Gloria Montenegro
Thomas W. Swetnam

Contents

Contributors

Ernesto Alvarado	Forestry Sciences Laboratory, University of Washington, Seattle, WA 98105, USA
R. Scott Anderson	Center for Environmental Sciences and Education, Northern Arizona University, Flagstaff, AZ 86011, USA
Juan Carlos Aravena	Departamento de Biología, Facultad de Ciencias, Universidad de Chile, Correo 653, Santiago, Chile. dendro@uchile.cl
Juan J. Armesto	Departamento de Biología, Facultad de Ciencias, Universidad de Chile, Correo 653, Santiago, Chile
Christopher H. Baisan	Laboratory of Tree-Ring Research, University of Arizona, Tucson, AZ 85721, USA
William L. Baker	Department of Geography and Recreation, University of Wyoming, Laramie, WY 82071, USA. bakerwl@uwyo.edu

Marco Cortés — Departamento de Ciencias Forestales, Universidad Catolica de Temuco, Casilla 151, Temuco, Chile

Guillermo E. Defossé — Consejo Nacional de Investigaciones Cientificas y Tecnicas, 9200 Esquel, Chubut, Argentina

Francisca Díaz — Departamento de Ciencias Vegetales, Facultad de Agronomía e Ingeniería Forestal, Pontificia Universidad Católica de Chile, Casilla 306, Campus San Joaquin, Santiago, Chile

Mark A. Finney — USDA Forest Service, Rocky Mountain Research Station, Fire Sciences Laboratory, Missoula, MT 59807, USA

Mike Flannigan — Canadian Forest Service, Edmonton T6H 3S5, Canada. MFlannig@NRCan.gc.ca

C.J. Fotheringham — Department of Organismic Biology, Ecology and Evolution, University of California, Los Angeles, CA 09995, USA

Shawn Fraver — Department of Forest Ecosystem Science, University of Maine, Orono, ME 04469-5755, USA

Rosanna Ginocchio — Departamento de Ecología, Facultad de Ciencias Biologicas, Pontificia Universidad Católica de Chile, Alameda 340, Santiago, Chile

Miguel Gómez — Departamento de Ciencias Vegetales, Facultad de Agronomía e Ingeniería Forestal, Pontificia Universidad Católica de Chile, Campus San Joaquin, Casilla 306, Santiago, Chile

Emily K. Heyerdahl — USDA Forest Service, Rocky Mountain Research Station, Fire Sciences Laboratory, Missoula, MT 59807, USA. eheyerdahl@fs.fed.us

Ulli M. Huber — Geobotanical Institute, Unversity of Bern, CH-3013 Bern, Switzerland. huberu@bluewin.ch

Héctor Jiménez — Departamento de Biología, Facultad de Ciencias, Universidad de Chile, Correo 653, Santiago, Chile

Robert E. Keane — USDA Forest Service, Rocky Mountain Research Station, Fire Sciences Laboratory, Missoula, MT 59807, USA. rkeane@fs.fed.us

Jon E. Keeley — Western Ecological Research Center, Sequoia National Parks, Three Rivers, CA 93271-9651, USA. jon_keeley@usgs.gov

Thomas Kitzberger — Laboratorio El Ecotono, Universidad Nacional del Comahue, E.P. Universidad, 8400 Bariloche, Argentina. tkitzber@crub.uncoma.edu.ar

Antonio Lara — Instituto de Silvicultura, Universidad de Austral, Casilla 567, Valdivia, Chile. alara@uach.cl

Carlos LeQuesne — Instituto de Silvicultura, Universidad de Austral, Casilla 567, Valdivia, Chile

Diane C. Lorenz — Geological Society of America, Boulder, CO 80301-9140, USA

Vera Markgraf — Institute of Arctic and Alpine Research, University of Colorado, Boulder, CO 80309-0450, USA

Carol Miller — USDA Forest Service, Rocky Mountain Research Station, Aldo Leopold Wilderness Research Institute, Missoula, MT 59807, USA. cmiller04@fs.fed.us

Gloria Montenegro — Departamento de Ciencias Vegetales, Facultad de Agronomía e Ingeniería Forestal, Pontificia Universidad Católica de Chile, Campus San Joaquin, Casilla 306, Santiago, Chile. gmonten@puc.cl

Penelope Morgan	College of Natural Resources, University of Idaho, Moscow, ID 83844-1133, USA. pmorgan@uidaho.edu
Estela Raffaele	Laboratorio El Ecotono, Universidad Nacional del Comahue, E.P. Universidad, 8400 Bariloche, Argentina
Norberto F. Rodríguez	Consejo Nacional de Investigaciones Cientificas y Tecnicas, 9200 Esquel, Chubut, Argentina
Fernando Silla	Departamento de Ecología, Universidad de Salamanca, Salamanca, Spain
Brian Stocks	Canadian Forest Service, Sault Ste. Marie, Ontario P6A 2E5, Canada
Thomas W. Swetnam	Laboratory of Tree-Ring Research, University of Arizona, Tucson, AZ 85721, USA. tswetnam@ltrr.arizona.edu
Thomas T. Veblen	Department of Geography, University of Colorado, Boulder, CO 80309-0260, USA. thomas.veblen@colorado.edu
Mike Weber	Canadian Forest Service, Edmonton T6H 3S5, Canada
Cathy Whitlock	Department of Geography, University of Oregon, Eugene, OR 97403, USA. whitlock@oregon.uoregon.edu
Alexia Wolodarsky-Franke	Instituto de Silvicultura, Universidad de Austral, Casilla 567, Valdivia, Chile

1. Methods and Models

1. Fire History Reconstructions Based on Sediment Records from Lakes and Wetlands

Cathy Whitlock and R. Scott Anderson

Fire-history reconstructions that extend beyond the age of living trees and subfossil wood are based on an analysis of particulate charcoal and other fire proxies preserved in the sediments of lakes and wetlands. The goal of such research is to document the long-term fire history with enough temporal and spatial resolution to complement and extend reconstructions provided by dendrochronological and historical records. Long-term records also provide an opportunity to examine how fire regimes were affected by periods of major climate change and vegetation reorganization in the past. Such insights are critical for understanding the legacy of past fires in present ecosystems, as well as the role of fire with projected climate changes as a result of increased greenhouse gases in the future (e.g., Overpeck, Rind, and Jones 1990; Price and Rind 1994; Bartlein, Whitlock, and Shafer 1997). In the last decade several advances have been made in the analysis of lake and wetland sediment records for fire history reconstructions. These advances reflect a growing interest within the paleoecological community to consider fire as an ecosystem process operating on long and short time scales, as well as an increasing need on the part of resource managers to understand prehistoric fire regimes. In this chapter we review the theoretical and empirical basis for charcoal analysis, including assumptions about the charcoal source area and the processes that transport and deposit charcoal into lakes and wetlands. We discuss issues of site selection, chronology, and data analysis. In an effort to standardize procedures and establish greater confidence in inter-site comparisons, we suggest a research protocol for long-term fire history studies in the western Americas based on our own work and the recommendations of a charcoal workshop held in Eugene, Oregon, in June 1996 that was sponsored by the Inter-American Institute and National Science Foundation. Finally, we present examples of three fire history reconstructions in the western United States using this protocol.

Fire reconstructions based on lake and wetland records are derived from (1) the analysis of particulate charcoal (both macroscopic and microscopic in size), which provides direct evidence of burning, (2) pollen evidence of fluctuations in vegetation that can be tied to disturbance, and (3) lithologic evidence of watershed adjustments caused by fire, such as erosion or the formation of fire-altered minerals. The first of these, charcoal analysis, is based on the accumulation of charcoal particles in sediments during and following a fire event. Stratigraphic levels with abundant charcoal (so-called charcoal peaks in the core) are inferred to result from past fire activity. The use of pollen analysis to detect periods of burning is based on the assumption that the pollen of disturbance-adapted species increases immediately following a fire, while that of fire-sensitive species decreases. For example, a grass-dominated assemblage in a period otherwise characterized by forest taxa might indicate a fire event. Lithologic analyses supplement charcoal data by detecting changes in the input of allochthonous sediment and alteration of soil minerals due to heating. The registration of fire-related lithologic changes varies among sites, but where present, the information helps constrain the fire location.

Our experience in conducting fire history studies comes from regions with natural lakes and wetlands. Lake sites are used for most stratigraphic fire history studies, and our understanding of charcoal deposition and burial (i.e., charcoal taphonomy) comes from such sites. Fire history studies from wetlands avoid some of the problems of sediment reworking found in lakes and offer a more local fire signal. Thus wetlands provide complementary information and an important alternative in regions where lakes are absent.

Charcoal Taphonomy

The rate at which charcoal accumulates in a lake or wetland depends on the characteristics of the fire (e.g., how much charcoal is produced) and the processes that transport and deliver charcoal to the lake (e.g., how far the charcoal is carried aloft; how much charcoal is introduced by streams and surface runoff in the years following a fire) (Fig. 1.1). Primary charcoal refers to the material introduced during or shortly after a fire event. Secondary charcoal is introduced to the sedimentary record during non-fire years, as a result of surface runoff and redeposition. Fire size, intensity, and severity all affect charcoal production and transport, and if these were the only processes at work, all sedimentary charcoal would be primary and thus a direct measure of biomass burning. However, studies have shown that the record reflects both primary and secondary sources, and estimating fire size, severity, or intensity has been possible only in the most general terms. In the forested regions of the western United States, for example, partic-

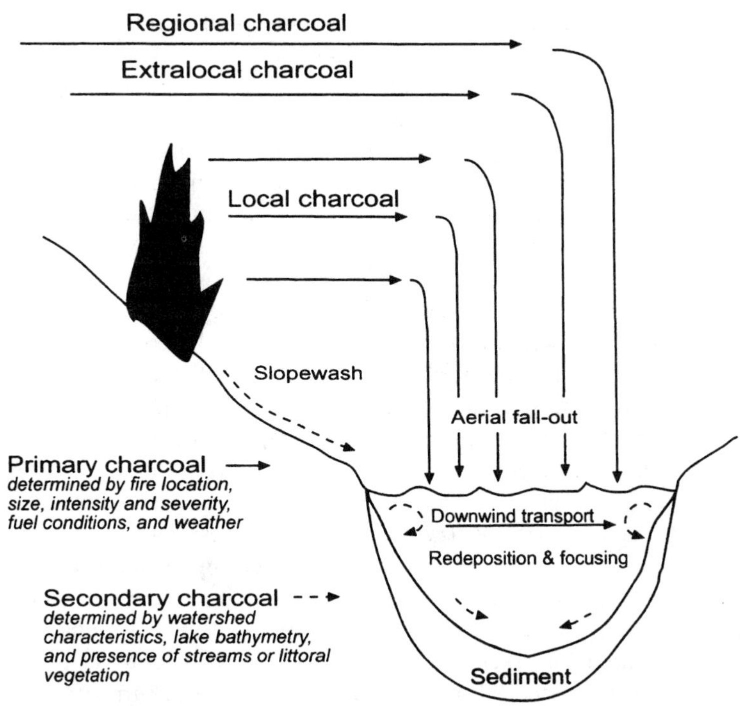

Figure 1.1. Schematic figure showing sources and pathways by which particulate charcoal is introduced into lake sediments.

ulate charcoal is composed of burned fragments of wood and needles (as opposed to grass cuticles), suggesting the charcoal was produced during high-severity or mixed-severity fires. Low-severity surface fires often do not produce much charcoal (Mohr, Whitlock, and Skinner 2000), with the possible exception of prairie fires (Umbanhower 1996; Pearl 1999).

Fire combustion products are carried aloft to great heights and transported long distances (Radtke et al. 1991; Andreae 1991), and the source of the charcoal may be from local watershed fires but also extralocal (i.e., nearby but outside the watershed) or regional (i.e., distant) fires. Charcoal in wetland sites may also record periods when the wetland itself was dry enough to burn (Huber and Markgraf, Chapter 13, this volume; Huber 2001). The distance that charcoal is carried during a fire has been discussed in several papers, including Swain (1978), Tolonen (1986), Patterson, Edwards, and MacGuire (1987), Clark (1988a), Clark and Royall (1995, 1996), Whitlock and Millspaugh (1996), Clark and Patterson (1997), Clark et al. (1998), Ohlson and Tryterud (2000), Whitlock and Millspaugh (1996), and Gardner and Whitlock (2001). Simple Gaussian plume models suggest that particles >1000 μm in diameter, if released relatively close to the ground, are deposited within <100 m of a fire (Clark and Patterson 1997). These models predict that particles <10 μm in size travel well beyond 100 m, and very small particles can be transported long distances.

Empirical studies are consistent with model projections by showing a decrease in charcoal particle size and abundance away from the source. A study of charcoal accumulation following the 1988 fires in Yellowstone National Park indicates that charcoal particles >125 μm diameter were abundant in sites <7 km from the fire (Whitlock and Millspaugh 1996); beyond that distance the accumulation of such particles declined sharply. In a study of 35 lakes following a 1996 fire in the Cascade Range, levels of >125 μm size charcoal were also highest in sites within the burned perimeter (Gardner and Whitlock 2001). Unburned sites located 100 ms beyond the burned area had significantly less charcoal, and nearby sites upwind of the fire had the lowest charcoal amounts of all. Clark et al. (1998) collected charcoal in a series of traps during a prescribed fire in Siberia in 1993. The distribution of particle sizes was the same for traps in the burned area as it was for those located 80 m beyond the burn. Again, charcoal abundance dropped off sharply at the edge of the fire. All of these results, as well as those of Clark and Hussey (1996) and Ohlson and Tryterud (2000), suggest that large charcoal particles provide a record of local fire activity. To reconstruct fire history at multiple spatial scales would require an analysis of several particle size ranges.

Studies of charcoal accumulation following modern fires also indicate that the deposition of charcoal in lakes can take place several years after the actual event. Whitlock and Millspaugh (1996) observed that lakes in both burned and unburned watersheds in Yellowstone received charcoal during the 1988 fires, but the amounts continued to increase significantly for five years in burned watersheds. Anderson et al. (1986) described accrual of charcoal into a lake in Maine for several decades following a 1910 fire. The secondary charcoal, in both cases, could have been introduced from wind erosion of standing burned snags, especially in winter, as well as from dead trees that eventually fell into the lake. Surface runoff may also have deposited charcoal in the lake during the first few years following a fire, but after that slopes become stabilized by vegetation. Another source, noted in the Yellowstone study, was the accumulation of particles that landed on the lake during the fire and were blown to the shore and deposited in the littoral zone. In the years after the fire, this material was refocused to deep water.

Bradbury (1996) documented the movement of the littoral charcoal in Elk Lake, a large lake in north-central Minnesota. By associating the charcoal peaks in the deep-water core with the seasonal flux of diatoms, he showed that shallow-water charcoal was mobilized in the lake during spring circulation. In both the Yellowstone and Elk Lake studies, the focusing of charcoal to deep water occurred within a few years of the fire event. Focusing of littoral charcoal is also blamed for variation in charcoal accumulation rates in sediment cores in different parts of a lake (Edwards and Whittington 2000). The cautionary note is that charcoal peaks are composed of particles deposited during and after the fire. For this reason it would be difficult to infer levels of charcoal production or biomass burning in the past based on charcoal abundance in lakes.

Larsen and MacDonald (1993) and Larsen et al. (1998) considered the characteristics of lakes best suited for paleoecological studies. Deep lakes were preferable because the sediments are less mixed by biological activity and less impacted by wind-driven currents. Lakes with steep-sided basins are less suitable because of the likelihood of subaqueous slumping. Studies in Yellowstone offer some support for these recommendations. Charcoal abundance in surface sediments was compared along a transect from shallow to deep water in eight lakes for several years after the 1988 Yellowstone fires (Whitlock and Millspaugh 1996). Charcoal accumulation was slowest and the year-to-year variation was less in the deep-water sediments of lakes with >10 m water depth. In contrast, shallow-water sites showed significant interannual variation in charcoal abundance.

To examine the patterns of charcoal accumulation in lakes in more detail, a transect of 42 short cores from shallow to deep water was collected from Duck Lake, Yellowstone National Park, in 1993 (Fig. 1.2). The small watershed was

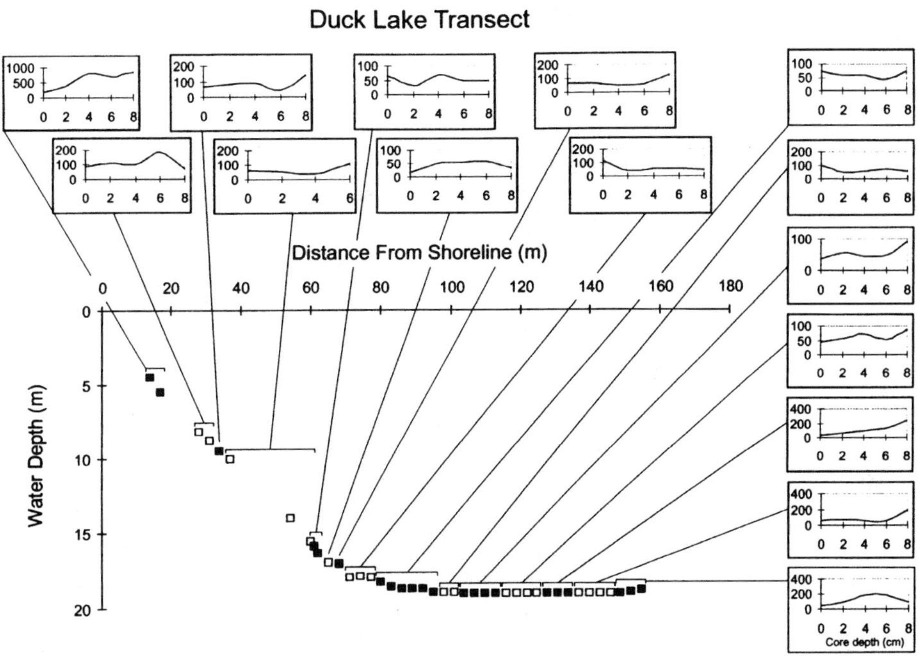

Figure 1.2. Charcoal abundance profile in a series of short cores from Duck Lake at Yellowstone National Park in 1993. Cores were collected from shallow to deep water as indicated by squares. The graphs show the charcoal abundance at 2-cm intervals to a core depth of 10 cm (each interval of the x-axis represents the top of a 2-cm sample, i.e., 0 = 0–2 cm, 2 = 2–4 cm). The y-axis shows number of charcoal particles >125 µm/gm dry weight. Adjacent cores with similar profiles are indicated by the series of black and white squares. The high abundance of charcoal in the uppermost samples is attributed to the 1988 fire. A high level of charcoal at depths >4 cm in some cores is attributed to a fire in 1889 or rapid deposition since the 1988 fires.

60% burned by the 1988 fires. In each core the charcoal accumulation was calculated for 2-cm-long intervals to a depth of 10 cm. The profiles indicate that charcoal from the 1988 fire was unevenly distributed across the lake. Shallow-water cores contained the most charcoal. The source is probably primary material that was blown to shore before sinking and secondary charcoal that was introduced by surface runoff and tree blowdown. Little charcoal was present in cores taken from the steepest slopes of the lake, perhaps because of slope instability. The amount of charcoal in the upper sediments of the deep-water cores was highly variable. Some cores contained a distinct charcoal peak, whereas others had very little charcoal. Two explanations may account for the pattern. First, charcoal might not have been deposited uniformly across the lake bottom during the 1988 fire, and postfire focusing of charcoal may have accentuated core-to-core variability. (Some cores also showed a peak in charcoal in the lower 4 cm that may represent a fire in 1889; however, no independent dating of the cores was undertaken.) Second, the charcoal variability might have been related to variations in sedimentation rates and bioturbation since 1988. Parts of the basin with higher sedimentation rates could have "buried" the charcoal peak. Again, without an independent chronology there is no way to choose between these explanations.

Both the Yellowstone and Elk Lake studies suggest that a charcoal peak represents accumulation occurring over a few years, and at any particular site, charcoal transport and deposition are affected by fire and fuel characteristics, weather conditions during and following the fire, surface runoff, and stream input. Although these processes lead to spatial variability in the abundance of charcoal across the lake bottom, charcoal samples from any single coring location yield similar results. In Yellowstone, for example, charcoal values of 30 surface cores taken from the same location fell within 10% of the mean charcoal value at that location. Thus analytical errors associated with field sampling and laboratory preparation are relatively small (Whitlock and Millspaugh 1996).

Fire-history information is also obtained from wetland deposits and soils, particularly in Europe (e.g., Iversen 1941; Tolonen 1985; O'Sullivan 1991; Odgaard 1992; Kuhry 1994; Carcaillet and Thinon 1996; Bradshaw, Tolonen, and Tolonen 1997; Pitkänen, Turunen, and Tolonen 1999; Innes and Simmons 2000). Wetland studies have also been undertaken in South America (e.g., Huber and Markgraf, Chapter 13, this volume; Heusser 1994; Markgraf and Anderson 1994; Huber 2001) and North America (e.g., Mehringer, Arno, and Petersen 1977; Terasmae and Weeks 1979; Wein et al. 1987; Anderson and Smith 1994, 1997; Brunner Jass 1999). Assumptions about charcoal accumulation in wetland sites are not well tested by models or empirical studies, but it seems clear that such sites avoid the problems of sediment focusing and mixing that complicate the interpretation of lake-sediment records. Close agreement has been found between the tree-ring record of known fires and the age of charred particles in bogs (Tolonen 1985; Bradshaw, Tolonen, and Tolonen 1997; Brunner Jass 1999).

In wetland sites, charcoal is introduced not only from upland fires, but also is produced in situ when the wetland surface burns (Huber and Markgraf, Chapter 13, this volume). Water levels likely determine the depth of *in situ* wetland burning, and so the thickness of the charred layer is an indication of effective moisture at the time of the fire. Wetland surfaces are uneven, and the lateral extent and thickness of a charcoal layer depend on spatial variations in flammability. Huber and Markgraf (Chapter 13, this volume) combined a fire history based on charcoal data with a drought record based on wetland-plant macrofossils to examine climate variability at the forest-steppe ecotone in southern Patagonia. They noted that charcoal layers were associated with sedge remains, indicating bog fires during dry periods, whereas little charcoal was found in sediments with abundant moss fragments, indicating wetter conditions. In Denmark, Odgaard (1992) combined charcoal and pollen analysis to reconstruct a local fire history of heathland fires. Charcoal peaks were associated with periods of *Calluna* pollen, implying an expansion of the bog as a result of anthropogenic burning of the watershed and forest clearance.

Methodological Issues

Site Selection and Field Methods

> There is no point in carrying out historical studies of fire from lake sediments if the sediment quality and coring sites do not fulfill the criteria for finely resolved pollen analysis.
>
> —Tolonen (1986)

In selecting a site for charcoal studies, several issues need to be addressed: What type of fires (surface, crown, or a combination) characterizes the present fire regime? How does topography influence fire patterns and the introduction of charcoal to the lake? How do lake or wetland characteristics influence charcoal accumulation and deposition? What is the desired spatial and temporal resolution of the fire history reconstruction—local or regional and annual, decadal, centennial, or millennial? The answers to these questions affect the choice of a site and the methods used.

Assuming that charcoal transport and deposition are not unlike that of pollen, regional records of fire can be obtained by looking at charcoal records from a large lake (sensu Jacobson and Bradshaw 1981) or by looking at small charcoal particles that might be transported long distances (Patterson, Edwards, and MacGuire 1987). In either case, the fire history integrates information from a large area. In general, small lakes (<10 ha) are selected when a local fire history is of interest. Whitlock and Millspaugh (1996) suggest that deep lakes (>10 m water depth) in steep catchments provide better charcoal records than lakes in low-gradient watersheds, since such sites increase the input of fire-related material (Meyer, Wells, and Tull 1995) and sediment focusing. Sites with a fringing margin of littoral vegetation may be less desirable because aquatic vegetation can entrap charcoal and mitigate charcoal reaching deep water. On the other hand, littoral vegetation may filter out local inputs, making such sites suitable for studies of regional fire history (Terasmae and Weeks 1979). Sites with significant stream activity are avoided because of the likelihood that secondary charcoal will be introduced from distal parts of the watershed long after the fire event. Lakes with large watersheds (e.g., >10× the size of the lake) are sometimes chosen on the assumption that they amplify the limnological signal of watershed disturbance through the greater input of allochthonous material (Rhodes and Davis 1995; Birks 1997). On the other hand, inputs from a large watershed limit the spatial specificity of the local fire reconstruction.

Local fire history information can also be obtained from charcoal preserved in wetlands. The best sites are small; have forest margins, rapid sedimentation rates, and little through-flow; and remain moist throughout the year. Such sites have the potential to incorporate charcoal particles from upland fires into the sediments as discrete layers. Anderson and Smith (1997) have shown that multiple cores from a single site are needed to capture all fire events because burned layers in wetlands are discontinuous. Suitable wetland areas with thick sediment accumulations are common in the narrow glaciated valleys of the western Cordillera (see photo in Anderson and Smith 1997). Reconstructions of *in situ* fire events that burn the wetlands themselves target sites that dry seasonally and thus have a greater potential to burn during the fire season (Huber and Markgraf, Chapter 13, this volume).

Site selection of lakes and wetlands should also accord with the availability of independent information on fire history against which to calibrate the charcoal data. This information includes documentary records of historic fires and dendrochronological data within and near the watershed. Analysis of the uppermost sediments of a core should reveal charcoal peaks that match known fire events, especially fires that were severe or near the lake or wetland margin. Sites with sedimentary records that do not register known fires, for whatever reason, will probably not provide a reliable record of older events, and it is best to find another, more sensitive site.

Magnetic measurements of lake sediments can complement the information obtained from charcoal analysis (Rummery et al. 1979; Thompson and Oldfield 1986; Gedye et al. 2000). The usefulness of such data depends on fire location, fire type and intensity, and soils and substrate type. Millspaugh and Whitlock (1995) examined magnetic susceptibility to detect periods of fire-related erosion or the formation of paramagnetic minerals due to soil heating. Lakes that recorded the highest sediment magnetism were located in steep-sided watersheds, where the potential for postfire erosion was greatest. Low-gradient watersheds, in comparison, showed no signal. Gedye et al. (2000) correlated the magnetic stratigraphy with pollen and charcoal evidence of fire in a Swiss lake. Long et al. (1998) found that magnetic susceptibility increased dramatically in the late Holocene, but that peaks of magnetic susceptibility did not match the charcoal peaks in most cases. Fire-induced erosion has also been inferred from increases in the content of aluminum, vanadium, and inorganic sediments immediately overlying charcoal peaks (Cwynar 1978) and from an increase in varve thickness (Larsen and MacDonald 1998a).

Most researchers collect cores for charcoal analysis from the deepest water or the center of the lake basin, or from the thickest section or center of the wetland, as is standard practice for pollen analysis. Whitlock and Millspaugh (1996) provide justification for this decision based on their studies of charcoal abundance in shallow- and deep-water sediments in Yellowstone (described above). They also found that more charcoal was deposited on the downwind shore of a lake than on the upwind shore. Thus it is likely that shallow-water areas under- or overrepresent charcoal compared to the center of the basin.

In most studies, a "long" core is obtained with a piston corer, vibracorer or percussion corer, and the cores are transported to the lab for further analysis. In addition a "short" core or a frozen core of the uppermost meter is collected for modern calibration purposes, including determining the size fraction most useful for identifying local fires in the long core. The short core is extruded in the field in 1-cm intervals and stored in plastic bags; frozen cores are sampled in the laboratory, also at a fine interval (Clark 1988b).

Fire History Reconstructions Based on Charcoal Accumulation Rates

Laboratory Methods

One issue in fire history studies has been the lack of a standardized methodology (see also Whitlock and Larsen, in press). Several methods have been proposed for generating charcoal time series and quantifying the results (Table 1.1). Methods concerned with general fire activity have been focused on pollen slide or microscopic charcoal with size fractions generally $<150\,\mu m$. In this approach the number or area of charcoal particles is determined along a series of traverses, and the data are expressed as a percentage of the pollen sum, as a ratio of the pollen count, or as charcoal accumulation rates (e.g., Swain 1973, 1978; Cwynar 1987; Smith and Anderson 1992; Bradshaw, Tolonen, and Tolonen 1997). The advantage is that microscopic charcoal is tallied on pollen slides and no further preparation is required. A concern, however, is that charcoal particles are broken during pollen preparation, thus creating artificially high abundances of microscopic charcoal. R. L. Clark (1982) modified the pollen slide method by determining charcoal area with a point count method. This faster method may underestimate charcoal when values are low (Patterson, Edwards, and MacGuire 1987). A technique that uses chemical digestion and loss on ignition has also been used to calculate charcoal abundance by weight (Winkler 1985), but some analysts have found the results unreliable (MacDonald et al. 1991). Laird and Campbell (2000) modified the Winker approach by using a total carbon analyzer rather than loss on ignition, and the results correlated fairly well with fires in the upper watershed but not those located at the lake margin.

The sampling interval in microscopic charcoal studies generally matches that of pollen analysis. Samples are taken centimeters apart in a core, which can represent a spacing of several centuries. Microscopic particles are generally considered to be evidence of regional or extralocal fires, but the exact source area is usually vague—somewhere in the area, but not necessarily within the watershed.

Table 1.1. Comparison of methods of charcoal analysis

Method	Procedure (P) and quantification (Q)	Objective	Advantages (Adv) and disadvantages (Dis)	References
Macroscopic sieving	P—Contiguous 1-cm core intervals are gently washed through analytical sieves (mesh sizes >0.125 mm). Sieved samples put in gridded petri dish (see Box). Q—Macroscopic charcoal (>125 μ) counted under stereomicroscope. Recorded as charcoal per volume.	To reconstruct history of local and extralocal fires on decadal to millennial time scales.	Adv—Easy, can be used for nonlaminated lake sediments, preserves macrofossils for AMS-dating. Dis—Time-consuming	Millspaugh and Whitlock 1995; Long et al. 1998.
Thin-section	P—Varved or nonlaminated sediments dehydrated with acetone, impregnated with epoxy, cured, then thin-sectioned. Q—A grid moved on traverses across each varve. Number and area of macroscopic charcoal (>50 μ) are recorded.	To reconstruct history of local and extralocal fires on annual to millennial time scales.	Adv—Provides record with annual or subdecadal resolution. Dis—Expensive, varved sediments are rare.	Clark 1988b; Anderson and Smith 1997.
Chemical Extraction	P—Digest sediment in nitric acid, then weigh sample. Ignite sample at 500°C, then weigh again or use total carbon analyzer to calculate carbon content. Q—To calculate % charcoal: subtract weight after nitric digestion from weight after ignition, multiply by 100, then divide by weight of sample or total carbon	To determine the importance of fire on millennial time scales.	Adv—Don't have to worry about visual misidentification of charcoal. Dis—Poor temporal resolution; record may be influenced by watershed processes.	Winkler 1985; Laird and Campbell 2000.
Image analysis	P—Uses a video camera, mounted on a microscope, to scan preparation for charcoal particles. Q—Scanner recognizes charcoal based on optical density. Number, area, and size-class distributions of charcoal recorded. Verification of each particle is required.	To quantify charcoal area for different size ranges.	Adv—Use of scanner is less time-consuming than visual counting. Dis—Scanner misidentifies other types of dark particles. Scanner doesn't focus on all particles.	MacDonald et al. 1991; Horn, Horn, and Byrne 1992; Earle, Brubaker, and Anderson 1996.
Pollen slide	P—Standard pollen preparation methods. Q—A grid (in microscope eyepiece) is moved on traverses across pollen slide. Number and area of charcoal particles recorded. Expressed as % of pollen sum or ratio of total pollen count. Q—A grid is moved step by step across a pollen slide and only charcoal particles that intersect a grid line are counted. Area of charcoal particles is estimated.	To determine the importance of fire in a region on centennial or millennial time scales.	Adv—Charcoal is counted on pollen slides without additional preparation. Dis—Spatial and temporal resolution of charcoal record is poor; difficult to identify breakage; influx problems with exotic.	Swain 1973; Cwynar 1978; Clark 1982; Patterson, Edwards, and MacGuire 1987.

Fire frequency per se cannot be calculated, because the source area is diffuse and the records are discontinuous. Nonetheless, the data are useful in that they disclose broad periods of burning in the past, and often the paleoclimatic inferences are consistent with those based on the pollen record, probably because the source areas of pollen and microscopic charcoal are similar in size. A common conclusion from studies that look at pollen and microscopic charcoal, for example, is that lots of fires occurred during periods when disturbance-adapted species were more prevalent; thus both charcoal and pollen suggest a drier climate and/or more climate variability.

Recent efforts have focused on extracting the local fire signal from charcoal data by examining macroscopic charcoal particles, generally defined as particles >60 to 100 μm in diameter (Clark 1988b; Millspaugh and Whitlock 1995). The most convincing demonstration that large particles indeed provide a record of local fires comes from comparing the charcoal from varved (annually laminated) lake sediments with known watershed fires (e.g., Clark 1990). In such sites, charcoal peaks can be dated to a particular year, and the accumulation of charcoal particles or charcoal area can be calculated for a particular fire.

Macroscopic charcoal is quantified from petrographic thin-sections or in sieved sediment fractions. Both methods of analysis yield comparable fire reconstructions, as long as contiguous samples are examined and the records are calibrated against known fires in an explicit way. Thin-section analysis is desirable for varved-sediment records, because it is possible to tally charcoal particles on an annual time scale. Anderson and Smith (1997) also used the thin-section method on wet meadow sites in the Sierra Nevada, California, which enabled them to tally charcoal particles at 1-mm intervals.

The sieving approach has yielded promising results in cases where contiguous, usually 1-cm-thick, core segments have been analyzed (see Box 1.1). Charcoal peaks in nonlaminated sediment records, dated by ^{210}Pb age determinations, have been shown to match fairly closely with the timing of known fire events within the watershed (Millspaugh and Whitlock 1995; Long et al. 1979; Mohr, Whitlock, and Skinner 2000). Methods of enumeration include simple counts of particles of different size (Millspaugh and Whitlock 1995; Mehringer, Arno, and Petersen 1977) and area measures (Horn, Horn, and Byrne 1992; Earle, Brubaker, and Anderson 1996; Clark 1990) (Table 1.1). Hallett and Walker (2000) compared macroscopic charcoal counts and charcoal area measurements in the same core and concluded that the approaches produced similar results.

In most lakes of the western United States, a single centimeter represents about 5 to 20 years, depending on the sedimentation rate. Where fires are infrequent, this sampling interval is short enough to discriminate particular fire events, but in regions of frequent burning, a single sample may represent one or more fires occurring several years apart. For that reason, the term "fire event" or "fire episode", rather than "fire," is more appropriate for the information provided by most charcoal studies. In our experience, subsampling lake-sediment cores at intervals of <1 cm (e.g., at 0.5 cm intervals) did not improve the temporal resolution because bioturbation blurred the charcoal signal at a finer scale. However,

Box 1.1. Macroscopic Sieving Method at the University of Oregon

This method provides a simple means of quantifying macroscopic charcoal in nonlaminated lake sediments to provide a record of past local fire events.

Equipment and Materials

Sodium hexametaphosphate
100-ml beakers
Diffused spray nozzle attachment for faucet (i.e., shower head type)
Metal sedimentology sieves (sizes 0.063, 0.125, and 0.250 mm)
Large wash bottle
Plastic petri dishes with grids etched into them
Stereomicroscope

Procedure

Sediment Sampling

Slice the core lengthwise and describe core lithology before subsampling. Take a known volume of sediment from contiguous intervals. The sample interval width should be chosen based on information on sediment accumulation rate of your core. We generally take samples at 1-cm intervals.

If analyzing for both charcoal and magnetic susceptibilty (using a cup sampling device), subsample ca. 8 or 10 cm^3 from each interval. Run each subsample through a magnetic susceptibility meter, and then subsample the sediment for charcoal analysis. Generally, we take 2.0 or 5.0 cm^3 of sediment for charcoal analysis based on the concentration of charcoal in the sediment. Soak each charcoal subsample in 60 ml of a solution of ca. 10% sodium hexametaphosphate and water (in a small beaker) for two to five days to deflocculate the sediment.

Sieving Subsamples

Gently wash each subsample through a set of nested sedimentology sieves. The number and sizes of sieves depends on the research design. We suggest using sieves with mesh sizes of 63, 125, and 250 μm to test whether the different size ranges show similar trends. Often the smallest size fraction (63–125 μm) is more abundant, but also more cumbersome and time-consuming to count. We have found that charcoal particles 125–250 μm in length were present in all samples, but not so abundant as to make analysis impractical. Particles >250 μm were present in low numbers in most samples. After this initial test, we chose to sieve for particles in the 125–250 μm, and >250 μm size ranges.

Using a spray nozzle attached to a faucet, gently spray the surface of the top sieve for 1.5 to 2 minutes so that the entire subsample is washed through the sieves. Separate the sieves, and then gently wash the sediment to one side of each sieve. Turn the sieve so that its surface is perpendicular to the counter top and the sediment is at the bottom (closest to the counter). Using a large wash bottle, direct a stream of water at the charcoal and remaining particles and wash them into a gridded plastic petri dish. It is best to use as little water as possible so that the charcoal and other particles do not float around as you try to count them.

Counting Charcoal Particles

Under a stereomicroscope at 50–100× magnification, count all charcoal particles. The gridded rows helps you keep track of your counting. Collect large pieces of charcoal (>500 μm) while you are counting for AMS radiocarbon dating. Save samples in plastic bags in case the charcoal needs to be recounted at a later date.

Data Analysis

This procedure gives number of charcoal particles (in a particular size range) for a volume of sediment. To calculate the charcoal concentration for each sample, divide the number of charcoal particles by the volume to get charcoal particles in cm^{-3}. Enter the charcoal concentration data and age-depth data (derived from radiocarbon dates) into a computer program such as TILIA (Grimm, ND). Calculate an age-depth curve, charcoal accumulation rates (pieces $cm^{-2} yr^{-1}$), and sediment-deposition time for each sample. Transfer information to CHAPS for decomposition approach (available from Department of Geography, University of Oregon).

Anderson and Smith (1997) used finer sampling in wetland sites where bioturbation is less of a problem. In the sieving method, the core is sampled at continuous 1-cm intervals and every sample is analyzed.

Sample volume is measured carefully, and it can be adjusted depending on the abundance of particulate charcoal. Between 2 and 5 cm^3 per sample is used in lake-sediment studies, as little as 0.5 to 1.0 cm^3 of sediment is used in wetland and lakes with abundant charcoal. Each sample is soaked in a deflocculant for a few days and then gently washed through a series of nested sieves (with mesh sizes of 250, 125, and 63 μm). Initially the amount of charcoal in the different size fractions is tallied or measured for several samples to ensure that the three fractions show similar trends. In the western United States, we have found that the smallest, 63–125 μm size fraction contains abundant charcoal in nearly every sample and is tedious and difficult to count accurately. The >250 μm fraction is not present in many samples, suggesting that the largest particle sizes may not be deposited evenly across the lake. Most of our studies use the 125–250 μm fraction or the >125 μm fraction as the most practical size range for analysis. In this range, a fire event is typically represented by >50 particles cm^{-3} and a nonfire event by substantially fewer particles. The resulting data set is converted to charcoal concentrations (number of charcoal particles cm^{-3}) and then to charcoal accumulation rates (CHAR = number of charcoal particles $cm^{-2} yr^{-1}$) by dividing by the deposition time ($yr\,cm^{-1}$).

Chronological Issues

Adequate chronological control is necessary for any high-resolution time series, and sediments that have annual laminations (varves) offer an opportunity for fire history reconstructions on annual time scales. In nonlaminated sediments, the chronology for the fire reconstruction is based on ^{210}Pb dating of sediments that span the last 200 years and AMS ^{14}C dating of charcoal and terrestrial macrofossils from the remainder of the core. Radiocarbon years are converted to calendar years using the calibration program of Stuiver et al. (1998) in order to calculate charcoal accumulation rates in calendar years. In developing an age model, it is desirable to use as smooth a regression curve as possible to calculate the deposition time of particular lithologic units. Sharp discontinuities in deposition time that are artificially imposed by using linear interpolation between dates will influence the charcoal accumulation rates.

Variations in sedimentation rate often make it difficult to sample a core at equally spaced time intervals. This is especially true for wet meadow records (Anderson and Smith 1997). For practical purposes and to facilitate comparison with other records, we convert our observations to regularly spaced time intervals. Because direct interpolation of CHAR to a constant time interval may not conserve the quantity of charcoal within the intervals, concentration values are first interpolated to pseudo-annual intervals, and those values are integrated over decadal or longer time intervals. The unit of aggregation is generally equal to the shortest deposition time; for example, Mohr, Whitlock, and Skinner (2000) aggregated samples at 12-year intervals and Long et al. (1998) and Millspaugh, Whitlock, and Bartlein (2000) used an aggregation of 10 years. This approach preserves the features of the raw charcoal accumulation rates but allows the data to be analyzed at evenly spaced time intervals (Fig. 1.3).

Decomposition Approach for Analyzing Charcoal Accumulation Rates. The purpose of the data-analytical phase is to separate the charcoal component related to the fire event from that related to variations in fuel biomass and depositional processes. Clark and Royall (1996) and Long et al. (1998) suggest that this separation can be accomplished statistically by decomposing the charcoal time series into separate series. Time series of the charcoal accumulation rate (CHAR) display a low-frequency or slowly varying component, called the *background*

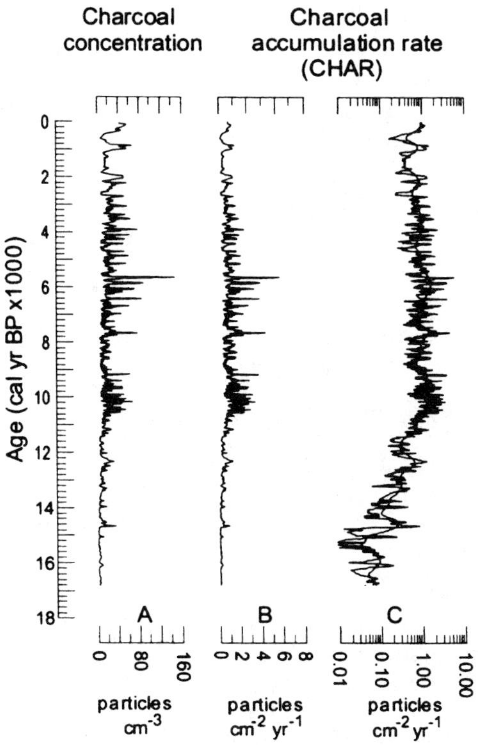

Figure 1.3. Charcoal data from Cygnet Lake at Yellowstone National Park showing the transformation of the data from charcoal concentrations (A) to charcoal accumulation rates (CHAR) at evenly spaced time intervals. CHAR are plotted on both normal (B) and logarithmic (C) scales (after Millspaugh, Whitlock, and Bartlein 2000).

component, and a higher-frequency or rapidly varying component, called the *peaks component*. This type of decomposition approach also assumes that the relationship between these two components stays constant throughout the record. The background component or general trend in the data arises from any of several sources, which are poorly understood and difficult to separate. A general time-varying level of background CHAR may be the result of changes in fuel accumulation and its influence on charcoal production. For example, Millspaugh, Whitlock, and Bartlein (2000) argue that the increase in background CHAR in Yellowstone lakes about 11,000 years ago occurred as a result of changes in fuel during the transition from open meadow to forest vegetation. Background CHAR has also been attributed to secondary charcoal, namely material stored in the watershed and littoral zone that is delivered to the lake over a long period of time. In this case, the charcoal is not directly related to a fire event. An increase in charcoal in late-Holocene sediments at Little Lake in the Coast Range was attributed to more mass movements occurring with the onset of a wetter climate (Long et al. 1998). This hypothesis was supported by the high magnetic susceptibility of late-Holocene sediments. A third contributor of background charcoal may be from extralocal or regional fires. This possibility, proposed by Clark and Royall (1996), needs further testing by comparing the background charcoal stratigraphy with that of a microscopic charcoal record. Of the three sources of background charcoal, both variations in charcoal production and secondary charcoal delivery are affected by changes in vegetation, climate, and fire weather, as well as by changes in hydrology, fluvial geomorphology, and lake conditions. The regional fire component also may have varied as the vegetation and climate changed.

The peaks component is evident when the charcoal record is compared with historical and dendrochronological records of fires (Clark 1990; Millspaugh and Whitlock 1995). The peak represents the contribution of charcoal from a fire event. As discussed above, this component has its source area within the watershed and sometimes from adjacent upwind basins. In addition to a particular fire event, it also represents "noise" from analytical error (Whitlock and Millspaugh 1996) and natural random variations in CHAR. In practice, the largest variations in the peaks component are attributed to fire events, and the minor "noise" component is disregarded.

Peaks of significance are identified by assigning a threshold value, such that CHAR higher than that value is assumed to represent a fire event. Depending on the deposition time, an event may comprise one or more fires occurring during the time span represented by the peak. In sites with fast deposition times, a peak is generally less than 20 years (1 or 2 cm thick) (Long et al. 1998; Millspaugh, Whitlock, and Bartlein 2000), whereas in sites with slow sedimentation, a comparable size peak may span several decades (Anderson and Smith 1997; Mohr, Whitlock, and Skinner 2000; Hallett and Walker 2000). To detect individual fires, it is necessary to have a sedimentary record that can be sampled at a shorter interval than the time between fires (Whitlock and Larsen, in press).

The decomposition approach has also been applied to magnetic susceptibility data. Background levels of magnetic minerals provide information on pedologic and geomorphic processes that operate within the basin over the long term. Peaks in magnetic susceptibility measurements indicate individual geomorphic events, such as landslides, similar to the CHAR peaks. In Yellowstone, peaks in magnetic susceptibility corresponded well with charcoal peaks, suggesting that they were fire-related erosion events (Millspaugh and Whitlock 1995). In other studies in Yellowstone, the Coast Range, the Sierra Nevada, and the Klamath Mountains, no direct relation between CHAR peaks and magnetic susceptibility peaks was noted, even when the possibility of a time lag was considered (Millspaugh 1997; Long et al. 1998; Brunelle 1997; Mohr, Whitlock, and Skinner 2000).

Charcoal data, like other paleoenvironmental records in lake sediments, are approximately lognormally distributed, in that most of the charcoal is deposited close to the site and the abundance declines exponentially away from the source area (Clark 1988a; Clark et al. 1998). Consequently, CHAR and magnetic susceptibility data are usually log transformed before analysis (Fig. 1.3). A locally weighted (moving) average is used to define the background component. It is calculated by moving a "window" along the CHAR series, and at each point determining a weighted average of CHAR values for the points contained in the window. The weight assigned to each point is based on the distance of the point from the center of the window so that points near the edge of the window have less influence than those near the center. This method of locally weighted averaging is related to the "lowess" approach for smoothing scatter diagrams (Cleveland 1979), and weights are determined using a tri-cube or approximately bell-shaped function. The width of the window affects the smoothness of the background component. If too wide, a window does not capture long-term variations in the data; if too narrow, a window produces a background trend that mimics the high-frequency or peaks component. In sites with fast sedimentation rates relative to the fire frequency, window widths of 500 to 1000 years have been used to convey the general trends in the data (e.g., Long et al. 1998). However, in sites with very slow sedimentation rates, a shorter window width is preferred because each interval of high CHAR spans several decades and is considered significant (Mohr, Whitlock, and Skinner 2000). In these cases, a broader backgrounds width would tend to smooth the data and not identify potentially significant peaks.

The CHAR threshold value is set or calibrated based on the timing of known fires evident in dendrochronological or historical records. The calibration determines what specific values of the peaks components correspond with a fire event. The threshold value is defined in terms of a threshold ratio, that is, a ratio of CHAR at a particular time relative to background. For example, a ratio of 1.00 would identify all peaks greater than background as a fire event. In the case of lake records, the peak begins at the *oldest* interval at which the CHAR threshold value is exceeded, and it is registered until CHAR drops below that value. The assumption is that the oldest date marks the fire event and the younger part of the peak is reworked or secondary charcoal. In wetland records the peak is marked at the *youngest* interval with CHAR greater or equal to the threshold value, on the ground that the fire burns the surface and penetrates some depth into the wetland sediment (Huber and Markgraf, Chapter 13, this volume).

Clark and Royall (1996) used a Fourier series filter (Press et al. 1986), based on the variance spectrum of the CHAR series, to describe the background component. The peaks component was defined as the positive deviations of the CHAR series from background. This approach assumes that the background series is composed of many sinusoidal components, and can be adjusted by the choice of the width of the "spectral window" used in constructing a variance spectrum either through smoothing the periodogram or transforming an autocovariance function. Clark and Royall (1996) do not explicitly define a CHAR threshold for identifying fire events but by plotting the positive residual from the background component, such a threshold is implicitly defined. The low values of the noise component are not separated from the horizontal axis of their plots of the peaks components. Because the variance spectrum and resulting filter are defined using the entire record, as opposed to locally as in our approach, their strategy assumes that the CHAR background does not change through time. The CHAR data at Little Lake (Fig. 1.4), for example, suggest that the variance spectrum did indeed vary over time in response to changing climate and vegetation. We favor an approach where the background component may adapt to changes in the variability of the CHAR data.

Window width and threshold-ratio parameters are selected by (1) examining the CHAR from the short core relative to the record of recent fires near the site, and (2) by testing a variety of values of the two parameters to decompose the long record. The results of the decomposition are compared with information on present-day fire regimes in the region. This iterative approach provides an opportunity to examine the robustness of the method and the sensitivity of the outcomes to the choice of parameter values (Fig. 1.4). We display the fire events as a locally weighted mean frequency of peaks (number of peaks/1000 years). This peak-frequency series was obtained by smoothing a binary series of peaks (1, peaks; 0, no peaks) using a locally weighted average with a 2000-year window width.

A software package (Charcoal Analysis Programs, or CHAPS, developed by P.J. Bartlein) is available from the University of Oregon to facilitate decomposition of the charcoal records. The program converts charcoal concentration data into concentration at pseudo-annual intervals and then into charcoal concentration and CHAR at decadal intervals. The program also allows consideration of different background and threshold values to produce a plot of peak frequency.

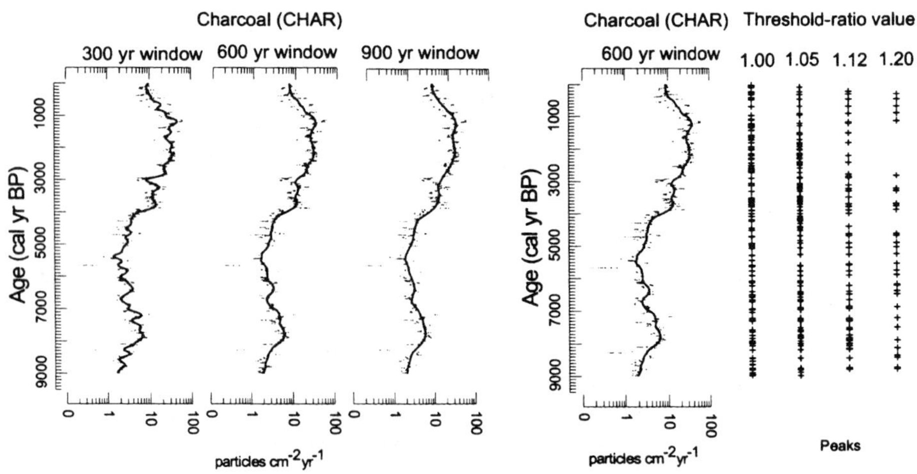

Figure 1.4. Comparison at Little Lake of different window widths to define background charcoal (left) and different threshold-ratio levels to identify significant peaks that represent fire events (right). In Long et al. (1998), a window width of 600 years and a threshold-ratio value of 1.12 was used to reconstruct the fire history.

Examples of High-Resolution Charcoal Studies

Charcoal and pollen data from Little Lake in the Oregon Coast Range (Long et al. 1998), charcoal records from Bluff and Crater lakes in the Klamath Mountains of northern California (Mohr, Whitlock, and Skinner 2000), and a charcoal study of wet meadows in the Sierra Nevada (Anderson and Smith 1997) illustrate the type of insights that can be gained from high-resolution fire history studies. In each case macroscopic charcoal was analyzed in contiguous intervals.

At Little Lake, an 11.33-m-long core was taken that spanned the last 9000 cal years. The chronology for this core was based on four AMS ^{14}C dates on charcoal particles, one conventional bulk-sediment ^{14}C date, and the age of the Mazama volcanic ash, which was identified in the core. A third-order polynomial was used to fit a smooth age-to-depth model. At the coring location, a 45-cm-long short core was also retrieved and dated by ^{210}Pb method. The cores were sliced into 1-cm-thick intervals, and from each sample, sediment was taken for magnetic susceptibility and charcoal analyses. The pollen stratigraphy had already been described in a previous study (Worona and Whitlock 1995). Charcoal samples (2.5 cm^3 volume) were washed through sieves of 63-, 125-, and 250-μm mesh diameters, and the particles were counted under a stereomicroscope and compared. As a result, only the two larger size fractions were examined, because they contained abundant charcoal but not so much that counting was impractical. Data were converted to concentration data and then to CHAR at decadal intervals, using CHAPS software.

Very little information was available on the modern fire history of the Little Lake watershed, because much of the area was logged and reforested in the twentieth century. The choice of parameters to assign for window-width and threshold-ratio values came from an understanding of the recent fire regime, as well as an inspection of the CHAR data. Long et al. (1998) identified eight large CHAR peaks in the last 1500 years, which seemed to represent fire events (Fig. 1.4). The temporal spacing of the peaks was consistent with the mean return interval of fires in the Coast Range at present based on dendrochronological studies. Different combinations of window-width and threshold-ratio values were considered in an effort to find parameters that would identify the eight peaks as fire events. A background window of 600 years and a threshold value of 1.12 were chosen, because they identified all eight peaks and no additional ones. These values also produced fire return intervals of $<$600 years in the rest of the record, which seemed reasonable given the return intervals of large fires in the wettest and driest parts of the Pacific Northwest rain forest suggested by dendrochronological data (Agee 1993).

Applying the peak and threshold parameters to the entire record produced a fire event frequency that showed variations throughout the Holocene. The background component at Little Lake was low for the first 5000 years, and then increased abruptly at 4000 cal yr BP (before present). The increase corresponds to a change in deposition time and implies an increase in secondary charcoal during non-fire years in the late Holocene. It was ascribed to more woody fuel biomass

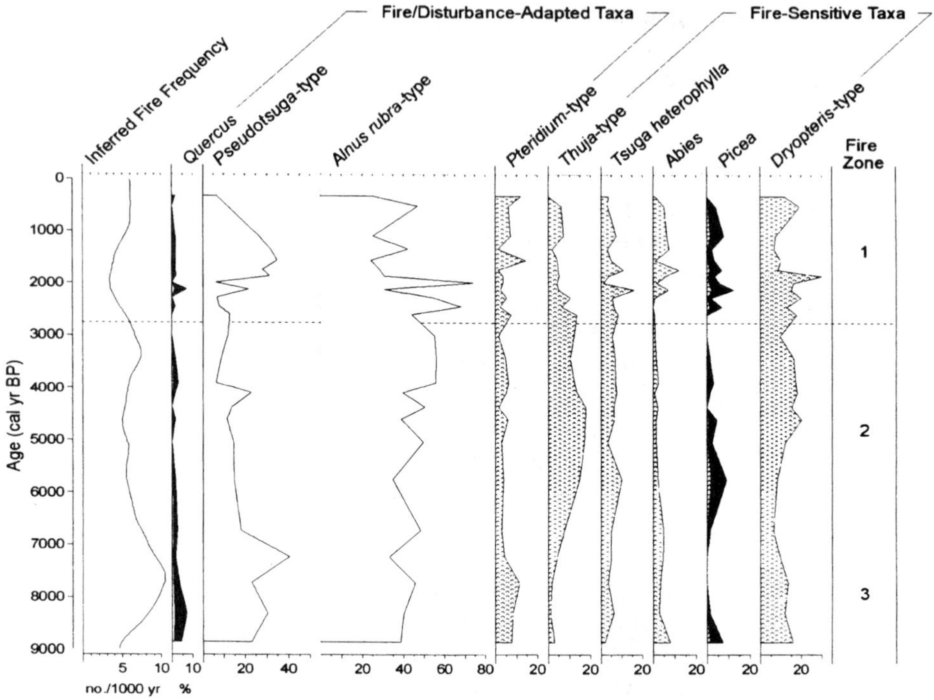

Figure 1.5. Little Lake fire history reconstruction and comparison with pollen data (Long et al. 1998).

with the development of closed rain forest, changes in sediment storage characteristics and mass wasting, and possibly an increase in the severity of fires with the onset of cool wet conditions.

The fire history was divided into three periods (Fig. 1.5): a period from 9000 to 8500 cal yr BP, when fire occurrence ranged from 5 to 8 events/1000 years, with a high of 10 events/1000 years at ca. 7500 cal yr BP; a middle-Holocene period (ca. 6850–2750 cal yr BP) when the record ranged from 6 to 8 events/1000 years, and a late-Holocene period (2750 cal yr BP to present) when the frequency has been about 5 events/1000 years. The fire record matched well with changes in the vegetation, inferred from the pollen data. In the early Holocene, when the climate was warm and dry, fires were frequent and disturbance-adapted taxa, such as *Alnus*, *Quercus*, and *Pseudotsuga*, were prevalent in the vegetation. In the middle Holocene, the fire frequency lengthened and regional paleoclimate records suggest a shift to cool wet conditions and more *Tsuga heterophylla* and *Thuja plicata* at Little Lake. In the late Holocene, the fire interval was longest and coincided with the establishment of cool wet conditions and a dominance of mesophytic species. The record suggested that the fire frequency has changed continuously with climate change and that the present fire regime has been in existence for only the last 1000 to 2000 years.

The second example combined pollen and high-resolution charcoal data from two lakes, Bluff and Crater, in the Klamath Mountains of northern California (Mohr, Whitlock, and Skinner 2000). In these mixed conifer forests the historic fire return interval is very short, on the order of decades (Skinner and Chang 1996). The same methodological approach was used as at Little Lake, but the Klamath lakes were located on serpentine substrates and had very slow sedimentation rates. Each 1-cm sample spanned 24 to 180 years at Bluff Lake, and 12 to 120 years at Crater Lake, which was longer than the average fire return interval based on tree-ring data. The charcoal record was decomposed using a background window width of 120 years and a threshold-ratio value of 1.00, based on a comparison with dendrochronological and documentary evidence of recent fires and their registration of specific charcoal peaks in a ^{210}Pb-dated short core. Because of the slow deposition time, the goal was to select a value that would correctly identify multiple decadal intervals with significant burning. Charcoal peaks represented one or more fires occurring over a time span of 12 to 180 years, and, as a result, the data were not directly comparable to the dendrochronological record.

The pollen and charcoal record considered together provided information on the postglacial vegetation and fire regimes near the lakes. The vegetation and fire history indicated more xerophytic vegetation and more frequent fire events during the early Holocene than at present. The early Holocene is widely documented as a period of intense summer drought in the Pacific Northwest based on several lines of evidence (Thompson et al. 1993). As the climate became cooler and wetter in the late Holocene, mesophytic taxa, such as *Tsuga heterophylla*, became more important and fire event frequencies decreased. The modern forest was established in the last 2000 years; fire event frequencies were high at ca. 1000 cal yr BP and have declined since then. Both watersheds experienced highest fire frequencies during dry periods. Fire events were frequent at 8300 cal yr BP, 4000 cal yr BP, and during the so-called Medieval Warm Period, ca. 1000 cal yr BP (Stine 1994) (Fig. 1.6). The synchroneity of the fire history at the sites implied a response to regional changes in climate on submillennial time scales.

The third example considered high-resolution charcoal records in several wet meadow cores in the Sierra Nevada of California (Anderson and Smith 1997). The meadows varied in elevation from 1786 to 2206 m. The goal of the study was to examine the broad-scale fire patterns within the montane forest during the last 10,000 years by comparing the charcoal record from several sites. Using the thin-section charcoal method (Clark 1988b), charcoal particles were identified and tallied in contiguous 1-mm-depth intervals for each core. A chronology was developed by assigning ages based on a suite of AMS and conventional radiocarbon dates. The data were aggregated into 25-year periods, because sedimentation rates within the wet meadows were assumed to be variable (Anderson and Smith 1994). More recently Anderson and Smith (1998)

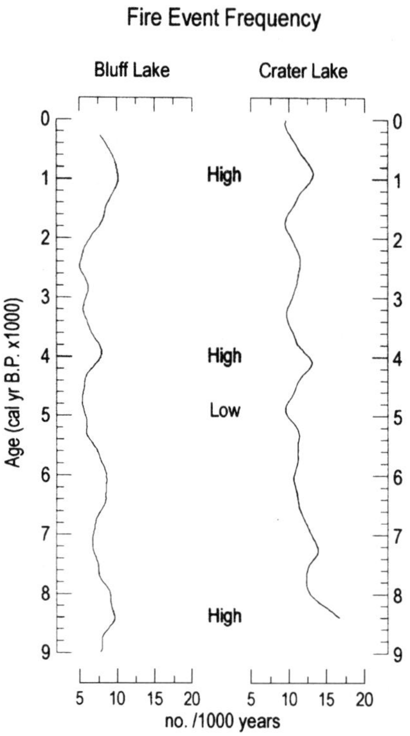

Figure 1.6. Comparison of two Klamath Lake records for the last 8500 years, showing synchronous periods of high and low fire occurrence (Mohr, Whitlock, and Skinner 2000; reprinted with permission from *The Holocene*, © Arnold Publishers).

refined the chronology with additional AMS dates and converted the time scale to calendar years.

Although charcoal is recovered in variable amounts throughout the profile (note logarithmic scale), distinct charcoal peaks were recorded at each of the sites, and are inferred to be local fire events (Fig. 1.7). Lesser amounts of charcoal could be attributed to regional as well as local fires. Over the last 1000 cal years, three sites showed a prominent peak between 550 and 400 cal yr BP. Three sites also registered a charcoal peak between 790 and 665 cal yr BP, while two displayed charcoal peaks between 985 and 935 cal yr BP. Higher concentrations of charcoal indicating increased fire activity occurred over the last ca. 1200 years, at ca. 2200, 2700, and 3000 cal yr BP, between ca. 3700 and 5200 cal yr BP, at ca. 7250 cal yr BP and between ca. 9300 and 9700 cal yr BP. In between these intervals were periods of considerably lower charcoal deposition (Anderson and Smith 1997, 1998).

Like the Coast Range and Klamath studies, the Sierran records suggest that climate changes were responsible for the long-term variations in fire occurrence. Warmer drier conditions in the early Holocene led to open forests with more pine and montane chaparral shrubs than today (Anderson 1990). With increased effective precipitation in the late Holocene, forests became more closed and fires were less frequent (Anderson 1990; Anderson and Smith 1994).

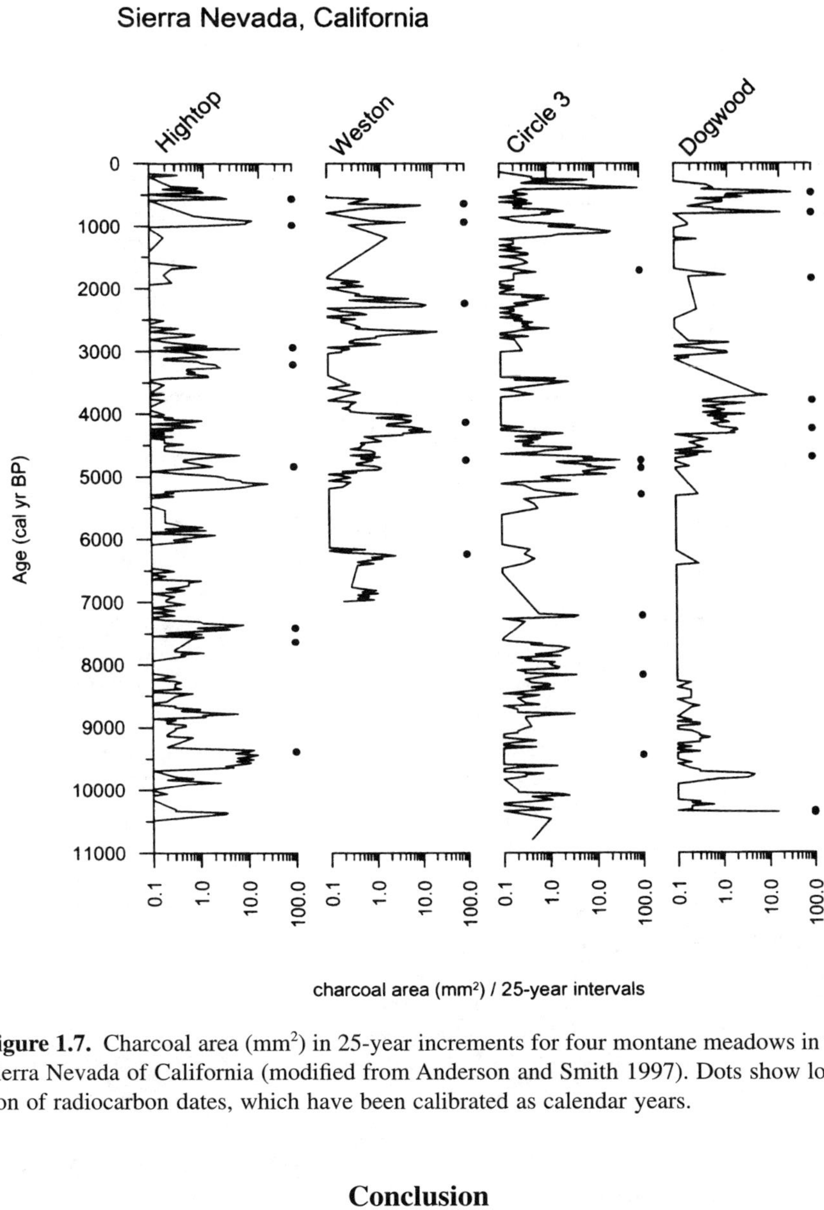

Figure 1.7. Charcoal area (mm^2) in 25-year increments for four montane meadows in the Sierra Nevada of California (modified from Anderson and Smith 1997). Dots show location of radiocarbon dates, which have been calibrated as calendar years.

Conclusion

As more charcoal records become available in the Americas, their value in paleoclimate reconstructions and in assessing the proximal causes of vegetation change will increase. Studies to date suggest that variations in fire frequency offer a more sensitive proxy of millennial-scale climate variations than do pollen data. The development of high-resolution charcoal records in both hemispheres offers an opportunity to examine climate variations and teleconnections associated with

ENSO and changes in the seasonal cycle of insolation. To realize the potential of charcoal studies as a paleoclimate proxy, however, requires that the paleoecological community standardize both the techniques and interpretation of such data. Too many charcoal studies are based on imprecise or unsubstantiated assumptions and analytical approaches.

Modern studies of charcoal transport and deposition are rare. Two have been undertaken in the western United Sates, and none are available from South America. Data on modern processes are needed to calibrate charcoal data and refine the interpretation of the stratigraphic record. Similarly, modeling efforts that focus on the relation between fire and charcoal production and transport are needed to verify the assumptions developed from empirical studies.

The language used to describe past fire regimes from charcoal data has been imprecise, and often does not convey information that is useful to fire ecologists. For example, the term "fire frequency" has been used in the paleoecological literature to describe everything from changes in abundance of microscopic charcoal in discontinuous records to peaks in high-resolution records of macroscopic charcoal. These two data sets do not describe the same phenomenon. Likewise, the term "regional fires" is seldom defined in a paper, and it is not clear whether it refers to distant fires, large widespread fires, or a more general attribute of a fire regime. The lack of standardization in terminology has limited our ability both to compare charcoal records and to link charcoal-based fire reconstructions with those provided by dendrochronological data. It is important that the charcoal and the dendrochronological communities work together to understand and describe fire regimes on multiple spatial and temporal scales. Charcoal analysis can contribute in significant ways toward assessing the representativeness of dendrochronological records on longer time scales, and tree-ring records can assist in refining the interpretation of charcoal data.

Several methods are now available for charcoal analysis, and while the choice may seem bewildering, we offer some recommendations:

1. Charcoal studies should routinely examine macroscopic charcoal in order to get a local fire reconstruction. The source area of macroscopic charcoal is much better known than that of microscopic charcoal, and fire location is an essential part of any fire reconstruction.

2. Contiguous sampling at a fine sampling interval is critical to calculate fire event frequency; discontinuous sampling misses charcoal peaks and often background trends are interpreted as fire events.

3. An adequate chronology is essential, as is some method of calibration to identify a significant threshold level. Thus, charcoal studies require varved-sediment records or calibrated AMS ^{14}C-dated and ^{210}Pb-dated records.

4. Macroscopic charcoal data are quantified in different ways, most commonly as particle counts or area measurements. Charcoal accumulation rates are calculated based on sediment weight or volume. These different methods seem

to give generally similar trends through time. More important is the decision to undertake high-resolution sampling by analyzing contiguous samples.

5. The time series consists of at least two components, a slowly varying background component, superimposed upon which is a peaks component. The information contained in these two components is different and should be interpreted separately. Periods with abundant charcoal may not necessarily represent times of more fires; they could be periods of high background charcoal as a result of a shift in fire severity or the introduction of secondary charcoal.

6. Each macroscopic charcoal record is a local reconstruction; to infer landscape, regional, or larger-scale patterns requires a network of sites, all done to a similar high standard, or calibration of the microscopic charcoal component.

7. In addition to charcoal data, other fire proxies are worth considering, especially as they can supplement the fire interpretations. Closely sampled pollen records have been used as a proxy of fire (Sugita et al. 1997; MacDonald et al. 1991; Larsen and MacDonald 1998b). The sensitivity of pollen records to a particular fire event should to be carefully tested in each locality. Similarly, lithologic analyses, in particular, magnetic susceptibility measurements, have proved to be a useful indicator of fire-related erosion in some regions, but the success varies among sites.

Acknowledgments. The research that motivated this chapter was supported by grants from the National Science Foundation (SBR9616951, EAR9906100, ATM0117160 to CW), USDA Forest Service (USFS PSW-95-0022CA, USFS PNW-98-5122-1CA to CW), U.S. Geological Survey Global Change Program (1434-WR-97-AG-00013 to RSA), and the National Park Service Global Change Program (CA 8000-7-0001, CA 8013-8-0002 to RSA).

References

Agee, J.K. 1993. *Fire Ecology of Pacific Northwest Forests*. Washington, DC: Island Press.

Anderson, R.S. 1990. Holocene forest development and paleoclimates within the central Sierra Nevada, *Cal. J. Ecol.* 78:470–489.

Anderson, R.S., Davis, R.B., Miller, N.G., and Stuckenrath, R. 1986. History of late- and post-glacial vegetation and disturbance around Upper South Branch Pond, northern Maine. *Can. J. Bot.* 64:1977–1986.

Anderson, R.S., and Smith, S.J. 1994. Paleoclimatic interpretations of meadow sediment and pollen stratigraphies from California. *Geology* 22:723–726.

Anderson, R.S., and Smith, S.J. 1997. The sedimentary record of fire in montane meadows, Sierra Nevada, California, USA: A preliminary assessment. In *Sediment Records of Biomass Burning and Global Change*, eds. J.S. Clark, H. Cachier, J.G. Goldammer, and B. Stocks, pp. 313–328. NATO ASI Series 1: Global Environmental Change, vol. 51. Berlin: Springer.

Andreae, M.O. 1991. Biomass burning: its history, use, and distribution and its impact on environmental quality and global climate. In *Global Biomass Burning: Atmospheric, Climatic, and Biospheric Implications*, ed. J.S. Levin, pp. 3–21. Cambridge: MIT Press.

Bartlein, P.J., Whitlock, C., and Shafer, S.L. 1997. Future climate in Yellowstone National Park region and its potential impact on vegetation. *Conservation Biol.* 11:782–792.

Birks, H.J.B. 1997. Reconstructing environmental impacts of fire from the Holocene sedimentary record. In *Sediment Records of Biomass Burning and Global Change*, eds. J.S. Clark, H. Cachier, J.G. Goldammer, and B. Stocks, pp. 295–312. NATO ASI Series 1: Global Environmental Change, vol. 51. Berlin: Springer-Verlag.

Bradbury, J.P. 1996. Charcoal deposition and redeposition in Elk Lake, Minnesota, USA. *Holocene* 6:339–344.

Bradshaw, R.H.W., Tolonen, K., and Tolonen, M. 1997. Holocene records of fire from the boreal and temperate zones of Europe. In *Sediment Records of Biomass Burning and Global Change*, eds. J.S. Clark, H. Cachier, J.G. Goldammer, and B. Stocks, pp. 347–366. NATO ASI Series 1: Global Environmental Change, vol. 51. Berlin: Springer-Verlag.

Brunelle, A. 1997. A post-glacial record of fire and vegetation from Siesta Lake, Yosemite National Park, California. M.S. thesis. Northern Arizona University, Flagstaff. 107p.

Brunner Jass, R.M. 1999. Fire occurrence and paleoecology at Alamo Bog and Chihuahueños Bog, Jemez Mountains, New Mexico. M.S. thesis. Northern Arizona University, Flagstaff. 140p.

Carcaillet, C., and Thinon, M. 1996. Pedoanthracological contribution to the study of the evolution of the upper treeline in the Maurienne Valley (North French Alps): Methodology and preliminary data. *Rev. Palaeobot. Palynol.* 91:399–416.

Clark, J.S. 1988a. Particle motion and the theory of stratigraphic charcoal analysis: Source area, transport, deposition, and sampling. *Quat. Res.* 30:67–80.

Clark, J.S. 1988b. Stratigraphic charcoal analysis on petrographic thin sections: Applications to fire history in northwestern Minnesota. *Quat. Res.* 30:81–91.

Clark, J.S. 1990. Fire and climate change during the last 750 years in northwestern Minnesota. *Ecol. Monogr.* 60:135–159.

Clark, J.S., and Hussey, T.C. 1996. Estimating the mass flux of charcoal from sedimentary records: Effects of particle size, morphology, and orientation. *Holocene* 6:129–145.

Clark, J.S., and Patterson, W.A. III. 1997. Background and local charcoal in sediments: Scales of fire evidence in the paleorecord. In *Sediment Records of Biomass Burning and Global Change*, eds. J.S. Clark, H. Cachier, J.G. Goldammer, and B. Stocks, pp. 27–48. NATO ASI Series 1: Global Environmental Change, vol. 51. Berlin: Springer.

Clark, J.S., and Royall, P.D. 1995. Particle size evidence for source areas of charcoal accumulation in late Holocene sediments of eastern North American lakes. *Quat. Res.* 43:80–89.

Clark, J.S., and Royall, P.D. 1996. Local and regional sediment charcoal evidence for fire regimes in presettlement northeastern North America. *J. Ecol.* 84:365–382.

Clark, J.S., Lynch, J., Stocks, B., and Goldammer, J. 1998. Relationships between charcoal particles in air and sediments in west-central Siberia. *Holocene* 8:19–29.

Clark, R.L. 1982. Point count estimation of charcoal in pollen preparations and thin sections of sediment. *Pollen Spores* 24:523–535.

Cleveland, W.S. 1979. Robust locally weighted regression and smoothing scatterplots. *J. Am. Stat. Assoc.* 74:829–836.

Cwynar, L.C. 1978. Recent history of fire and vegetation from annually laminated sediment of Greenleaf Lake, Algonquin Park, Ontario. *Can. J. Bot.* 56:10–12.

Cwynar, L.C. 1987. Fire and the forest history of the north Cascade Range. *Ecology* 68:791–802.

Earle, C.J., Brubaker, L.B., and Anderson, P.M. 1996. Charcoal in north central Alaskan lake sediments: Relationships to fire and late-Quaternary vegetation history. *Rev. Palaeobot. Palynol.* 92:83–95.

Edwards, K.J., and Whittington, G. 2000. Multiple charcoal profiles in a Scottish lake: Taphonomy, fire ecology, and human impact and interference. *Palaeogeogr. Palaeoclim. Palaeoecol.* 164:67–86.

Gardner, J., and Whitlock, C. 2001. Charcoal accumulation following a recent fire in the Cascade Range, northwestern USA, and its relevance for fire-history studies. *Holocene* 11:541–549.

Gedye, S.J., Jones, R.T., Tinner, W., Ammann, B., and Oldfield, F. 2000. The use of mineral magnetism in the reconstruction of fire history: A case study from Lago di Origlio, Swiss Alps. *Palaeogeogr. Palaeoclim. Palaeoecol.* 164:101–110.

Grimm, E.C., ND. Tilia Software Package. Illinois State Museum, Springfield, IL.

Hallett, D.J., and Walker, R.C. 2000. Paleoecology and its application to fire and vegetation management in Kootenay National Park, British Columbia. *J. Paleolim.* 24:401–414.

Heusser, C.J. 1994. Paleoindians and fire during the late Quaternary in southern South America. *Rev. Chilena Hist. Nat.* 67:435–443.

Horn, S.P., Horn, R.D., and Byrne, R. 1992. An automated charcoal scanner for paleoecological studies. *Palynology* 16:7–12.

Huber, U. 2001. Holocene variations among fire, climate, and vegetation in southern Patagonia. Ph.D. dissertation. University of Colorado, Boulder.

Innes, J.B., and Simmons, I.G. 2000. Mid-Holocene charcoal stratigraphy, fire history, and palaeoecology at North Gill, North York Moors, UK. *Palaeogeogr. Palaeoclim. Palaeoecol.* 164:151–165.

Iversen, J. 1941. Land occupation in Denmark's Stone Age. *Danmarks Geologiske Forenhandlungen II* 66:1–126.

Jacobson, G.L. Jr., and Bradshaw, R.H.W. 1981. The selection of sites for paleovegetational studies. *Quat. Res.* 16:80–96.

Laird, K.D., and Campbell, I.D. 2000. High resolution palaeofire signals from Christina Lake, Alberta: A comparison of charcoal signals extracted by two different methods. *Palaeogeogr. Palaeoclim. Palaeoecol.* 164:111–123.

Larsen, C.P.S., and MacDonald, G.M. 1993. Lake morphology, sediment mixing and the selection of sites for fine resolution palaeoecological studies. *Quat. Sci. Rev.* 12:781–792.

Larsen, C.P.S., and MacDonald, G.M. 1998a. An 840-year record of fire and vegetation in a boreal white spruce forest. *Ecology* 79:106–118.

Larsen, C.P.S., and MacDonald, G.M. 1998b. Fire and vegetation dynamics in a jack pine and black spruce forest reconstructed using fossil pollen and charcoal. *J. Ecol.* 86:815–828.

Larsen, C.P.S., Peinitz, R., Smol, J.P., Moser, K.A., Cumming, B.F., Blais, J.M., MacDonald, G.M., and Hall, R.I. 1998. Relations between lake morphometry and the presence of laminated lake sediments: A reexamination of Larsen and MacDonald (1993). *Quat. Sci. Rev.* 17:711–717.

Long, C.J., Whitlock, C., Bartlein, P.J., and Millspaugh, S.H. 1998. A 9000-year fire history from the Oregon Coast Range, based on a high-resolution charcoal study. *Can. J. For. Res.* 28:774–787.

MacDonald, G.M., Larsen, C.P.S., Szeicz, J.M., and Moser, K.A. 1991. The reconstruction of boreal forest fire history from lake sediments: a comparison of charcoal, pollen, sedimentological, and geochemical indices. *Quat. Sci. Rev.* 10:53–72.

Markgraf, V., and Anderson, L. 1994. Fire history of Patagonia: Climate versus human cause. *Rev. Instituto Geologico*, Sao Paulo 15:35–47.

Mehringer, P.J., Arno, S.F., and Petersen, K.L. 1977. Postglacial history of Lost Trail Pass Bog, Bitterroot Mountains, Montana. *Arct. Alp. Res.* 9:345–368.

Meyer, G.A., Wells, S.G., and Tull, A.J.T. 1995. Fire and alluvial chronology in Yellowstone National Park: Climatic and intrinsic controls on Holocene geomorphic processes. *Geol. Soc. Am. Bull.* 107:1211–1230.

Millspaugh, S.H. 1997. Late-glacial and Holocene variations in fire frequency in the Central Plateau and Yellowstone-Lamar Provinces of Yellowstone National Park. Ph.D. dissertation. University of Oregon, Eugene.

Millspaugh, S.H., and Whitlock, C. 1995. A 750-year fire history based on lake sediment records in central Yellowstone National Park, USA. *Holocene* 5:283–292.

Millspaugh, S.H., Whitlock, C., and Bartlein, P.J. 2000. Variations in fire frequency and climate over the last 17,000 years in central Yellowstone National Park. *Geology* 28:211–214.

Mohr, J.A., Whitlock, C., and Skinner, C.J. 2000. Postglacial vegetation and fire history, eastern Klamath Mountains, California. *Holocene* 10:587–601.

Odgaard, B.V. 1992. The fire history of Danish heathland areas as reflected by pollen and charred particles in lake sediments. *Holocene* 2:218–226.

Ohlson, M.C., and Tryterud, E. 2000. Interpretation of the charcoal record in forest soils: forest fires and their production and deposition of macroscopic charcoal. *Holocene* 10:519–525.

O'Sullivan, A. 1991. Historical and contemporary effects of fire on the native woodland vegetation of Killarney, S.W. Ireland. Ph.D. dissertation. Trinity College, Dublin.

Overpeck, J.T., Rind, D., and Goldberg, R. 1990. Climate-induced changes in forest disturbance and vegetation. *Nature* 343:51–53.

Patterson, W.A. III, Edwards, K.J., and MacGuire, D.J. 1987. Microscopic charcoal as a fossil indicator of fire. *Quat. Sci. Rev.* 6:3–23.

Pearl, C.A. 1999. Holocene environmental history of the Willamette Valley, Oregon: Insights from an 11,000-year record from Beaver Lake. M.S. thesis. Environmental Studies Program. University of Oregon, Eugene.

Pitkänen, A. Turunen, J., and Tolonen, K. 1999. The role of fire in the carbon dynamics of a mire, eastern Finland. *Holocene* 9:453–462.

Press, W.H., Flannery, B.P., Teukolsky, S.A., and Vetterling, W.T. 1986. *Numerical Recipes*. Cambridge: Cambridge University Press.

Price, C., and Rind, D. 1994. The impact of a 2 × CO_2 climate on lightning caused fires. *J. Clim.* 7:1484–1494.

Radtke, L.F., Hegg, D.A., Hobbs, P.V., Nance, J.D., Lyons, J.H., Laursen, K.K., Weiss, R.E., Riggan, P.J., and Ward, D.E. 1991. Particulate and trace gas emissions from large biomass fires in North America. In *Global Biomass Burning: Atmospheric, Climatic, and Biospheric Implications*, ed. J.S. Levin, pp. 209–224. Cambridge: MIT Press.

Rhodes, T.E., and Davis, R.B. 1995. Effects of late Holocene forest disturbance and vegetation change on acidic Mud Pond, Maine, USA. *Ecology* 76:734–746.

Rummery, T.A., Bloemendal, J., Dearing, J., Oldfield, F., and Thompson, R. 1979. The persistence of fire-induced magnetic oxides in soils and lake sediments. *Ann. Geophys.* 35:103–107.

Skinner, C.N., and Chang, C. 1996. Fire regimes, past and present. In Sierra Nevada Ecosystem Project: Final Report to Congress, Vol. II, Assessments and Scientific Basis for Management Options. University of California at Davis, Centers for Water and Wildland Resources 1041–1069.

Smith, S.J., and Anderson, R.S. 1992. Late Wisconsin paleoecologic record from Swamp Lake, Yosemite National Park, California. *Quat. Res.* 38:91–102.

Stine, S. 1994. Extreme and persistent drought in California and Patagonia during mediaeval time. *Nature* 369:546–549.

Stuiver, M., Reimer, P.J., Bard, E., Beck, J.W., Burr, G.S., Hughen, K.A., Kromer, B., McCormac, F.G., van der Plicht, J., and Spurk, M. 1998. INTCAL 98 Radiocarbon age calibration 24,000–0 cal BP. *Radiocarbon* 40:1041–1083.

Sugita, S., MacDonald, G.M., and Larsen, C.P.S. 1997. Reconstruction of fire disturbance and forest succession from fossil pollen in lake sediments: Potential and limitations. In *Sediment Records of Biomass Burning and Global Change*, eds. J.S. Clark, H. Cachier, J.G. Goldammer, and B. Stocks, pp. 387–412. NATO ASI Series 1: Global Environmental Change, vol. 51. Berlin: Springer.

Swain, A.M. 1973. A history of fire and vegetation in northeastern Minnesota as recorded in lake sediments. *Quat. Res.* 3:383–396.

Swain, A.M. 1978. Environmental changes during the past 2000 yr in north-central Wisconsin: Analysis of pollen, charcoal and seeds from varved lake sediments. *Quat. Res.* 10:55–68.

Terasmae, J., and Weeks, N.C. 1979. Natural fires as an index of paleoclimate. *Can. Field Nat.* 93:116–125.

Thompson, R., and Oldfield, F. 1986. *Environmental Magnetism.* London: Allen and Unwin.

Thompson, R.S., Whitlock, C., Bartlein, P.J., Harrison, S.P., and Spaulding, W.G. 1993. Climatic changes in western United States since 18,000 yr BP. In *Global Climates since the Last Glacial Maximum*, eds. H.E. Wright Jr., J.E. Kutzbach, T. Webb III, W.F. Ruddiman, and F.A. Street-Perrot, pp. 468–513. Minneapolis: University of Minnesota Press.

Tolonen, K. 1986. Charred particle analysis. In *Handbook of Holocene Palaeoecology and Palaeohydrology*, ed. B.E. Berglund, pp. 485–496. New York: Wiley.

Tolonen, M. 1985. Paleoecological record of local fire history from a peat deposit in SW Finland. *Ann. Bot. Fenn.* 15:177–209.

Umbanhowar, C.E., Jr. 1996. Recent fire history of the northern Great Plains. *Am. Midl. Nat.* 135:115–121.

Wein, R.W., Burzynski, M.P., Screenivasa, B.A., and Tolonen, K. 1987. Bog profile evidence of fire and vegetation dynamics since 3000 years BP in the Acadian Forest. *Can. J. Bot.* 65:1180–1186.

Whitlock, C., and Larsen, C.P.S., in press. Charcoal as a fire proxy. In *Tracking Environmental Change Using Lake Sediments*: Terrestrial, Algal, and Siliceous Indicators, eds. J.P. Smol, H.J.B. Birks, and W.M. Last, vol. 3. Dordrecht: Kluwer Academic.

Whitlock, C., and Millspaugh, S. 1996. Testing assumptions of fire history studies: An examination of modern charcoal accumulation in Yellowstone National Park. *Holocene* 6:7–15.

Winkler, M.G. 1985. Charcoal analysis for paleoenvironmental interpretation: a chemical assay. *Quat. Res.* 23:313–326.

Worona, M.A., and Whitlock, C. 1995. Late-Quaternary vegetation and climate history near Little Lake, Central Coast Range, Oregon. *Geol. Soc. Am. Bull.* 107:867–876.

2. The Simulation of Landscape Fire, Climate, and Ecosystem Dynamics[1]

Robert E. Keane and Mark A. Finney

Wildland fire is a critical disturbance process in many ecosystems worldwide, yet it is difficult to study fire and its relationship to climate across large landscapes over long time periods (Crutzen and Goldammer 1993). Most field studies evaluate the effects of fire at the stand level, and usually after only one fire event. Field investigation of the cumulative effects of many fires across an entire landscape would require exorbitant amounts of time and money not available to many fire scientists. Simulation modeling, however, provides an alternative tool to investigate and understand how landscapes respond to changes in fire brought on by climate change across large spatial and temporal domains. Simulation approaches allow the integration of diverse scientific information into a comprehensive framework to explore more complex problems, such as the effects of climate change on fire dynamics. An advantage of simulation modeling is that it enables us to determine the relative importance of one factor in regulating ecosystem behavior by holding other factors constant.

There are many types of simulation models, and they are usually categorized into four groups (Table 2.1). *Empirical models* are primarily built on statistical relationships derived from actual data. *Deterministic models* use generalized

[1] The use of trade or firm names in this paper is for reader information and does not imply endorsement by the U.S. Department of Agriculture of any product or service. This paper was written and prepared by U.S. Government employees on official time, and therefore is in the public domain and not subject to copyright.

Table 2.1. Contrasting aspects of the four simulation approaches discussed in this chapter

Attribute	Empirical	Deterministic	Stochastic	Mechanistic
Complexity	Low	Low	Moderate	High
Parameter requirements	Low	Moderate	Moderate	High
Accuracy	High	Variable	Low	Low
Exploratory uses	Low	Moderate	Moderate	High
Management application	High	High	Low	Low
Portability to other situations	Low	Moderate	Moderate	High
Expandability	Low	Moderate	Moderate	High
Computer requirements	Low	Moderate	Moderate	High
Preparation time	Low	Low	Moderate	High

functions to represent the relationships that drive simulation dynamics. *Stochastic models* use probability distributions to represent primary ecosystem processes. And last, *mechanistic models* use fundamental biological and physical relationships to simulate the underlying processes or causal mechanisms that dictate system behavior (Gay 1989). While all of these model approaches have their various advantages and disadvantages (Table 2.1), the best simulation models are often combinations of all four types.

This chapter presents a general review of spatially explicit fire and fire effects simulation modeling, but uses a conceptual fire simulation system called FESM (fire effects simulation model) as the context for this review. The primary focus of this conceptual framework is to investigate the interaction of fire and climate with ecosystem dynamics at landscape scales as it relates to management issues or research goals. This conceptual framework or general model is presented only to organize and integrate the discussion of the important processes needed to simulate fire and its immediate and long-term effects on ecosystems and landscapes (Fig. 2.1). The lack of computer resources, limited research findings, complex interrelationships, and incomplete scientific expertise prohibit the inclusion of all processes into FESM as yet, so not all processes shown in Figure 2.1 will be discussed.

Terminology used in this chapter must be clearly defined to avoid confusion. *Fire behavior* is defined as the quantification of the physical properties of a fire. Descriptors of fire behavior include spread rate ($m\,s^{-1}$), fire line intensity ($kW\,m^{-1}$), and flame length (m) (Anderson 1969; Rothermel 1972; Albini 1976a). *Fire effects* are the direct and indirect consequences of a fire on ecosystem components. These effects may or may not be correlated to fire behavior. Direct or first-order fire effects include fuel consumption, tree mortality, and smoke generation. Indirect or second-order fire effects include plant succession, soil erosion, and landscape pattern. A model *component* is the abstract representation of a simulated process or characteristic used for descriptive purposes, whereas a *module* is the quantification and representation of that process into a computer algorithm. Model *compartments* are the state variables that represent characteristics of an

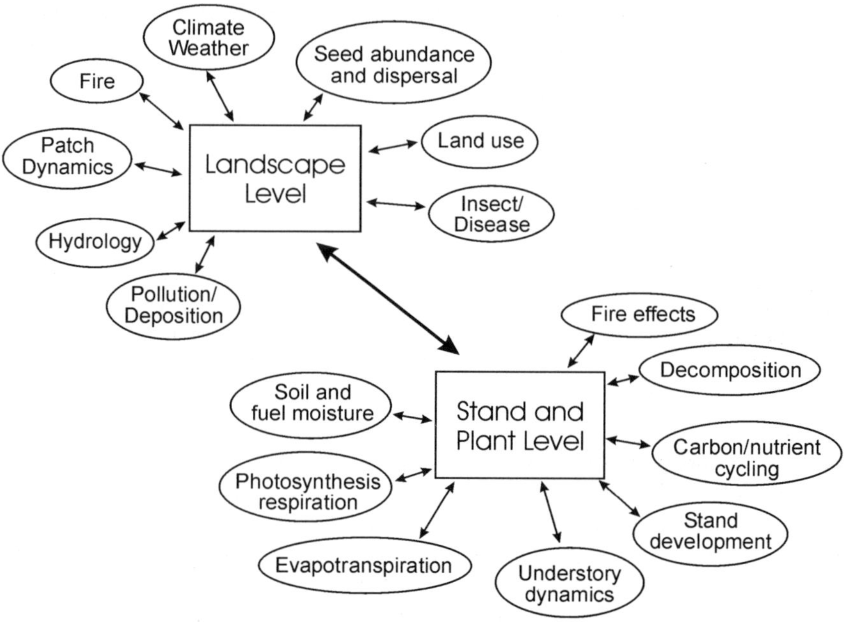

Figure 2.1. Critical processes needed to simulate fire and landscape interactions in the FESM construction.

ecosystem, such as leaf carbon or soil nitrogen (Swartzman 1979). *Processes* are the dynamic exchange of energy across the landscape, such as photosynthesis and respiration (Forman and Godron 1986). *Mechanisms* refer to the factors, such as temperature and radiation, that influence the flow of energy across model components.

The Fire Effects Simulation Model (FESM) Overview

It is necessary to specify the critical assumptions, goals, and objectives of FESM design. FESM is a not a prognostic model. It is, instead, primarily used to explore the interactions of fire, climate, and ecosystems on a landscape. FESM is mechanistic in design because exploratory models require explicit representations of the causal mechanisms that affect fire and landscapes (Bossel 1991). However, empirical, deterministic, and stochastic methods can substituted when underlying physical processes are unknown, inherently complex, or not required. FESM should be spatially explicit to address the effect of fire severity on the pattern, composition, and structure of landscapes (Forman and Godron 1986; Goodchild, Parks, and Steyaert 1993). Further, since FESM has a landscape focus, ecosystem processes must be integrated at appropriate time and space scales (Ball and Gimblett 1992). FESM must also have outputs that are applicable to both research and land management (Korzukhin, Ter-Mikaelian, and Wagner 1996). Last, FESM's design must maintain its flexibility so that empirical modules can be replaced with complex mechanistically driven modules as new research becomes available.

Four hierarchically nested organizational levels corresponding to an appropriate spatial scale are useful for FESM design (Simard 1996) (Fig. 2.2). The *landscape* is defined by spatial extent and is usually at least 5 to 10 times the size of the largest fire (Knight 1987) or target area (Keane et al. 2002), or 50 to 100 times the average fire size (Baker 1992a). Hydrologic boundaries (i.e., watersheds) are often used to define landscapes because most hydrological processes are completely represented within a watershed, and a watershed usually contains a fair representation of the ecosystems and topography that compose the surrounding areas (Forman and Godron 1986). The landscape is then divided into *biophysical settings* (Fig. 2.2), which define areas of similar soils, topography, land form, and hydrology that do not change throughout a simulation. Habitat types or potential vegetation types can be used to define biophysical settings in lieu of explicitly mapped weather, soils, and topography characteristics (Keane, Morgan, and Running 1996; Pfister et al. 1977). Each biophysical setting is then divided into *stands* that are best described as successional communities. Stand boundaries are dynamic because they are created by fire and other disturbances. Within the stand are the *organisms* that represent the successional community. This chapter will limit the discussion of organisms to plants and predominantly trees.

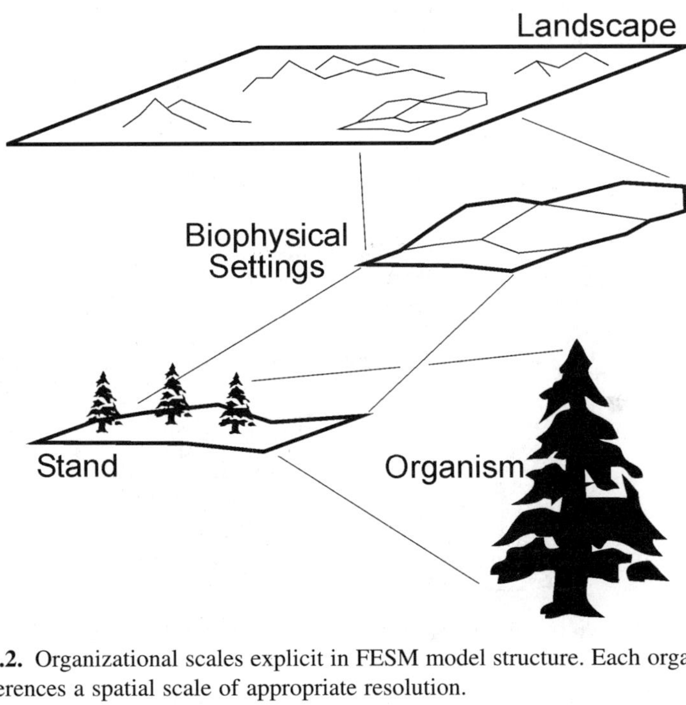

Figure 2.2. Organizational scales explicit in FESM model structure. Each organizational scale references a spatial scale of appropriate resolution.

Landscape and Biophysical Setting Processes

There are essentially five major processes that should be included in landscape-scale architecture of FESM: climate, fire, insects and disease, seed dispersal, and hydrology (Fig. 2.3). *Landscape* is the only state variable shown in Figure 2.3, and it represents the collective characteristics of a spatial setting such as pattern, biomass, water, and nitrogen. Simulation of the landscape processes in Figure 2.3 are reviewed next. Insects and diseases are beyond the scope of this chapter, although they are important disturbance processes on the landscape and their simulation should be included in FESM. Especially important is the interrelationship of insects and disease with climate and fire regimes. A thorough discussion of human impacts, such as forestry practices, human settlement, grazing, and fire suppression, on fire dynamics is also beyond the scope of this chapter.

Climate and Weather

Climate plays a critical role across all ecosystems and scales, and its simulation is critical to understand fire and succession dynamics. It is represented as weather at a daily time step at the biophysical setting and stand spatial scales. Climate and weather are important at the coarse scale for ecosystem processes such as species distributions, hydrologic cycles, and fire regimes; at the midscale for plant growth, decomposition, and fire patterns, and at the fine scales for plant regeneration, mortality, and fire spread. Conversely, the effect of fire on landscape structure and composition can influence regional climate and weather (Segal et al. 1988; Pielke and Avissar 1990).

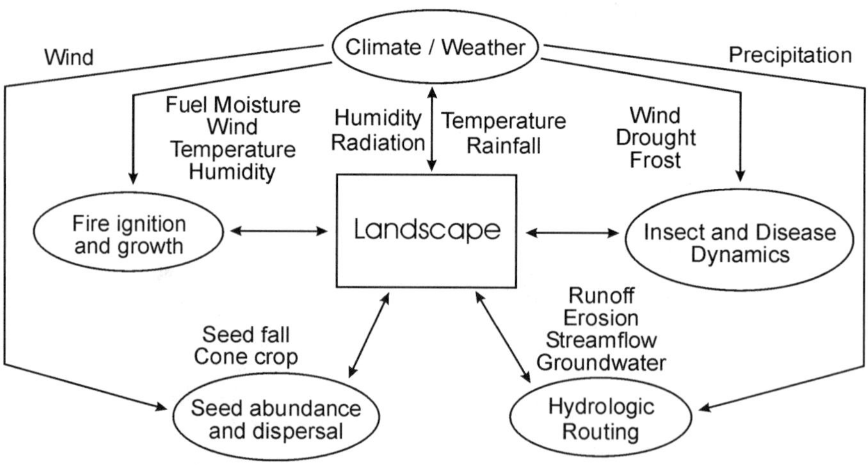

Figure 2.3. Diagram of FESM processes at the scale of the landscape and biophysical setting. Circles indicate processes while squares indicate states. Flows of energy are depicted by the arrows.

A minimum of seven daily measurements would be needed to quantify weather for all FESM components: maximum and minimum temperature, humidity, precipitation, wind direction and speed, and solar radiation. Most fire models require weather estimates at still smaller time steps, usually hourly but at least daily (Rothermel et al. 1986). But Finney (1998) uses a cosine function to derive hourly values from these daily estimates to compute fire behavior. Vegetation succession models usually need weather data at daily (Friend, Schugart, and Running 1993; Keane, Morgan, and Running 1996), weekly, monthly (Kercher and Axelrod 1984; Pastor and Post 1985), or yearly (Reed and Clark 1979; Mohren, Van Gerwen, and Spitters 1984; Bossel and Shafer 1988) time steps, depending on the detail of the simulation approach. Species dispersal models may use significantly longer time steps for simulation, often monthly or yearly. Each landscape process requires different lengths and temporal resolution of weather records.

In general, three methods can be used to simulate weather for input to FESM components across all organizational scales. The most complicated method uses a top-down approach where simulated climate from global circulation models (GCMs) are obtained for very coarse scale grids, usually 1 to 5 degrees in size. These grids can then be used as boundary conditions for modeling the mesoscale climate in gridded mechanistic models such as RegCM2 (Giorgi et al. 1993a; Luce, Kluzek, and Bingham 1995), MM4 (Hsie 1987), or RAMS (Pielke et al. 1992). Mesoscale models can compute the seven weather variables (listed above) on smaller grids of about 10 to 50 km square at time steps compatible with some fire simulations (Pinty et al. 1992). These gridded weather estimates can then be used to compute site-specific weather for biophysical settings through various extrapolation and interpolation techniques (Luce, Kluzek, and Bingham 1995) or finer-scale mechanistic modeling (Running, Nemani, and Hungerford 1987). Dickinson et al. (1989) took this approach in simulating the regional climate from global climate models. This multiple-scale approach requires an inordinate amount of computer resources and expertise, but it does provide a consistent scaling of weather information across a landscape (Blyth, Dolman, and Noihan 1994).

A second method employs a bottom-up approach where empirical weather data, measured at weather stations scattered across the landscape, are extrapolated to biophysical settings by empirical and process-based relationships. Many modelers have used MTCLIM (Running, Nemani, and Hungerford 1987; Hungerford et al. 1989) to extrapolate daily weather measurements taken at a base station to various sites on the landscape for ecosystem modeling (White 1996; Keane, Morgan, and Running 1996). Thornton, Running, and White (1997) improved MTCLIM algorithms and implemented them in a spatial domain to generate maps of daily weather for input to coarse-scale biogeochemical models (Running and Coughlan 1988; Thornton, Running, and White 1997). Everham, Wooster, and Hall (1991) built the TOPOCLIM model to simulate landscape climate at hourly time steps from empirical and mechanistic relationships. Bottom-up methods usually produce short weather records because the base station record is limited. Although there are many scaling problems, such as complex topography, with this extrapolation method, it remains perhaps the most widely used and accurate of all methods.

The last method involves simulating weather streams from historical weather station data or simulated GCM inputs using stochastic or empirical methods. Pastor and Post (1985, 1986) simulated monthly weather variations by way of a normal distribution with stochastic parameters quantified by actual weather data. Synthetic weather records were stochastically simulated from monthly weather summaries by Strandman, Vaisanen, and Kellomaki (1993) for input to climate change models in boreal ecosystems. A problem with creating daily stochastic weather streams is that daily observations are autocorrelated in time, space, and across the seven measurements. Failure to account for this correlation usually produces unrealistic weather patterns, such as high temperatures occurring on rainy days. These errors are then magnified because fire and ecosystem dynamics models will translate effects of odd weather trends onto the state variables. Lall and Sharma (1996) and Rajagopalan et al. (1997) used nonparametric resampling and time series to create statistically driven weather streams from historical data. Next-day weather is computed by matching the recent weather pattern to historical sequences. Desanker and Reed (1991) stochastically generate autocorrelated weather streams for ecological models using Markov chains and multivariate techniques. An advantage to this stochastic method is that it produces a seemingly endless weather record that is invaluable for century and millennial simulations.

Fire

The simulation of fire across the multiple scales implemented in FESM will require at least two landscape-scale modules. First, a *fire ignition* module is needed to start a fire on the landscape, and then a *fire growth* module is needed to spread that fire across the landscape. These two modules must be linked in space and time to the landscape and climate components for realistic simulations. This is an extremely difficult task. A mechanistic fire ignition module will require daily or hourly weather data from century-long records across an entire landscape defined at a resolution fine enough to distinguish lightning strikes on fuel beds. A detailed mechanistic fire growth model also requires hourly weather data and high-resolution spatial data layers that define fuel characteristics, topography, and vegetation. Obviously a compromise must be made between model resolution and algorithmic realism given the state of available research and current computer technology.

Integrated mechanistic fire models require many specialized parameters to calculate fire ignition, behavior, and growth. First, a detailed expression of the fuel bed is essential to fire modeling (Anderson 1982; Burgan and Rothermel 1984). Fuels must be described by type (live or dead), size (diameter of fuel particle), loading ($kg\,m^{-2}$), depth (m), heat content ($J\,kg^{-1}$), surface area-to-volume ratio ($m^2 m^{-3}$), mineral content (%), and moisture (%) (Brown 1970, 1981; Anderson

Table 2.2. Fuel components needed for fire simulation

Fuel component	Diameter (cm)	Material name	Material type
Duff	Very small	Decomposing Material	Humus
Litter	<1	Foliage	Leaves, needles, grass
1-h time lag	0–1	Twig	Small wood
10-h time lag	1–3	Branch	Wood
100-h time lag	3–7	Large branch	Wood
1000-h time lag	>7	Log	Wood

Notes: See Fosberg 1972 for details.

1982; Brown and Bevins 1986). Of these characteristics, usually only fuel loadings and moistures by size class are simulated in most ecosystem fire models. Their values would be obtained from other FESM modules at the stand level (see later sections) (Keane, Morgan, and Running 1996). The other fuel characteristics can be generally quantified for each biophysical classification category. Most fire models require fuel loadings by size classes that are based on relative drying rates (Fosberg 1970) and are defined by the diameter classes shown in Table 2.2. So litter and woody debris shed from vegetation compartments must be stratified by these size classes, which is rarely done in most ecosystem dynamics models. Pastor and Post (1985) stratify woody fuel by species and size class, but only to more accurately simulate decomposition. Fire models also require a detailed description of the topography to simulate weather and fire spread. This is usually taken from digital elevation models (DEMs) (U.S. Geological Survey 1987). Last, the four stand characteristics of canopy cover, crown bulk density, stand height, and crown height will be needed to compute crown fire dynamics and surface fuel moistures (Finney 1995).

Fire Ignition

Perhaps the most difficult and least understood challenge in any fire simulation is predicting when and where a fire actually starts on the landscape. Human ignitions can be somewhat easy to model using a stochastic approach that is dependent on Julian date, fire danger, and distance from developed areas (Martell, Bevilacqua, and Stocks 1989; Garcia et al. 1995). But natural ignitions, especially those resulting from lightning, are much more difficult to simulate using a mechanistic approach. Their prediction requires a fundamental understanding of lightning dynamics, ignition processes, smoldering processes, and combustion physics coupled with extensive weather data sets.

Predicting the timing and location of lightning strikes is a complex task. It involves a multiple-scale approach that links regional weather to local lightning activity and site-specific lightning strikes to point-level fuel ignition (Barrows, Sandberg, and Hart 1977; Fuquay, Baughman, and Latham 1979). At a coarse scale, thunderstorm direction and intensity vary by geographical region, year, and season (Barrows, Sandberg, and Hart 1977; Uman 1987). Lightning activity within a single storm, such as the number of cloud-to-cloud and cloud-to-ground strikes, also changes during the life of the storm as it passes over a region (Uman 1987). Some lightning dynamics can be simulated by mesoscale climate models (Pielke et al. 1992), but the majority of factors that affect lightning activity are still unknown. Land form, aspect, elevation, and slope position are major landscape characteristics that can influence lightning strikes (Fowler and Asleson 1984). And, at an organism level, Knight (1987) mentions that dead trees may be more susceptible to lightning ignition but that live trees may be struck more often by lightning. Lightning strike locations and strike characteristics for cloud-to-ground lightning strikes are available across the United States using directional-finding and time-of-arrival technologies (Graham, Holle, and Lopez 1997). These data can be used for model building, but the geo-referenced inaccuracies (0.4–4 km) and limited sampling periods may prevent their use at fine scales.

The ignition of the fuel bed by the predicted lightning strike must also be explicitly simulated in FESM (Hartford 1990). Electric energy from the lightning discharge must be translated to heat of ignition for a given fuel bed, and this complex process depends on average peak current, period of continuing current, charge, and fuel-bed characteristics (Latham 1983). Only a portion of lightning strikes have the characteristics needed to start a fire (Fuquay et al. 1972; Latham and Schlieter 1989). Positive charges start most fires because they have higher peak currents and longer periods of continuing current (Fuquay 1980; Latham and Schlieter 1989; Flannigan and Wotton 1991). Fuels must have high surface areas, low moistures, and sufficient fuel loading to sustain active burning and provide the heat needed to start a wildfire (Fowler and Asleson 1984). Fuel moisture is greatly dependent on whether the lightning storm produced any rain. Latham and Schlieter (1989) describe stochastic approaches to simulating ignition on various fuel beds where arc duration, fuel moisture, and fuel-bed depth were major factors in ignition success.

Initiation of smoldering and flaming combustion processes from lightning ignitions will need to be simulated next in FESM (Hungerford, Frandsen, and Ryan 1995). Smoldering rates and heat of combustion depend on the bulk density, moisture content, and inorganic concentrations of the fuels (Latham and Schlieter 1989; Flannigan and Wotton 1991; Frandsen 1991a,b). The transition of smoldering combustion sufficient to flaming combustion sufficient to start an active wildland fire is also difficult to simulate because the fire could smolder for weeks and even months before fuel and weather conditions are conducive for the initiation of a flaming surface fire. Excessive computer resources would be required to simulate smoldering combustion processes at the requisite fine scale given the complexity of the models used to simulate this phenomenon (Frandsen 1991a; Hungerford, Frandsen, and Ryan 1995). Clearly, since the smoldering combustion is a fine-scale process acting on very small pieces of ground (1–5 m^2) (Frandsen 1991a), a detailed resolution of the input data layers describing fuels and moistures would be prohibitive for development and simulation efficiency.

The entire complex ignition process is often modeled using stochastic approaches where the probability of a fire start is approximated from fire history or stand structure data (see Johnson 1992; Johnson and Gutsell 1994; Boychuk et al. 1997). Weibull functions and their derivatives are used to generate probabilities of fire occurrence for only stand-replacement fires (Van Wagner 1978; Johnson and Van Wagner 1985; Baker 1989a, 1993; Baker, Egbert, and Frazier 1991). Reed (1994) used likelihood functions to estimate annual probability of stand-replacement fire and Baker, Egbert, and Frazier (1991) used probabilities in the DISPATCH model to simulate the interaction of climate change on fire regime and landscape dynamics. Li et al. (1996) investigated the sensitivity of four fire probability functions and their parameters on fire rotation periods. Unfortunately, it is difficult to parameterize these functions because fire history and tree age data are unavailable, inadequate, or inappropriate for many stands (Fox 1989; Baker 1989a; Finney 1995; Boychuk et al. 1997). Moreover frequent, nonlethal fire regimes (i.e., not stand replacement) are often described from fire-scar evidence that does not accurately represent fire size (Marsden 1983; Fox 1989; Baker 1989a; Finney 1995).

Perhaps the most efficient way to simulate fire starts is by a melding of approaches where mechanistic variables, dynamically linked to other FESM compartments, are used to drive stochastic functions. But it is problematic to simulate fire ignition stochastically, and then subsequent fire behavior mechanistically, to produce results that compare well to fires observed on the landscape (McKenzie, Peterson, and Alvarado 1996; Keane and Long 1997). Incompatibilities across module linkages may result when weather and spatial databases are inconsistent in time and space scales across the simulation of both processes. For example, the fire ignition module could start fires in wet periods if ignition probability functions do not include daily or monthly weather (Keane et al. 1997; Keane and Long 1997). This inconsistency is common in many spatially explicit fire succession models, and more research is needed to pioneer comprehensive and complementary techniques to link ignition with behavior.

Fire Growth Simulation

There are many fire behavior simulation models available for research and management applications, but only a few are compatible with the mechanistic FESM design strategy. Fire growth models can be grouped by mechanistic and nonmechanistic approaches in a spatial or nonspatial implementation. Mechanistic, nonspatial models developed by Albini (1976a), Rothermel (1972, 1991), McArthur (1967), and Noble, Bary, and Gill (1980) are used extensively in land and fire management programs around the world. Computer programs containing these models, such as BEHAVE (Andrews 1986), are the backbone of many fire management programs. Nonspatial models with empirical approaches include the Australian model developed by MacArther (1967) and quantified by Noble, Bary, and Gill (1980). Although these models are not as robust and sometimes have only local applications, they are useful because they require minimal data for input and generate somewhat accurate predictions.

Spatially explicit fire behavior models have recently become important in landscape ecology and fire management simulations (Andrews 1990). Accurate predictions of fire sizes and fire spread speeds and directions are critical to researching and managing both prescribed fire and wildfire. These models are reviewed and categorized by Gardner, Romme, and Turner (1999) and somewhat by McCarthy and Gill (1997). We did not find any purely empirical spatial fire models, but there are many mechanistic fire models that use a variety of stochastic, fractal, and geometric techniques to spread fire across landscapes.

The most common approach to mechanistic spatial fire modeling is to simulate fire growth as a discrete process of cell-to-cell ignitions across a regularly spaced landscape grid (i.e., cellular models). Kourtz and O'Reagan (1971) calculated fire arrival time for distances between the eight neighboring cells in a grid to compute which cell the fire spreads to first. Some cellular techniques use templates of varying shapes and sizes to circumscribe fire perimeters (Green 1983), while others use stochastic percolation procedures (Von Niessen and Blumen 1988; Beer and Enting 1990; Gardner et al. 1996), biased percolation (Ohtsuki and Keyes 1986), or fractal algorithms (Clarke, Brass, and Riggan 1994) to simulate the uncertainty in fire spread across a landscape. Fire behavior in nonuniform fuels was spatially simulated using hexagon-shaped pixels and mechanistic fire spread equations by Frandsen and Andrews (1979). Vasconcelos and Guertin (1992) linked a cellular fire spread model to a Geographic Information System (GIS) for fire management applications. Karafyllidis and Thanailakis (1997) incorporate weather and topography in their forest fire cell automata model. Turner et al. (1994) use a cellular model to investigate the effects of fire on landscape pattern in Yellowstone National Park. Cellular models can produce unrealistic fire shapes when environmental conditions and landscapes become complex and heterogeneous because they do not respond well to subtle changes in diurnal wind speed, topography, wind direction, and fuel moisture (French 1992).

Vector or wave approaches to fire growth modeling treat the fire front as a continually expanding polygon using discrete timesteps (Anderson et al. 1982). Fire polygon boundaries are defined by a series of vertices (x, y coordinates) determined from a computation of the spread rate and direction for the time interval (Richards 1995; Finney 1998). Huygens's principle, which states that growth along boundaries can be modeled as a progression of elliptical wavelets, was used by Sanderlin and Sunderson (1975) and Finney (1998) to spatially grow fires. Richards (1990, 1995) analytically derived a differential equation to propagate fire from various points using elliptical and other fire shapes. Many others have also developed procedures for computing fire perimeter positions based on Huygens's principle (Anderson et al. 1982; Catchpole, Alexander, and Gill 1982; Dorrer 1993; Knight and Coleman 1993; Wallace 1993).

Many spatial fire models used in ecosystem simulations do not explicitly simulate fire behavior, but rather infer subsequent fire effects from vegetation and fuel characteristics inside a simulated burn perimeter. The EMBYR model (Gardner et al. 1996) spreads fire across the landscape using a stochastic cell automata approach where probabilities are related to vegetation and fuel attributes. Burn severity is calculated as a linear combination of fuel type, fuel moisture, wind speed, and cell burn rate. McCarthy and Gill (1997) use probability distributions to compute fire ignition and size, and then used a cell automata model to spread this fire across the landscape. Green (1989) simulated landscape fires by drawing the number of fires from a Poisson distribution and the area burnt from a geometric distribution. These types of cell automata models may be more appropriate for midscale to coarse time and space scale applications (McKenzie, Peterson, and Alvarado 1996; Keane and Long 1997). Other landscape models confine the spread process to mapped polygon boundaries and select the polygon to burn using probability distributions. Roberts and Betz (1999) simulated landscape fire dynamics using fuzzy systems theory at the stand level. The SIMPPLLE (Chew 1997) and CRBSUM (Keane et al. 1996) models simulate fire by selecting stands to be burned based on fire interval probabilities, and then modeling fire effects based on cover type and biophysical setting.

Specific properties and linkages are required to mechanistically simulate fire effects on the landscape in FESM. Most important, the model must simulate fire behavior from attributes computed in other FESM components. For example, combustion processes are simulated from weather derived in the climate module and downed organic biomass (i.e., fuels) derived in the ecosystem dynamics module to compute fire behavior characteristics. Next, FESM must have a spatially explicit fire spread simulation to incorporate large-scale influences (e.g., topography) on fine-scale fire behavior so that realistic fire patterns are generated across the landscape. The model must also have the capability of simulating the transition from surface to crown fire (Van Wagner 1977) and crown fire spread (Van Wagner 1977; Rothermel 1991). This is important for the computation of smoke, crown fuel consumption, and postfire tree mortality. Ember spotting, or the ignition of additional fires downwind, must also be included in this fire model because many fire patterns are a result of complex spot fires rather than the direct spread of the main fire (Albini 1979). Next, the model might have the ability to use weather and vegetation characteristics computed from other FESM components (e.g., shading, wind damping) to dynamically compute fuel moistures at hourly time steps (Rothermel et al. 1986).

Seed Abundance and Dispersal

Seed crop abundance and subsequent seed dispersal are needed to simulate the migration of plant species across a landscape. The amount and distribution of plant propagules across landscapes play important roles in postfire successional dynamics and subsequent landscape composition and structure. Seed crop abundance has rarely been mechanistically modeled because it is dependent on many cross-scale factors including species, plant health, long-term weather trends (drought), short-term weather disturbances (winds, hail storms, early frosts), and animal predation (Eis and Craigdallie 1983; Shearer 1985). The frequency and intensity of seed crops is usually simulated by species at the landscape level using stochastic approaches where probabilities of seed crop classes (e.g., good, fair, and poor) are taken from field studies (Kercher and Axelrod 1984; Keane, Arno, and Brown 1989). Mechanistic approaches may be possible as ecophysiological research efforts quantify the relationships between plant carbon allocation to reproductive organs and growing environment (Landsberg and Gower 1997).

Seed dispersal depends on many biotic and abiotic factors including propagule release height, topography (e.g., slope, elevation), wind speed and direction, tree density, and seed morphology (Johnson et al. 1981; Van der Pijl 1982; Greene and Johnson 1996). In addition there are many vectors besides wind that disperse seeds across a landscape, including birds, rodents, water, and large mammals (Van der Pijl 1982). Many plant species regenerate primarily from organs that survive the fire and these will be discussed in later sections.

Seed dispersal models have been developed for diverse spatial applications and most simulate only tree species. Clark et al. (1998) developed a set of general dispersal probability density functions (i.e., kernels) using the gamma function to study long-distance seed dispersal of many forest species. Malanson and Armstrong (1996) modeled seed dispersal as a negative algebraic decay away from source pixels using a Monte Carlo approach implemented in the JABOWA-II tree growth model (Botkin 1993; Malanson 1996). Keane, Morgan, and Running (1996) used a hybrid approach where empirical dispersal equations for wind-dispersed species (McCaughey, Schmidt, and Shearer 1985) and bird-dispersed species (Tomback, Hoffman, and Sund 1990) generate probabilities of tree seed landing on any given landscape pixel. Animal and wind dispersal vectors are included in the McClanahan (1986) simulation of seedflow across vegetation islands. Andersen (1991) uses a more complex, physical approach to simulate seed shadows of plants whose seeds are dispersed by the wind. In a highly mechanistic approach, Greene and Johnson (1989) simulated the effect of flight morphology on seed dispersal for wind-dispersed plant species, and then improved the model to include microclimate effects (Greene and Johnson 1996).

A mechanistic simulation of seed dispersal may need to be simplified in the multiple-scale approach of FESM to increase computer efficiency. For example, dispersal could be simulated at the stand level instead of the organism level by using average species height, wind direction and speed, elevation, and slope. These factors may then be integrated into a spatially explicit stochastic simulation of dispersal across the landscape (Keane, Morgan, and Running 1996a). Mladenoff et al. (1996) used forest age class structure rather that individual trees to simulate seed dispersal across large landscapes in the LANDIS model. Propagule dispersal from non-tree species may also need to be simplified for computational efficiency because of the vast number of plant species and dispersal vectors on the landscape (Van der Pijl 1982).

Hydrologic Processes

The routing of water as it flows across the landscape is a critical link to the understanding of stand-level water cycling, which can determine unique compositions and processes for specific areas in the landscape. Many hydrologic models have been developed using varied approaches for diverse purposes. Band et al. (1991) partitioned the landscape into "hillslopes" to more effectively simulate overland and subsurface water flow in the TOPMODEL. Beven and Kirkby (1979) constructed a basin hydrology model using a physical approach stratified by contributing landscape areas. Boumans and Sklar (1990) simulate hydrologic drainage and its effect on forest succession in a Louisiana wetland. Narasimhan (1995) reviews hydrogeologic process-based models and approaches. The selection of which hydrologic routing model to include in FESM would ultimately depend on modeling objectives. A simple and less comprehensive hydrologic routing module might be suitable if only upland stand dynamics are important. If riparian stand dynamics or fire's effect on streamflow is of concern, then a detailed representation of hydrologic processes should be included. However, detailed, physically based hydrology models have many parameters that require intensive quantification and calibration, and model outputs can have a high level of predictive uncertainty (Binley et al. 1991).

Stand and Organism Processes

FESM must contain specific model components and compartments at the stand and organism simulation level to achieve the stated objective of exploring fire's role in landscape dynamics (Fig. 2.4). Vegetation, fuels, and soils must be defined by an appropriate set of compartments that allow the application of model results to management issues and research problems (Fig. 2.4). The level of stratification of stand components again depends on simulation objective and desired outputs. A workhorse FESM module, called the *ecosystem dynamics module*, is where all stand and organism processes are simulated. Plant growth, regeneration, and mortality are simulated from climate drivers using mechanistic approaches linked to landscape-level simulation results (Landsberg and Gower 1997). The cycling of organic matter and nutrients is explicitly modeled from the processes of plant litterfall, atmospheric deposition, and decomposition (Waring and Schlesinger 1985). The effects of fire on abiotic and biotic components of the stand are also included in this module design.

Simulation Compartments

At least six types of organic material (detailed in Table 2.2) are needed to adequately represent forest floor dynamics at the stand level (Fig. 2.4) for linkage to fire behavior and ecosystem dynamics calculations. Woody material fallen from the canopy must be placed into four carbon pools depending on size of fallen

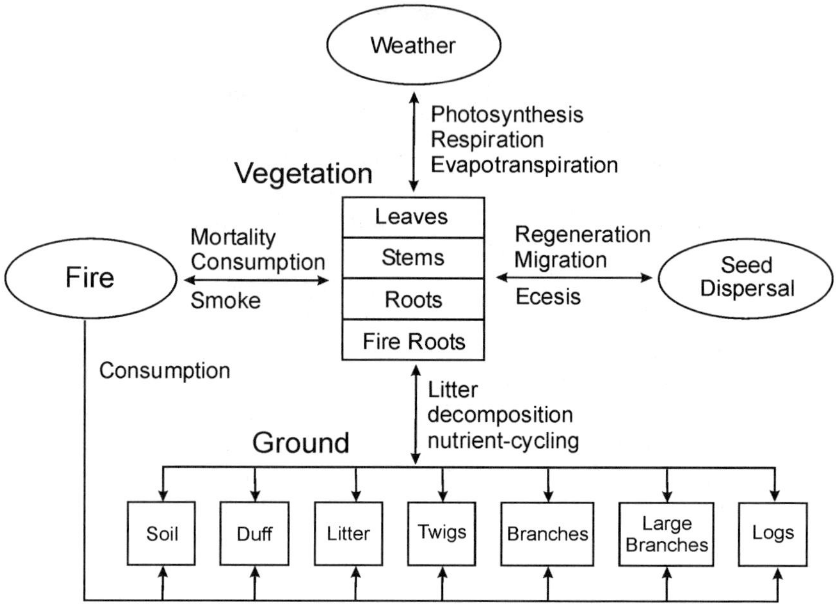

Figure 2.4. Important stand-level compartments needed for FESM simulation. Boxes represent model compartments or state variables with the arrows indicating the flows of carbon, water, and nutrients across the state variables. Circles indicate processes simulated at other scales.

particle (see Table 2.2). Needlefall is placed into the litter or duff compartment depending on the lignin concentration or level of decomposition (Meetenmeyer 1978). The woody pools and the litter compartment can then be passed to the fire growth model to compute fire intensity and spread. Carbon is transferred from woody and litter carbon pools to the soil and duff as decomposition advances. The duff compartment is necessary because its thickness influences some plant regeneration processes (Boyce 1985), and its consumption by smoldering combustion can generate high temperatures in the soil profile (Hungerford 1990) and smoke (Brown et al. 1985). The soil compartment is needed because it provides a carbon and nitrogen sink for duff decomposition and root mortality.

Plants, simulated as individuals or species groups (i.e., guilds or functional groups; see Diaz and Cabido 1997), should be explicitly represented in the model architecture. It is necessary to simulate large plants as individuals, especially trees, because the differential effect of fire on plants of different sizes and species will directly dictate postfire community and landscape composition and structure. For example, low-intensity fires in dry, montane Rocky Mountain ecosystems often maintain the dominance of ponderosa pine because they kill small Douglas-fir trees, which are more susceptible to fire mortality due to their low crown heights (Arno, Simmerman, and Keane 1985). Further the structural characteristics of individual trees describe the vertical structure of the stand, so it is important to accurately simulate sunlight and rainfall attenuation through the canopy and to model surface-to-crown fire transitions (Van Wagner 1977; Finney 1998). It is probably not necessary, nor practical, to spatially locate each tree across the simulated landscape or stand due to computer limitations and inadequate research in mechanistic spatial plant interactions at fine scales.

Since there can be tens of thousands of individual plants within a stand, there must be some simplification of plant representation in model structure to more efficiently manage computer resources. Many models simulate only individual trees and then represent other vascular plants by species or by groups of species (Shugart and West 1977; Keane, Morgan, and Running 1996). Moreover many models only simulate a small portion or vignette of a stand to increase simulation time, but this vignette must be large enough to adequately represent all ecosystem processes and small enough to ensure efficient use of computer resources (Dale and Hemstrom 1984; Botkin 1993). FESM design should probably limit individual plant simulation to only trees and represent the remaining plants by species, guilds, life-forms, or functional groups.

All plants or plant guilds should be represented by leaf, stem, coarse root, and fine root carbon and nitrogen compartments, such as in mechanistic stand-level ecosystem process models of FOREST-BGC (Running and Coughlan 1988; Running and Gower 1991) and CENTURY (Parton et al. 1987; Parton, Stewart, and Cole 1988) (Fig. 2.4). This way an efficient simulation can be obtained of fire's effects on many ecosystem processes, including photosynthesis, respiration, transpiration, and carbon allocation (Dixon et al. 1990; Bossel 1991; Mohren, Bartelink, and Lansen 1994). Ecophysiological single-tree models, such as Fire-BGC (Keane, Ryan, and Running 1996), TREE-BGC (Korol et al. 1991; Korol, Running, and Milner 1995) and HYBRID (Friend, Schugart, and Running 1993), require separate leaf, stem, coarse root, and fine root carbon and nitrogen compartments for each tree. In addition plants must be described by the structural characteristics that dictate fire, light, and water dynamics, such as height, age, diameter, and live crown height. Structural characteristics are also needed because they are used to initialize state variables, to compute intermediate variables from allometric equations, and to summarize simulation results in a form useful to management. For instance, the FARSITE model uses stand height, average live crown base height, crown bulk density (derived from leaf area), and crown closure to compute the transition and spread of crown fires. FESM design must include these compartments for a comprehensive biogeochemical landscape simulation so that the full range of fire and landscape interactions can be explored across various ecosystems and scales.

Ecosystem Dynamics Modeling

Ecosystem models can be classified into combinations of four categories: (1) stand level or plant level, (2) stand based or plant based, (3) mechanistic and non-mechanistic, and (4) spatial and nonspatial. A *stand-level* model simulates all ecosystem processes across a homogeneous piece of ground, whereas a *plant-level* model simulates the dynamics of only one plant. Plant-level models are difficult to scale up to stand level because of their inherent complexity and detail. As a result they are primarily used to investigate carbon and nutrient cycling patterns on plant growth (Bassow, Ford, and Kiester 1990; Host and Isebrands 1994; Zhang et al. 1994). ECOPHYS is a mechanistic whole-tree growth model for juvenile poplar that includes morphological, phenological, and physiological interactions in the simulation (Rauscher et al. 1990; Host and Isebrands 1994). Plant-level models are probably not appropriate for FESM. *Stand-based* models simulate stand characteristics as one entity instead of a collection of individual plants. For example, growth and yield models used in forestry simulate changes in stand basal area over time (Mohren, Bartelink, and Lansen 1994). *Plant-based* models simulate interactions between and within individual plants and their environment across a stand to investigate ecosystem dynamics. *Mechanistic* models attempt to simulate basic biogeochemical processes from fundamental physical relationships and relate them to ecosystem dynamics. Ecosystem models that directly simulate spatial interactions are called *spatial models* (Busing 1991). A summary of these models is provided in Hunsaker et al. (1993) and Baker (1989b). We believe that only stand-level, and stand- and plant-based mechanistic models are appropriate for inclusion into FESM.

Stand-based, stand-level, mechanistic ecosystem models are commonly used to explore the role of changing environment on ecosystem productivity. Grassland ecosystem models are reviewed by Hanson, Parton, and Innis (1985), while models of forest ecosystems are discussed by Dixon et al. (1990), Dale and Rauscher (1994), and Ågren et al. (1991). The "big-leaf" models such as Forest-BGC (Running and Coughlan 1988; Running and Gower 1991), Biome-BGC (Hunt et al. 1996), CENTURY (Parton et al. 1987; Parton, Stewart, and Cole 1988), and BIOMASS (McMurtrie et al. 1992) simulate fluxes of carbon, nitrogen, and water across stand-level stem, leaf, and root compartments to describe ecosystem productivity (Ågren et al. 1991). The lumped-parameter PnET model of Aber and Federer (1992) computes production from photosynthesisis, respiration, and evapotranspiration in temperate and boreal forests. Rastetter et al. (1991) built a highly aggregated biogeochemical model to investigate effects of CO_2 and other environmental factors on plant and soil carbon and nitrogen.

Unfortunately, these standlevel models are rarely directly applicable to landscape fire simulations because they fail to recognize those components that influence fire occurrence, behavior, and effects. For example, fuel loadings by size class categories (Table 2.2) are rarely simulated in these models, and none of these models explicitly represents the vegetation characteristics needed to compute fire effects. The IMAGE 2.0 model developed by Goldewijk et al. (1994) does stratify organic matter pools by woody size classes but does not simulate fire and its effects on these compartments. Most stand-based models have been implemented in a spatial domain because they do not have the tremendous number of computations required by individual plant-based models. White (1996) integrated the Biome-BGC model into a spatial implementation and linked it to the hydrology model TOPMODEL (Band et al. 1991) and other landscape modules to study landscape carbon and water dynamics.

Many tree-based, stand-level, nonspatial mechanistic models are currently available for implementation into FESM (Dale, Doyle, and Shugart 1985), but only a few have all compartments and processes critical for FESM construction. The gap-phase models, originating with JABOWA (Botkin, Janak, and Wallis 1972) and FORET (Shugart and West 1977), are a special class of mechanistic models because they abstractly simulate the effect of environmental processes on tree regeneration, growth, and mortality using growth response functions (see reviews by Botkin and Schenk 1996; Dale and Rausher 1994; Dale, Doyle, and Shugart 1985; Shugart and West 1980; Urban and Shugart 1992). Pacala, Canham, and Silander (1993) modified the JABOWA-FORET model to simulate spatial interactions between individual trees in a stand-level simulation. Some gap-phase models have the compartments and characteristics needed to simulate fire and fire effects, but very few include an explicit simulation of fire. For example, Dale, Hemstrom, and Franklin (1986) simulated fire as a probability of tree mortality on a simulation plot without regard to fuel loading, fire intensity, or topography. Examples of gap-phase fire models include BRIND (Shugart and Noble 1981), SILVA (Kercher and Axelrod 1984), FIRESUM (Keane, Arno, and Brown 1989), and ZELIG (Miller and Urban 1999a), where fire modules were added to compute fire effects at the stand level. A few gap-phase models have been implemented in landscape applications (Shugart and Seagle 1985; Urban et al. 1991; Mladenoff and Baker 1999).

The investigative power of gap-phase models was greatly improved when the growth modules were refined using more mechanistic approaches (Reed 1980; Huston, DeAngelis, and Post 1988; Dixon et al. 1990; Levine et al. 1993; Mohren, Bartelink, and Lansen 1994). For example, Friend, Shugart, and Running (1993) created the HYBRID model by merging the gap-phase model ZELIG (Burton and Urban 1990; Urban et al. 1991) with the stand-based mechanistic model FOREST-BGC (Running and Coughlan 1988). Leemans and Prentice (1989) based the growth algorithm in FORSKA on photosynthesis calculations to simulate forest succession in a Swedish broad-leaved forest for seven species (Leemans 1992). Kimmins (1993) included nutrient dynamics, photosynthetic functions, and carbon allocation in the hybrid model FORCYTE-11, which combines the historical bioassay modeling approach with tree-level process-based simulation. Battaglia and Sands (1998) and Sharp (1986) provide a review of factors, scale, and objectives of the application of several process-based forest productivity models.

Few individual tree gap models contain the compartments needed for multiscale FESM simulations for two reasons. First, they rarely include a comprehensive simulation of non-tree plant species; the exceptions being FORSKA, which simulates species in the understory layer of broadleaf European forests (Leemans and Prentice 1989) and the Kellomäki and Vaissane (1991) model that simulates the understory in boreal ecosystems. Most simulate the undergrowth as life-forms or species guilds independent of tree-level processes (Kercher and Axelrod 1984;

Keane, Arno, and Brown 1989). Second, gap-phase models seldom simulate the forest floor dynamics in compartments directly useful for fire modeling. For example, Kercher and Axelrod (1984) use only litter to carry the fire in their SILVA model so there is no treatment of woody fuel dynamics. Miller and Urban (1999a,b) used downed, dead woody and herbaceous fuels to define a fuel bed.

Regeneration is the most difficult process to simulate in most plant-based models (Pukkala 1987; Blake and Hoogenboom 1988). Once a seed has landed on the ground, it is subject to many environmental factors that may prevent or delay its germination and ultimate development to a tree, including soil moisture, temperatures, seed bed condition, nutrient status, and competition from other tree seedlings. Because of the inherent complexity in plant regeneration processes, many models establish trees as saplings and ignore germination and seedling dynamics (Blake and Hoogenboom 1988; Groot 1988; Botkin 1993). However, it is important that the effects of fire on regeneration processes be included in this simplification. Keane, Arno, and Brown (1989) linked duff reduction by fire to regeneration success using empirical relationships from Boyce (1985). Fire-caused seedling mortality is important to understory dynamics (Kercher and Axelrod 1984). The postfire flush of nutrients to the soil from combustion processes can enhance seedling survival and germination for several years after the fire and must be accounted for in FESM model design (Grier 1975).

Fire Effects

Most fire effects would be computed at the stand and organism scale in FESM design but would translate upward in scale to affect landscape composition and structure. It is impossible to simulate the full extent of fire's influence on all ecosystem components because of the complexity of fire processes and the lack of long-term field studies that take a comprehensive approach to the fire effects research. Most research efforts study only one aspect of fire's aftermath, such as fuel consumption, and do not link that effect to changes in ecosystem processes across spatial and temporal scales. Currently there are five major fire effects that should be included, at a minimum, in FESM design—fuel consumption, plant mortality, soil heat pulse, smoke, and nutrient cycling. Other second-order fire effects, such as soil erosion, are important but can be added to FESM as needed and will not be discussed here.

Fuel consumption is important because it affects carbon and nitrogen cycling, soil heat pulse, and smoke generation. Downed woody and litter consumption can be empirically calculated using consumption equations from the FOFEM (Reinhardt, Keane, and Brown 1997) or CONSUME (Ottmar et al. 1990) fire effects models. Fuel consumption regression models exist for many areas of the United States, but these equations are often limited in scope and application (e.g., Norum 1974; Sanberg 1980; Brown et al. 1985; Reinhardt, Keane, and Brown 1997). An alternative is to use the mechanistic BURNOUT model (Albini et al. 1995; Albini and Reinhardt 1995) to directly simulate fuel consumption from active and smoldering combustion, but the required extensive inputs of fuel loadings, moistures, and fuel characteristics may prohibit its use until computer technology and research findings are sufficiently advanced.

A stochastic approach is usually employed for the computation of fire-caused tree mortality because of the complex interactions involved when a fire kills a tree. First, fire can kill the tree's cambium across all or some of the circumference of the stem, resulting in a wide array of mortality responses across a wide variety of tree species (Ryan, Peterson, and Reinhardt 1987; Ryan and Reinhardt 1988). Next, heat generated from smoldering combustion can kill fine and coarse roots in the soil. And last and more common, fire can scorch all or part of the crown to kill the tree outright or over the next few years (Van Wagner 1973; Peterson 1985). It is usually a combination of these factors that contributes to the eventual demise of a tree over short to long time periods. Ryan and Reinhardt (1988) developed a robust set of stochastic equations to predict fire-caused tree mortality after one year for seven Rocky Mountain conifers. Variables in these equations act as surrogates for fire mortality vectors discussed above, and they include bark thickness (cambium insulation), percentage of crown scorched (loss of photosynthetic tissue), and species (fire resistance).

Evaluation of fire-caused mortality for non-tree plants is more difficult because many species in fire-dominated ecosystems have developed diverse and specialized strategies to survive fires (Flinn and Wein 1977; Noble and Slatyer 1977; Grime 1979; Canham and Marks 1985). Shrubs often regenerate from rhizomes or root sprouts beneath the soil surface that are protected from fire. Grasses can sprout from deep-rooted rhizomes and corms, while forbs can regenerate from bulbs, corms, or caudices (Fischer and Bradley 1987; Stickney 1990). Some plants germinate from light seeds that are dispersed great distances to burn areas, while other plants germinate from seeds that have survived the fire either in the canopy or in the soil (Stickney 1990). Still other species rely on birds to disperse seeds into recent burns (Tomback 1982; Tomback, Hoffman, and Sund 1990). FESM design should account for the diverse regenerative strategies that directly influence successional dynamics, especially if management issues such as species biodiversity and migration are important simulation objectives.

A simplified, computationally efficient method is needed to simulate undergrowth fire successional dynamics. Cattelino et al. (1979) simulated succession after fire using a vital attributes approach pioneered by Noble and Slatyer (1977). The FATE model also uses the vital attributes approach to simulate successional dynamics of Australian forest and woodland communities (Moore and Noble 1990). Roberts (1996) expanded the vital attributes concept into a more comprehensive spatial fire succession model for Bryce National Park. Both Keane (1987) and Moeur (1985) used a regression techniques and vital attributes to predict post-fire species coverage. The Fire Effects Information System (Fischer et al. 1996) is a valuable source of input parameters and species descriptions needed in the simulation of plant species dynamics. None of these models allow the direct linkage of climate, water, and light on understory plant survival and subsequent growth. Additional research is needed to link understory dynamics with the

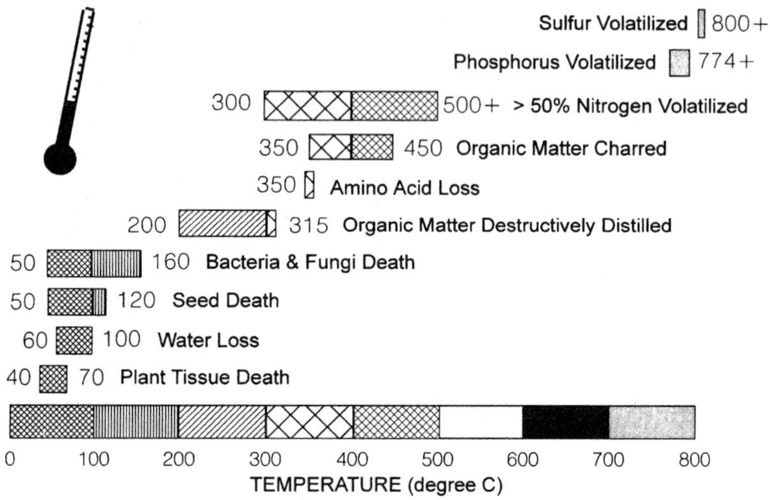

Figure 2.5. Critical temperatures for changes in important soil characteristics caused by the heat pulse from fire.

tree and climate compartments to comprehensively and mechanistically study postfire succession.

The smoldering combustion of large woody fuel and duff after the flaming front of the fire has passed causes a pulse of heat to move through the soil profile (Campbell et al. 1995). This heat pulse can kill plant roots, soil organisms, plant reproductive parts, and it can also alter important soil properties such as fertility and texture (Flinn and Wein 1977; Levitt 1980) (Fig. 2.5). The duration and intensity of this heat pulse depends on many properties of the duff and soil such as depth, inorganic content, moisture, texture, and temperature (Hungerford 1990). The heat pulse phenomenon can be simulated by a variety of mechanistic soil heat transfer models (see the review by Albini et al. 1996). Campbell et al. (1995) used a soil heat and moisture transport model to simulate soil temperatures at various soil depths under differing soil moisture conditions. Hungerford (1990) presents a conceptual model containing the major components needed to mechanistically simulate soil heat pulse and its effect on plant tissue mortality. The Aston and Gill (1976), Philip and deVries (1957), and deVries (1958) models of soil heat and moisture transport were used as a basis for many other modeling efforts (Schroeder 1974; Jury 1973; Peter 1992). Unfortunately, very little is known of the transfer of heat from the soil to living plant parts and the translation of that heat to plant mortality or reduction in vigor (Flinn and Wein 1977).

Smoke production is becoming an increasingly important issue in fire research and management, and its inclusion in a simulation of wildland fire is essential for three reasons. First, smoke emissions from wildland fires are important greenhouse gases, and any comprehensive investigation of the effect of global climate change on ecosystem and carbon dynamics must include smoke inputs to the atmosphere (Ward 1990). Second, current smoke emission regulations often restrict prescribed burning on many public lands because of the adverse effects on human health. It makes no sense to simulate detailed fire management scenarios if the smoke generated by simulated fires would prevent their implementation. Last, important minerals and nutrients are lost from the ecosystem in the smoke plume. Nitrogen is volatilized during the combustion process and transported away from the fire in smoke, as are other elements such as phosphorous, magnesium, and potassium (Ward 1990). It may be important to account for this loss in the nutrient cycle, especially where they are limited. Nutrient losses are often computed from the proportion of the fuel consumed and the fuel type (Little and Ohmann 1988; Keane, Morgan, and Running 1996). Kutiel and Shaviv (1992) found postfire nutrient dynamics dictated successional trajectory and understory dynamics. These conclusions demonstrate that the direct linkage of fire-caused changes in nutrient pools to ecosystem dynamics is essential for long-term successional simulations.

Smoke is directly related to the amount of fuel consumed and the rate or efficiency of combustion (Ward 1990). Smoke is easily computed by multiplying an emission factor by the amount of woody biomass consumed based on combustion efficiency (Reinhardt, Keane, and Brown 1997). Emission factors are available for many forest species and woody fuel size classes (Ward, Peterson, and Hao 1993). Especially important in smoke management issues is the dispersal of smoke over large land areas, which requires the rate of smoke production over time. This can be computed in FESM by linking the Albini et al. (1996) BURNOUT model with any of the current smoke dispersion models (see Breyfogle and Ferguson 1996) such as CALPUFF (Scire et al. 1995), VALBOX (Sestak, Marlatt, and Riebau 1989), PLUMP (Latham 1994), and SASEM (Sestak and Riebau 1988). This linkage would require comprehensive representation of wind speed and direction at different atmospheric heights over several days from the climate module.

FESM Implementation

The acute complexity and detail involved in a strict mechanistic FESM design would ultimately prevent its development because of previously mentioned reasons. However, it is possible to create a spatially explicit landscape fire model using parts or simplifications of the FESM structure presented here. Empirical and stochastic modules can be substituted for some mechanistic components until adequate research and computer technology become available. We have several recommendations for the successful construction of such a model. First, it is imperative that the simulation objectives be clearly defined prior to construction so that appropriate modules can be designed and included in FESM structure (Korzukhin, Ter-Mikaelian, and Wagner 1996). The "shotgun" approach of simulating every ecosystem process may allow investigation of many aspects of ecosystem dynamics, but ultimately it will lead to large, unwieldy computer programs that require abundant computer resources and are only moderately useful. Second, it is more efficient if the model is constructed from a set of linked computer programs that can be easily added or removed from the simulation like building blocks or tinker toys (Bevins and Andrews 1994; Bevins, Andrews and Keane 1995). Comprehensive simulation of ecosystem processes and their interactions is an extremely complex task and the development of detailed mechanistic models to simulate these processes is often best left to the appropriate discipline. It may be more practical if simulation modules are simply taken from previously tested models and modified for inclusion in linked fire simulations. Last, there must be comprehensive verification and testing of intermediate and final simulation results to assess model behavior, sensitivity, accuracy, and precision (Rastetter 1996). This requires extensive field data sets to validate internal algorithms and parameters so that simulation computations can be interpreted in the right context (Turner, Costanza, and Sklar 1989).

Simulation platforms can be useful tools for linking many simulation programs with databases, GIS, and other software to form a comprehensive computer application. Many complex simulation efforts use this integrated approach to link programs into one application. The Loki system (Bevins and Andrews 1994) allowed Keane et al. (1996, 1997) to link the FARSITE fire behavior model with the seed dispersal model SEEDER, the Fire-BGC ecosystem process model, the FIRESTART model, and the GRASS GIS system (USA CERL 1990) to simulate the role of fire on Glacier National Park ecosystems. Ford, Running, and Nemani. (1994) developed a modular system to link mechanistic ecosystem process models across several scales. An object-oriented, event-driven simulation system was developed by Bolte, Fisher, and Ernst (1993) to investigate the complexities of biological systems. Lauenroth et al. (1993) coupled four models to investigate interactions between vegetation structure and ecosystem processes along environmental gradients.

Models of Landscape Change

Landscape change models investigate the role of disturbance, primarily fire, on landscape dynamics, and several existing models provide examples of how FESM components can be developed using different approaches integrated into a comprehensive application. Baker (1989a) examines several models of landscape change and groups them into whole, distributional, and spatial landscape models depending on the level of data aggregation. Details of some landscape models are presented in Mladenoff and Baker (1999). We only review landscape models where spatial interactions are directly simulated in model design.

The LANDIS model was used to evaluate fire, wind throw, and harvest disturbance regimes on landscape pattern and structure (Mladenoff et al. 1996). Fire is indirectly simulated at the standlevel by quantifying fire effects based on age class structure, and succession is simulated as a competitive process driven by species life history parameters. Roberts (1996) used life history parameters or vital attributes (Noble and Slatyer 1977) to drive succession in his polygon-based model LANDSIM, which also simulates fire effects at the polygon level without a fire spread model (Roberts and Betz 1999). The DISPATCH model of Baker (1992b, 1993) and Baker, Egbert, and Frazier (1991) stochastically simulates fire occurrence and spread based on the dynamically simulated weather, fuel loadings, and topographic setting, and then simulates subsequent forest succession as a change in cover type and stand age. The SIMPPLLE model (Chew 1997) uses a multiple-pathway approach to simulate succession on landscape polygons. It uses a stochastic approach to determine when a polygon will burn. Miller (1994) implemented a spatial application of fire in the Zelig model to assess the interaction of fire, climate, and pattern in Sierra Nevada forests (Miller and Urban 1999a, 199b). Ratz (1995) used a single-pathway succession model linked to a cellular automata fire spread model to simulate long-term fire patterns in boreal forests. Keane and Long (1997) used a multiple-pathway succession model to simulate coarse-scale fire succession, where fire is simulated as an independent stochastic process that can burn across polygon boundaries.

A major problem with many landscape fire models is the forcing of fire spread along polygon boundaries when actual fire growth depends on many factors other than vegetation. Weather, topography, wind, and landform all influence fire pattern. As a result fire spread must be modeled independent of vegetation layer delineations and allowed to transect and divide polygons to achieve realistic simulations. Another limitation of current landscape fire simulation efforts is their inability to recognize the range of severities within and across stands. Daily fluctuations in wind and humidity can cause fine-scale differences in fire effects within a stand. Tree mortality, fuel consumption, and smoke can vary a great deal within a stand boundary because of changes in fire weather. Future models would need to incorporate fine-scale fire severity patterns into the spatial design.

Conclusion

A comprehensive, mechanistic simulation of wildland fire and ecosystem dynamics across a landscape may not be possible because of computer limitations, inadequate research, inconsistent data, and extensive parameterization. Therefore empirical and stochastic approaches must be substituted for many mechanistic modules until research and technology improve. Unfortunately, nonmechanistic approaches limit the scope and applicability of spatial ecosystem process models. Ecosystem dynamics models need to be refined so that appropriate compartments that model fire spread and effects are explicitly represented in their structure. Landscape disturbance models need to simulate fire growth unconstrained by vegetation polygon delineations. Most important, these models must be designed in the context of the simulation objective to ensure that appropriate simulation modules are included.

References

Aber, J.D., and Federer, C.A. 1992. A generalized, lumped-parameter model of photosynthesis, evapotranspiration, and net primary production in temperate and boreal forest ecosystems. *Oecologia* 92:463–474.

Ågren, G.I., and Axelsson, B. 1980. PT-A tree growth model. *Ecol. Bull.* (Stockholm) 32: 525–536.

Ågren, G.I., McMurtrie, R.E., Parton, W.J., Pastor, J., and Shugart, H.H. 1991. State-of-the-art of models of production-decomposition linkages in conifer and grassland ecosystems. *Ecol. Appl.* 1(2):118–138.

Albini, F.A. 1976a. *Computer-Based Models of Wildland Fire Behavior: A User's Manual.* Ogden, UT: USDA Forest Service, Intermountain Forest and Range Experiment Station. 68p.

Albini, F.A., and Reinhardt, E.D. 1995. Modeling ignition and burning rate of large woody natural fuels. *Int. J. Wildl. Fire* 5(2):81–91.

Albini, F.A., Brown, J.K., Reinhardt, E.D., and Ottmar, R.D. 1995. Calibration of a large fuel burnout model. *Int. J. Wildl. Fire* 5(3):173–192.

Albini, F.A., Amin, M.R., Hungerford, R.D., Frandsen, W.H., and Ryan, K.C. 1996. Models for fire-driven heat and moisture transport in soils. USDA Forest Service Gen. Tech. Rep. INT-GTR-335. 16p.

Andersen, M. 1991. Mechanistic models for the seed shadows of wind-dispersed plants. The *Am. Naturalist* 137(4):476–497.

Anderson, D.G., Catchpole, E.A., DeMestre, N.J., and Parkes, T. 1982. Modeling the spread of grass fires. *J. Austral. Math. Soc.* (ser. B.) 23:451–466.

Anderson, H.E. 1969. Heat transfer and fire spread. Res. Pap. INT-69. Ogden, UT: USDA, Forest Service, Intermountain Forest and Range Experiment Station. 20p.

Anderson, H.E. 1982. Aids to determining fuel models for estimating fire behavior. Gen. Tech. Rep. INT-122. Ogden, UT: USDA Forest Service, Intermountain Forestry and Range Experimental Station. 22p.

Andrews, P.L. 1986. BEHAVE: Fire behavior prediction and fuel modeling system—BURN subsystem. USDA Forest Service Gen. Tech. Rep. INT-194. 130p.

Andrews, P.L. 1990. Application of fire growth simulation models in fire management. In *Proceedings of the 10th Conference on Fire and Forest Meteorology*, pp. 317–321. April 17–21, Ottawa, Canada. Society of American Foresters, Washington, DC.

Arno, S.F., Simmerman, D.G., and Keane, R.E. 1985. Forest succession on four habitat types in western Montana. Gen. Tech. Rep. INT-177. Ogden, UT: USDA Forest Service, Intermountain Forest and Range Experiment Station. 74p.

Aston, A.R., and Gill, A.M. 1976. Coupled soil moisture, heat, and water vapor transfers under simulated fire conditions. *Austral. J. Soil Res.* 14:55–66.

Baker, W.L. 1989a. Effect of scale and spatial heterogeneity on fire-interval distributions. *Can. J. For. Res.* 19:700–706.

Baker, W. L. 1989b. A review of models of landscape change. *Landscape Ecol.* 2(2): 111–133.

Baker, W.L. 1992a. Effects of settlement and fire suppression on landscape structure. *Ecology* 73:1879–1887.

Baker, W.L. 1992b. The landscape ecology of large disturbances in the design and management of nature reserves. *Landscape Ecol.* 7(3):181–194.

Baker, W.L. 1993. Spatially heterogeneous multi-scale response of landscapes to fire suppression. *Oikos* 66:66–71.

Baker, W.L., Egbert, S.L., and Frazier, G.F. 1991. A spatial model for studying the effects of climatic change on the structure of landscapes subject to large disturbances. *Ecol. Model.* 56:109–125.

Ball, G.L., and Gimblett, R. 1992. Spatial dynamic emergent hierarchies simulation and assessment system. *Ecol. Model.* 62:107–121.

Band, L.E., Peterson, D.L., Running, S.W., Coughlan, J., Lanners, R., Dungan, J., and Nemani, R. 1991. Forest ecosystem processes at the watershed scale: basis for distributed simulation. *Ecol. Model.* 56:171–196.

Barrows, J.S., Sandberg, D.V., and Hart, J.D. 1977. Lightning fires in Northern Rocky Mountain forests. USDA Forest Service Final Report for Contract Grant 16-440-CA. On file, USDA Forest Service, Intermountain Fire Sciences Laboratory, P.O. Box 8089, Missoula, MT. 221p.

Bassow, S.L., Ford, E.D., and Kiester, A.R. 1990. A critique of carbon-based tree growth models. In *Process Modeling of Forest Ggrowth Responses to Environmental Stress*, eds. R.K. Dixon, R.S. Meldahl, G.A. Ruark, and W.G. Warren, pp. 50–57. Portland, OR: Timber Press.

Battaglia, M., and Sands, P.J. 1998. Process-based forest productivity models and their application in forest management. *For. Ecol. Manag.* 102:13–32.

Beer, T., and Enting, I.G. 1990. Fire spread and percolation modelling. *Mathl. Comput. Model.* 13(11):77–96.

Beven, K.J., and Kirkby, M.J. 1979. A physically based, variable contributing area model of basin hydrology. *Hydrol. Sci. Bull.* 24(1):43–69.

Bevins, C.D., and Andrews, P.L. 1994. The Loki software architecture for fire and ecosystem modeling: A tinker toy approach. In *12th Conference on Fire and Forest Meteorology*, pp. 252–260, October 26–28, Jekyll Island, GA. Society of American Foresters, Washington, DC.

Bevins, C.D., Andrews, P.L., and Keane, R.E. 1995. Forest succession modelling using the Loki software architecture. *Lesnictvi-Forestry* 41(4):158–162.

Binley, A.M., Beven, K.J., Calver, A., and Watts, L.G. 1991. Changing responses in hydrology: Assessing the uncertainty in physically based model predictions. *Water Resources Res.* 27(6):1253–1261.

Blake, J.I., and Hoogenboom, G. 1988. A dynamic simulation of loblolly pine (*Pinus taeda*) seedling establishment based upon carbon and water balances. *Can. J. For. Res.* 18:833–850.

Blyth, E.M., Dolman, A.J., and Noilhan, J. 1994. The effect of forest on mesoscale rainfall: An example from HAPEX-MOBILHY. J. Appl. Meteorol. 33(4):445–454.

Bolte, J.P., Fisher, J.A., and Ernst, D.H. 1993. An object-oriented, message-based environment for integrating continuous, event-driven and knowledge-based simulation. In *Proceedings of Conference on Application of Advanced Information Technologies: Effective Management of Natural Resources*, pp. 290–308, June 18–19, Spokane, WA. American Society of Agricultural Engineers.

Bossel, H. 1991. Modeling forest dynamics: Moving from description to explanation. *Forest Ecol. Manag.* 42:129–142.

Bossel, H., and Schäfer, H. 1988. Eco-physiological dynamic simulation model of tree growth, carbon, and nitrogen dynamics. In *Forest Simulation Systems: Proceedings of the IUFRO Conference*, eds. L.C. Wensel, G.S. Biging. November 2–5, 420p. Berkeley, CA: University of California, Division of Agriculture and Natural Resources. pp. 121–122.

Botkin, Daniel B. 1993. *Forest Dynamics: An Ecological Model.* New York: Oxford University Press.

Botkin, D.B., and Schenk, H.J. 1996. Review and analysis of JABOWA and related forest models and their use in climate change studies. NCASI Tech. Bull. 717. 62p.

Botkin, D.B., Janak, J.F., and Wallis, J.R. 1972. Some ecological consequences of a computer model of forest growth. *J. Ecol.* 60:849–872.

Boumans, R.M.J., and Sklar, F.H. 1990. A polygon-based spatial (PBS) model for simulating landscape change. *Landscape Ecol.* 4(2/3):83–97.

Boyce, R.B. 1985. Conifer germination and seedling establishment on burned and unburned seedbeds. MS thesis. University of Idaho, Moscow.

Boychuk, D., Perera, A.H., Ter-Mikaelian, M.T., Martell, D.L., and Li, C. 1997. Modelling the effect of spatial scale and correlated fire disturbances on forest age distribution. *Ecol. Model.* 95:143–162

Breyfogle, S., and Ferguson, S.A. 1996. User assessment of smoke-dispersion models for wildland biomass burning. USDA Forest Service Gen. Tech. Rep. PNW-GTR-379. 30p.

Brown, J.K. 1970. Ratios of surface area to volume for common fire fuels. *For. Sci.* 16: 101–105.

Brown, J.K. 1981. Bulk densities of nonuniform surface fuels and their application to fire modeling. *For. Sci.* 27:667–683.

Brown, J.K., and Bevins, C.D. 1986. Surface fuel loadings and predicted fire behavior for vegetation types in the Northern Rocky Mountains. Res. Note INT-358. Ogden, UT: USDA Forest Service, Intermountain Forest and Range Experiment Station. 9p.

Brown, J.K., Marsden, M.A., Ryan, K.C., and Reinhardt. E.D. 1985. Predicting duff and woody fuel consumed by prescribed fire in the Northern Rocky Mountains. Res. Pap. INT-337. Ogden, UT: USDA Forest Service, Intermountain Forest and Range Experiment Station. 23p.

Burgan, R.E., and Rothermel, R.C. 1984. BEHAVE: Fire behavior prediction and fuel modeling system—FUEL subsystem. USDA Forest Service Gen. Tech. Rep. INT-167. 126p.

Burton, P.J., and Urban, D.L. 1990. An overview of ZELIG, a family of individual-based gap models simulating forest succession. In *Symposia Proceedings Vegetation Management: An Integrated Approach*, E. Hamilton, (compiler), November, 14–16, pp. 92–96, Victoria, BC. Forestry Canada Pacific Forestry Centre FRDA Rep. 109.

Busing, R.T. 1991. A spatial model of forest dynamics. *Veg. Sci.* 92:167–179.

Campbell, G.S., Jungbauer, J.D., Bristow, K.L., and Hungerford, R.D. 1995. Soil temperature and water content beneath a surface fire. *Soil Sci.* 159(6):363–374.

Canham, C.D., and Marks, P.L. 1985. The response of woody plants to disturbance: patterns of establishment and growth. In *The Ecology of Natural Disturbance and Patch Dynamics*, pp. 197–216. S.T.A. Piclult and P.S. White, San Diego, CA: Academic Press

Catchpole, E.A., Alexander, M.E., and Gill, A.M. 1982. Elliptical fire perimeter and area intensity distributions. *Can. J. For. Res.* 22:968–972.

Cattelino, P.J., Noble, I.R., Slatyer, R.O., and Kessell, S.R. 1979. Predicting multiple pathways of plant succession. *Environ. Manag.* 3:41–50.

Chew, J.D. 1997. Simulating landscape patterns and processes at landscape scales. In *Proceedings of the 11th Annual Symposium on Geographic Information Systems*, pp. 287–291. Vancouver, B.C. GIS World Publications, Fort Collins, CO.

Cipollini, M.L., Wallace-Senft, D.A., and Whigham, D.F. 1994. A model of patch dynamics, seed dispersal, and sex ratio in the dioecious shrub *Lindera benzo*in (Lauraceae). *J. Ecol.* 82:621–633.

Clark, J.S., Fastie, C., Hurtt, G., Jackson, S.T., Johnson, C., King, G., Lewis, M., Lynch, J., Pacala, S., Prentice, C., Schupp, E.W., Webb, T., and Wyckoff, P. 1998. Reid's Paradox of rapid plant migration. *Biosci.* 48:13–18.

Clarke, K.C., Brass, J.A., and Riggan, P.J. 1994. A cellular automaton model of wildfire propagation and extinction. *Photogramm. Eng. Rem. Sens.* 60(11):1355–1367.

Crutzen, P.J., and Goldammer, J.G. 1993. *Fire* in *the Environment: The Ecological, Atmospheric and Climatic Importance of Vegetation Fires*. New York: Wiley.

Dale, V.H., Doyle, T.W., and Shugart, H.H. 1985. A comparison of tree growth models. *Ecol. Model.* 29:145–169.

Dale, V.H., and Hemstrom, M. 1984. CLIMACS: A computer model of forest stand development for western Oregon and Washington. USDA Forest Service Res. Pap. PNW-327. Portland, OR: USDA Forest Service, Pacific Northwest Forest and Range Experiment Station. 60p.

Dale, V.H., Hemstrom, M., and Franklin, J. 1986. Modeling the long-term effects of disturbances on forest succession, Olympic Peninsula, Washington. *Can. J. For. Res.* 16:56–67.

Dale, V.H., and Rauscher, H.M. 1994. Assessing impacts of climate change on forests: The state of biological modeling. *Clim. Change* 28:65–90.

Desanker, P.V., and Reed, D.D. 1991. A stochastic model for simulating daily growing season weather variables for input into ecological models. In: M.A. Buford (compiler). *Proceedings of the 1991 Symposium on Systems Analysis* in *Forest Resources*, pp. 1–11, March 3–6, Charleston, SC. USDA Forest Service Gen. Tech. Rep. SE-74.

DeVries, D.A. 1958. Simultaneous transfer of heat and moisture in porous media. *Trans. Am. Geophys. Union* 39:909–916.

Diaz, S., and Cabido, M. 1997. Plant functional types and ecosystem function in relation to global change. *J. Veg. Sci.* 8:121–133.

Dickinson, R.E., Erroco, R.M., Giorgi, F., and Bates, G.T. 1989. A regional climate model for the western United States. *Clim. Change* 15:383–422.

Dixon, R.K., Meldahl, R.S., Ruark, G.A., and Warren, W.G., eds. 1990. *Process Modelling of Forest Growth Responses to Environmental Stress*. Portland, OR: Timber Press.

Dyer, J.M. 1995. Assessment of climatic warming using a model of forest species migration. *Ecol. Model.* 79:199–219.

Eis, S., and Craigdallie, D. 1983. Reproduction of conifers: A handbook for cone crop assessment, pp. 12–27. Canadian Forest Service Tech. Rep. 31.

Everham, E.M., Wooster, K.B., and Hall, C.A.S. Forest landscape climate modeling. In: M.A. Buford (compiler). *Proceedings of the 1991 Symposium on Systems Analysis* in *Forest Resources*, March 3–6, pp. 11–16, Charleston, SC. USDA Forest Service Gen. Techn. Rep. SE-74.

Fall, J., and Fall, A. 1996. SELES: A spatially explicit landscape event simulator. In *Proceedings of the NCGIA Conference on GIS and Environmental Modeling*, January 12, pp. 1–12, Santa Fe, NM.

Finney, M.A. 1995. The missing tail and other considerations for the use of fire history models. *Int. J. Wildl. Fire* 5(4):197–202.

Finney, M.A. 1998. FARSITE: Fire area simulator—Model development and evaluation. USDA Forest Service Gen. Tech. Rep. RMRS-GTR-4. 47p.

Fischer, W.C., and Bradley, A.F. 1987. Fire ecology of western Montana forest habitat types. Gen. Tech. Rep. INT-223. Intermountain Research Station. USDA Forest Service. 95p.

Fischer, W.C., Miller, M., Johnston, C.M., Smith, J.K., Simmerman, D.G., and Brown, J.K. 1996. Fire effects information system: User's guide. USDA Forest Service Gen. Tech. Rep. INT-GTR-327. 131p.

Flannigan, M.D., and Wotton, B.M. 1991. Lightning-ignited forest fires in northwestern Ontario. *Can. J. For. Res.* 21:277–287.

Flinn, M.A., and Wein, R.W. 1977. Depth of underground plant organs and theoretical survival during fire. *Can. J. Bot.* 55:2550–2554.

Ford, R., Running, S.W., and Nemani, R. 1994. A modular system for scalable ecological modeling. *IEEE Comp. Sci. Eng.* 10:32–44.

Forman, R.T.T., and Godron, M. 1986. *Landscape Ecology*. New York: Wiley.

Fosberg, M.A. 1970. Drying rates of heartwood below fiber saturation. *For. Sci.* 16:57–63.

Fowler, P.M., and Asleson, D.O. 1984. The location of lightning-caused wildland fires, Northern Idaho. *Phys. Geogr.* 5(3):240–253.

Fox, J.F. 1989. Bias in estimating forest disturbance rates and tree lifetimes. *Ecology* 70(5): 1267–1272.

Frandsen, W.H., and Andrews, P.L. 1979. Fire behavior in nonuniform fuels. Res. Pap. INT-232. Ogden, UT: USDA Forest Service, Intermountain Forest and Range Experiment Station. 34p.

Frandsen, W.H. 1991a. Heat evolved from smoldering peat. *Int. J. Wild. Fire* 1:197–204.

Frandsen, W.H. 1991b. Burning rate of smoldering peat. *Northwest Sci.* 65(4):166–172.

French, I.A. 1992. Visualization techniques for the computer simulation of bushfires in two dimensions. M.S. thesis. University of New South Wales, Australian Defence Force Academy. 140p.

Friend, A.D., Shugart, H.H., and Running, S.W. 1993. A physiology-based gap model of forest dynamics. *Ecology*. 74(3):792–797.

Fuquay, D.M. 1980. Lightning that ignites forest fires. In *Proceedings, Sixth Conference on Fire and Meteorology*, pp. 109–112. April 22–24, Seattle, WA. Society of American Foresters, Washington, DC.

Fuquay, D.M., Baughman, R.G., and Latham, D.J. 1979. A model for predicting lightning-fire ignition in wildland fuels. USDA Forest Service Res. Pap. INT-217. 21p.

Fuquay, D.M., Taylor, A.R., Hawe, R.G., and Schmid, C.W. 1972. Lightning discharges that caused forest fires. *J. Geophy. Res.* 77:2156–2158.

Garcia, C.V., Woodard, P.M., Tinus, S.J., Adamowicz, W.L., and Lee, B.S. 1995. A logit model for predicting the daily occurrence of human caused forest fires. *Int. J. Wild. Fire* 5(2):101–111.

Gardner, R.H., Hargrove, W.W., Turner, M.G., and Romme, W.H. 1996. Climate change, disturbances and landscape dynamics. pp. 149–172. In *Global Change and Terrestrial Ecosystems*, ed. B.H. Walker and W.L. Steffen. Cambridge: Cambridge University Press.

Gardner, R.H., Romme, W.H., and Turner, M.G. 1999. Predicting forest fire effects at landscape scales. In *Spatial Modeling of Forest Landscape Change: Approaches and Applications*, eds. D.J. Mladenoff and W.L. Baker, pp. 163–185. Cambridge: Cambridge University Press.

Gay, C.A. 1989. Modeling tree level processes. In *Proceedings of the Second US–USSR Symposium Air Pollution Effects on Vegetation Including Forest Ecosystems*, eds. I. Nobel and D. Reginald, pp. 143–155. Broomall, PA, September 1989.

Giorgi, F., Marinucci, M.R., Bates, G.T., and Canio, G.D. 1993a. Development of a second generation regional climate model (RegCM2). I. Boundary-layer and radiative transfer processes. *Mon. Wea. Rev.* 121:2794–2813.

Giorgi, F., Marinucci, M.R., Bates, G.T., and Canio, G.D. 1993b. Development of a second generation regional climate model (RegCM2). II. Convective processes and assimilation of lateral boundary conditions. *Mon. Wea. Rev.* 121:2813–2832.

Goldewijk, K.K., van Minnen, J.G., Kreileman, G.J., Vloedbeld, M., and Leemans, R. 1994. Simulating the carbon flux between the terrestrial environment and the atmosphere. *Water Air Soil Pollut.* 76:199–230.

Goodchild, M.F., Parks, B.O., and Steyaert, L.T. 1993. *Environmental Modeling with GIS*. New York: Oxford University Press.

Graham, B.L., Holle, R.L., and Lopez, R.E. 1997. Lightning detection and data use in the United States. *Fire Manag. Notes* 57(2):4–9.

Green, D.G. 1983. Shapes of simulated fires in discrete fuels. *Ecol. Model.* 20:21–32.

Green, D.G. 1989. Simulated effects of fire, dispersal and spatial pattern on competition within forest mosaics. *Vegetatio* 82:139–153.

Greene, D.F., and Johnson, E.A. 1989. A model of wind dispersal of winged or plumed seeds. *Ecol.* 70:339–347.

Greene, D.F., and Johnson, E.A. 1996. Wind dispersal of seeds from a forest into a clearing. *Ecol.* 77(2):595–609.

Grier, C.C. 1975. Wildfire effects on nutrients distribution and leaching in a coniferous ecosystem. *Can. J. For. Res.* 5:599–607.

Grime, J.P. 1979. *Plant Strategies and Vegetation Processes*. New York: Wiley.

Groot, A. 1988. Methods for estimating seedbed receptivity and for predicting seedling stocking and density in broadcast seeding. *Can. J. For. Res.* 18:1541–1549.

Hanson, J.D., Parton, W.J., and Innis, G.S. 1985. Plant growth and production of grassland ecosystems: a comparison of modelling approaches. *Ecol. Model.* 29:131–144.

Hartford, R.A. 1990. Smoldering combustion limits in peat as influenced by moisture, mineral content, and organic bulk density. In *Proceedings of the 10th Conference on Fire and Forest Meteorology*, eds. D.C. MacIver, H. Auld, R. Whitewood, pp. 282–286. April 17–21, Ottawa, Ontario. Forestry Canada, Petawawa National Forestry Institute, Chalk River, Ontario.

Host, G.E., and Isebrands, J.G. 1994. An interregional validation of ECOPHYS, a growth process model of juvenile poplar clones. *Tree Physiol.* 14:933–945.

Hsie, E.Y. 1987. MM4 (Penn State/NCAR) mesoscale model version 4 documentation. NCAR Tech. Note, NCAR/TN294 + STR, National Center for Atmospheric Research, P.O. Box 3000, Boulder, CO 80307. 215p.

Hungerford, R.D. 1990. Modeling the downward heat pulse from fire in soils and in plant tissue. In *Proceedings of the 10th Conference on Fire and Forest Meterology*, eds. D.C. MacIver, H. Auld, R. Whitewood, pp. 148–151. April 17–21, Ottawa, Ontario. Forestry Canada, Petawawa National Forestry Institute, Chalk River, Ontario.

Hungerford, R.D., Frandsen, W.H., and Ryan, K.C. 1995. Ignition and burning characteristics of organic soils. In *Fire* in *Wetlands: A Management Perspective. Proceedings of the Tall Timbers Fire Ecology Conference No. 19*, eds. S.I. Cerulean and R.T. Engstrom, pp. 78–91. Tallahasee, FL. Tall Timbers Research Station.

Hungerford, R.D., Nemani, R.R., Running, S.W., and Coughlan, J.C. 1989. MTCLIM: A mountain microclimate simulation model. Research Paper INT-414. Ogden, UT: USDA Forest Service, Intermountain Research Station. 52p.

Hunsaker, C.T., Nisbet, R.A., Lam, D.C., Brower, J.A., Baker, W.L., Turner, M.G., and Botkin, D.B. 1993. Spatial models of ecological systems and processes: The role of GIS. In *Environmental Modeling with GIS*, eds. M.F. Goodchild, B.O. Parks, L.T. Steyaert, pp. 248–264. New York: Oxford University Press.

Hunt, E.R., Piper, S.C., Nemani, R., Keeling, C.D., Otto, R.D., and Running, S.W. 1996. Global net carbon exchange and intra-annual atmospheric CO_2 concentrations predicted by an ecosystem process model and three-dimensional atmospheric transport model. *Global Biogeochem. Cycles* 10(3):431–456.

Huston, M., DeAngelis, D., and Post, W. 1988. New computer models unify ecological theory. *BioScience* 38(10):682–691.

Johnson, E.A. 1992. *Fire and Vegetation Dynamics: Studies from the North American Boreal Forest*. New York: Cambridge University Press.

Johnson, E.A., and Gutsell, S.L. 1994. Fire frequency models, methods, and interpretations. *Adv. Ecol. Res.* 25:239–285.

Johnson, E.A., and Van Wagner, C.E. 1985. The theory and use of two fire history models. *Can. J. For. Res.* 15:214–220.

Johnson, W.C., Sharpe, D.M., DeAngelis, D.L., Fields, D.E., and Olson, R.J. 1981. Modeling seed dispersal and forest island dynamics. In *Forest Island Dynamics* in *Man-Dominated Landscapes*, eds. R.L. Burgess and D.M. Sharpe, pp. 215–239. New York: Springer.

Jury, W.A. 1973. Simultaneous transport of heat and moisture through a medium sand. Ph.D. dissertation. Physics Department, University of Wisconsin, Madison. 19p.

Karafyllidis, I., and Thanailakis, A. 1997. A model for predicting forest fire spreading using cellular automata. *Ecol. Model.* 99:87–97.

Keane, R.E. 1987. Forest succession in western Montana—A computer model designed for resource management. Res. Note INT-376. USDA Forest Service, Intermountain Research Station. 8p.

Keane, R.E., Arno, S.F., and Brown, J.K. 1989. FIRESUM—An ecological process model for fire succession in Western conifer forests. Gen. Tech. Rep. INT-266. Ogden, UT: USDA Forest Service, Intermountain Research Station. 76p.

Keane, R.E., Arno, S.F., and Brown, J.K. 1990. Simulating cumulative fire effects in ponderosa pine/Douglas-fir forests. *Ecology* 71(1):189–203.

Keane, R.E., Morgan, P., and Running, S.W. 1996. Fire-BGC—A mechanistic ecological process model for simulating fire succession on coniferous forest landscapes of the Northern Rocky Mountains. USDA Forest Service Res. Pap. INT-484. 122p.

Keane, R.E., Long, D.G., Menakis, J.P., Hann, W.J., and Bevins, C. 1996. Simulating coarse scale vegetation dynamics with the Columbia River Basin succession model CRBSUM. USDA Forest Service Gen. Tech. Rep. INT-GTR-340. 50p.

Keane, R.E., Hardy, C.C., Ryan, K.C., and Finney, M.A. 1997. Simulating effects of fire on gaseous emissions from future landscape of Glacier National Park, Montana, USA. *World Resources Rev.* 9(2):177–205.

Keane, R.E., and Long, D.G. 1997. A comparison of coarse scale fire effects simulation strategies. *Northwest Sci.* 72(2):76–90.

Keane, R.E., Rarsons, R., and Hessburg, P. 2002. Estimating historical range and variation of landscape path dynamics: Limitations of the simulation approach. *Ecol. Model.* (In press).

Kellomäki, S., and Väisänen, H. 1991. Application of a gap model for the simulation of forest ground vegetation in boreal conditions. *For. Ecol. Manag.* 42:35–47.

Kercher, J.R., and Axelrod, M.C. 1984. A process model of fire ecology and succession in a mixed-conifer forest. *Ecology* 65(6):1725–1742.

Kessell, S.R., Good, R.B., and Hopkins, A.J.M. 1984. Implementation of two new resource management information systems in Australia. *Environ. Manag.* 8:251–270.

Kimmins, J.P. 1993. Scientific foundations for the simulation of ecosystem function and management in FORCYTE-11. Inf. Rep. NOR-X-328. Edmonton, Alberta: Forestry Canada, Northwest Region, Northern Forestry Centre. 88p.

Knight, D.H. 1987. Parasites, lightning and the vegetation mosaic in wilderness andscapes. In *Landscape Heterogeneity and Disturbance*, ed. M.G. Turner, pp. 59–83. New York: Springer-Verlag.

Knight, I., and Coleman, J. 1993. A fire perimeter expansion algorithm based on Huygens' wavelet propagation. *Int. J. Wild. Fire* 3(2):73–84.

Korol, R.L., Running, S.W., Milner, K.S., and Hunt, E.R. 1991. Testing a mechanistic carbon balance model against observed tree growth. *Can. J. For. Res.* 21:1098–1105.

Korol, R.L., Running, S.W., and Milner, K.S. 1995. Incorporating intertree competition into an ecosystem model. *Can. J. For. Res.* 25:413–424.

Korzukhin, M.D., Ter-Mikaelian, M.T., and Wagner, R.G. 1996. Process versus empirical models: which approach for forest ecosystem management? *Can. J. For. Res.* 26: 879–887.

Kourtz, P., and O'Reagan, W.G. 1971. A model for a small forest fire to simulate burned and burning areas for use in a detection model. *For. Sci.* 17(2):163–169.

Kutiel, P., and Shaviv, A. 1992. Effects of soil type, plant composition and leaching on soil nutrients following a simulated forest fire. *For. Ecol. Manag.* 53:329–343.

Lall, U., and Sharma, A. 1996. A nearest neighbor bootstrap for time series resampling. *Water Resources Res.* 32(3): 679–693.

Landsberg, J.J., and Gower, S.T. 1997. *Applications of Physiological Ecology to Forest Management*. San Diego, CA: Academic Press.

Latham, D. 1983. LLAFFS-A lightning-locating and fire-forecasting system. USDA Forest Service Res. Pap. INT-315. 44p.

Latham, D., Burgan, R., Chase, C., and Bradshaw, L. 1997. Using lightning location in the Wildland Fire Assessment System. USDA Forest Service Gen. Tech. Rep. INT-GTR-349. 5p.

Latham, D.J., and Schlieter, J.A. 1989. Ignition probabilities of wildland fuels based on simulated lightning discharges. USDA Forest Service Res. Pap. INT-411. 16p.

Latham, D. 1994. PLUMP: A plume predictor and cloud model for fire managers. USDA Forest Service INT-GTR-314. 15p.

Lauenroth, W.K., Urban, D.L., Coffin, D.P., Parton, W.J., Shugart, H.H., Kirchner, T.B., and Smith, T.M. 1993. Modeling vegetation structure-ecosystem process interactions across sites and ecosystems. *Ecol. Model.* 67:49–80.

Leemans, R. 1992. Simulation and future projection of succession in a Swedish broad-leaved forest. *For. Ecol. Manag.* 48:305–319.

Leemans, R., and Prentice, I. 1989. FORSKA, a general forest succession model. Gen. Rep. 89/2, Institute of Ecological Botany, Uppsala, Sweden. 45p.

Levitt, J. 1980. *Responses of Plants to Environmental Stresses: Chilling, Freezing, and High Temperature Stresses*, vol. 1. New York: Academic Press.

Levine, E.R., Ranson, K.J., Smith, J.A., Williams, D.L., Knox, R.G., Shugart, H.H., Urban, D.L., and Lawrence, W.T. 1993. Forest ecosystem dynamics; linking forest succession, soil process and radiation models. *Ecol. Model.* 75:199–219.

Li, C., Ter-Mikaelian, M., and Perera, A. 1996. Temporal fire disturbance patterns on a forest landscape. *Ecol. Model.* 99(2, 3):137–150.

Little, S.N., and Ohmann, J.L. 1988. Estimating nitrogen lost from the forest floor during prescribed fires in Douglas-fir/western hemlock clearcuts. *For. Sci.* 34(1):152–164.

Luce, C.H., Kluzek, E., and Bingham, G.E. 1995. Development of a high resolution climatic data set for the Northern Rockies. In *Interior West Global Change Workshop*, ed. R. Tinus, pp. 106–111. USDA Forest Service Gen. Tech. Rep. RM-GTR-262.

Malanson, G.P. 1996. Effects of dispersal and mortality on diversity in a forest stand model. *Ecol. Model.* 87:103–110.

Malanson, G.P., and Armstrong, M.P. 1996. Dispersal probability and forest diversity in a fragmented landscape. *Ecol. Model.* 87:91–102.

Marsden, M.A. 1983. Modeling the effect of wildfire frequency on forest structure and succession in the Northern Rocky Mountains. *J. Environ. Manag.* 16(1):45–62.

Martell, D.L., Bevilacqua, E., and Stocks, B.J. 1989. Modelling seasonal variation in daily people-caused forest fire occurrence. *Can. J. For. Res.* 19:1555–1563.

McArthur, A.G. 1967. Fire behavior in eucalypt forests. Commonwealth of Australia Forestry and Timber Bureau Leaflet No. 107. 80p.

McCarthy, M.A., and Gill, A.M. 1997. Fire modeling and biodiversity. In *Natural and Altered Landscapes: Disturbance Ecology of Ecosystems*, pp. 79–88. Amsterham: Elsevier.

McCaughey, W.W., Schmidt, W.C., and Shearer, R.C. 1985. Seed dispersal characteristics of conifers of the Inland Mountain West. In *Proceedings of Symposium on Conifer Seed* in *Inland Mountain West*, ed. R.C. Shearer (compiler), pp. 50–61. April 5–6, Missoula, MT.

McClanahan, T.R. 1986. Seed dispersal from vegetation islands. *Ecol. Model.* 32:301–309.

McKenzie, D., Peterson, D.L., and Alvarado, E. 1996. Extrapolation problems in modeling fire effects at large spatial scales: A review. *Int. J. Wild. Fire* 6(4):165–176.

McMurtrie, R.E., Leuning, R., Thompson, W.A., and Wheeler, A.M. 1992. A model of canopy photosynthesis and water use incorporating a mechanistic formulation of leaf CO_2 exchange. *For. Ecol. Manag.* 52:261–278.

Meetenmeyer, V. 1978. Macroclimate and lignin control of decomposition rates. *Ecology* 59:465–472.

Miller, C. 1994. A model of the interactions among climate, fire, and forest pattern in the Sierra Nevada. MS thesis. Department of *Forest Sciences*, Colorado State University, Fort Collins. 77p.

Miller, C., and Urban, D.L. 1999a. A model of surface fire, climate and forest pattern in the Sierra Nevada, California. *Ecol. Model.* 114:113–135.

Miller, C., and Urban, D.L. 1999b. Interactions between forest heterogeneity and surface fire regimes in the southern Sierra Nevada. *Can. J. For. Res.* 29:202–212.

Mladenoff, D.J., and Baker, W.L. 1999. *Spatial Modeling of Forest Landscape Change*. Cambridge: Cambridge University Press.

Mladenoff, D.J., Host, G.E., Boeder, J., and Crow, T.R. 1996. LANDIS: A spatial model of forest landscape disturbance, succession and management. In *GIS and Environmental Modeling*, pp. 175–181. NCGIA, Santa Barbara, CA.

Moeur, M. 1985. COVER: A user's guide to the CANOPY and SHRUBS extension of the Stand Prognosis Model. USDA Forest Service Gen. Tech. Rep. INT-190. 49p.

Mohren, G.M.J., Van Gerwen, C.P., and Spitters, C.J.T. 1984. Simulation of primary production in even-aged stands of Douglas-fir. *For. Ecol. Manag.* 9:27–49.

Mohren, G.M.J., Bartelink, H.H., and Lansen, J.J., eds. 1994. Contrasts between biologically based process models and management-oriented growth and yield models. Special issue—*For. Ecol. Manag.* 69(1–3):1–350.

Moore, A.D., and Noble, I.R. 1990. An individual model of vegetation stand dynamics. *J. Environ. Manag.* 31:61–81.

Narasimhan, T.N. 1995. Models and modeling of hydrogeologic processes. *Soil Sci. Soc. Am. J.* 59:300–306.

Noble, I.R., Bary, G.A.V., and Gill, A.M. 1980. McArthur's fire danger meters expressed as equations. *Austral. J. Ecol.* 5:201–203.

Noble, I.R., and Slatyer, R.O. 1977. Postfire succession of plants in Mediterranean ecosystems. In *Proceedings of Symposium on the Environmental Consequences of Fire and Fuel Management in Mediterranean Ecosystems*, eds. H.A. Mooney and C.E. Lowrad, pp. 27–36. USDA Forest Service Gen. Tech. Rep. WO-3.

Ohtsuki, T., and Keyes, T. 1986. Biased percolation: Forest fires with wind. *J. Phys. Advanus Math. Gen.* 19:L281–L287.

Ottmar, R.D., Burns, M.F., Hall, J.N., and Hanson, A.D. 1993. CONSUME users guide. USDA Forest Service Gen. Tech. Rep. PNW-GTR-304. 118p.

Pacala, S.W., Canham, C.D., and Silander, J.A. 1993. Forest models defined by field measurements: I. The design of a northeastern forest simulator. *Can. J. For. Res.* 23: 1980–1988.

Parton, W.J., Schimel, D.S., Cole, C.V., and Ojima, D. 1987. Analysis of factors controlling soil organic levels of grasslands in the Great Plains. *Soil Sc. Soc. Am. J.* 51: 1173–1179.

Parton, W.J., Stewart, J.W.B., and Cole, C.V. 1988. Dynamics of C, N, P, and S in grassland soils: A model. *Biogeochemistry* 5:109–131.

Pastor, J., and Post, W.M. 1985. Development of a linked forest productivity-soil process model. Environmental Sciences Division Publication No. 2455. Oak Ridge, TN: Martin Marietta Energy Systems, Inc. for the U.S. Department of Energy, Environmental Sciences Division. 162p.

Pastor, J., and Post, W.M. 1986. Influence of climate, soil moisture, and succession on forest carbon and nitrogen cycles. *Biogeochemistry* 2:3–27.

Peter, S.J. 1992. Heat transfer in soils beneath a surface fire. Ph.D. dissertation. Development of Chemical Engineering, University of New Brunswick, Fredericton. 479p.

Peterson, D.L. 1985. Crown scorch volume and scorch height: Estimates of post-fire tree condition. *Can. J. For. For. Res.* 15:596–598.

Pfister, R.D., Kovalchik, B.L., Arno, S.F., and Presby, R.C. 1977. Forest habitat types of Montana. Gen. Tech. Rep. INT-34. Ogden, UT: USDA Forest Service, Intermountain Forest and Range Experiment Station. 174p.

Philip, J.R., and DeVries, D.A. 1957. Moisture movement in porous materials under temperature gradients. *Trans. Am. Geophys. Union* 38:222–232.

Pielke, R.A., and Avissar, R.A. 1990. Influence of landscape structure on local and regional climate. *Landscape Ecol.* 4:133–155.

Pielke, R.A., Cotton, W.R., Walko, R.L., Tremback, C.J., Nicholls, M.E., Moran, M.D., Wesley, D.A., Lee, T.J., and Copland, J.H. 1992. A comprehensive meteorological modeling system—RAMS. *Meteorol. Atmos. Phys.* 49:69–91.

Pinty, J.P., Mascart, P., Bechtold, P., and Rosset, R. 1992. An application of the vegetation-atmosphere coupling concept to the HAPEX-MOBILHY experiment. *Agricult. For. Meteorol.* 61:253–279.

Pukkala, T. 1987. Simulation model for natural regeneration of *Pinus sylvestris*, *Picea abies*, *Bedtula pendula* and *Betula pubescens*. *Silva Fennica* 21(1):37–53.

Rajagopalan, B., Lall, U., Tarboton, D.G., and Bowles, D.S. 1997. Multivariate non-parametric resampling scheme for generation of daily weather variables. *Stochast. Hydrol. Hydraul.* 11(1):65–95.

Rastetter, E.B., Ryan, M.G., Shaver, G.R., Melillo, J.M., Wadelhoffer, K.J., Hobbie, J.E., and Aber, J.D. 1991. A general biogeochemical model describing the responses of the C and N cycles in terrestrial ecosystems to changes in CO_2, climate, and N deposition. *Tree Physiol.* 9:101–126.

Rastetter, E. B. 1996. Validating models of ecosystem response to global change. *Bioscience* 46(3):190–197.

Ratz, A. 1995. Long-term spatial patterns created by fire: A model oriented towards boreal forests. *Int. J. Wildland Fire* 5(1):25–34.

Reed, K.L. 1980. An ecological approach to modeling growth of forest trees. *For. Sci.* 26: 33–50.

Reed, K.L., and Clark, S.G. 1979. SUCcession SIMulator: A coniferous forest simulator. Model documentation. Bulletin No. 11. Seattle: University of Washington, Coniferous Biome Ecosystem Analysis. 96p.

Reed, W.J. 1994. Estimating the historic probability of stand-replacement fire using age-class distribution of undisturbed forest. *For. Sci.* 40(1):104–119.

Reinhardt, E.D., Keane, R.E., and Brownm, J.K. 1997. First order fire effects model: FOFEM 4.0, user's guide. USDA Forest Service Gen. Tech. Rep. INT-GTR-344. 65p.

Richards, G.D. 1990. An elliptical growth model of forest fire fronts and its numerical solution. *Int. J. Numer. Meth. Eng.* 30:1163–1179.

Richards, G.D. 1995. A general mathematical framework for modeling two-dimensional wildland fire spread. *Int. J. Wildl. Fire* 5(2):63–72.

Roberts, D.W. 1996. Landscape vegetation modeling with vital attributes and fuzzy ststems theory. *Ecol. Model.* 90:175–184.

Roberts, D.W., and Betz, D.W. 1999. Simulating landscape vegetation dynamics of Bryce Canyon National Park with the vital attributes/fuzzy systems model VAFS.LANDSIM. In *Spatial Modeling of Forest Landscape Change: Approaches and Applications*, eds. D.J. Mladenoff and W.L. Baker, pp. 99–123. Cambridge: Cam bridge University Press.

Rothermel, R.C. 1972. A mathematical model for predicting fire spread in wildland fuels. Res. Pap. INT-115. Ogden, UT: USDA Forest Service, Intermountain Forest and Range Experiment Station. 40p.

Rothermel, R.C. 1991. Predicting behavior and size of crown fires in the Northern Rocky Mountains. USDA Forest Service Res. Pap. INT-438. 46p.

Rothermel, R.C., Wilson, R.A., Morris, G.A., and Sackett, S.S. 1986. Modeling moisture content of fine dead wildland fuels: Input to the BEHAVE fire prediction system. USDA Forest Service *Res. Pap.* INT-359. 61p.

Running, S.W., and Coughlan, J.C. 1988. A general model of forest ecosystem processes for regional applications. I. Hydrologic balance, canopy gas exchange and primary production processes. *Ecol. Model.* 42:125–154.

Running, S.W., and Gower, S.T. 1991. FOREST-BGC, a general model of forest ecosystem processes for regional applications. II. Dynamic carbon allocation and nitrogen budgets. *Tree Physiol.* 9:147–160.

Running, S.W., Nemani, R.R., and Hungerford, R.D. 1987. Extrapolation of synoptic meteorological data in mountainous terrain and its use for simulating forest evapotranspiration and photosynthesis. *Can. J. For. Res.* 17:472–483.

Ryan, K.C., Peterson, D.L., and Reinhardt, E.D. 1987. Modeling long-term fire-caused mortality of Douglas-fir. *For. Sci.* 34(1):190–199.

Ryan, K.C., and Reinhardt, E.D. 1988. Predicting postfire mortality of seven western conifers. *Can. J. For. Res.* 18:1291–1297.

Sandberg, D.V. 1980. Duff reduction by prescribed underburning in Douglas-fir. USDA Forest Service Res. Pap. PNW-272. 18p.

Sanderlin, J.C., and Sunderson, J.M. 1975. A simulation for wildland fire management planning support (FIREMAN). In *Volume II. Prototype Models for FIREMAN (Part II): Campaign Fire Evaluation.* Mission Research Corp. Contract No. 231–343, Spec. 222. 249p.

Schroeder, C.N. 1974. The development of an optimized computer simulation model for heat and moisture transfer in soils. Ph.D. dissertation. Texas A&M University, College Station. 318p.

Scire, J., Strimaitis, D.G., Yamartino, R.J., and Xiamong, Z. 1995. A user's guide for CALPUFF dispersion model. Document 1321–2. Concord, MA: Sigma Research/Earth Tech. 315p.

Segal, M., Avissar, R., McCumber, M.C., and Pielke, R.A. 1988. Evaluation of vegetation effects on the generation and modification of mesoscale circulation. *J. Atmos. Sci.* 45: 2268–2292.

Sestak, M.L., Marlatt, W.E., and Riebau, A.R. 1989. VALBOX: Ventilated valley box model. Unpublished report on file with Michael Sestak, USDI Bureau of Land Management and Colorado State University, Environmental Science and Technology Center, 2401 Research Blvd., Suite 205, Fort Collins, CO 80526.

Sestak, M.L., and Riebau, A.R. 1988. SASEM: Simple approach smoke estimation model. Tech. Note 382. USDI Bureau of Land Management, Fort Collins, CO 80526. 31p.

Sharpe, P.J.H., Walker, J., Penridge, L.K., Wu, H., and Rykiel, E.J. 1986. Spatial considerations in physiological models of tree growth. *Tree Physiol.* 2:403–421.

Shugart, H.H., and Noble, I.R. 1981. A computer model of succession and fire response of the high-altitude Eucalyptus forest of the Brindabella Range, Australian Capital Territory. *Austral. J. Ecol.* 6:149–164.

Shugart, H.H., and Seagle, S.W. 1985. Modeling forest landscapes and the role of disturbance in ecosystems and communities. In *The Ecology of Natural Disturbance and Patch Dynamics*, eds. S.T.H. Pickett and P.S. White, pp. 353–368. San Diego, CA: Academic Press.

Shugart, H.H., and West, D.C. 1980. Forest succession models. *Bioscience* 30(5):308–313.

Shugart, H.H., and West, D.C. 1977. Development of an Appalachian deciduous forest succession model and its application to assessment of the impact of the chestnut blight. *J. Environ. Manag.* 5:161–179.

Sievänen, R., and Burk. T.E., 1993. Adjusting a process-based growth model for varying site conditions through parameter estimation. *Can. J. For. Res.* 23:1837–1851.

Sievänen, R., Hari, P., Orava, P.J., and Pelkonen, P. 1988. A model for the effect of photosynthate allocation and soil nitrogen on plant growth. *Ecol. Model.* 41:55–65.

Simard, A.J. 1996. Fire severity, changing scales, and how things hang together. *Int. J. Wildl. Fire* 1(1):23–34.

Sirois, L., Bonan, G.B., and Shugart, H.H. 1994. Development of a simulation model of the forest-tundra transition zone of northeastern Canada. *Can. J. For. Res.* 24:697–706.

Stickney, P.F. 1990. Early development of vegetation following holocaustic fire in Northern Rocky Mountain Forests. *Northwest Sci.* 64(5):243–249.

Strandman, H., Vaisanen, H., and Kellomaki, S. 1993. A procedure for generating synthetic weather records in conjunction of climatic scenario for modelling of ecological impacts of changing climate in boreal conditions. *Ecol. Model.* 70:195–220.

Swartzman, G.L. 1979. Simulation modeling of material and energy flow through an ecosystem: methods and documentation. *Ecol. Model.* 7:55–81.

Thornton, P.E., Running, S.W., and White, M.A. 1997. Generating surfaces of daily meteorological variables over large regions of complex terrain. *J. Hydrol.* 190:214–251.

Tomback, D.F. 1982. Dispersal of whitebark pine seeds by Clark's nutcracker: A mutualism hypothesis. *J. Animal Ecol.* 51:451–467.

Tomback, D.F., Hoffman, L.A., and Sund, S.K. 1990. Coevolution of whitebark pine and nutcrackers: Implications for forest regeneration. In *Proceedings of the Symposium: Whitebark Pine Ecosystems: Ecology and Management of a High Mountain Resource*, pp. 118–130. March 29–31. Bozeman, MT. USDA Forest Service Gen. Tech. Rep. INT-270.

Turner, M.G., Costanza, R., and Sklar, F.H. 1989. Methods to evaluate the performance of spatial simulation models. *Ecol. Model.* 48:1–18.

Turner, M.G., Hargrove, W.W., Gardner, R.H., and Romme, W.H. 1994. Effects of fire on landscape heterogeneity in Yellowstone National Park, Wyoming. *J. Veg. Sci.* 5: 731–742.

Uman, M.A. 1987. *The Lightning Discharge*. Orlando, FL. Academic Press.

Urban, D.L., Bonan, G.B., Smith, T.M., and Shugart, H.H. 1991. Spatial applications of GAP models. *For. Ecol. Manag.* 42:95–110.

Urban, D.L., and Miller, C. 1996. Modeling Sierran forests: Capabilities and prospectus for gap models. In Final Report to Congress, Status of the Sierra Nevada Volume III, Assessments, Commissioned Reports, and Background Information. University of California, Davis, CA, Centers for Water and Wildland Resources. pp. 733–744.

Urban, D.L., and Shugart, H.H. 1992. Individual-based models of forest succession. In *Plant Succession Theory and Prediction*, eds. D.C. Glenn-Lewin, R.K. Peet, T.T. Veblen, pp. 249–292. London: Chapman and Hall.

USA CERL. 1990. GRASS 4.0 Reference Manual. United States Army Corps of Engineers, Construction Engineering Research Laboratory, Champaign, IL. 208p.

U.S. Geological Survey. 1987. *Digital Elevation Models Data User's Guide*. U.S. Department of the Interior. 38p.

Van der Pijl, L. 1982. *Principles of Dispersal in Higher Plants*. Berlin: Springer-Verlag.

Van Wagner, C.E. 1973. Height of crown scorch in forest fires. *Can. J. For. Res.* 3: 373–378.

Van Wagner, C.E. 1977. Conditions for the start and spread of crownfire. *Can. J. For. Res.* 3:373–378.

Van Wagner, C.E. 1978. Age-class distribution and the forest fire cycle. *Can. J. For. Res.* 8:220–227.

Vasconcelos, M.J., and Guertin, D.P. 1992. FIREMAP—Simulation of fire growth with a geographic information system. *Int. J. Wildl. Fire* 2:87–98.

Von Niessen, W., and Blumen, A. 1988. Dynamic simulation of forest fires. *Can. J. For. Res.* 18:805–812.

Wallace, G. 1993. A numerical fire simulation model. *Int. J. Wildl. Fire* 3(2):111–116.

Wang, Y.P., and Jarvis, P.G. 1990. Description and validation of an array model-MAESTRO. *Agric. For. Meteorol.* 51:257–280.

Ward, D.E. 1990. Factors influencing the emissions of gases and particulate matter from biomass burning. In *Fire in the Tropical Biota*, ed. J.G. Goldammer, pp. 418–436. Berlin: Springer.

Ward, D.E., Peterson, J., and Hao, W.M. 1993. An inventory of particulate matter and air toxic emissions from prescribed fires in the USA for 1989. In *Proceedings of the Air and Waste Management Association 1993 Annual Meeting and Exhibition*. Denver, CO, June 14–18, pp. 1–19.

Waring, R.H., and Schlesinger, W.H. 1985. *Forest Ecosystems Concepts and Management*. San Diego, CA: Academic Press.

Wiens, J.A. 1989. Spatial scaling in ecology. *Funct. Ecol.* 3:385–397.

White, J.D. 1996. Spatial, and temporal scale effects on assessment of a regional ecosystem model: Modeling climate change in Glacier National Park, USA. Ph.D. dissertation. University of Montana, Missoula. 191p.

Zhang, Y., Reed, D.D., Cattelino, P.J., Gale, M.R., Jones, E.A., Liechty, H.O., and Mroz, G.D. 1994. A process-based growth model for young red pine. *For. Ecol. Manag.* 69: 21–40.

3. Simulation of Effects of Climatic Change on Fire Regimes

Carol Miller

Increasing concentrations of greenhouse gases in the atmosphere will likely be accompanied by substantial warming of the earth's surface (1.5–4.5°C), altered precipitation patterns, and an increase in climate variability (Houghton et al. 2001). The impact of such climatic change on vegetation and fire activity is of great economic, social, and ecological interest across the globe. Plant species have evolved under a range of environmental conditions that have occurred in the past, including the timing and severity of fires, and some species may not be able to persist in a new climatic regime if the changes in environmental conditions exceed pre-adapted tolerances. Changes in plant species distributions brought on by global warming could seriously affect biotic diversity and ecosystem function (Peters and Lovejoy 1992).

Of particular concern are changes in climate that result in an increase in the size, severity, or frequency of wildfires because the social and economic consequences of such changes are likely to compound existing management challenges (Arno and Brown 1991). For example, in western North America the accumulation of hazardous fuels due to decades of fire exclusion already poses serious threats to human life and property (Hardy et al. 1999; U.S. General Accounting Office 1999), and climate-mediated changes in the fire regime could serve to exacerbate the situation (Riggan et al. 1994). Changes in the size, severity, or frequency of fires could also have important ecological consequences such as changes in vegetation structure, species composition and native plant diversity (Christensen 1988; Brown 2000).

Fire Regimes and Climate

A fire regime is a generalized description of the role fire plays in an ecosystem (Agee 1993). A generic set of descriptors of disturbance regimes (White and Pickett 1985) have been applied specifically to fire regimes and include descriptors such as frequency, spatial extent, rotation period, intensity, severity, and seasonality. The specific combinations of measures that are used to describe fire regimes vary greatly in the literature. Fire regimes have been classified according to fire frequency and intensity (Heinselman 1973), potential vegetation types (Davis, Clayton, and Fischer 1980), and the effect of fire on dominant vegetation (Agee 1990). In this chapter only two aspects of the fire regime are discussed: frequency and spatial extent. Furthermore the discussion is limited to surface fires, which spread by flaming combustion through fuels at or near the surface (Pyne, Andrews, and Laven 1996); crown fires, which burn through the crowns of trees, are not considered here.

Climate influences the environment under which fires burn across a wide range of temporal scales. Temperature, relative humidity, and precipitation influence how a wildland fire burns on a time scale of hours to days by influencing the moisture content of the live and dead vegetation (i.e., fuel) and the amount of heat transfer required for combustion of those fuels (Pyne, Andrews, and Laven 1996). On a time scale of weeks to months, climatic variables combine to influence drought and the duration of the fire season (Pyne, Andrews, and Laven 1996). A short fire season provides fewer opportunities for fires to occur than does a long fire season, thereby influencing the fire frequency at a site. On longer time scales of years to decades, climate can influence fire regimes by governing plant distributions, growth rates, and the type and amount of fuel that result (Christensen 1993).

Describing the variability of climate and its influence on fire regimes is critical for establishing reference conditions as targets for management (Landres, Morgan, and Swanson 1999; Swetnam, Allen, and Betancourt 1999) and paleoecological studies can provide useful insights into the interconnections between climate and fire regimes. Very long term records of the relationship between fire and climate have been inferred from the charcoal content in lake sediments and from proxy data such as pollen (Whitlock and Anderson, Chapter 1, this volume). The impacts of global climate cycles on regional fire regimes have been demonstrated using networks of tree-ring studies (Swetnam and Baisan, Chapter 6, this volume; Heyerdahl and Alvarado, Chapter 7, this volume). Tree-ring studies can also provide information on the influence of interannual and seasonal variations on fire regimes (Swetnam and Baisan, Chapter 6, this volume). Our understanding of how fire and climate interact to affect vegetation patterns in many different ecosystems has been improved through the use of paleoecological data. Such data provide us with valuable insights about the interactions between climate and fire, but because the data correspond to a particular time period (and associated climate) from the past, our ability to use these data to infer the impacts from future climatic change may be limited (Millar and Woolfenden 1999). Furthermore temperature and precipitation may vary independently from one another, perhaps even in different directions, and a future climate may not have an analog from the past with which to compare.

Interactions with Vegetation

The interaction of vegetation, climate, and fire adds to the challenge of using paleoecological data for understanding the effect of climatic change on fire regimes. Past changes in vegetation composition have been inferred from pollen and macrofossil records, and changes in some aspects of fire regimes can be inferred from charcoal in lake sediments and fire-scarred trees (Betancourt, Van Devender, and Martin 1990; Clark 1990). The relative timing of changes are often difficult to discern in paleoecological records (e.g., Clark, Royall, and Chumbley 1996), in part because vegetation response can significantly lag changes in environmental conditions (Davis and Botkin 1985). A change in climate might invoke a change in vegetation type and subsequent changes in associated fuels, thereby altering patterns of fire occurrence and fire spread. Alternatively, a change in the climate might directly affect fire frequency or severity, subsequently altering the vegetation. If future climatic change results in an altered fire regime, will that new fire regime precede or follow a change in vegetation? To understand how climatic change might influence fire regimes, we need to also understand the complexities of vegetation response (Clark 1993).

The vegetation that occurs on any site is a collection of individual plants, each responding to its environment on an individual basis. Different plant species have minimum and/or maximum tolerances for environmental factors such as temperature, soil moisture, light, and nutrients; multiple species may be competing for resources within these tolerances. When climatic change causes an environmental factor, such as temperature or soil moisture, to exceed the tolerance limit of a species, that species can become locally extinct. Furthermore a change in environmental conditions can cause a shift in the competitive balance among species and result in the increase of some species over others (Urban, Harmon, and Halpern 1993). The ability of a plant species to take advantage of new environmental conditions over the long term depends on its opportunities for reproduction on the site. Because disturbances often create new establishment sites for plants (e.g., by leaving bare soil and/or an opening in a forest canopy), the rate at which vegetation responds to climatic change might be accelerated by increased fire occurrence (Overpeck, Rind, and Goldberg 1990). Therefore vegetation response to climatic change will depend on the species present, the site conditions, and the creation of establishment sites by disturbances.

How can we explore these complex interactions and interrelationships and anticipate ecosystem response to future climatic change? Simulation modeling is a viable option for expanding our understanding of the site-specific and species-specific responses to climatic change and the important interactions between fire and vegetation.

Models for Understanding

At continental and global scales, dynamic global vegetation models have been used to predict the changes in the distribution of general vegetation types that would occur in a new climate (Prentice et al. 1992; Nielsen 1995; Haxeltine, Prentice, and Cresswell 1996). These models simulate vegetation in terms of life-form (e.g., evergreen vs. deciduous trees, shrubs, and C3 *versus* C4 grasses) and cannot simulate changes in species composition that might result from climatic change. At landscape scales (10^4–10^6 ha), dynamic landscape simulation models have been used to explore the effects of fire on vegetation pattern (Gardner et al. 1999), and several of these models have been used to simulate the implications of climatic change over time periods of centuries to millennia (Baker, Egbert, and Frazier 1991; Keane, Ryan, and Running 1996; He and Mladenoff 1999; Hargrove et al. 2000). Most models are not designed to explore how climatic change might alter fire regimes, but instead make explicit assumptions about how climatic change will affect fire frequency, and in some cases, fire size (e.g., Baker, Egbert, and Frazier 1991). However, because of the feedback of vegetation on fire regimes, it can be difficult to determine what effect a particular climatic change scenario might have on fire regimes. For the purpose of investigating future vegetation response to climatic change, a model needs to simulate a realistic disturbance regime (Overpeck, Rind, and Goldberg 1990); for fire, this means simulating a fire regime coupled to both climate and vegetation dynamics.

This chapter discusses insights on climate-fire interactions provided by a model that was developed for the Sierra Nevada in California to study the complex linkages among climate, fire, and forest dynamics. The Sierra Nevada is a particularly appropriate region for studying the implications of climatic change on fire regimes for two major reasons. First, there is a strong connection between climate and fire regimes in the Sierra Nevada. Climate and fire histories reconstructed from multimillenial tree-ring chronologies have established that fire regimes are directly related to climatic factors in the Sierra Nevada (Swetnam 1993; Caprio and Swetnam 1995). In addition, data from vegetation inventory plots in Sequoia National Park, as well as results from simulation models, reveal that the soil water balance strongly governs tree species distributions (Stephenson 1988; Urban et al. 2000), which in turn dictate the type and amount of fuel that accumulate (Miller and Urban 1999b). Second, the fire management issues and climatic change concerns are typical of much of western North America. Fire exclusion during the twentieth century has resulted in an abundance of dead surface fuels, an increase in forest stand densities, and a shift in species composition toward shade-tolerant tree species (Vankat and Major 1978; Parsons and DeBenedetti 1979; van Wagtendonk 1985). Although the importance of restoring the natural role of fire to these forests is widely acknowledged (Kilgore 1973; Graber 1985; Parsons et al. 1986; Biswell 1989; Husari and McKelvey 1996), there is substantial disagreement about how to accomplish this restoration (e.g., Bonnicksen and Stone 1982; Stephenson 1999). Global climatic change could amplify these management challenges by contributing to changes in species distributions, loss of biotic diversity, increased frequency or severity of wildfires, and increased tree mortality (Stephenson and Parsons 1993).

The model discussed here simulates surface fire regimes from climate and vegetation. Because the predominant fire regimes in the Sierra Nevada are surface fire regimes (McKelvey et al. 1996; Skinner and Chang 1996), this model was not designed to simulate crown fire behavior. In the model, temperature and precipitation directly influence fuel moisture, and indirectly influence fuel loads and fuel type via their influence on forest productivity and tree species distribution. The model can be used to examine how some aspects of fire regimes vary across climatic gradients, such as those that occur with elevation, as well as to investigate feedbacks that may accompany transient (short-lived) responses to climatic change. The purpose of this model is not to predict fine-scale patterns in vegetation for fire regimes, nor is it to predict what will happen under a particular climatic change scenario. Unlike the modeling approach introduced by Keane and Finney (Chapter 2, this volume), this model does not simulate detailed fire behavior and is not expected to be able to reproduce actual fire perimeters. Instead, the value of this model is in its ability to generate some aspects of fire regimes that are realistic and influenced by climate and forest properties.

Model Description

Gap Models

The model described here is an extension of the forest gap model ZELIG (Smith and Urban 1988; Urban et al. 1991). All forest gap models simulate the establishment, growth, and death of individual trees on a tree-sized plot about the size of a typical canopy gap that would be created by the death of an overstory tree. This tree-sized plot is considered to be homogeneous and representative of an entire forest stand. Tree growth is usually specified as a maximum potential which is then reduced to reflect suboptimal environmental conditions (e.g., low light, low temperature, or drought). A key characteristic of gap models is that they simulate system feedbacks: not only are trees affected by their environment, but each tree exerts an influence on its environment (e.g., through shading and water use). Different species are often ranked by their tolerances to environmental conditions, and these rank tolerances are used to simulate species replacement during succession. For example, as trees grow and increase leaf area, less light reaches the forest floor, ultimately allowing only shade-tolerant species to establish.

A spatially explicit variant of ZELIG arrays the individual treesized plots in a rectangular grid and these plots interact with one another via shading (i.e., tall trees on one plot may shade out smaller trees on neighboring plots). This raster configuration relaxes the assumption of a homogeneous forest stand and allows for the investigation of causes and consequences of spatial pattern within a stand (Smith and Urban 1988; Urban et al. 1991). The model

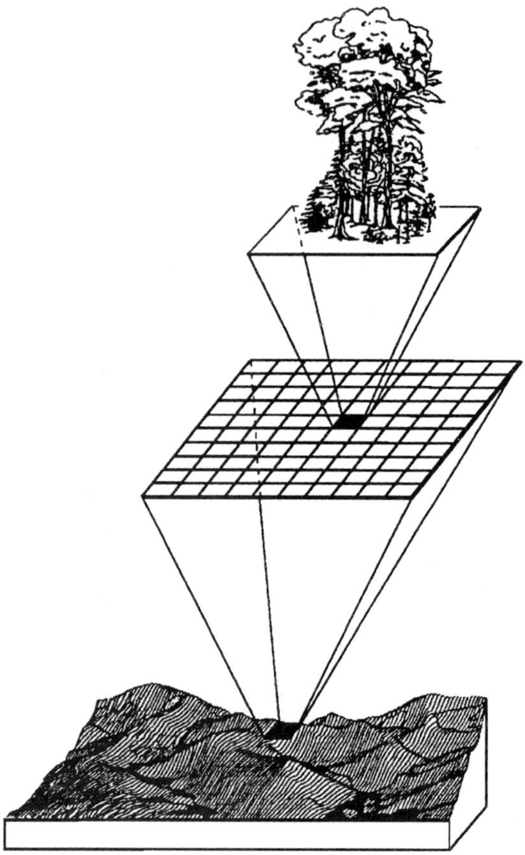

Figure 3.1. FM simulates a grid of 15 × 15 m forest plots as a landscape "facet" that has an assigned elevation, slope, and aspect. In the simulation experiments the landscape is 20 × 20 cells (300 × 300 m = 9 ha).

version discussed here is FM 97.5 (FACET Model version 97.5). The functional unit is the slope-facet (Daly, Neilson, and Phillips 1994), which is defined in the model as a grid of cells with homogeneous slope and aspect, and with one edge of the grid located at a specified elevation (Fig. 3.1). Elevation, slope, and aspect are used in FM's weather model to adjust incoming solar radiation (Nikolov and Zeller 1992), and to adjust estimates of monthly temperature and precipitation according to lapse rates (Running, Nemani, and Hungerford 1987). For the Sierran version of FM, the lapse rates were derived from seven meteorological stations in Sequoia National Park (Urban et al. 2000). In the experiments discussed in this chapter, a 9-ha forest stand is represented by 15 × 15 m cells arrayed in a 20 × 20 cell grid. FM is a spatial model at the stand scale but not at the landscape scale because only one slope-facet is run at a time. It does not simulate fire spread across a landscape of multiple slope-facets. To apply the 9-ha slope-facet to the larger landscape scale, a simple sampling approach was used—each simulation was assigned a unique combination of elevation, slope, and aspect—effectively distributing the model grid over topographic gradients (Urban et al. 2000). Therefore the landscape-scale patterns presented in this chapter are the result of a simple aggregation of stand-scale results.

Climate, Vegetation, and Fire

Monthly temperature and precipitation affect tree growth in two main ways in FM. First, species have a temperature tolerance whereby cold temperatures constrain tree growth. This sorts out species abundance along temperature gradients that exist with latitude or elevation (Urban et al. 2000). Second, soil moisture is sensitive to both temperature and precipitation (Urban et al. 2000) and is indexed as the number of drought-days per year. Species tolerances to drought govern each tree's growth response to soil moisture. Nine tree species were simulated here: *Quercus kelloggii* (black oak), *Calocedrus decurrens* (incense cedar), *Pinus ponderosa* (ponderosa pine), *Pinus jeffreyii* (Jeffrey pine), *Pinus lambertiana* (sugar pine), *Abies concolor* (white fir), *Abies magnifica* (red fir), *Pinus contorta* (lodgepole pine), and *Pinus monticola* (western white pine).

In FM, forest dynamics are coupled to the fire regime because trees produce surface fuels (Fig. 3.2). Each year in FM, a portion of each tree's foliage and

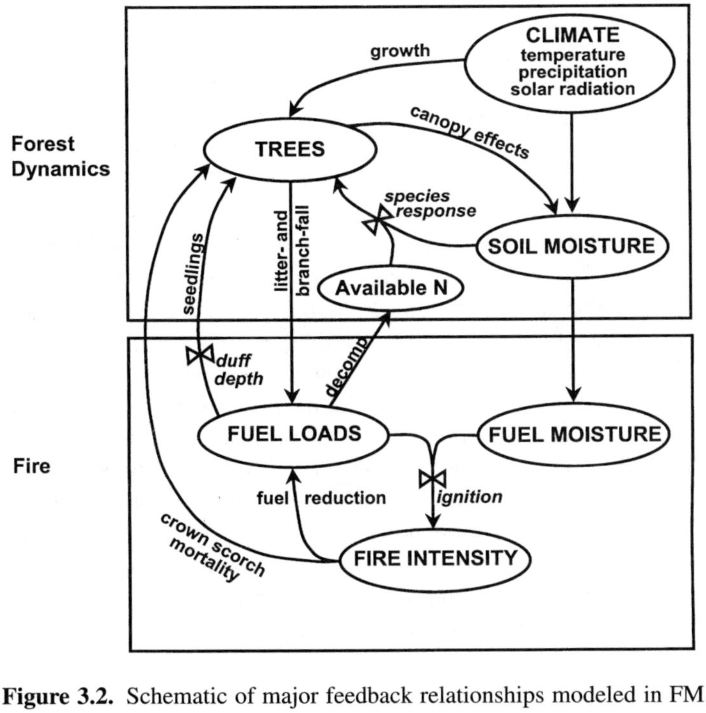

Figure 3.2. Schematic of major feedback relationships modeled in FM.

branch biomass is shed according to tree size and species. In addition, when an individual tree dies, its bole, bark, branch, and foliage biomass is added to the fuel load. Thus fuels respond to temporal changes in forest structure and composition that occur during successional development of the forest and to differences in forest structure and composition that exist across topographic gradients. Fuels are partitioned into separate fuel classes that accumulate and decay independently. The fuel classes that represent "dead and down" fuels are litter, fine wood (<7.5 cm diameter, in three size classes corresponding to 1-, 10-, and 100-hour time lag fuels), and coarse wood (>7.5 cm diameter, or 1000-hour fuels). These fuel classes are kept as separate compartments in the model, making FM compatible with mechanistic surface fire behavior equations and models (Keane and Finney, Chapter 2, this volume). In addition to these dead fuel classes, FM simulates a duff layer representing the compact, partially decomposed layer of litter. Although the duff layer is not considered important in prediction of fire behavior, it is stored in the model's fuel array for convenience. Additionally it functions as the surface soil layer in FM's soil moisture model (described below) and can limit tree seedling establishment.

Because fine herbaceous fuels can be an important factor in Sierra Nevada fire regimes, particularly at lower elevations where oak-pine woodlands can occur, FM includes a grass component in its fuel bed. Grass production is simulated as a function of precipitation, temperature, shade from overstory trees, and the depth of the duff layer (Miller and Urban 1999a). As FM does not currently simulate crown fire behavior, no other living fuels (e.g., live tree foliage or branches) have been included in the simulation of fire behavior.

Climate directly affects the fire regime in FM by influencing fuel moisture (Fig. 3.2). Fuel moisture is estimated from FM's soil moisture model (Urban et al. 2000), whereby the uppermost soil layer of a multilayer soil profile comprises the duff layer (the partially decomposed portion of foliage litter). The moisture content of this layer is used to derive fuel moisture (Cohen and Deeming 1985) for the different fuel classes for each grid cell. Surface evaporation reduces the moisture content of the duff layer as a function of temperature and incoming solar radiation, and therefore this effect varies with elevation and topographic position. Elevation and topographic position also affect rainfall and snowmelt, both of which affect the moisture content of the duff layer. In this way FM simulates fuel moistures that vary across topographic gradients and that vary with monthly temperature and precipitation. Fuel moisture in the model does not vary at time scales shorter than one month and therefore reflects seasonal drought conditions as opposed to diurnal or weekly fluctuations in weather conditions.

The types of fuels comprising the fuel bed may influence flammability as much as the gross amount of fuel mass that exists. The loosely packed litter of long-needled ponderosa pine forests will burn more readily than a tightly packed short-needled fir forest floor, and a fuel bed with a large grass component may burn more readily than a fuel bed comprised only of forest fuels. To capture these differences, FM simulates bulk density of the fuel bed for each grid cell as a function of species composition and grass content. Fuel-bed bulk density is directly related to the fuel-bed depth used for mechanistic simulation of fire behavior. In FM simulations, fuel-bed bulk density tends to increase with elevation as species composition shifts from pine to fir and as grass production declines (Miller and Urban 1999a).

Although FM does not simulate detailed fire spread behavior, it does simulate fire frequency, area burned, and fire intensity. Fire ignition events are possible every year, but for fire to occur, low fuel moisture and suitable fuels must exist (Fig. 3.2). Each year, the fireline intensity (i.e., the amount of energy released along a linear fire front) for each of the grid cells is computed from the accumulated fuels and fuel moisture conditions following equations for surface fire behavior (Rothermel 1972; Albini 1976). Cells are considered to be burnable if the fireline intensity is at least $45\,kW\,m^{-1}$ (roughly equivalent to a scorch height of 0.5 m); fire effects are not simulated for grid cells with intensities less than this. A simple algorithm is used to address the contagious nature of fire. Fires initiate from a randomly located ignition point on the grid and spread to any adjoining cells that are burnable. Thus fires are restricted to a contagious cluster of burnable cells, and on average, fires tend to burn the largest cluster of burnable cells. Although this method does not simulate detailed fire spread patterns as a function of fine-scale weather data, it is a simple way to approximate the extent of burning using coarser-scale monthly weather information.

The fireline intensity estimated from the fire behavior model equations is used to estimate fire effects for each cell that burns (Fig. 3.2). Fuels are reduced as a function of pre-fire fuel load (Brown et al. 1985), scorch height is calculated as a function of mean daytime temperature and fireline intensity (Van Wagner 1973) and fire mortality is computed as a function of crown damage (Ryan and Reinhardt 1988; Stephens 1995; Mutch and Parsons 1998). FM calculates only those fire effects that have a significant feedback to forest condition, and therefore, a feedback to the fire regime.

Model Validation

The climatic variables that drive forest processes and the fire regime in FM vary at large landscape scales and the model is not expected to be able to reproduce forest pattern or fire history at finer scales. Rather, landscape-scale patterns simulated by the model must be compared to similarly scaled data. For example, simulated tree species distributions were verified using plot-level data from 280 quadrats widely distributed throughout the west slope of Sequoia National Park (Stephenson 1988; Graber, Haultain, and Fessenden 1993). The model reproduces the gradient pattern in species composition that occurs with elevation and reproduces overall patterns of species abundance across a wide range of site conditions (Urban et al. 2000). Elevation gradient patterns in fuel loads observed in independent field data are also reproduced by the model (Miller and Urban 1999b).

Paleoecological fire history data are often summarized and used to parameterize the inputs for model simulations. However, because FM generates fire frequency and fire size as output, paleoecological data from the Sierra Nevada were used for model validation instead of model parameterization. Two sets of appropriately scaled fire history data were used to this end. First, fire-scar data from an elevation transect on the western slope of Sequoia National Park provide empirical evidence for decreasing average fire frequencies with increasing elevation (Caprio and Swetnam 1995). Because the soil moisture simulated by FM—and thus fuel moisture—varies with elevation, FM also generates decreasing fire frequencies (i.e., increasing intervals between fires) with elevation (Miller and Urban 1999b). The second set of fire history data comprises very long fire-scar chronologies from giant sequoia groves in Sequoia and Yosemite National Parks, which provide evidence that larger (more widely spreading) fires occurred during periods of less frequent fire in the past (Swetnam 1993). Model simulations also produce this result due to the greater amounts of fuel that may accumulate when fires are less frequent (Miller and Urban 1999b). Both sets of results provided some confidence that the model generates fire regimes that are realistic and sensitive to forest condition and climate, but a more rigorous comparison of model results with data could not be done for several reasons. First, data on monthly temperature and precipitation are not available for the time period represented by the fire scar record (AD 1700–1900). Model simulations used mean monthly temperature and precipitation derived from data from the latter half of the twentieth century, which may be quite different from the historical period. Second, ignition rates for the historical time period are completely unknown. In the absence of such information, a uniform ignition rate (once per year) across the elevation gradient was assumed in the simulations. Third, neither the area burned in FM nor the area inferred from the fire scar represent actual fire extent, and they cannot be directly compared. The measure derived from the fire-scar data is the percent of sampled trees scarred and can be interpreted only as a coarse index of area burned. In FM the simulated area burned depends partly on the burnability threshold of $45\,kW\,m^{-1}$, which is a somewhat arbitrary number.

Fire Regimes and Climate

Simulation results from FM illustrate the direct and indirect influences of climate on fire frequency and area burned. Two sets of model results are discussed here. The first represents a steady-state examination of the connectivity of the fuel bed (Miller and Urban 2000). The second set of results is from specific hypothetical climatic change scenarios (Miller and Urban 1999c).

Fuel-Bed Connectivity

The spread of fire is a contagious process, and therefore the area that burns during a fire is very sensitive to the connectivity of fuels across a landscape. Understanding how climate affects this connectivity can lead to a better understanding of how climatic change might affect future trends of area burned. This is particularly relevant in forests in western North America because decades of fire exclusion may have already increased fuel-bed connectivity, consequently increasing the likelihood of extremely large fires.

FM's explicit spatial design is useful for examining connectivity and its role in the fire regime. For the purposes here, an area is defined as connected if fire will burn through it under a given set of conditions. One way to index connectivity is with the correlation length, which is a measure of the average within-cluster distances for a map (Stauffer 1985). As applied here, it is the average distance that fire can spread in a randomly selected direction without encountering an unburnable cell. Burnability is a function of the same variables that predict fireline intensity, such as fuel moisture, fuel loads (or fuel mass), and fuel-bed bulk density.

Each of these variables (fuel moisture, fuel loads, and fuel-bed bulk density) can vary greatly from grid cell to grid cell in FM, making the direct and indirect effects of climate on connectivity difficult to discern. Fortunately, the relative influence of fuel moisture, fuel loads, and fuel-bed bulk density on connectivity of the fuel bed can be isolated by postprocessing output from FM and recomputing which grid cells are burnable when one or more of these variables are held constant across the grid. For example, a single map of fuels was postprocessed to generate the three maps shown in Figure 3.3. In each case, fuel moisture was held constant at a different value (1, 5, and 10%) when calculating burnability. The average connectivity of burnable area was computed for each of these moisture levels and for simulations run at elevations ranging from 1000 to 3000 m (Fig. 3.4a). When fuel moisture is moderate, the pattern of fuels dictates the burnability and connectedness of the map, but under extremely dry conditions (e.g., fuel moisture = 1%), most of the map is burnable, and consequently connectivity is very high. Indeed, events such as the Yellowstone fires in 1988 demonstrate that fire pays no heed to the existing vegetation mosaic or to topographic features when wind and moisture conditions are extreme (Turner and Romme 1994).

Results from this model also illustrate how the connectivity of a fuel bed can be quite sensitive to other less-studied properties of the fuel bed. The bulk density of the fuel bed, for example, greatly affects the connectivity of the fuel bed. The presence of grass in the fuel bed at elevations below 1500 m contributes to a low fuel-bed bulk density, resulting in relatively high connectivity (Fig. 3.4). The increase in forest productivity and accompanying fuel mass that occurs above 2500 m (Fig. 3.4c) should serve to increase connectivity along this gradient. But as species composition shifts from grass and long-needled pine at low elevations to short-needled fir at higher elevations, fuel-bed bulk density increases (Fig. 3.4b), thereby countering the effect of increased fuel mass on the connectivity of the fuel bed. The zone of low connectivity simulated between 1550 and 1850 m elevation (Fig. 3.4a) is the result of low grass production and low litter produc-

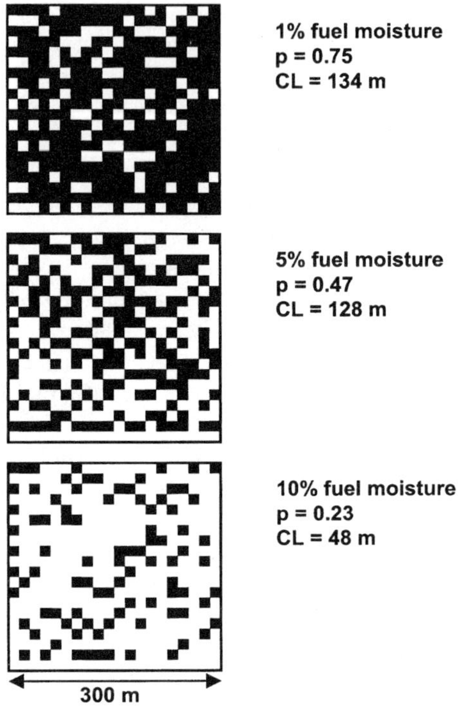

Figure 3.3. Maps of burnable area for three levels of fuel moisture. The fraction of the map that is burnable, *p*, and the correlation length, *CL*, are given.

tion from forest trees simulated by the model at these elevations. The connectivity of the fuel bed varies with elevation in complex ways because fuel moisture, fuel mass, and fuel-bed bulk density all vary independently across the environmental gradient.

In the past many fires burned across all elevations in the Sierra Nevada (Caprio and Swetnam 1995), perhaps linking disparate vegetation types along the elevation gradient. The fire regime and vegetation pattern in one elevation zone could influence the fire regime and vegetation pattern in other zones. Although FM does not explicitly link forest stands from one site to another across the elevation gradient, the suggestion that connectivity varies with elevation may have important implications for fires that spread throughout the landscape. For example, if there is an elevation zone of low connectivity, such as that simulated between 1550 and 1850 m (Fig. 3.4a), this zone could act as a natural fire break for fires burning upslope from sites below, except, of course, during conditions of extremely dry weather. If climatic change results in extended and more severe droughts, larger fires might occur due to low fuel moisture.

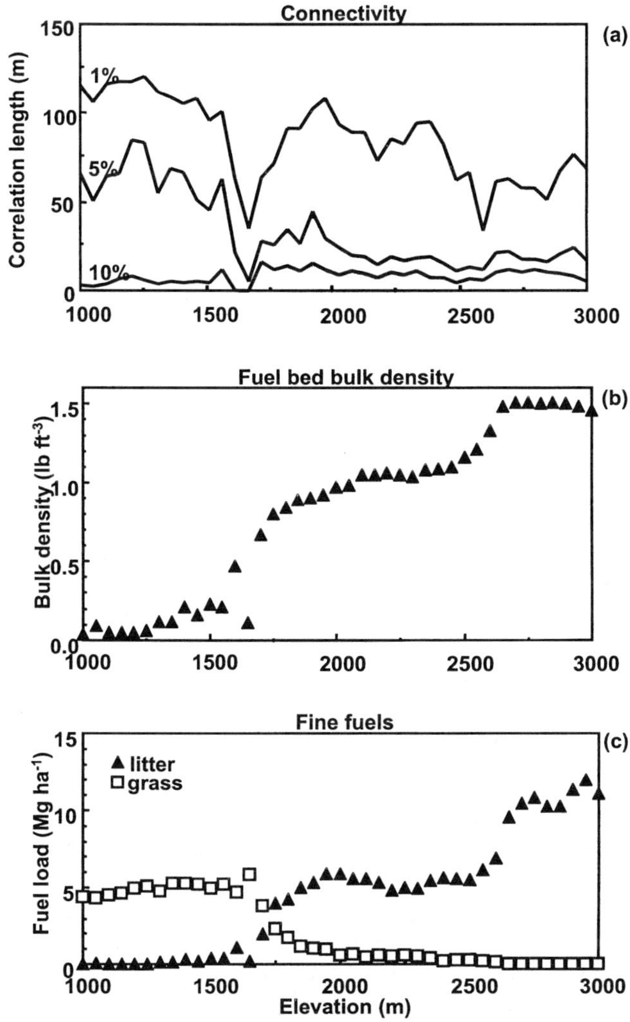

Figure 3.4. Elevational gradients in connectivity, fuel-bed bulk density and fuel loads. (a) Connectivity measured by correlation length. Isolines represent different levels of fuel moisture; fuel-bed bulk density was allowed to vary within the stand and with elevation. (b) Mean fuel-bed bulk density simulated during fire years. (c) Mean fuel loads for litter and grass fuels simulated during fire years.

Climatic Change Experiments

FM was used to investigate the impacts that climatic change might have on Sierra Nevada forests and fire regimes. Several simulation experiments were conducted to investigate the sensitivity of the fire regime and forest vegetation to departures from baseline (i.e., current) mean temperature and precipitation. These experi-

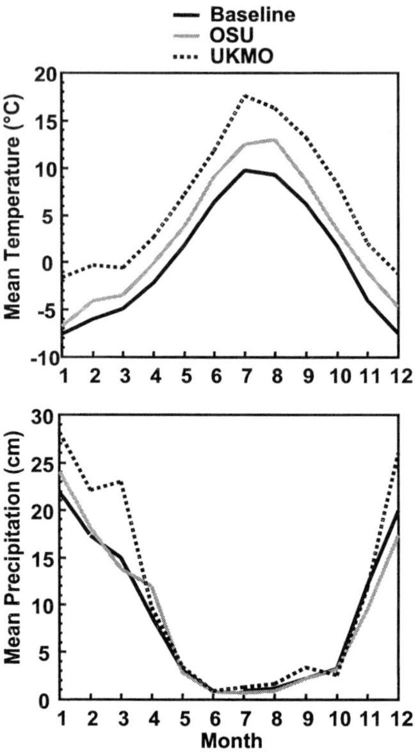

Figure 3.5. Mean monthly temperatures and precipitation predicted by the OSU and UKMO GCMs compared to baseline climate. Data are for the nominal average elevation in the GCM grid cells corresponding to Sequoia National Park (ca. 2400 m).

ments do not represent predictions of a future climate, but rather are designed to develop an understanding of how the complex linkages among forest pattern, fire, and climate might interact to elicit responses in forest structure, species composition, and fire regimes. A warm-dry case (+2°C and –20% precipitation), a cool-wet case (–2°C and +20% precipitation), and two 2 × CO_2 predictions of temperature and precipitation from general circulation models (GCMs) were simulated. The 2 × CO_2 predictions for the Sierra Nevada region were generated by the Oregon State University (OSU) and the United Kingdom Meteorological Office (UKMO) GCMs. These represent the most conservative and most extreme GCM predictions, respectively, that were available for the Sierra Nevada region (UCAR 1997). Both GCMs predict warmer temperatures, but they differ somewhat in their prediction of precipitation relative to baseline climate (Fig. 3.5). The simulations were run from bare ground for 800 years. The first 200 years were run without fire to allow successional dynamics to stabilize. Climate changes were applied gradually with the temperature and precipitation changes occurring linearly over 100 years from simulations years 501–600.

To illustrate the effect of altered temperature and precipitation on the climatic environment, the weather model in FM was used to simulate 31 sites ranging from 500 to 3000 m elevation. Growing degree-days per year (a temperature index) is plotted against drought-days per year (a drought index) for the five simulation experiments (Fig. 3.6). The set of curves in Figure 6 describe the gradi-

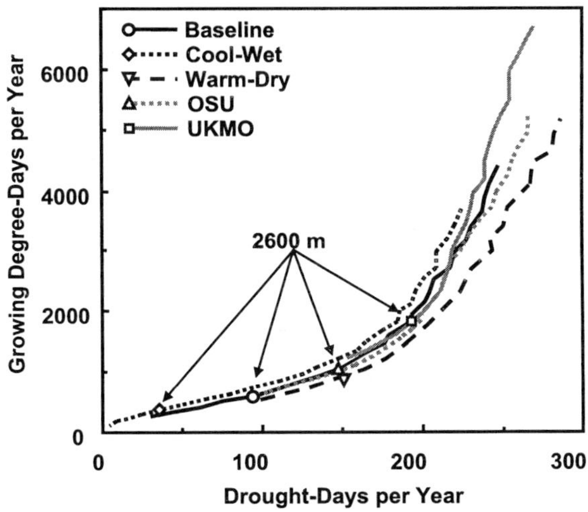

Figure 3.6. The climatic environment simulated by FM's weather model for baseline conditions and three of the climate scenarios. The uppermost right-hand point on each curve is a site at 500 m and the lowest left-hand point is a site at 3000 m. Pointers indicate the same site at 2600-m elevation under the different climate scenarios.

ents that result under the different experiments with the lower left corner of the graph representing cool and wet conditions (and higher elevations), and the upper right corner representing warm and dry environments (and lower elevations). The symbol on each curve represents a site at 2600 m elevation. A comparison of the relative position of these points illustrates the different climatic environments experienced by trees in each simulation experiment.

Climatic change affects simulated forest structure and composition across the elevation gradient. Generally speaking, tree species migrate either upslope or downslope, following the environmental conditions under which they best compete with other species. When climatic change results in a net increase in available water (i.e., a decrease in drought-days), water-limited sites experience an increase in woody biomass. This is the case in the cool-wet experiment at 1800, 2200, and 2600 m elevation (Fig. 3.7). At sites that are limited by the length of the growing season (e.g., 3000 m elevation), the response to a warmer and drier climate is also an increase in woody biomass, such as in warm-dry and OSU experiments (Fig. 3.7). When the new climate exceeds the temperature or drought tolerances of all tree species, the forest is converted to grassland or other nonforest type, a result seen in the warm-dry and the two GCM experiments at 1800 m (Table 3.1). The climatic change experiments simulated here alter conditions severely enough to produce dramatic shifts in species composition as well as conversion to nonforest conditions at certain sites.

The mean fire interval (average time between fires) for the 9-ha model grid (Fig. 3.8a) and the mean area burned by each fire (Fig. 3.8b) were computed for each of the climatic change experiments and averaged over 10 replicate

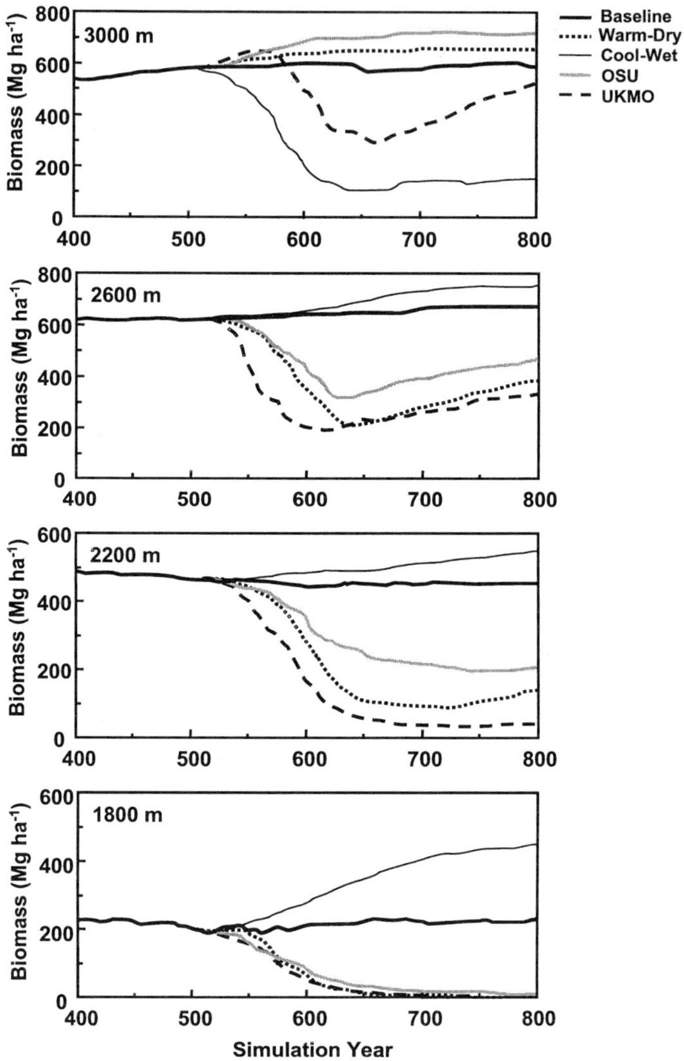

Figure 3.7. Total woody biomass simulated at four elevations for the four climatic change scenarios and baseline conditions. Values were averaged over 10 replicate simulations. The climate transient from baseline conditions occurred from simulation year 501–600.

simulations. Warmer and drier climates tend to generate more frequent fires than cooler and wetter climates, and in general, the average area burned by each fire is inversely related to fire frequency. An important point, however, is that more precipitation does not necessarily mean "wetter" with respect to the annual water balance. For example, the UKMO GCM predicts higher precipitation than baseline climate, but the warmer temperatures in the UKMO experiment create a

Table 3.1. Species basal areas ($m^2 ha^{-1}$) in the final simulation year for the baseline and four climate change experiments

	1800 m				
Species name	Baseline	Warm-dry	Cool-wet	OSU	UKMO
White fir	1	0	16	0	0
Red fir	0	0	0	0	0
Incense cedar	9	1	15	1	0
Lodgepole pine	0	0	0	0	0
Jeffrey pine	3	1	5	1	0
Sugar pine	0	0	2	0	0
Western white pine	0	0	0	0	0
Ponderosa pine	15	1	15	1	0
Black oak	1	1	0	1	0
	2200 m				
Species name	Baseline	Warm-dry	Cool-wet	OSU	UKMO
White fir	45	2	46	6	0
Red fir	0	0	14	0	0
Incense cedar	2	11	0	13	3
Lodgepole pine	0	0	0	0	0
Jeffrey pine	1	3	1	2	1
Sugar pine	4	0	1	0	0
Western white pine	0	0	0	0	0
Ponderosa pine	0	9	0	11	3
Black oak	0	0	0	0	1
	2600 m				
Species name	Baseline	Warm-dry	Cool-wet	OSU	UKMO
White fir	9	41	0	48	13
Red fir	51	0	63	0	0
Incense cedar	0	0	0	0	13
Lodgepole pine	0	0	0	0	0
Jeffrey pine	1	3	0	1	2
Sugar pine	0	2	0	4	0
Western white pine	0	0	0	0	0
Ponderosa pine	0	0	0	0	14
Black oak	0	0	0	0	0
	3000 m				
Species name	Baseline	Warm-dry	Cool-wet	OSU	UKMO
White fir	0	0	0	2	49
Red fir	53	59	0	61	2
Incense cedar	0	0	0	0	2
Lodgepole pine	1	0	20	0	0
Jeffrey pine	0	0	0	0	0
Sugar pine	0	0	0	0	11
Western white pine	1	1	8	0	0
Ponderosa pine	0	0	0	0	0
Black oak	0	0	0	0	0

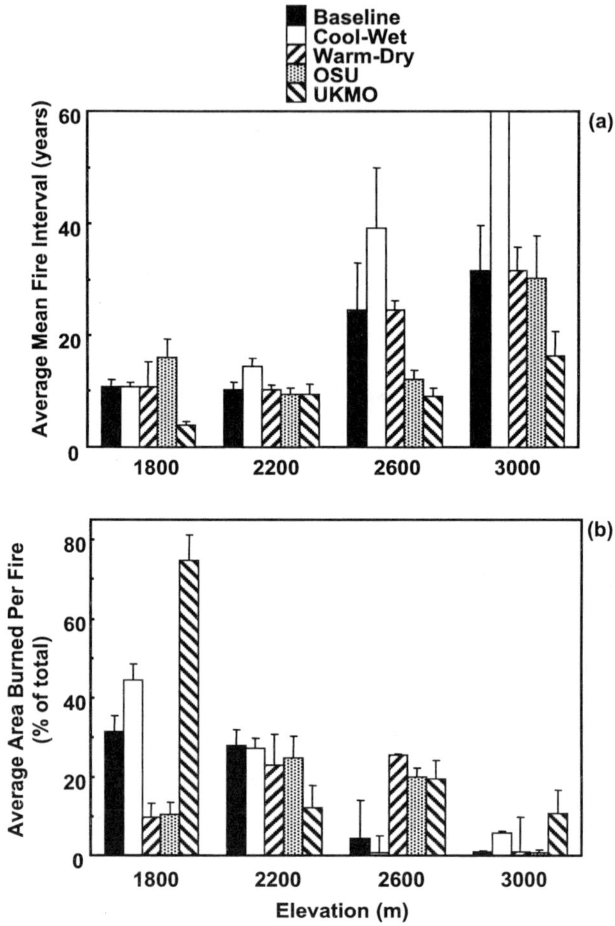

Figure 3.8. Summary of two aspects of the fire regime for four climatic change scenarios and baseline conditions. (a) Average mean fire interval during simulation years 501–800 and (b) average percent of the total area burned per fire during simulation years 501–800. Values were averaged over 10 replicate simulations and error bars are ±1 standard deviation. The average mean fire interval for the cool-wet scenario at 3000 m is not shown but was 202 years.

water demand that exceeds the increase in precipitation. As a result conditions in the UKMO simulation experiment are actually droughtier (i.e., more drought-days per year) than current baseline climate (Fig. 3.6) and the mean fire interval decreases at all sites in this experiment. The most significant differences in fire frequency occurred at the highest elevation sites.

More intriguing, perhaps, are the strong *indirect* effects of climate on the fire regime. Climatic change can influence the spatial extent of fires indirectly because altered forest structure and composition affect both the amount and type of fuel that are available for combustion. The simulations at 2600 m illustrate the effect of species composition on fire extent. In the warm-dry experiment at 2600 m, each fire burns an average of 25% of the model grid during years 501–800, whereas under baseline climate conditions, each fire burns only an average of 4% of the area (Fig. 3.8b). The difference is due to a shift in forest composition from red fir to white fir that occurs in the warm-dry experiment (Fig. 3.9a) and the influence that this change in species composition has on the properties of the fuel bed (Fig. 3.9b). The fuel bed that develops under red fir forests has a tightly packed

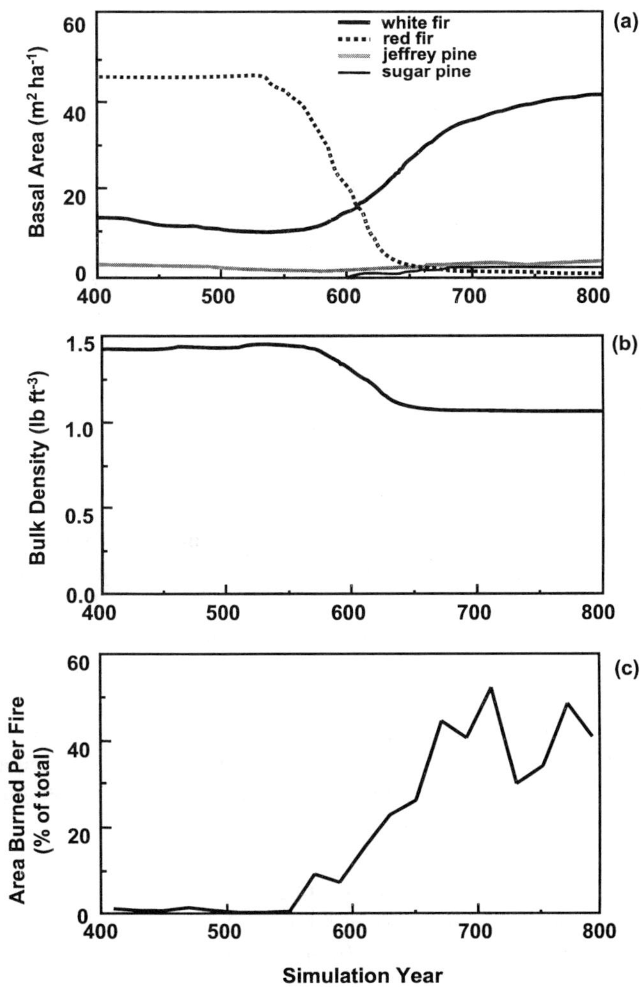

Figure 3.9. The effect of species composition on fire extent: (a) species composition, (b) fuel-bed bulk density, and (c) area burned per fire at 2600-m elevation for the warm-dry scenario averaged over 20-year intervals. All values were averaged over 10 replicate simulations. The climate transient occurred during simulation years 501–600.

litter component that does not burn readily. The fuel bed produced under a white fir forest, however, is less compact (van Wagtendonk, Benedict, and Sydoriak 1998), more burnable, and results in more area burned (Fig. 3.9c).

Limitations

The experiments simulated here are intended to demonstrate the sensitivity of these forests to climatic change and are not intended to make predictions of forest response to a particular climatic change scenario. Rather than looking at the model results for a single experiment or scenario, it is more instructive to compare the range of responses that may be possible. Climate predictions from GCMs carry substantial uncertainties, especially with respect to precipitation patterns, and it is noteworthy to point out that the GCM predictions used here predict *increases* in precipitation for the Sierra Nevada. If these predictions are wrong and precipitation instead decreases in the Sierra, we might expect even greater impacts on forest structure and composition and the fire regime.

The global climate may be responding to increasing greenhouse gas concentrations at a much faster rate than the climatic changes observed in the earth's paleoecological record (Houghton et al. 2001). The future sustainability of forests and ecosystems depends on whether plant migrations can keep pace with this rapid change (Davis 1990; Solomon and Kirilenko 1997). Unfortunately, we still do not understand many of the mechanisms responsible for plant migration (Pitelka 1997). FM, like most forest gap models, assumes that all species can disperse to all sites. Therefore it is possible that the model overpredicts the rate that forests can respond to climatic change (Loehle and LeBlanc 1996).

On the other hand, FM may actually underestimate the rate of forest response to climatic change because it tends to underestimate the severity of fires. For example, the model does not simulate crown fire behavior or the contribution of live fuels other than grass to fire behavior. Live foliage and branches from trees and shrubs in the subcanopy can serve as fuel ladders and promote crown fires, thereby increasing tree mortality. Higher mortality rates would (1) open up the forest canopy more than is currently simulated by the model, (2) provide more establishment opportunities for species better suited to the new climate, and (3) increase the rate of forest response to the simulated climatic change (Overpeck, Rind, and Goldberg 1990). FM also does not simulate extreme weather and wind conditions, and as a result probably underestimates the overall extent of fires and severity of fire effects.

Fire is not the only disturbance that interacts with vegetation and climate. For example, the frequency and severity of insect and disease outbreaks may increase in the Sierra Nevada under altered climatic conditions (Ferrell 1996). The associated tree mortality would provide increased dead fuels, potentially increasing fire frequency and area burned. Furthermore fire-damaged trees may be more susceptible to bark beetle attack (Ryan and Amman 1996), especially during droughts (Ferrell 1996), potentially resulting in a positive feedback cycle between drought, fire, and insect outbreaks.

Results from the simulations discussed here demonstrate the fire regime's sensitivity to fuel-bed bulk density which is a function of species composition and grass content in FM. However, other physical environmental factors, such as the depth and duration of snow pack, could affect fuel-bed bulk density. Furthermore fuel-bed bulk density is probably not the only fuel property that varies with elevation in the Sierra Nevada. For example, surface-area-to-volume ratios for woody fuels vary among species (van Wagtendonk, Benedict, and Sydoriak 1996), but these differences were not simulated here. Elevation gradients in other fuel properties could either enhance or offset the influence of fuel-bed bulk density on connectivity of burnable area that was demonstrated here.

Beyond the Sierra Nevada

Although this version of FM was developed for the Sierra Nevada, the model can be used to investigate climate-fire-forest relationships in other forests where surface fires dominate the fire regime. To apply the model to other forests, however, substantial data for parameterization are needed. This version uses life history information and tree-size allometric data from a variety of sources, including unpublished data sets, for the nine tree species simulated here (see Miller and Urban 1999b for a detailed list of parameters and data sources); similar information would be required to include other tree species. FM also requires considerable information about fuels. Data from several field studies of fuel accumulation rates, fuel input rates, and species-specific fuel characteristics were used to parameterize the model (e.g., Parsons 1978; van Wagtendonk, Benedict, and Sydoriak 1996, 1998; J. van Wagtendonk, USGS Yosemite Field Station, unpublished data). The success in applying FM to other forests may depend on the availability of these data.

Although the details of simulation results will differ if FM is applied to other forests, some general results are expected to hold. Because the model uses fuel dynamics to link forest dynamics with fire, fuels act as an intermediary for climatic change effects on the fire regime. This result is likely to be borne out in other forests as well. The simulation experiments highlight the influence of fuel-bed characteristics (e.g., fuel-bed bulk density) on the fire regime, but other variables may be important in other forests. Other aspects of fuel dynamics, such as fuel input rates or decay rates, or other fuel-bed characteristics (e.g., particle size relationships, heat content of fuels) may be critical variables in other forests. A thorough understanding of fuel dynamics is needed to predict climatic change effects on fire regimes.

Conclusion

The response of ecosystems and fire regimes to global climatic change will depend on a host of site-specific factors and a complex set of interactions among climate, fire, and vegetation. If our society is to anticipate the potential consequences of climatic change, we must improve our understanding of the complex interactions between climate and fire regimes. To study the impacts of climatic change on fire regimes, the effects of climatic change on vegetation must be simultaneously studied because vegetation affects fire regimes through its influence on the fuels that accumulate. Although paleoecological data have been used to reconstruct linkages among past changes in fire regimes, climate, and vegetation, the use of these retrospective data for projecting future consequences is limited because the climate of the future may not be analogous to any climate in the past. Simulation models may help us understand how fire-dependent ecosystems might respond to climatic change, particularly if the models simulate fire regimes that are driven by both climate and vegetation dynamics.

The simulation model FM was used to investigate how climatic change might impact surface fire regimes and forests in the Sierra Nevada. Simulated changes in temperature and precipitation affected forest biomass, forest species composition, fire frequency, and area burned. These effects were site specific and varied in direction and magnitude depending on the elevation of the simulated site. Total forest biomass and species composition changed in response to the change in site conditions relative to the tolerances of tree species simulated in the model. In some cases changes in temperature and precipitation affected the fuel moisture simulated by the model, thus directly affecting fire frequency and area burned. In other cases the fire regime was affected by changes in the amount and type of fuel that resulted from changes in forest structure or composition. These results suggest that the fire regime may be impacted most by climatic change where species composition shifts significantly and alters the flammability of the fuel bed.

Results from the simulation experiments suggest future directions for research on fire and climatic change. Vegetation response to climatic change will occur at the species level with individual species responding to environmental conditions as their life history adaptations allow, and it's possible that even subtle changes in the composition of species could dramatically influence the flammability of the fuel bed, thereby altering the fire regime. Therefore climatic change research would benefit from the use of models that simulate species-specific responses to climatic factors and that simulate fire regimes that respond to climate and vegetation processes. Fuels are at the heart of climate-fire-vegetation interactions, and we need to improve our understanding of fuel dynamics and the properties of fuel beds that influence flammability. Additional field studies of the biological and physical factors that govern fuel dynamics and fuel-bed characteristics will be essential for increasing our understanding of potential climatic change effects on fire regimes.

References

Agee, J.K. 1990. The historical role of fire in Pacific Northwest forests. In *Natural and Prescribed Fire in Pacific Northwest Forests*, eds. J.D. Walstad, S.R. Radosevich, and D.B. Sandberg, pp. 25–38. Corvallis, OR: Oregon State University Press.

Agee, J.K. 1993. *Fire Ecology of Pacific Northwest Forests*. Washington, DC: Island Press.

Albini, F.A. 1976. Estimating wildfire behavior and effects. USDA Forest Service Gen. Tech. Rep. INT-30.

Arno, S.F., and Brown, J.K. 1991. Overcoming the paradox in managing wildland fire. *Western Wildlands* (Spring):40–46.

Baker, W.L., Egbert, S.L., and Frazier, G.F. 1991. A spatial model for studying the effects of climatic change on the structure of landscapes subject to large disturbances. *Ecol. Model.* 56:109–125.

Betancourt, J.L., Van Devender, T.R., and Martin, P.S., eds. 1990. *Packrat Middens: The Last 40,000 Years of Biotic Change*. Tucson: University of Arizona Press.

Biswell, H.H. 1989. Prescribed burning in California wildlands vegetation management. Berkeley: University of California Press.

Bonnicksen, T.M., and Stone, E.C. 1982. Managing vegetation within U.S. national parks: A policy analysis. *Environ. Manag.* 6:101–102, 109–122.

Brown, J.K. 2000. Ecological principles, shifting fire regimes and management considerations. In *Wildland Fire in Ecosystems: Effects of Fire on Flora*, eds. J.K. Brown and J. Smith, pp. 185–203. USDA Forest Service Gen. Tech. Rep. RMRS-42-vol. 2.

Brown, J.K., Marsden, M.A., Ryan, K.C., and Reinhardt, E.D. 1985. Predicting duff and woody fuel consumed by prescribed fire in the northern Rocky Mountains. USDA Forest Service Res. Pap. INT-337.

Caprio, A.C., and Swetnam, T.W. 1995. Historic fire regimes along an elevational gradient on the west slope of the Sierra Nevada, California. In *Proceedings of Symposium on Fire in Wilderness and Park Management*, tech. coords. J.K. Brown, R.W. Mutch, C.W. Spoon, and R.W. Wakimoto, pp. 173–179. Missoula, MT, March 30–April 1, 1993.

Christensen, N.L. 1988. Succession and natural disturbance: paradigms, problems, and preservation of natural ecosystems. In *Ecosystem Management for Parks and Wilderness*, eds. J.K. Agee and D.R. Johnson, pp. 62–86. Seattle: University of Washington Press.

Christensen, N.L. 1993. Fire regimes and ecosystem dynamics. In *Fire in the Environment*, eds. P.J. Crutzen and J.G. Goldammer, pp. 233–244. New York: Wiley.

Clark, J.S. 1990. Twentieth-century climate change, fire suppression, and forest production and decomposition in northwestern Minnesota. *Can. J. For. Res.* 20:219–232.

Clark, J.S. 1993. Paleoecological perspectives on modeling broad-scale responses to global change. In *Biotic Interactions and Global Change*, ed. P.M. Kareiva, J.G. Kingsolver, and R.B. Huey, pp. 315–332. Sinauer Associates, Sunderland, MA.

Clark, J.S., Royall, P.D., and Chumbley, C. 1996. The role of fire during climate change in an eastern deciduous forest at Devil's Bathtub, New York. *Ecology* 77:2148–2166.

Cohen, J.D., and Deeming, J.E. 1985. The national fire-danger rating system: basic equations. USDA Forest Service Gen. Tech. Rep. PSW-82.

Daly, C., Neilson, R.P., and Phillips, D.L. 1994. A digital topographic model for distributing precipitation over mountainous terrain. *J. Appl. Meteorol.* 33:140–158.

Davis, M.B. 1990. Climatic change and the survival of forest species. In *The Earth in Transition: Patterns and Processes of Biotic Impoverishment*, ed. G.M. Woodwell, pp. 99–111. Cambridge: Cambridge University Press.

Davis, M.B., and Botkin, D.B. 1985. Sensitivity of cool-temperate forests and their fossil pollen record to rapid temperature change. *Quat. Res.* 23:327–340.

Davis, K.M., Clayton, B.D., and Fischer, W.C. 1980. Fire ecology of Lolo National Forest habitat types. USDA Forest Service Gen. Tech. Rep. INT-79.

Ferrell, G.T. 1996. The influence of insect pests and pathogens on Sierra forests. In *Sierra Nevada Ecosystem Project: Final Report to Congress, Vol. II: Assessments and Scientific Basis for Management Options*, pp. 1177–1192. Davis: University of California, Centers for Water and Wildland Resources.

Gardner, R.H., Romme, W.H., and Turner, M.G. 1999. Predicting forest fire effects at landscape scales. In *Spatial Modeling of Forest Landscape Change—Approaches and Applications*, eds. D.J. Mladenoff and W.L. Baker, pp. 163–185. Cambridge: Cambridge University Press.

Graber, D.M. 1985. Coevolution of National Park Service fire policy and the role of national parks. In *Proceedings of Symposium and Workshop on Wilderness Fire*, tech. coords., J.E. Lotan, B.M. Kilgore, W.C. Fischer, and R.W. Mutch, pp. 345–349. USDA Forest Service Gen. Tech. Rep. INT-182.

Graber, D.M., Haultain, S., and Fessenden, J.E. 1993. Conducting a biological survey: a case study from Sequoia and Kings Canyon National Parks. In *Proceedings of Fourth Conference on Research in California's National Parks*, eds. S.D. Veirs Jr., T.J. Stohlgren, and C. Schonewald-Cox, pp. 17–35. Transactions and Proceedings Series 9, U.S. Department of the Interior, National Park Service.

Hardy, C.C., Bunnell, D.L., Menakis, J.P., Schmidt, K.M., Long, D.G., Simmerman, D.G., and Johnston, C.M. 1999. *Coarse-scale Spatial Data for Wildland Fire and Fuel Management*. World Wide Web site: *www.fs.fed.us/fire/fuelman*

Hargrove, W.W., Gardner, R.H., Turner, M.G., Romme, W.H., and Despain, D.G. 2000. Simulating fire patterns in heterogeneous landscapes. *Ecol. Model.* 135:243–263.

Haxeltine, A., Prentice, I.C., and Cresswell, I.D. 1996. A coupled carbon and water flux model to predict vegetation structure. *J. Vegetation Sci.* 7:651–666.

He, H.S., and Mladenoff, D.J. 1999. Spatially explicit and stochastic simulation of forest-landscape fire disturbance and succession. *Ecol.* 80:81–99.

Heinselman, M.L. 1973. Fire in the virgin forests of the Boundary Waters Canoe Area, Minnesota. *Quat. Res.* 3:329–382.

Houghton, J.T., Ding, Y., Griggs, D.J., Noguer, M., van der Linden, P.J., and Xiasou, D., eds. 2001. *Climate Change 2001: The Scientific Basis*. Cambridge: Cambridge University Press.

Husari, S.J., and McKelvey, K.S. 1996. Fire-management policies and programs. In *Sierra Nevada Ecosystem Project: Final Report to Congress, Vol. II, Assessments and Scientific Basis for Management Options*, pp. 1101–1117. Davis: University of California, Centers for Water and Wildland Resources.

Keane, R.E., Ryan, K.C., and Running, S.W. 1996. Simulating effects of fire on northern Rocky Mountain landscapes with the ecological process model FIRE-BGC. *Tree Physiol.* 16:319–331.

Kilgore, B.M. 1973. The ecological role of fire in Sierran conifer forests: Its application to national park management. *J. Quat. Res.* 3:496–513.

Landres, P.B., Morgan, P., and Swanson, F.J. 1999. Overview of the use of natural variability concepts in managing ecological systems. *Ecol. Appl.* 9:1179–1188.

Loehle, C., and LeBlanc, D. 1996. Model-based assessments of climate change effects on forests: a critical review. *Ecol. Model.* 90:1–31.

McKelvey, K.S., Skinner, C.N., Chang, C., Erman, D.C., Husari, S.J., Parsons, D.J., van Wagtendonk, J.W., and Weatherspoon, C.P. 1996. An overview of fire in the Sierra Nevada. In *Sierra Nevada Ecosystem Project: Final Report to Congress, Vol. II: Assessments and Scientific Basis for Management Options*, pp. 1033–1040. Davis: University of California, Centers for Water and Wildland Resources.

Millar, C.I., and Woolfenden, W.B. 1999. The role of climate change in interpreting historical variability. *Ecol. Appl.* 9:1207–1216.

Miller, C., and Urban, D.L. 1999a. Forest heterogeneity and surface fire regimes. *Can. J. For. Res.* 29:202–212.

Miller, C., and Urban, D.L. 1999b. A model of surface fire, climate and forest pattern in Sierra Nevada, California. *Ecol. Model.* 114:113–135.

Miller, C., and Urban, D.L. 1999c. Forest pattern, fire, and climatic change in the Sierra Nevada. *Ecosystems* 2:76–87.

Miller, C., and Urban, D.L. 2000. Connectivity of forest fuels and surface fire regimes. *Landscape Ecol.* 15:145–154.

Mutch, L.S., and Parsons, D.J. 1998. Mixed conifer forest mortality and establishment before and after prescribed fire in Sequoia National Park, California. *For. Sci.* 44: 341–355.

Nielsen, R.P. 1995. A model for predicting continental-scale vegetation distribution and water balance. *Ecol. Appl.* 5:362–385.

Nikolov, N.T., and Zeller, K.F. 1992. A solar radiation algorithm for ecosystem dynamic models. *Ecol. Model.* 61:149–168.

Overpeck, J.T., Rind, D., and Goldberg, R. 1990. Climate-induced changes in forests disturbance and vegetation. *Nature* 343:51–53.

Parsons, D.J. 1978. Fire and fuel accumulation in a giant sequoia forest. *J. For.* 76: 104–105.

Parsons, D.J., and DeBenedetti, S.H. 1979. Impact of fire suppression on a mixed-conifer forest. *For. Ecol. Manag.* 2:21–33.

Parsons, D.J., and van Wagtendonk, J.W. 1996. Fire research and management in the Sierra Nevada National Parks. In *Science and Ecosystem Management in the National Parks*, eds. W.L. Halvorson and G.E. Davis, pp. 25–48. Tucson: University of Arizona Press.

Parsons, D.J., Graber, D.M., Agee, J.K., and van Wagtendonk, J.W. 1986. Natural fire management in national parks. *Environ. Manag.* 10:21–24.

Peters, R.L., and Lovejoy, T.E., eds. 1992. *Global Warming and Biological Diversity*. New Haven: Yale University Press.

Pitelka, L.F. 1997. Plant migration and climate change. *Am. Scientist* 85:464–473.

Prentice, I.C., Cramer, W., Harrison, S.P., Leemans, R., Monserud, R.A., and Solomon, A.M. 1992. A global biome model based on plant physiology and dominance, soil properties and climate. *J. Biogeogr.* 19:117–134.

Pyne, S.J., Andrews, P.A., and Laven, R.D. 1996. *Introduction to Wildland Fire*. New York: Wiley.

Riggan, P.J., Franklin, S.E., Brass, J.A., and Brooks, F.E. 1994. Perspectives on fire management in Mediterranean ecosystems of southern California. In *The Role of Fire in Mediterranean-Type Ecosystems*, eds. J.M. Moreno and W.C. Oechel, pp. 140–162. New York: Springer-Verlag.

Rothermel, R.C. 1972. A mathematical model for predicting fire spread in wildland fuels. USDA Forest Service Res. Pap. INT-115.

Running, S.W., Nemani, R., and Hungerford, R.D. 1987. Extrapolation of meteorological data in mountain terrain, and its use for simulating forest evapotranspiration and photosynthesis. *Can. J. For. Res.* 17:472–483.

Ryan, K.C., and Amman, G.D. 1994. Bark beetle activity and delayed tree mortality in the Greater Yellowstone Area following the 1988 fires. In *The Ecological Implications of Fire in Greater Yellowstone: Proceedings of the Second Biennial Conference on the Greater Yellowstone Ecosystem*, ed. J. Greenlee, pp. 151–158. Fairfield, WA: International Association of Wildland Fire.

Ryan, K.C., and Reinhardt, E.D. 1988. Predicting postfire mortality of seven western conifers. *Can. J. For. Res.* 18:1291–1297.

Skinner, C.N., and Chang, C. 1996. Fire regimes, past and present. In *Sierra Nevada Ecosystem Project: Final Report to Congress, Vol. II: Assessments and Scientific Basis for Management Options*, pp. 1041–1069. Davis: University of California, Centers for Water and Wildland Resources.

Smith, T.M., and Urban, D.L. 1988. Scale and resolution of forest structural pattern. *Vegetatio* 74:143–150.

Solomon, A.M., and Kirilenko, A.P. 1997. Climate change and terrestrial biomass: What if trees do not migrate? *Global Ecol. Biogeogr. Letts.* 6:139–148.

Stauffer, D. 1985. *An Introduction to Percolation Theory*. London: Taylor and Francis.

Stephens, S.L. 1995. Effects of prescribed and simulated fire and forest history of giant sequoia (*Sequoiadendron giganteum* [Lindley] Buccholz.)-mixed conifer ecosystems of the Sierra Nevada, California. Ph.D. dissertation. University of California, Berkeley.

Stephenson, N.L. 1988. Climatic control of vegetation distribution: the role of the water balance with examples from North America and Sequoia National Park, California. Ph.D. dissertation. Cornell University, Ithaca.

Stephenson, N.L. 1999. Reference conditions for giant sequoia forest restoration: structure, process, and precision. *Ecol. Appl.* 9:1253–1265.

Stephenson, N.L., and Parsons, D.J. 1993. A research program for predicting the effects of climatic change on the Sierra Nevada. In *Proceedings of the Fourth Conference on Research in California's National Parks*, eds. S.D. Veirs Jr., T.J. Stohlgren, and C. Schonewald-Cox, pp. 93–109. U.S. Department of the Interior National Park Service Transactions and Proceedings Series 9.

Swetnam, T.W. 1993. Fire history and climate change in giant sequoia groves. *Science* 262:885–889.

Swetnam, T.W., Allen, C.D., and Betancourt, J.L. 1999. Applied historical ecology: Using the past to manage for the future. *Ecol. Appl.* 9:1189–1206.

Turner, M.G., and Romme, W.H. 1994. Landscape dynamics in crown fire ecosystems. *Landscape Ecol.* 9:59–77.

United States General Accounting Office. 1999. *Western National Forests: A Cohesive Strategy Is Needed to Address Catastrophic Wildfire Threats.* General Accounting Office Report GAO/RCED-99–65.

University Corporation for Atmospheric Research (UCAR). 1997. World Wide Web site: *www.cgd.ucar.edu/vemap/scenario.html*

Urban, D.L., Bonan, G.B., Smith, T.M., and Shugart, H.H. 1991. Spatial applications of gap models. *For. Ecol. Manag.* 42:95–110.

Urban, D.L., Harmon, M.E., and Halpern, C.B. 1993. Potential response of Pacific northwestern forests to climatic change, effects of stand age and initial composition. *Clim. Change* 23:247–266.

Urban, D.L., Miller, C., Halpin, P.N., and Stephenson, N.L. 2000. Forest gradient response in Sierran landscapes: the physical template. *Landscape Ecol.* 15:603–620.

Vankat, J.L., and Major, J. 1978. Vegetation changes in Sequoia National Park, California. *J. Biogeogr.* 5:377–402.

Van Wagner, C.E. 1973. Height of crown scorch in forest fires. *Can. J. For. Res.* 3: 373–378.

van Wagtendonk, J.W. 1985. The role of fire in the Yosemite wilderness. In *Proceedings of the National Wilderness Research Conference*, Fort Collins, CO, July 23–26, pp. 2–9.

van Wagtendonk, J.W., Benedict, J.M., and Sydoriak, W.M. 1996. Physical properties of woody fuel particles of Sierra Nevada conifers. *Int. J. Wildland Fire* 6:117–123.

van Wagtendonk, J.W., Benedict, J.M., and Sydoriak, W.M. 1998. Fuel bed characteristics of Sierra Nevada conifers. *Western J. Appl. For.* 13:73–84.

White, P.S., and Pickett, S.T.A. 1985. Natural disturbance and patch dynamics, an introduction. In *The Ecology of Natural Disturbances and Patch Dynamics*, eds. S.T.A. Pickett and P.S. White, pp. 3–13. New York: Academic Press.

2. North America

4. Fire Regimes and Climatic Change in Canadian Forests

Mike Flannigan, Brian Stocks, and Mike Weber

Forest fire is the dominant disturbance regime in circumboreal forests, burning an average of 5 to 10 million ha annually (Stocks 1991; Weber and Stocks 1998), almost exclusively in Canada, Alaska, and Russia. Forest fire is the primary process organizing the physical and biological attributes of the boreal biome over most of its range, shaping landscape diversity and influencing energy flows and biogeochemical cycles, particularly the global carbon cycle. Settlement and exploitation of the boreal zone has been accomplished in conjunction with the development of sophisticated fire management systems designed to suppress unwanted fires that threaten public and industrial interests while permitting natural forest cycling through fire where possible.

Fire and climate/weather are intimately linked (Johnson 1992; Swetnam 1993), which means the fire regime will respond rapidly to changes in climate. For the purposes of this chapter we define the fire regime as having six components; frequency, size, intensity, seasonality, type, and severity (cf. Flannigan 1993; Malanson 1987; Merrill and Alexander 1987). The ecological importance of some of these components of a fire regime has been put into perspective by Malanson (1987) and Whelan (1995). Fire frequency affects ecosystems by interrupting or terminating individual life cycles. If fires recur more or less regularly, selection pressure will favor those organisms that better take advantage of the recurrence at a given interval. Fire size determines landscape patchiness and determines the distance seed will have to travel for regeneration. Fire intensity is equivalent to the amount of energy released per unit length of fireline. Intensity, within the confines of a single burn, can vary greatly depending on the fuel type and loading, topography, meteorological influences, and characteristics of the previous disturbance, among other factors. Season of the year at which fire occurs is one of the determinants of the successional trajectories on which ecosystems embark after fire. Time of year may affect fire intensity through differences in surface and crown fuel moisture contents. Seasonal phenological state of the plants burned will determine the characteristics of the vegetative or seed reproductive response and have a pronounced effect on the structure of postfire ecosystems and landscapes. Fire type refers to crown, surface, and ground fires, which are largely controlled by fire intensity and fuel characteristics (structure, load, and moisture). Fire type, like intensity, can vary across the area of the burn, giving rise to a mosaic of postfire plant communities that might be initiated by crowning, surface fires, intermittent crowning, or a combination thereof. Fire severity is a description of the depth of burn into the surface soil organic layers and therefore another important controlling factor of postfire ecosystem structure and function through direct impacts on underground plant root and reproductive tissues, soil seed bank, and forest floor microbial populations. These component parts of a fire regime with their intricate linkage to Canadian forest ecosystem structure and function are, in turn, highly dependent on climate (Kirschbaum and Fishlin 1996).

Since 1980 there has been an annual average of over 10,000 fires with an area burned close to 3 million ha in Canada. Approximately 3% of the fires are larger than 200 ha, but these fires are responsible for 97% of the area burned (Weber and Stocks 1998). In the boreal forest the dominant fire type is stand-replacing crown fires. Typical fire cycles range from 30 to 500 years for most of the boreal forest (Flannigan et al. 1998). Species regenerate vegetatively or through serotinous cones within the area burned, or they can regenerate from seeds from adjacent unburned stands. Fire is a critical aspect in the regeneration of many forests as it removes competition, allows sunlight to reach the forest floor, and prepares the seed bed by removing organic matter. Figure 4.1 shows the forest regions of Canada along with the location of large fires (≥200 ha) during the 1980 to 1989 period (Stocks et al. 1998).

Discussion in this chapter is restricted to changes in climate and vegetation during the last 10,000 years. Ten thousand years ago northern latitudes were still greatly influenced by the continental ice sheet. The climate warmed to a point where it was warmer than the present day for the period 7000 to 3000 years before present (Intergovernmental Panel on Climate Change [IPCC] 2001). A general cooling trend has been experienced in the last 3000 years with relatively short periods of warming, such as the recent warming since the end of the Little Ice Age (ca. AD 1850).

There is consensus that human activities are responsible for recent changes in the climate (IPCC 2001). Specifically, increases in radiatively active gases such as carbon dioxide, methane, and the chlorofluorocarbons in the atmosphere are causing a significant warming of the earth's surface. Significant increases in temperature are anticipated in this century and beyond, with general

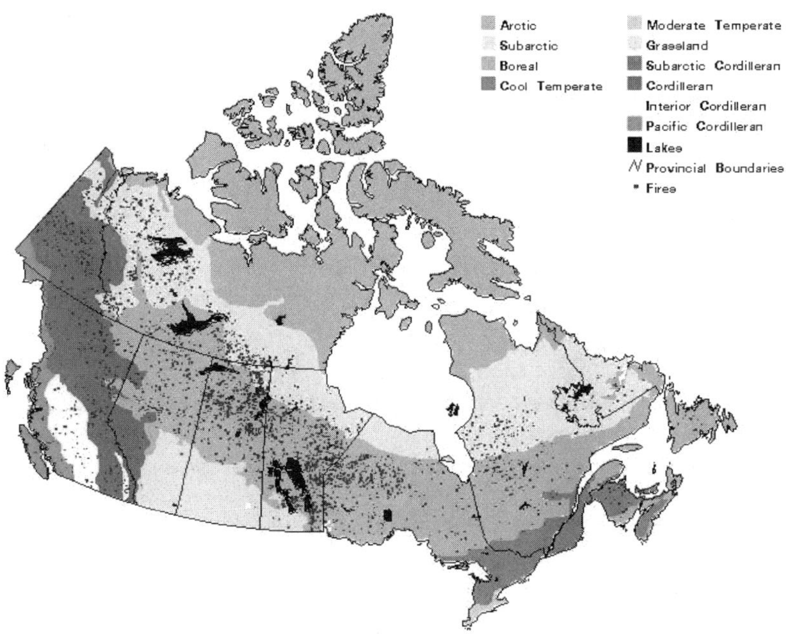

Figure 4.1. Ecoclimatic regions of Canada with the 1980 to 1989 fires comprising >200 ha.

circulation models (GCMs) projecting a mean global temperature increase of 1.4–5.8°C by AD 2100, an increase greater than any observed in the last 10,000 years.

Weather and climate are crucial to the occurrence and growth of forest fires. Lightning is the key ignition agent for naturally caused forest fires. Lightning is the result of an electrical discharge from a thunderstorm, which itself is a result of the appropriate meteorological conditions, namely atmospheric instability, moisture, and a lifting agent. The weather prior to ignition is important in determining the fuel moisture, which in turn will determine if ignition will occur and if the fire will grow. These weather conditions that influence fuel moisture include temperature, precipitation, wind speed, and atmospheric moisture (vapor pressure deficient). Fire growth is a function of a number of variables, but if fuels are available and dry, then wind speed is the key factor.

When studying the role of the weather or climate on the area burned by forest fires several meteorological parameters are important. Temperature, precipitation, wind, atmospheric moisture, upper atmospheric features, teleconnections, vertical structure of the atmosphere, drought indexes, and components of fire weather index systems have all been used to elucidate the relationships between the weather/climate and area burned by forest fire (Flannigan and Wotton 2001). Mean and maximum temperature are frequently used in studies. Precipitation measures include, amount, frequency, and duration. Wind speed and direction are often used, as well as the dew point, relative humidity, or other measures of the moisture in the atmosphere near the earth's surface. Features such as upper-level ridges and the stability of the atmosphere have been addressed in some studies relating fire to climate and weather (Flannigan and Harrington 1988; Skinner et al. 1999). Often the term blocking ridges has been associated with fire outbreaks. These are persistent ridges in the upper atmosphere (usually at the 500-mb level which is approximately 5600 m above sea level) that last a week or longer. These ridges tend to block or divert precipitation-bearing systems to the north or south of the ridge; thus dry and warm weather at the surface is typically associated with these upper ridges.

Drought indexes such as the Palmer Drought Index and components of fire weather index systems like the Canadian Forest Fire Weather Index (FWI) System (Van Wagner 1987) have been employed in investigations between fire and weather. Fire danger rating systems integrate daily weather information into qualitative outputs that describe the relative fire danger across an area. Fire danger systems are typically designed to suit fuel types in a specific region (Deeming, Burgan, and Cohen 1977; Van Wagner 1987; Stocks et al. 1989). These systems vary greatly around the world in terms of input and output complexity but, in general, use daily air temperatures, relative humidities, wind speeds, and precipitation amount (and perhaps rate) to calculate a series of cumulative fuel moisture indicators. The fuel moisture indicators are then used as relative danger indexes or are combined to give a more general index of fire potential. In some systems differences in fuel type and topographical effects are also taken into account, though these inputs are more typically used for site-specific fire behavior prediction. With regard to the relative importance of vegetation and weather on fire behavior, research has shown that weather is the most important factor by far (Bessie and Johnson 1995; Hely et al. 2001).

There are numerous other factors such as ignition agents, topography, vegetation, landscape fragmentation, and fire management activities that could influence the fire activity in a region. The agent of ignition can be lightning, or ignition can be human caused by a wide variety of activities. In some cases fire can be a cultural practice such as burning fields or conversion of forest to agriculture using slash and burn practices (Pyne 1997). Topography, slope, and orientation can significantly influence fire behavior (Van Wagner 1977). Vegetation can also play an important role as aspects of fuel amount, continuity, moisture, arrangement, and structure are key determinants in fire occurrence and spread. The fragmentation of the landscape through natural features such as lakes or via human activities including roads, agriculture, and settlements can influence the area burned (Weir and Johnson 1998; Weir, Johnson, and Miyanishi 2000). The influence of fire management on area burned is a function of the effectiveness of the fire crews and the suppression policy in place.

The objective of this chapter is to estimate how climate change will influence the fire regime across Canada in the twenty-first century and, in turn, how this change in fire regime will impact Canadian forests. We will begin by reviewing connections between climate and fire that have been elucidated by paleo studies, fire-history studies, and fire-weather studies. Those studies that encompass warm periods in the past might be analogues to future warming. Predictions of the climate derived from GCMs will be used to estimate the fire weather in this century. The implications of climate change on Canadian forests will be discussed with regard to changes in the fire regime.

Fire-Climate Interactions

This section will address fire-climate interactions during the Holocene (ca. 10,800 years ago to the present) which represents the present interglacial period. There are several methods to determine the long-term fire regime data (Tolonen 1983). These approaches include fire scars, time-since-fire maps, charcoal in peat, and laminated lake sediments. Short-term fire-weather studies will also be discussed in this section along with an overview of modeling efforts of fire activity in the future.

Fire Scars and Time-Since-Fire Maps

Trees that survive fires often scar, which allows a reconstruction of the fire history at that location. Multiple scarring is possible, and for longer-lived tree species such as sequoia and bristlecone pine, a long fire history (1000s of years) may be available (Swetnam 1993). Additionally fire scars on snags can be used to extend the fire history even further by using a master chronology. However, for much of the North American boreal forest, fire scars allow a reconstruction of fire activity for only the last 100 to 300 years, as this method is limited by the longevity of the trees. This same limitation is in effect for the time-since-fire map method. A time-since-fire map depicts regions of vegetation that are delineated by the last year in which they burned. Time-since-fire maps were introduced by Heinselman (1973). Using likelihood inference historical fire frequency can be obtained from these maps (Reed et al. 1998).

Payette et al. (1989) studied fire at the treeline in northern Quebec, and found that fire was related to vegetation and climate. In shrub tundra, fires are small and infrequent. In the open forest, fire is more frequent and sizes are larger, whereas in the closed forest, fire activity was the greatest. They conclude that the gradient of fire regimes is partly the result of climate and vegetation type. Sirois and Payette (1991) found that recent fires in the forest tundra have contributed to deforestation with the burned areas reverting to tundra. This is in agreement with results from Payette and Gagnon (1985), who found that increased fire activity was the immediate cause of a tree-line recession in northern Quebec starting around 3000 years ago. The southward retreat of vegetation is consistent with the overall cooling of the climate during the last few thousand years. Fire can be an agent of change that hastens the rate of vegetation change associated with a change in climate.

Bergeron (1991) found that fire frequency in western Quebec has decreased since the end of the Little Ice Age (ca. AD 1850) despite warmer temperatures.

Bergeron and Archambault (1993) attribute the decrease in fires to reduced drought frequency which might be the result of the region being under the influence of a warmer and moister tropical air mass during the fire season. Foster (1983) found that years of high fire activity in southeastern Labrador were associated with low summer precipitation.

Johnson, Fryer, and Heathcott (1990) used fire scars and time-since-fire maps to examine the fire regimes of Glacier National Park in British Columbia. They discovered a decrease in the fire frequency after AD 1760, which they associate with moister conditions. Johnson and Larsen (1991) found that the fire cycle was about 50 years, prior to AD 1730, before increasing to 90 years for the 1730 to 1980 period for the Kananaskis Watershed in the southern Canadian Rockies. They attribute this change in fire cycle to a change in climate as determined by dendroclimatological studies. These studies suggest warm and dry conditions prior to 1730 becoming cooler and moister conditions thereafter.

Flannigan et al. (1998) surveyed the recent fire history studies in Canada and northern Europe. They found that fire frequency has decreased at almost every site in the last 150 years despite a warming since the end of the Little Ice Age. Some of these sites are influenced by human activities, including landscape fragmentation and fire suppression, which have a direct effect on fire frequency. Payette (1992) also displays a table of recent fire-history studies for several boreal forest areas in North America. These studies demonstrate the need for caution when extrapolating results from an individual study to infer a trend over a region or continent.

Charcoal Studies

The analysis of charcoal is often done by using slides prepared for pollen analysis. The charcoal is counted with the aid of a microscope or the charcoal abundance is determined by a chemical assay method (Winkler 1985). Charcoal is often described as macrofossil or microfossil. The microfossil charcoal or fine fraction charcoal reflects regional fire activity, whereas macroscopic charcoal reflects fire activity near the site of collection. Clark (1988) employed a technique that requires thin-sectioned lake sediments from varved (annually laminated) lakes. Although this method gives an improved temporal resolution because of the annual varves, it is limited in that varved lakes are not available in all regions. Charcoal beds in peat (Khury 1994) and charcoal abundance from lake sediments (Clark 1988; Winkler 1985) have also been used to reconstruct fire history. Fire activity in Quebec has been reconstructed using charcoal preserved in sand dunes (Filion et al. 1991). MacDonald et al. (1991) provides an excellent discussion on charcoal analysis.

Studies of charcoal beds in peat deposits in central Canada suggest a peak in fire frequency around 3500 to 4000 years before present (BP) (Bryson, Irving, and Larsen 1965; Nichols 1967). Nichols (1967) suggests that this increased fire activity may be related to a cooling from 6000 to 1500 years BP which brought central Canada under the influence of the cold and dry Arctic air mass. However, additional cooling since 3500 to 4000 years BP has not corresponded to increased fire activity. Khury (1994) studied charcoal beds in Canadian peatlands in Alberta, Saskatchewan, and Manitoba. He found that fire was more frequent by a factor of two to one during a mid-Holocene warm period prior to 5000 years BP. Vance, Emerson, and Habgood (1983), using charcoal from pollen slides obtained from sites in central Alberta, found charcoal influx was greater during the mid-Holocene warm period as opposed to after about 4000 years ago. Hu, Brubaker, and Anderson (1993) found that charcoal abundance was low at Wien Lake in central Alaska during the mid-Holocene warm period but that charcoal increased during the cooling after the mid-Holocene.

Future areas of research might use a regional approach to look at whether there is spatial synchrony in the fire-climate record. Ideally one technique would be to find and core a number of varved lakes across a large region and reconstruct the fire regime and the vegetation present. Finally warmer periods of the past such as the mid Holocene warm period can be used as analogues for future warming. Flannigan et al. (2001) use charcoal data from sites across Canada in combination with a specially modified GCMs that was run for 6000 years ago to suggest that previous warm periods may be analogues of future warming. In that study there was good agreement between the charcoal data and model data, which tends to validate the results, as these two data sources can be treated as independent.

Fire-Weather Studies

Day to day weather can dramatically influence fire behavior and area burned. This has lead to many studies over various spatial and temporal scales that try to relate the weather to fire. There have been numerous case studies that address the weather associated with an individual fire or an outbreak of fires. Schaefer (1957) addressed the relationship with the upper-level jet stream on forest fires. Turner (1970) studied the effect of hours of sunshine on fire season severity. The synoptic weather types associated with critical fire weather were studied by Schroeder et al. (1964). Other studies (Flannigan and Harrington 1987; Hirsch and Flannigan 1990; Quintilio, Fahnestock, and Dube 1977; Stocks and Walker 1973; Stocks 1975) have documented the weather prior to and during major fire runs. These studies have shown that fire spread rapidly when the fuels were dry and the weather conditions were warm to hot, dry, and windy. These studies by themselves have limited application because of the narrow scope in terms of temporal and spatial scales used. However, these studies are of value in identifying the most likely meteorological predictors related to fire activity that can be used in studies with a larger time and space domain.

Harrington, Flannigan, and Van Wagner (1983) related the monthly provincial area burned in Canada to components of the Canadian Fire Weather Index (FWI) System for 1953 to 1980. Results showed that the monthly means and extreme maximum values of the Duff Moisture Code (DMC) and the daily severity rating (DSR) were the best predictors of area burned. In western Canada, with the exception of the Yukon and Northwest Territories, explained variance averaged 33%.

In the territories and eastern Canada the explained variance averaged 12%. Using the same data set, Flannigan and Harrington (1988) studied the relation between meteorological variables and monthly area burned by wildfire from May to August 1953–80 for nine provincial sized regions in Canada. They found that bad fire months were independent of rainfall amount but significantly dependent on rainfall frequency, temperature, and relative humidity. Results were similar to those obtained by Harrington, Flannigan, and Van Wagner (1983), except the meteorological variables did better in the Territories and eastern Canada than did the FWI System. The most important predictors were long sequences of days with less than 1.5 mm of precipitation and long sequences of days with relative humidity below 60%. These long sequences were assumed to be associated with blocking highs in the upper atmosphere.

Newark (1975) discovered that 500-mb longwave ridging was related to forest fire occurrence in northwestern Ontario during the summer of 1974. Nimchuk (1983) related two episodes of catastrophic burning during the Alberta 1981 fire season to the breakdown of the upper ridge over Alberta. These episodes, which lasted eight days, accounted for about 1 million ha burned. The breakdown of these upper ridges is often accompanied by increased lightning activity as upper disturbances (shortwaves) move along the west side of the ridge. Additionally, as the ridge breaks down, strong and gusty surface winds are common. Brotak and Reifsnyder (1977) also studied the upper air conditions associated with 52 major wildland fires (area burned 5000 acres or more) in the eastern United States from 1963 to 1973. They found that the vast majority of the fires were associated with the eastern portion of a small but intense shortwave trough at 500 mb. Despite the difference in geographical location, the Brotak and Reifsnyder study and the work by Nimchuk may both be discussing the same situation, though the emphasis changes from trough to ridge breakdown from the former to the latter. Cold fronts are often associated with the breakdown of the ridge or the passing of a shortwave trough, which are also important in terms of major wildland fires (Brotak and Reifsnyder 1977). In addition to strong, and at times gusty, surface winds associated with these upper features, it is also important to note that a wind shift from southwest to northwest occurs with the passage of the shortwave trough aloft and the cold front at the surface. This is important in that the flank of a fire with a southwest wind will become the head of the fire with a northwest wind.

Flannigan and Harrington (1988) found that the 700-mb-height anomaly for the forested regions of their provincial areas was the predictor that was selected the most when relating meteorological variables to monthly provincial area burned in Canada 1953–1980. Johnson and Wowchuk (1993) found that midtropospheric positive anomalies (blocking ridges) were related to large-fire years in the southern Canadian Rocky Mountains, whereas as negative anomalies were related to small-fire years. They observed that these blocking ridges associated with the large-fire years were teleconnected, both spatially and temporally correlated with respect to 500-mb heights, to upper-level troughs in the North Pacific and eastern North America which is the positive mode of the Pacific North America (PNA) pattern. The PNA teleconnection is really a triple connection among an anticyclonic circulation over the North Pacific, a cyclonic circulation over western Canada, and a second anticyclonic circulation over the southeastern United States (Horel and Wallace 1981). Skinner et al. (1999) found that 500-mb-height anomalies were well correlated with seasonal area burned over various large regions of Canada. They also found a structure similar to the PNA pattern for the extreme fire seasons in western and west-central Canada Current research suggests that blocking frequency is related to the wave number (the number of longwaves in the westerlies—typically 3–5) with blocking ridges being more frequent with higher wave numbers (Weeks et al. 1997). Also research has suggested that the persistence of blocking ridges in the upper atmosphere will increase in a $2 \times CO_2$ climate (Lupo, Oglesby, and Mokhov 1997). This could have significant impact on fire activity as these upper ridges are associated with dry and warm conditions at the surface that are conducive to forest fires.

Dry and unstable air enhances the growth of forest fires. Unstable air is a layer of air that is characterized by a vertical temperature gradient such that when air parcels are displaced upward, they will accelerate upward and away from their original altitude. Haines (1988) developed a lower-atmosphere severity index (LASI) for wildland fires to account for temperature stability and the amount of moisture in the lower-atmosphere. He determined that only 6% of all fire season days fell into the high-index class for the western United States. However, 45% of days with large and/or erratic wildfire occurred during those high-index class days. Potter (1996) examined atmospheric properties associated with large wildfires (over 400 ha) in the United States from 1971 to 1984. He compared the lower-atmosphere moisture, temperature, wind, and lapse rate for the 339 large fires in the data set with climatology using the same 14-year period. The results show that the fire day surface air temperature and moisture differ from climatology at the 0.001 significance level. There was no difference in wind shear between fire days and climatology days. Results from wind speed and lapse rate were inconclusive. To date, research like that conducted by Haines (1988) and Potter (1996) on the vertical structure of the lower atmosphere has not been applied in Canada.

Models of the Future Climate

There are many General Circulation Models that enable researchers to simulate the future climate. Although there are a number of shortcomings associated with the GCMs, they provide the best means available to estimate the impact of changes in the future climate on the fire regime. Most models are in agreement in predicting the greatest warming at high latitudes and over land. In Canada, winter temperatures are expected to increase by 6–10°C, while summer temperatures increase by 4–6°C for a doubling of carbon dioxide in the middle of this century. The confidence is lower for estimates of precipitation, but many models suggest an increased moisture deficit, particularly in the center of continents during the summer. Recent transient GCMs, which include ocean-atmosphere coupling and aerosols, support these findings (Flato et al. 2000). In addition to

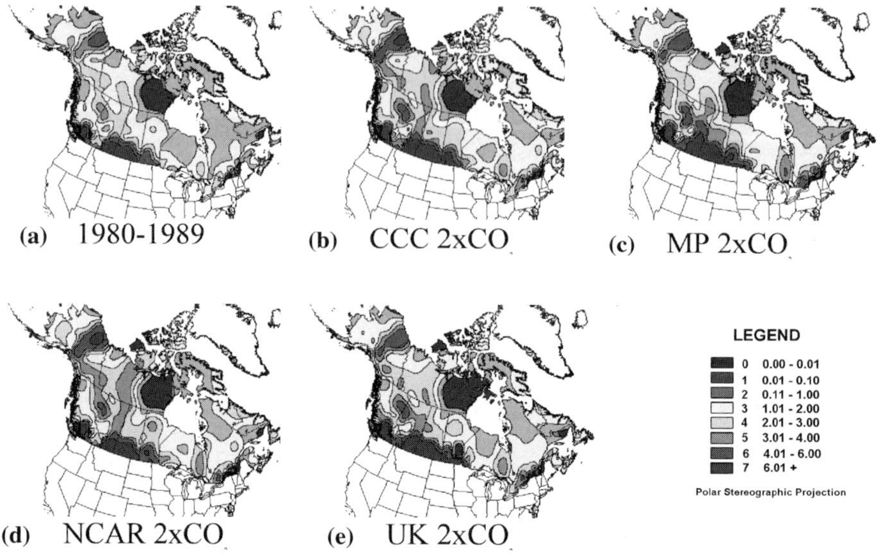

Figure 4.2. Average seasonal severity rating (SSR) maps for Canada showing (a) the 1980 to 1989 baseline SSR data and projected 2 × CO_2 SSR maps using the (b) Canadian, (c) United Kingdom, (d) German, and (e) U.S. GCMs (Stocks et al. 1998; reprinted with permission from *Climatic Change*, © Kluwer Academic Publishers).

temperature, other weather variables will be altered in the new climate such as precipitation, wind, and cloudiness. The variability of extreme events may be altered as well with increased variability anticipated (Mearns et al. 1989; Solomon and Leemans 1997).

Some studies suggest universal increases in fire frequency with climatic warming (Overpeck, Rind, and Goldberg 1990; IPCC 2001). The universality of these results is questionable because an individual fire is a result of the complex set of interactions that include ignition agents, fuel conditions, topography, and weather variables such as temperature, relative humidity, wind velocity, and the amount and frequency of precipitation. Increasing temperature alone does not necessarily translate into greater fire disturbance as assumed in these studies.

Studies that integrate several of the weather variables that influence forest fires provide better estimates than do simpler temperature-based models. Flannigan and Van Wagner (1991), for example, compared the seasonal fire severity rating (SSR, seasonal average of the Daily Severity rating which is devised from the FWI) from a $2 \times CO_2$ scenario (ca. AD 2050) versus the $1 \times CO_2$ scenario approximating the present day across Canada. Their study used monthly anomalies from three GCMs: Geophysical Fluid Dynamics Laboratory (GFDL), Goddard Institute for Space Studies (GISS), and Oregon State University (OSU). The results show increases in the SSR all across Canada with an average increase of nearly

50%, which they suggest would translate roughly into an increase of area burned by 50%. Stocks et al. (1998) used monthly data from four GCMs to examine climate change and forest fire potential in Russian and Canadian boreal forests. Forecast seasonal fire weather severity was similar for the four GCMs, indicating large increases in the areal extent of extreme fire danger (SSR values above 7) under a 2 × CO_2 scenario (Fig. 4.2). Stocks et al. (1998) also conducted a monthly analysis using the Canadian GCM, which showed an earlier start to the fire season and significant increases in the area experiencing high to extreme fire danger (monthly severity rating greater than 3) in Canada, particularly during June and July (Figs. 4.3 and 4.4). Wotton and Flannigan (1993) also found that the fire season length in Canada on average will increase by 22% or 30 days under a 2 × CO_2 climate. Flannigan et al. (1998) used daily output from the Canadian GCM to model the FWI for both the 1 × CO_2 and 2 × CO_2 scenarios for North America and Europe. Figure 4.5 shows the ratio of the 2 × CO_2 to 1 × CO_2 values for both mean FWI and maximum FWI for northern North America. There is a great deal of regional variation between areas where FWI decreases in a 2 × CO_2 scenario (values below 1.00) to areas where the FWI increases greatly in the warmer climate. There are significant increases in FWI for both mean and maximum over central Canada which is the region where most of the large fires

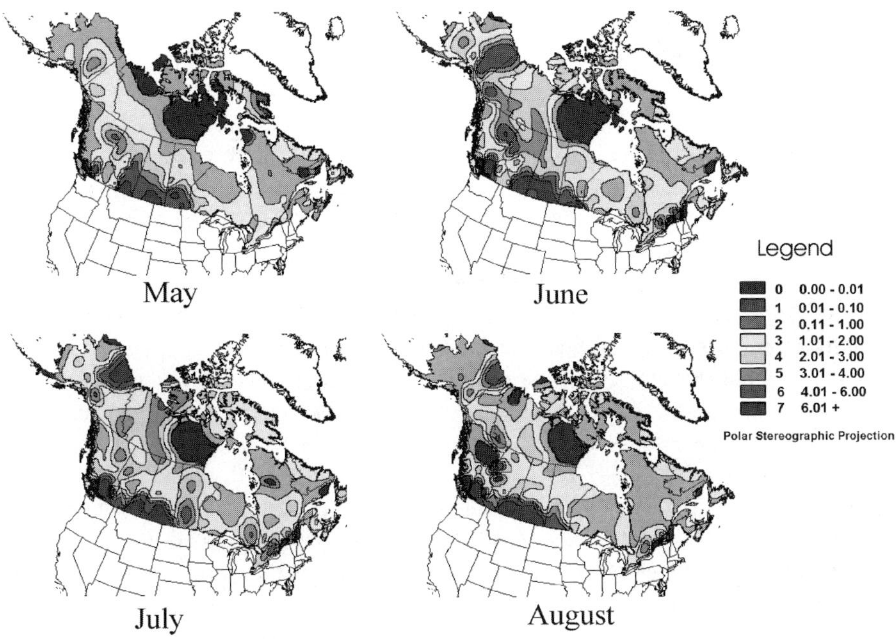

Figure 4.3. Average monthly severity rating (MSR) maps for Canada, based on 1980–1989 daily weather (Stocks et al. 1998; reprinted with permission from *Climatic Change*, © Kluwer Academic Publishers).

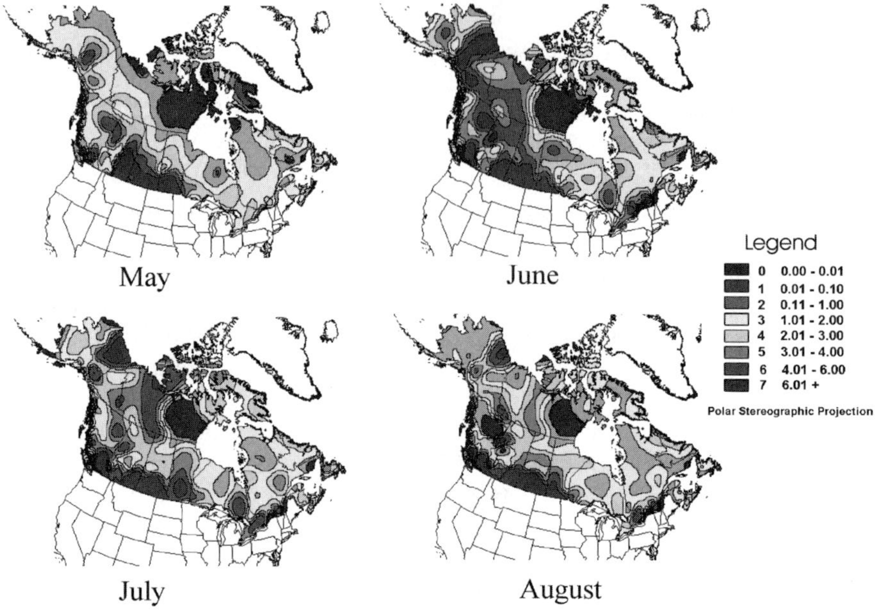

Figure 4.4. Average monthly severity rating (MSR) maps for Canada under a $2 \times CO_2$ climate using the Canadian GCM (Stocks et al. 1998; reprinted with permission from *Climatic Change*, © Kluwer Academic Publishers).

have occurred recently (Fig. 4.1). However, much of eastern Canada and northwestern Canada has ratios below 1.00, indicating that the FWI will decrease despite the warmer temperatures associated with a $2 \times CO_2$ climate. Noteworthy is the area of decreased FWI over western and northwestern sections of Canada where historically large portions of the landscape have been burned. However, due to the coarse spatial resolution of the GCM (~400 km) confidence in the results over complex, mountainous terrain is low. In such areas a Regional Climate Model (RCM) should be used (Caya and Laprise 1999) where the finer spatial resolution (ca. 40 km) can resolve mountain ranges. Significant increases in the FWI are evident over parts of central North America. The ratio of extreme maximum values of the FWI show a similar pattern, with higher ratios over central continental areas and lower values over portions of eastern Canada. On the other hand, there are increases in the maximum FWI over portions of western Canada. Consequences of climate change on fire disturbance must be viewed in a spatially dependent context.

Flannigan et al. (1998) suggest the reason for the decreased FWI despite the increasing temperature is due primarily to changes in the precipitation regime, and in particular to increases in precipitation frequency. These models results (Fig. 4.5) are in good agreement with recent fire-history studies, which cover

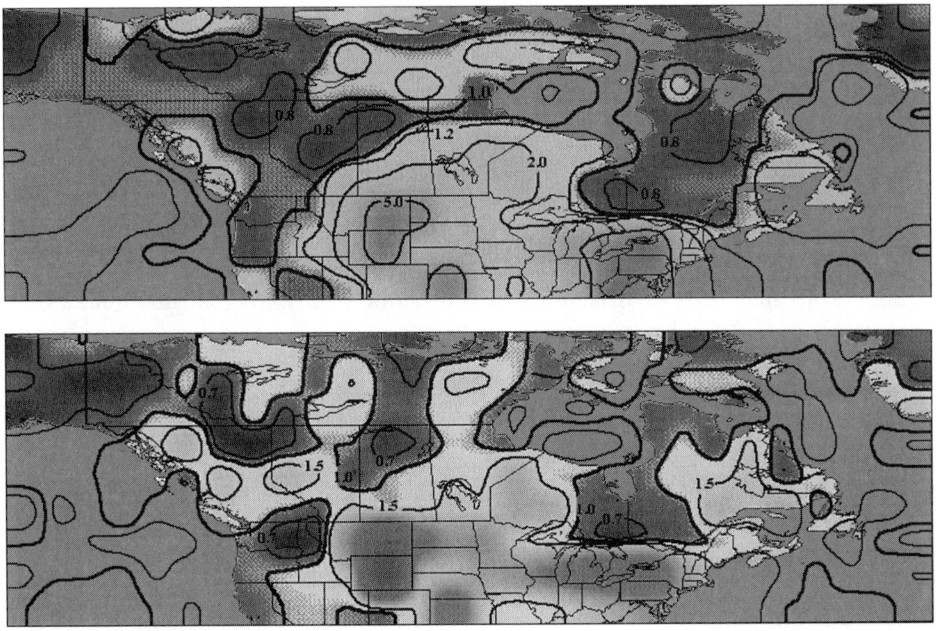
(a)
(b)

Figure 4.5. Mean (a) and maximum (b) FWI ratios ($2 \times CO_2/1 \times CO_2$) for North America (Flannigan et al. 1998; reprinted with permission from *Journal of Vegetation Science*, © Opulus Press).

roughly the last 200 years (Flannigan et al. 1998; Larsen 1996). Many of these studies show decreasing fire activity despite the warming since the end of the Little Ice Age (ca. AD 1850). These modeled results are also consistent with the modeled fire weather and charcoal record anomalies for a warm period during the mid-Holocence about 6000 years BP which was about 1°C warmer than present for Canada (Flannigan et al. 2001).

What will the fire regime be like for this century? Most studies in Canada suggest an overall increase in fire weather severity, although some areas of decreased fire weather severity are possible. Combine this with increasing fire season length and the increased cloud-to-ground lightning with a corresponding increase in ignitions (Price and Rind 1994), and greater fire activity is likely.

Climate Change: Impact on Canadian Forests

The forests of Canada will respond to changes in the climate over time. However, the almost instantaneous response of the fire regime to changes in the climate has the potential to overshadow importance of direct effects of global warming on species distribution, migration, substitution, and extinction. Thus fire is a catalyst for vegetation change.

In addition to climate's influence on the fire regime, other factors such as vegetation characteristics and human activities, fire management policies, and landscape fragmentation may greatly influence the fire regime in this century. Vegetation type, amount, and structure influence fire regime characteristics; thus any changes in vegetation due to changes in climate or fire regime have a feedback effect on the fire regime. Human activities such as fire management policies and effectiveness will continue to change. Other human activities such as conversion of forest lands to agriculture or urban areas along with the fragmentation of the landscape will influence the fire regime. These are confounding effects that may dampen or amplify the impact of a changing climate on the fire regime.

Fire may be more important than the direct effects of climate change for species distribution, migration, substitution, and extinction (Weber and Flannigan 1997). Fire can hasten the modification of the vegetation landscape into an new equilibrium with the climate if species are able to migrate fast enough. This would be true where the fire activity is expected to increase in this century. For example, increased fire frequency at the grassland–aspen parkland–boreal forest transition in western Canada (Fig. 4.1) may hasten the conversion of boreal forest to aspen parkland and aspen parkland to grassland. In those areas of Canada that experience a reduced fire frequency, in contrast, the transition of vegetation types may be retarded. For example, as the climate warms, the southern boreal forest in eastern Canada may be replaced by more southern species from the mixed wood region (Great Lakes–St. Lawrence Forest). This poleward migration of southern species would be enhanced by the presence of disturbed areas such as burns. In the absence of fire, existing shade-tolerant species such as balsam fir (*Abies balsamea* (L.) Mill.) and black spruce (*Picea mariana* (Mill.) B.S.P.) would dominate the landscape and would be hard to displace, retarding the poleward migration of southern species. Of course, increases in other disturbance regimes such as pests, diseases, and blowdown could offset decreases in area burned. Changes in climate and disturbance regimes may lead to assemblages of species that have never been encountered before (Martin 1993).

Vegetation models using GCM input have projected a large poleward shift in vegetation (Solomon and Leemans 1989; Rizzo and Wilken 1992; Smith and Shugart 1993a; IPCC 1998). However, most of these models have not incorporated forest fires.

Carbon and Nitrogen Cycling and Budgets

Changes in climate and the fire regime will impact on carbon and nitrogen cycling and budgets. Disturbances such as fire could be a critical factor in determining if Canadian forests are a carbon sink or source on a year-to-year basis.

Recent estimates are that 714 petagrams (Pg) of carbon ($1\,Pg = 10^{15}$ grams or 1 billion tonnes) are stored in the boreal forest region (Apps et al. 1993), and this represents about 37% of the total amount of carbon in the global terrestrial biosphere (Smith et al. 1993). The potential effects of climate change on levels of carbon storage in boreal forests has been estimated using changes in temperature and precipitation projected by GCMs to estimate changes in terrestrial biomes (Smith and Shugart 1993a, 1993b; Solomon et al. 1993). Fire has been shown to have a major effect on boreal carbon storage (Kasischke, Christensen, and Stocks 1995), but has been largely ignored in these models and even in the international Boreal Ecosystem–Atmosphere Study (BOREAS) conducted in Canada in 1994 and 1995 (*BOREAS* Special Issue 1998). With pervasive influence of fire across the boreal zone, and the strong likelihood of increased fire activity/severity under a warming climate, an improved understanding of the influence of fire on carbon cycling is essential.

Kasischke, Christensen, and Stocks (1995) described six ways that fire affects carbon storage in boreal forests: by directly releasing carbon to the atmosphere through combustion, through the conversion of plant material to charcoal, by strongly influencing the pattern of secondary succession on fire-disturbed landscapes, by altering the thermal regime of the forest floor and enhancing decomposition in these layers, by increasing the amount of soil nutrients available for plant growth, and by directly influencing the age-class distribution of forest stands. Amiro et al. (2001) found that direct emissions of carbon from forest fires in Canada from 1959 to 1999 averaged 27 Tg a year, which represents about 20% of the current carbon dioxide emissions from the Canadian energy sector. Kasischke, Christensen, and Stocks (1995) conducted a sensitivity analysis of the relationship between fire and carbon storage in the living-biomass and ground-layer compartments of boreal forests. They found that an increase in the occurrence and severity of fires under a warming climate would cause a net loss of carbon, as rapid loss of forest floor carbon would outpace carbon sequestration through plant regrowth. They concluded that because large amounts of carbon are stored in the ground layer of boreal forests, and fire significantly influences carbon storage in this area, any climate-induced changes in fire regimes will have major impacts.

The Carbon Budget Model of the Canadian Forest Sector (CFS-CBM) is a dynamic simulation model that accounts for carbon pools and fluxes in Canadian forest ecosystems and forest products (Kurz et al. 1992). The CBM-CFS has been used to analyze carbon flows both retrospectively (Kurz and Apps 1996) and to project future carbon budgets of Canadian boreal forests (Kurz and Apps 1995). In both cases the carbon sink/source strength of Canadian forests was determined to be significantly influenced by disturbance regimes, particularly fire and insects. Climate variation over the past two decades appears to have increased fire frequency, leading to a net carbon release from Canadian boreal forests. Periods of high fire activity were found to result in reduced carbon accumulation in biomass carbon pools, and a corresponding increase in soil carbon pools. The increase in dead organic matter associated with disturbance results in higher carbon loss from decomposition in the years following periods of high disturbance. The CBM-CFS results support the conclusion that fire activity is the major influence on the carbon budget of Canadian boreal forests. Apps et al. (2000) state that increased fire protection can perhaps delay, but not prevent, eventual carbon release from the ecosystem. If protection is not maintained, or the risk exceeds the protection measures, fire disturbance rates will again increase and the forest will become a carbon source.

Given that fire is natural and essential to boreal forest maintenance and productivity, large regions of Canada's boreal forest cannot and should not be protected from fire. Furthermore, given that economically feasible levels of fire protection in Canada's managed forests may delay but not prevent eventual fire impacts, and that projected climate change will result in more frequent and severe fires across much of Canada, it is difficult to avoid the conclusion that the impact of fires on the Canadian carbon budget will continue to increase. This increase in fire activity would result in shorter fire-return intervals, a skewing of forest age–class distribution toward younger stands, and a decrease in terrestrial carbon. This would also likely result in a positive feedback between boreal fires and climate change, exacerbating the problem (Kurz et al. 1995).

The very close coupling of nitrogen and carbon cycles within the plant and the ecosystem as a whole makes them particularly susceptible to modifications under global change as one or the other cycle may be altered by elevated CO_2 and climate change. Alteration in one of these two cycles can be expected to have immediate repercussions for the other because of the interaction and feedbacks between the two (Pastor and Post 1986; Reynolds et al. 1996). For example, the ability for increased carbon acquisition by plants in a higher CO_2 atmosphere could be limited by available soil N, which is in turn controlled by decomposition rates. The main avenue for interaction between C and N cycles may actually be via decomposition and litter quality. The reciprocal linkage between ecosystem cycles (N and C) and attributes (decomposition rates and litter quality) assumes added importance in the boreal forest biome for several reasons: (1) greater temperature impacts are predicted for northern latitudes under climate change, affecting all temperature sensitive processes, including decomposition and nutrient cycling (Anderson 1992); (2) boreal forest ecosystems are uniformly nitrogen limited and can be expected to respond to ameliorated nutrient conditions (Van Cleve et al. 1986); (3) the boreal forest's historical role as a carbon sink and likely reduction in sink strength under climate change (Kurz et al. 1995; Kurz and Apps 1993); and (4) effects of altered decomposition rates on fire regime via fire severity and changed organic layer thickness.

The principal pathway whereby elevated CO_2 interacts with decomposition is through effects on litter quality (O'Neill 1994). Litter characteristics, such as lignin and nutrient content, and most important, C/N ratios, strongly influence decay patterns and N availability, which in turn control the rate of biomass accumulation (Pastor and Post 1986; Reynolds et al. 1996). Therefore elevated CO_2 can alter ecosystem litter quality directly by affecting the C/N ratios of the plant material periodically deposited on the forest floor or indirectly, by changes in species composition of plant communities and their associated litter characteristics (O'Neill 1994). Evidence for direct effects of CO_2 enrichment on C/N ratios in plant tissue is inconclusive, especially because C/N ratios of living tissue may not be the same as senescent tissue shed as litter (Reynolds et al. 1996). In the case of a CO_2-caused shift in plant community species composition, changes in litter quality are expected because the type of carbon compounds in litter, and hence C/N ratios, are species specific (Pastor and Post 1986). C/N ratios control N availability; that is to say, the narrower the litter C/N ratios, the more rapid are the microbial decomposition rates, which in turn increase nitrogen availability for plant uptake and biomass production (Reynolds et al. 1996; Ryan 1991). Any time forest floor decomposition rates are altered due to soil warming, increased depth of active layer over permafrost, improved soil drainage, or accelerated substrate microbial activity, direct impacts on the fire regime are probable via fire severity (depth of burn). Improved soil drainage as a result of soil warming at northern latitudes is an important consideration for any climate change scenario (e.g., Anderson 1992; Bonan 1989; Dang and Lieffers 1989; Lashof 1989) because of the implications for organic layer drying and hence fire severity.

Combining increased fire severity in a changing climate with increased fire frequency, could accelerate carbon mineralization rates in arctic and subarctic soils underlying most of the boreal forests of North America (Anderson 1991). These faster carbon mineralization rates under warmer and drier conditions are due to low stabilization of soil organic matter and enhanced microbial responses to small changes in soil moisture and temperature (Anderson 1991). Accelerated C mineralization eventually feeds back to atmospheric CO_2 loading, possible biomass production impacts, litterfall quality, and quantity and decomposition rates. As a point of departure, for further information on the implications of such a scenario for global carbon cycling, mobilization of carbon stores from boreal forests, the carbon source/sink controversy, and feedback to global climate change, the reader is referred to Anderson (1992), Apps, Price, and Wisniewski (1995), Kasischke, Christensen, and Stocks (1995), Kurz et al. (1995), Oechel et al. (1993), and Thomas and Rowntree (1992).

Most of the atmospheric change-generated impacts are actually environmental stresses and may therefore predispose individuals and ecosystems to secondary stressors, such as insect and disease attack and drought (cf. Jones et al. 1993). Should this dynamic result in increased above-ground mortality and stand breakup, the fire regime may be affected immediately and in the short term because of to increased surface fuel loading and, hence, increased fire intensity (Stocks 1987).

Conclusion

Recently the climate has been warming over most of Canada (Gullett and Skinner 1992), and the warming is expected to continue throughout the twenty-first century (IPCC 2001). This warming and changes in other meteorological variables will alter the fire regime. Significant increases in fire weather indexes are anticipated over central sections of Canada where much of the current fire activity occurs. We believe that this increase in fire weather indexes will translate into significant increases in area burned in this century. Changes in the fire regime may have a significant impact on the composition, structure, and functioning of Canadian forests. Because the fire regime responds almost immediately to changes in the climate, the fire regime may act as a catalyst for change in Canadian forests. Therefore the rate and magnitude of fire-induced changes to Canadian forests could greatly exceed changes due directly to a changing climate. These changes would be most pronounced over regions where fire is prominent, such as in the boreal forest.

References

Amiro, B.D., Todd, J.B., Wotton, B.M., Logan, K.A., Flannigan, M.D., Stocks, B.J., Mason, J.A., Skinner, W.R., Martell, D.L., and Hirsch, K.G. 2001. Direct carbon emissions from Canadian forest fires, 1959 to 1999. *Can. J. For. Res.* 31:512–525.

Anderson, J.M. 1991. The effects of climate change on decomposition processes in grassland and coniferous forests. *Ecol. Appl.* 1:326–347.

Anderson, J.M. 1992. Response of soils to climate change. *Adv. Ecol. Res.* 22:163–210.

Apps, M.J., Price, D.T., and Wisniewski, J. 1995. *Boreal Forests and Climate Change.* Dortrecht: Kluwer Academic.

Apps, M.J., Bhatti, J.S., Halliwell, D.H., Jiang, H., and Peng, C.H. 2000. Simulated carbon dynamics in the boreal forest of central Canada under uniform and random disturbance regimes. In *Global Climate Change and Cold Regions Ecosystems*, eds. R. Lal, J. Kimble, and B. Stewart, pp. 107–121. Boca Raton: CRC Press.

Apps, M.J., Kurz, W.A., Luxmoore, R.J., Nilsson, L.O., Sedjo, R.A., Schmidt, R., Simpson, L.G., and Vinson, T.S. 1993. Boreal forests and tundra. *Water, Air Soil Pollut.* 70:39–53.

Bergeron, Y. 1991. The influence of island and mainland lakeshore landscapes on boreal forest fire regimes. *Ecology* 72:1980–1992.

Bergeron, Y., and Archambault, S. 1993. Decreasing frequency of forest fires in the southern boreal zone of Québec and its relation to global warming since the end of the "Little Ice Age." *Holocene* 3:255–259.

Bessie, W.C., and Johnson, E.A. 1995. The relative importance of fuels and weather on fire behavior in a subalpine forest. *Ecology* 76:747–762.

Bonan, G.B. 1989. A computer model of the solar radiation, soil moisture, and soil thermal regimes in boreal forests. *Ecol. Model.* 45:275–306.

BOREAS Special Issue 1997, *J. Geophys. Res.* 102 (D24): 28731–29745.

Brotak, E.A., and Reifsnyder, W.E. 1977. An investigation of the synoptic situations associated with major wildland fires. *J. Appl. Meteorol.* 16:867–870.

Bryson, R.A., Irving, W.N., and Larsen, J.A. 1965. Radiocarbon and soil evidence of former forest in the southern Canadian tundra. *Science* 147:46–48.

Caya, D., and Laprise, R. 1999. A semi-implicit semi-lagrangian regional climate model: The Canadian RCM. *Mon. Wea. Rev.* 127:341–362.

Clark, J.S. 1988. Particle motion and the theory of charcoal analysis: Source area, transport, deposition, and sampling. *Quat. Res.* 30:67–80.

Dang, Q.L., and Lieffers, V.J. 1989. Assessment of patterns of response of tree ring growth of black spruce following peatland drainage. *Can. J. For. Res.* 19:924–929.

Deeming, J.E., Burgan, R.E., and Cohen, J.D. 1977. The National Fire-Danger Rating System—1978. USDA Forest Service Gen. Tech. Rep. INT-39 63p. Intermountain Forest and range Experiment station, Ogden Utah, 84401.

Filion, L., Saint-Laurent, D., Desponts, M., and Payette, S. 1991. The late Holocene record of aeolian and fire activity in northern Quebec, Canada. *Holocene* 1:201–208.

Flannigan, M.D. 1993. Fire regime and the abundance of red pine. *Int. J. Wildl. Fire* 3: 241–247.

Flannigan, M.D., and Harrington, J.B. 1987. Synoptic conditions during the Porter Lake burning experiment. *Climatol. Bull.* 21:19–40.

Flannigan, M.D., and Harrington, J.B. 1988. A study of the relation of meteorological variables to monthly provincial area burned by wildfire in Canada 1953–80. *J. Appl. Meteorol.* 27:441–452.

Flannigan, M.D., and Van Wagner, C.E. 1991. Climate Change and wildfire in Canada. *Can. J. For. Res.* 21:66–72.

Flannigan, M.D., and Wotton, B.M. 2001. Connections—Climate/weather and area burned. In *Forest Fires: Behavior and Ecological Effects*, eds. E.A. Johnson, and K. Miyanishi, pp. 335–357. San Diego, CA: Academic Press.

Flannigan, M.D., Bergeron, Y., Engelmark, O., and Wotton, B.M. 1998. Future wildfire in circumboreal forests in relation to global warming *J. Veg. Sci.* 9:469–476.

Flannigan, M.D., Campbell, I., Wotton, B.M., Carcaillet, C., Richard, P., and Bergeron, Y. 2001. Future fire in Canada's boreal forest: Paleoecology results, and GCM/RCM simulations. *Can. J. For. Res.* 31:854–864.

Flato, G.M., Boer, G.J., Lee, W.G., McFarlane, N.A., Ramsden, D., Reader, M.C., and Weaver, A.J. 2000. The Canadian Centre for Climate Modelling and Analysis Global Coupled Model and its Climate. *Clim. Dyn.* 16:451–467.

Foster, D.R. 1983. The history and pattern of fire in the boreal forest of southeastern Labrador. *Can. J. Bot.* 61:2459–2471.

Gullett, D.W., and Skinner, W.R. 1992. The state of Canada's climate: Temperature change in Canada 1895–1991. A state of the Environment Report No. 92-2, Environ. Canada, Ottawa. Ontario.

Haines, D.A. 1988. A lower atmosphere severity index for wildland fires. *Nat. Wea. Digest* 13:23–27.

Harrington, J.B., Flannigan, M.D., and Van Wagner, C.E. 1983. A study of the relation of components of the Fire Weather Index System to monthly provincial area burned by wildfire in Canada 1953–80. Can. For. Serv., Petawawa Natl. For. Inst., Inf. Rep. PI-X-25.

Heinselman, M.L. 1973. Fire in the virgin forests of the Boundary Waters Canoe Area, Minnesota. *Quat. Res.* 3:329–382.

Hely, C., Flannigan, M.D., Bergeron, Y., and McRae, D. 2001. Role of vegetation and weather on fire behavior in the Canadian Mixedwood boreal forest using two fire behavior prediction systems. *Can. J. For. Res.* 31:430–441.

Hirsch, K.G., and Flannigan, M.D. 1990. Meteorological and fire behavior characteristics of the 1989 fire season in Manitoba, Canada. *International Conference on Forest Fire Research*, Coimbra, Portugal. pp. B.06-1–B.06-16.

Horel, J.D., and Wallace, J.M. (1981). Planetary-scale atmospheric phenomena associated with the southern oscillation. *Mon. Wea. Rev.* 109:813–829.

Hu, F.S., Brubaker, L.B., and Anderson, P.M. 1993. A 12,000 year record of vegetation change and soil development from Wien Lake, central Alaska. *Can. J. Bot.* 71: 1133–1142.

Intergovernmental Panel on Climate Change (IPCC). 2001. *Climate Change 2001: Impacts, Adaptation, and Vulnerability*, eds. J.J. McCarthy, O.F. Canziani, N.A. Leary, D.J. Dokken, and K.S. White. Cambridge: Cambridge University Press.

Intergovernmental Panel on Climate Change (IPCC). 1998. *The Regional Impacts of Climate Change: An Assessment of Vulnerability*. Cambridge: Cambrige University Press.

Johnson, E.A. 1992. *Fire and Vegetation Dynamics: Studies from the North American Boreal Forest*. Cambridge: Cambridge University Press.

Johnson, E.A., and Larsen, C.P.S. 1991. Climatically induced change in fire frequency in the southern Canadian Rockies. *Ecol.* 72:194–201.

Johnson, E.A., and Wowchuk, D.R. 1993. Wildfires in the southern Canadian Rocky Mountians and their relationship to mid-tropospheric anomalies. *Can. J. For. Res.* 23: 1213–1222.

Johnson, E.A., Fryer, G.I., and Heathcott, M.J. 1990. The influence of man and climate on frequency of fire in the interior wet belt forest, British Columbia. *J. Ecol.* 78: 403–412.

Jones, E.A., Reed, D.D., Mroz, G.D., Liechty, H.O., and Cattelino, P.J. 1993. Climate stress as a precursor to forest decline: Paper birch in northern Michigan, 1985–1990. *Can. J. For. Res.* 23:229–233.

Kasischke, E.S., Christensen, N.L., and Stocks, B.J. 1995. Fire, global warming, and the carbon balance of the boreal forests. *Ecol. Appl.* 5:437–451.

Khury, P. 1994. The role of fire in the development of *Sphagnum*-dominated peatlands in western boreal Canada. *J. Ecol.* 82:899–910.

Kirschbaum, M.U.F., and Fishlin, A. 1996. Climate change impacts on forests. In *Climate Change 1995. Contributions of Working Group II to the Second Assessment Report of the Intergovernmental Panel of Climate Change*, eds. R. Watson, M.C. Zinyowera, and R.H. Moss, pp. 93–129. Cambridge: Cambridge University Press.

Kurz, W.A., and Apps, M.J. 1993. Contribution of northern forests to the global C cycle: Canada as a case study. *Water Air Soil Pollut.* 70:163–176.

Kurz, W.A., and Apps, M.J. 1995. An analysis of future carbon budgets of Canadian boreal forests. *Water Air Soil Pollut* 82:321–331.

Kurz, W.A., and Apps, M.J. 1996. Retrospective assessment of carbon flows in Canadian boreal forests. In *Forest Ecosystems, Forest Management, and the Global Carbon Cycle*, eds. M.J. Apps and D.T. Price, pp. 173–182. Berlin: Springer-Verlag.

Kurz, W.A., Apps, M.J., Stocks, B.J., and Volney, J.A. 1995. Global climate change: Disturbance regimes and biospheric feedbacks of temperate and boreal forests. In *Biotic Feedbacks in the Global Climatic System. Will the Warming feed the Warming?* eds. G.M. Woodwell and F.T. Mackenzie. pp. 119–133. New York: Oxford University Press.

Kurz, W.A., Apps, M.J., Webb, T.M., and McNamee, P.J. 1992. The carbon budget of the Canadian forest sector: phase 1. For. Can., North. For. Cent. Inf. Rep. NOR-X-326, Edmonton, AB.

Larsen, C.P.S. 1996. Fire and climate dynamics in the boreal forest of northern Alberta, Canada from AD 1850 to 1989. *Holocene* 6:449–456.

Lashof, D.A. 1989. The dynamic greenhouse: Feedback processes that may influence future concentrations of atmospheric trace gases and climate change. *Clim. Change* 14:213–242.

Lupo, A.R., Oglesby, R.J., and Mokhov I.I. 1997. Climatological features of blocking anticyclones: A study of Northern Hemisphere CCM1 model blocking events in present-day and couble CO_2 concentration atmosphere. *Clim. Dyn.* 13:181–195.

MacDonald, G.M., Larsen, C.P.S., Szeicz, J.M., and Moser, K.A. 1991. The reconstruction of boreal forest fire history from lake sediments: A comparison of charcoal, pollen, sedimentological, and geochemical indices. *Quat. Sci. Rev.* 10:53–71.

Malanson, G.P. 1987. Diversity, stability, and resilience: effects of fire regime. In *The role of Fire in Ecological Systems*. ed. L. Trabaud, pp. 49–63. The Hague: SPB Academic Publishing.

Martin, P. 1993. Vegetation responses and feedbacks to climate: A review of models and processes. *Clim. Dyn.* 8:201–210.

Mearns, L.O., Schneider, S.H., Thompson, S.L., and McDaniel, L.R. 1989. Climate variability statistics from General Circulation Models as applied to climate change analysis. In *Natural Areas Facing Climate Change*, ed. G.P. Malanson, pp. 51–73. The Hague: SPB Academic Publishing.

Merrill, D.F., and Alexander, M.E. 1987. *Glossary of Forest Fire Management Terms*, 4th ed. National Research Council of Canada, Canadian Committee on Forest Fire Management. NRCC No. 26516.

Newark, M.J. 1975. The relationship between forest fire occurrence and 500 mb ridging. *Atmos.* 13:26–33.

Nichols, H. 1967. Pollen diagrams form sub-Arctic central Canada. *Science* 155: 1665–1668.

Nimchuk, N. 1983. Wildfire behavior associated with upper ridge breakdown. Alta. Energy and Nat. Resour., For. Serv,. Edmonton, Alta. ENR Rep. No. T/50.

Oechel, W.C., Hastings, S.J., Vourlitis, G., Jenkins, M., Riechers, G., and Grulke, N. 1993. Recent changes of arctic tundra ecosystems from a net carbon sink to a source. *Nature* 361:520–526.

O'Neill, E.G. 1994. Responses of soil biota to elevated atmospheric carbon dioxide. *Plant Soil* 165:55–65.

Overpeck, J.T., Rind, D., and Goldberg, R. 1990. Climate-induced changes in forest disturbance and vegetation. *Nature* 343:51–53.

Pastor, J., and Post, W.M. 1986. Influence of climate, soil moisture, and succession on forest carbon and nitrogen cycles. *Biogeochemistry* 2:3–27.

Payette, S. 1992. Fire as a controlling process in the North American boreal forest. In *A Systems Analysis of the Global Boreal Forest*, eds. H. Shugart, R. Leemans, and G.B. Bonan, pp. 144–169. Cambridge: Cambridge University Press.

Payette, S., and Gagnon, R. 1985. Late Holocene deforestation and tree regeneration in the forest tundra of Québec. *Nature* 313:570–572.

Payette, S., Morneau, C., Sirois, L., and Desponts, M. 1989. Recent fire history of the northern Québec biomes. *Ecol.* 70:656–673.

Potter, B.E. 1996. Atmospheric properties associated with large wildfires. *Int. J. Wildl. Fire* 6:71–76.

Price, C., and Rind, D. 1994. The impact of a 2 × CO_2 climate on lightning-caused fires. *J. Clim.* 7:1484–1494.

Pyne, S.J. 1997. *Vestal Fire: An Environmental History, Told through Fire, of Europe and Europe's Encounter with the World.* Seattle: University of Washington Press.

Quintilio, D., Fahnestock, G.R., and Dube, D.E. 1977. Fire behaviour in upland Jack Pine: The Darwin Lake Project. Environ. Can.. Can. For. Serv., Northern For. Res. Centre, Inf. Rep. NOR-X-174.

Reed, W.J., Larsen, C.P.S., Johnson, E.A., and MacDonald, G.M. 1998. Estimation of temporal variations in historical fire frequency from time-since-fire map data. *For. Sci.* 44:465–475.

Reynolds, J.F., Kemp, P.R., Acock, B., Chen, J.-L., and Moorhead, D.L. 1996. Progress, limitations, and challenges in modeling the effects of elevated CO_2 on plants and ecosystems. In *Carbon Dioxide and Terrestrial Ecosystems*, eds. G.W. Koch, and H.A. Mooney, pp. 347–380. San Diego, CA: Academic Press.

Rizzo, B., and Wilken, E. 1992. Assessing the sensitivity of Canada's forests to climatic change. *Clim. Change* 21:37–55.

Ryan, M.G. 1991. Effects of climate change on plant respiration. *Ecol. Appl.* 1:157–167.

Schaefer, V.J. 1957. The relationship of jet streams to forest wildfires. *J. For.* 55:419–425.

Schroeder, M.J., and others. 1964. Synoptic weather types associated with critical fire weather. USDA Forest Service, Pacific Southwest Forest Exp. Stn., Berkeley, CA, 492p.

Sirois, L., and Payette, S. 1991. Reduced postfire tree regeneration along a boreal forest–forest-tundra transect in northern Quebec. *Ecology* 72:619–627.

Skinner, W.R., Stocks, B.J., Martell, D.L., Bonsal, B., and Shabbar, A. 1999. The association between circulation anomalies in the mid-troposphere and area burned by wildland fire in Canada. *Theor. App. Clim.* 63:89–105.

Smith, T.M., and Shugart, H.H. 1993a. The transient response of carbon storage to a perturbed climate. *Nature* 361:523–526.

Smith, T.M., and Shugart, H.H. 1993b. The potential response of global terrestrial carbon storage to a climate change. *Water Air Soil Pollut.* 70:629–642.

Smith, T.M., Cramer, W.P., Dixon, R.K., Neilson, R.P., and Solomon, A.M. 1993. The global terrestrial carbon cycle. *Water Air Soil Pollut.* 70:19–37.

Solomon, A.M., and Leemans, R. 1989. Forest dieback inevitable if climate changes. Int. Inst. Appl. Syst. Anal., Luxemburg, Austria. IIASA Options.

Solomon, A.M., and Leemans, R. 1997.Boreal forest carbon stocks and wood supply: Past, present and future responses to changing climate, agriculture and species availability. *Agric. For. Met.* 84:137–151.

Solomon, A.M., Prentice, I.C., Leemans, R., and Cramer, W.P. 1993. The interaction of climate and land use in future terrestrial carbon storage and release. *Water Air Soil Pollut.* 70:595–614.

Stocks, B.J. 1975. The 1974 wildfire situation in northwestern Ontario. Can. For. Serv., Great Lakes Forest Res. Centre, Inf. Rep. O-X-232.

Stocks, B.J. 1987. Fire potential in the spruce-budworm damaged forests of Ontario. *For. Chron.* 63:8–14.

Stocks, B.J. 1991. The extent and impact of forest fires in northern circumpolar countries. In *Global Biomass Burning: Atmospheric, Climatic, and Biospheric Implications*, ed. J.S. Levine, pp. 197–202. Cambridge: MIT Press.

Stocks, B.J., and Walker, J.D. 1973. Climatic conditions before and during four significant forest fire situations in Ontario. Can. For. Serv., Great Lakes Forest Res. Centre, Inf. Rep. O-X-187.

Stocks, B.J., Lee, B.S., and Martell, D.L. 1996. Some potential carbon budget implications of fire management in the boreal forest. In *Forest Ecosystems, Forest Management and Global Carbon Cycle*, eds. M.J. Apps, and D.T. Price, pp. 89–96. NATO ASI Series Vol. I 40. Berlin: Springer.

Stocks, B.J., Fosberg, M.A., Lynham, T.J., Mearns, L., Wotton, B.M., Yang, Q., Jin, J.-Z., Lawrence, K., Hartley, G.R., Mason, J.A., and McKenney, D.W. 1998. Climate change and forest fire potential in Russian and Canadian boreal forests. *Clim. Change* 38:1–13.

Stocks, B.J., Lawson, B.D., Alexander, M.E., Van Wagner, C.E., McAlpine, R.S., Lynham, T.J., and Dubé, D.E. 1989. The Canadian Forest Fire Danger Rating System: an Overview. *For. Chron.* 65:450–457.

Swetnam, T.W. 1993. Fire history and climate change in giant sequoia groves. *Science* 262:885–889.

Thomas, G., and Rowntree, P.R. 1992. The boreal forests and climate. *Q.J.R. Meteorol. Soc.* 118:469–497.

Tolonen, K. 1983. The post-glacial fire record. In *The Role of Fire in Northern Circumpolar Ecosystems*, eds. W.R. Wein, and D.A. MacLean, pp. 21–44. New York: Wiley.

Turner, J.A. 1970. Hours of sunshine and fire season severity over the Vancouver Forest District. *For. Chron.* 46:106–111.

Vance, R.E., Emerson, D., and Habgood, T. 1983. A mid-Holocene record of vegetative change in central Alberta. *Can. J. Earth Sci.* 20:364–376.

Van Cleve, K., Chapin, F.S., III., Flanagan, P.W., Viereck, L.A., and Dyrness, C.T. 1986. *Forest Ecosystems in the Alaskan Taiga.* Ecological Studies 57. New York: Springer.

Van Wagner, C.E. 1977. Effect of slope on fire spread. *Can. For. Serv., Bi-Mon. Res. Notes* 33:7–8.

Van Wagner, C.E. 1987. The development and structure of the Canadian Forest Fire weather index system. Canadian Forest Service, Forest Tech. Rep. 35, Ottawa, Ontario.

Weber, M.G., and Flannigan, M.D. 1997. Canadian boreal forest ecosystem structure and function in a changing climate: impact on fire regimes. *Environ. Rev.* 5:145–166.

Weber, M.G., and Stocks, B.J. 1998. Forest fires and sustainability in the boreal forests of Canada. *Ambio* 27:545–550.

Weeks, E.R., Tian, Y., Urbach, J.S., Ide, K., Swinney, H.L., and Ghil, M. 1997. Transitions between blocked and zonal flows in a rotating annulus with topography. *Science* 278:1598–1601.

Weir, J.M.H., and Johnson, E.A. 1998. Effects of escaped settlement fires and logging on forest composition in the mixedwood boreal forest. *Can. J. For. Res.* 28:459–467.

Weir, J.M.H., Johnson, E.A., and Miyanishi, K. 2000. Fire frequency and the spatial age mosaic of the mixed-wood boreal in western Canada. *Ecol. Appl.* 10:1162–1177.

Whelan, R.J. 1995. *The Ecology of Fire*. Cambridge: Cambridge University Press.

Winkler, M.G. 1985. Charcoal analysis for paleoenvironmental interpretation: a chemical assay. *Quat. Res.* 23:313–326.

Wotton, B.M., and Flannigan, M.D. 1993. Length of the fire season in a changing climate. *For. Chron.* 69:187–192.

5. Fires and Climate in Forested Landscapes of the U.S. Rocky Mountains

William L. Baker

Scattered reports indicate that the number of fires or area burned has increased recently in parts of the northern temperate zone, but is climatic change responsible? Annual number of fires and area burned have generally increased since about 1950 in Canada and Sweden (Stocks 1991), the Rocky Mountains (Qu and Omi 1994; Fig. 5.1a) and the western United States (Arno 1996; Fig. 5.1b). However, trends in fire statistics may in part reflect increasing ability to monitor fires (Ryan 1976; Qu and Omi 1994). Moreover, in Canada and in Yellowstone National Park, trends are dominated by a few exceptional fire years in the 1980s (Stocks 1991; Balling, Meyer, and Wells 1992a). Also suppression of fires decades ago may have increased fuel loads, leading to the larger fires seen now (Covington and Moore 1994). Finally the landscape may shape potential responses to climatic change, leading to disequilibrium between climate and fires (Baker 1995). Identifying a climatic signal in historical fire data may thus require more understanding of how climate, fuels, the landscape, and land-use practices separately and jointly shape fire regimes.

To organize a discussion of the present state of understanding in the Rocky Mountains, I contrast a view that emphasizes how broad-scale patterns of climate and fuels control fire regimes, with a contingent view in which local spatial constraints and historical legacies may limit general trends. While these perspectives on what is important underlie models, empirical studies, and theories, they are seldom explicit. Models that represent the broad-scale view, for example, suggest that fires may hasten the response of vegetation to climatic

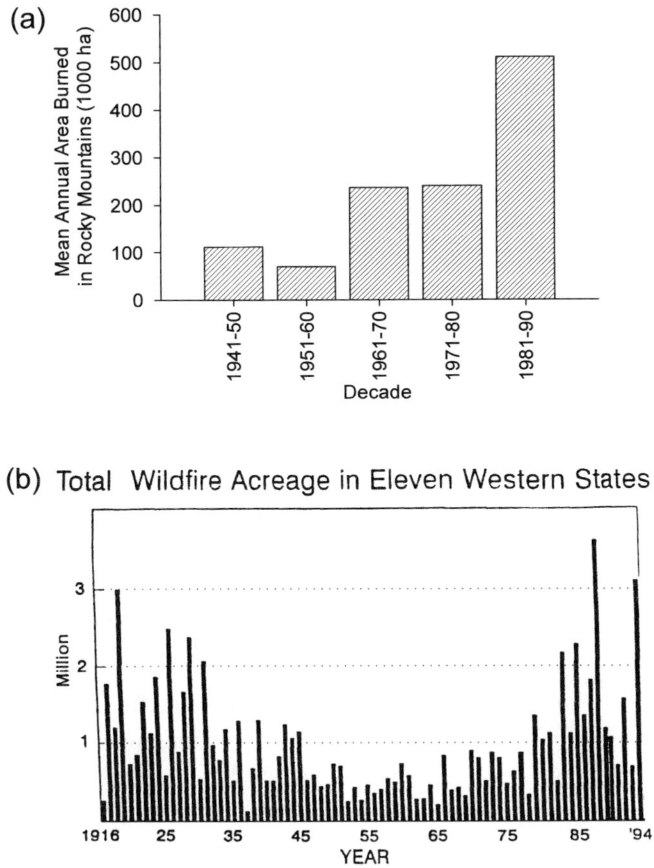

Figure 5.1. Trends in the occurrence of fires in (a) the Rocky Mountains (Qu and Omi 1994), and (b) the western United States (Arno 1996; reproduced with permission from the USDA Forest Service).

change by removing vegetation that may otherwise persist after climate is no longer favorable (Overpeck, Rind, and Goldberg 1990). This view is of a rapidly responding, climatically controlled fire regime affecting a passive and independent vegetation in a featureless landscape. The contingent view suggests that fire regimes are inherently spatial, are constrained by the physical landscape, and are shaped by climate and vegetation as well as by historical legacies. Fire regimes thus typically require decades to centuries to adjust to new climates (Baker 1995).

In this chapter I review the broad-scale and contingent views in the context of the U.S. Rocky Mountains. These mountains extend from northern Montana to the Sangre de Cristo Mountains of New Mexico and the San Francisco Peaks of Arizona (Peet 1988). The Rockies can be divided into the northern Rocky

Mountains in Montana, the central Rocky Mountains from southern Montana into central Wyoming, and the southern Rocky Mountains from southern Wyoming to northern New Mexico and Arizona.

The Broad-Scale View

Relative Roles of Climate and Fuels

The prevailing climate affects the probability of weather conducive to fire initiation and spread and affects fuel buildup and fuel moisture (Fig. 5.2). Fires are primarily ignited by lightning and humans, with lightning ignitions more probable during certain weather episodes, particularly thunderstorms (Price and Rind 1994a). Ignitions often do not spread significantly unless followed by weather that promotes spread, such as droughts and strong winds. However, the moisture content and abundance of fuel can also significantly constrain or promote fire spread.

The relative importance of fire weather and fuels in shaping fire regimes varies geographically (Table 5.1). In a generally warm, dry climate (e.g., Baja California) where fuel moisture is often low enough to carry a fire, and weather is often conducive to ignition and spread of fire, the primary limitation on fires may be the time required for fuel to build to levels sufficient to carry a fire (Minnich et al. 1993). In contrast, in the colder, more humid climate of western Canada, where suitable fire weather is rare, fuel buildup may be of little importance, and the fire regime is more strongly controlled by fire-initiation and spread weather (Bessie and Johnson 1995). Variation in the relative importance of

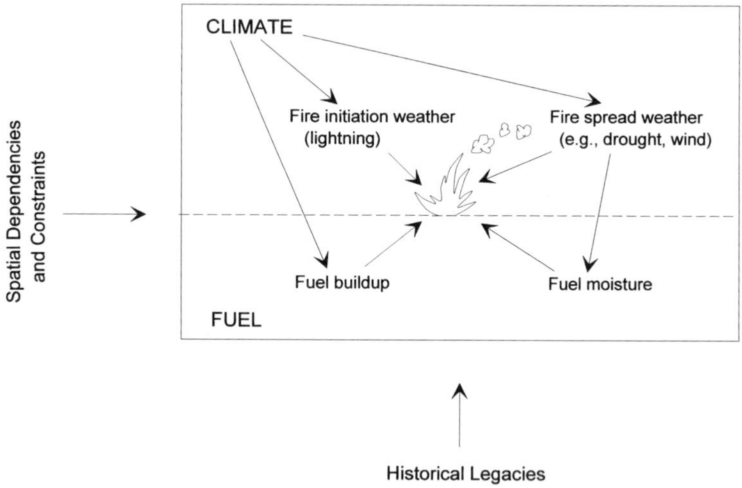

Figure 5.2. Major influences of climate on the occurrence of fires.

Table 5.1. Two ends of a continuum between fuel control of the fire regime and fire-weather control of the fire regime

Minnich et al. (1993)	Bessie and Johnson (1995)
Baja California chaparral and mixed conifer	Western Canadian subalpine forests
Fire regime predominantly fuel driven	Fire regime predominantly weather driven
Climatic change primarily affects fuel buildup; increased ignitions and fire weather irrelevant	Climatic change primarily affects fire weather; increased fuel buildup of minor importance

weather and fuels affects the potential response of the fire regime to changes in climate. Thus it is important to consider how climate, weather, and fuels may individually affect fires before their joint effects can be understood.

Climatic Setting of the Rocky Mountains

Air Mass Boundaries, Droughts, and Teleconnections

The central and southern Rocky Mountains are separated by a comparatively low-lying shrub steppe landscape between the Wind River Mountains and Medicine Bow Mountains in Wyoming. This is the location of a significant winter boundary between predominantly east–west airflow from the Pacific to the north and predominantly southerly flow, associated with an anticyclone over southern Nevada, to the south (Mitchell 1976; Adams and Comrie 1997). This boundary periodically breaks down, allowing Pacific cyclones to enter the southern Rockies. During summer, a monsoon boundary runs southwest to northeast across northwestern Colorado not far from the winter boundary (Mitchell 1976; Adams and Comrie 1997). To the north the Rockies are under predominantly westerly flow from the Pacific, which in summer results in prevailing warm, dry conditions. However, the northern Rockies are also an area of summer cyclogenesis (Changnon 1985). To the south, the southern Rockies are dominated in summer by the North American monsoon, which brings warm, moist tropical air from the gulfs of California and Mexico (Adams and Comrie 1997), and regular afternoon thunderstorms and lightning (Carleton 1985).

Low winter snowpacks, which may contribute to summer fire occurrence (e.g., Balling et al. 1992b), link to sea surface temperatures in the tropics and North Pacific. The El Niño phenomenon, associated with anomalous warming of eastern Pacific sea surface waters, may affect U.S. winter weather through extratropical teleconnections at periods concentrated in the four-year frequency band (Diaz and Markgraf 1992; Stahle et al. 1998). At the southern end of the southern Rockies (e.g., southern Colorado), winter precipitation and snowpack are enhanced during El Niños and lowered during La Niñas (Ropelewski and Halpert 1986; D'Arrigo and Jacoby 1991; Cayan 1996; Stahle et al. 1998; Kunkel and Angel 1999). Further north in the southern and central Rockies, El Niños and La Niñas may have less effect (Ropelewski and Halpert 1986; Woodhouse 1993). However, both

El Niño and La Niña increase snowfall in Wyoming (Smith and O'Brien 2001). In the western part of the northern Rockies, El Niño leads to average snowfall decreases of about 20%, and La Niña leads to similar increases in snowfall (Kunkel and Angel 1999; Smith and O'Brien 2001). Thus the effects of El Niño/La Niña vary along the Rockies, apparently with the greatest, but opposite, effects at the southern and northern ends. Another factor affecting low winter snowpacks is a strong Pacific North America (PNA) pattern, which consists of a deep Aleutian low and a blocking ridge over the northwestern United States and western Canada. This leads to low winter snowpack in the central and northern Rockies and often enhanced snowpack in the southern Rockies, as occurred from about 1977 to 1989 (Changnon, McKee, and Doesken 1993; Cayan 1996). Since 1989 the relationship between El Niño and the PNA pattern has broken down, probably because the North Atlantic Oscillation (NAO) and Pacific Decadal Oscillation (PDO) became locked in phase (Watanabe and Nitta 1999).

The PDO is an index of decadal variability in the climate of the Pacific Ocean (Mantua et al. 1997), with a period of about 23 years, but varying from 17 to 28 years (Biondi, Gershunov, and Cayan 2001). The PDO affects the strength of El Niño and La Niña, as well as the El Niño related PNA pattern. When the PDO is generally in a cold or low phase (e.g., 1947–1977) the effect of El Niño on U.S. climate is weakened, while the effect of La Niña is enhanced (Gershunov and Barnett 1998). The 1952 to 1956 central U.S. drought is an example (Barlow, Nigam, and Berbery 2001). Conversely, when the PDO is in a warm or high phase (e.g., 1977–1989) the effect of El Niño on U.S. climate is enhanced and the effect of La Niña is weakened. In addition to its effect on El Niño and La Niña, the PDO itself is associated with drought in the mid-Atlantic states and the extreme Northwest (Cole and Cook 1998), and also with dry summers in the central and northern Rockies and wet summers in the southern part of the southern Rockies (Barlow, Nigam, and Berbery 2001).

Summer droughts have less clear teleconnections with El Niño or La Niña in the central and southern Rockies, but they may also be linked to the winter PNA pattern. The 1988 drought that led to extensive fires in and near Yellowstone National Park in the northern central Rockies was associated with a teleconnection from the eastern Pacific following an El Niño early in 1988. But this teleconnection was unlike the typical winter PNA pattern, and it was probably only reinforced by, rather than caused by, La Niña (Trenberth, Branstator, and Arkin 1988; Palmer and Branković 1989). Major droughts in the western United States have often had a teleconnection to the North Pacific (Namias 1982). The 1997–98 drought, however, was broadly linked to exceptionally warm global sea surface temperatures, and not just El Niño or Pacific temperatures (Kumar et al. 2001).

The North American monsoon is influenced by the PNA pattern, El Niño/La Niña, and the PDO. A weak North American monsoon and dry conditions in the southern Rockies are associated with southward displacement of the summer subtropical ridge. Both tend to occur after a zonal or weak PNA pattern in winter, with no ridge over the western United States (Carleton, Carpenter, and Weser 1990; Higgins, Mo, and Yao 1998). A strong winter PNA pattern may lead to a wet monsoon the following summer, as the subtropical ridge is displaced north, allowing moist tropical air to flow into the southwest (Carleton, Carpenter, and Weser 1990; Higgins, Mo, and Yao 1998). El Niño (La Niña) is associated with late (early) onset of the monsoon (Higgins and Shi 2001). Monsoon intensity appears to be more controlled by intraseasonal effects, particularly the tropical Madden-Julian oscillation, and local influences, such as spring snow cover (Anderson, Roads, and Chen 2000; Higgins and Shi 2001). A high or warm PDO may also enhance monsoon strength (Barlow, Nigam, and Berbery 2001).

Droughts that promote fires in the Rocky Mountains also have statistical linkages to solar and lunar phenomena and, potentially, both oceans. A 22-year cycle of drought in the western United States correlates with the Hale sunspot cycle over long time periods (Mitchell, Stockton, and Meko 1979). Other strong sunspot-weather correlations have been found (van Loon and Labitzke 1988). The sunspot-weather relationship is weaker since 1895 (Diaz 1983), and may even be more strongly correlated with an 18.6-year lunar nodal-tide effect (Currie 1984). Tree rings reveal bi-decadal (20–23 year) and 7.8-year frequencies of drought in the western United States from 1700 to 1978 (Cook, Meko, and Stockton 1997). These authors found that both the Hale sunspot cycle and lunar tidal cycle are significantly correlated with drought, although an internally driven ocean-atmosphere oscillation in the North Pacific (e.g., PDO) is also a possible explanation (see also Woodhouse and Overpeck 1998). In the plains adjoining the Rocky Mountains, there is similar evidence of the Hale sunspot cycle in historical air temperatures (Chang and Smith 2001), but also a possible influence of the North Atlantic Oscillation on the bi-decadal drought cycle (Hu, Woodruff, and Mudrick 1998; Woodhouse and Overpeck 1998). Thus the bi-decadal drought cycle appears significant in the Rocky Mountain region, but there remain several hypotheses about the source of the cycle.

Influences of solar variation on fires have seldom been analyzed, since a compelling physical link with climate is lacking, but mechanisms have recently been proposed. These include cosmic ray influences on clouds (Wagner et al. 2001) and absorption of ultraviolet radiation by stratospheric ozone (Shindell et al. 1999). These explanations require further resolution, as do possible effects on fires. In the northwestern United States, including Idaho and Montana, the number of lightning fires correlates with sunspot numbers over the period from 1915 to 1939 (Bumstead 1943). In bristlecone pine forests in the southern Rockies, many stand origins, likely caused by fire, coincided with the Maunder sunspot minimum (Baker 1992).

Teleconnections with the tropics and the Pacific Ocean appear to influence Rocky Mountain climate, and there is now clear evidence of influence on Rocky Mountain fires. Many fire years in subalpine forests in the Rocky Mountains appear to have been regional in extent, suggesting a strong regional synoptic climatic control (Veblen 2000; Kipfmueller and Baker 2000). An early study found that above- and below-average fire years bear no relation to El Niño events in the Rocky Mountains (Simard, Haines, and Main 1985). However, the "Rocky Mountains" in this study include some Great Plains and southwestern states, clouding relationships in the mountains. In the first study, to clearly demonstrate an effect on fires in the Rockies, wet episodes associated with El Niño one to three years prior to drought were found to enhance fuel buildup that increases fires during La Niña-related drought in the Colorado Front Range (Veblen et al. 2000). This pattern of large fires occurring after a sequence of strong El Niño and La Niña years was also found to occur synchronously in the southwestern United States and Argentina (Kitzberger, Swetnam, and Veblen, in press). Decadal and centennial trends in fire occurrence are also approximately synchronous in the Colorado Front Range and Argentina (Veblen and Kitzberger, in press), and are probably related to variations in the strength of El Niño and La Niña.

Fire regimes also change in response to longer-term climatic trends. On the millennial time scale, fire frequency in Yellowstone National Park during the last 17,000 years increased as July insolation increased under the influence of variations in the earth's tilt and the timing of the perihelion (Millspaugh, Whitlock, and Bartlein 2000). The onset of warmer and drier conditions about 2600 years BP may have increased fire frequency in subalpine forests in central Colorado (Fall 1997).

Climate research continues to alter our understanding of Rocky Mountain climate, and some potential effects on fires have yet to be studied. For example, the effect of the PDO on U.S. climate has been elucidated (Mantua et al. 1997), but PDO effects on fires are unstudied. Fire research may always be awaiting further clarification of sources of variability in climate. This is particularly so in the Rocky Mountains, a complex meeting place for multiple climatic influences.

Lightning and Ignitions

Climatic episodes that lead to dry conditions are insufficient for fires, since ignition also is required, and lightning may be limiting. Lightning density is comparatively low in the Rockies, especially in the northern Rockies. The density of cloud-to-ground lightning averages 0.5 to 1.0 flashes $km^{-2} yr^{-1}$ in the central and northern Rockies to 1 to 3 flashes $km^{-2} yr^{-1}$ in the southern Rockies, compared to 9 to 13 flashes $km^{-2} yr^{-1}$ in parts of the midwest and southeast (Orville 1994; Orville and Silver 1997; Orville and Huffines 2001). In the southern Rockies, highest lightning densities are in July and August from noon to midnight, peaking in late afternoon, with much less in June and September, and very little in other months (López and Holle 1986).

Lightning-strike density gradually decreases by 50% from southwestern Colorado to southern Wyoming, while strike density remains about a third of that in southwestern Colorado across the central Rockies in Wyoming (Reap 1986). This north–south gradient is strongly related to thunderstorm density associated with moist tropical air from the North American monsoon (Reap 1986; Watson et al. 1994). However, in the northern Rockies a much larger percentage of thunderstorms is associated with summer cold fronts than with local convection and moist tropical air (Colson 1957). Lightning density in the western United States a little more than doubles from 1000 to 3000 m in elevation (Reap 1986), but precipitation also increases.

Lightning-strike density is often well correlated with thunderstorms and rainfall (e.g., Tapia, Smith, and Dixon 1998), but storms that start fires have less rain and more cloud-to-ground lightning. A study of 14,754 reports of thunderstorms from 270 or more fire lookouts stationed on high mountains in the northern Rockies over a five-year period linked lightning, rain, and ignitions (Gisborne 1931). Lightning storms that cover larger areas lead to more fires per unit area. The average lightning storm that does (does not) start a fire has about 9 (15) minutes of rain before the lightning starts and about 31 (44) minutes after the lightning ends. Six percent to 10% of thunderstorms have lightning with no rain. Gisborne found that these dry storms ignite fires no better than wet storms, but Rorig and Ferguson (1999) link dry lightning with increased fire starts in the northern Rockies. Storms that start fires have a high percentage of cloud-to-ground (as opposed to cloud-cloud) lightning and long-continuing current strokes (Fuquay et al. 1967a, b; Latham and Schlieter 1989). One thunderstorm ignited 335 fires in the northern Rockies in a day (Barrows 1951a).

Lags are common between ignition and fire detection or spread. Gisborne (1931) found that about 8% of fires were not detected until more than 48 hours after the storm that led to ignition. After ignition a fire may smolder for weeks before spreading; the Ouzel fire in 1978 in Rocky Mountain National Park, for example, ignited August 9 but did not begin its major spread until dry conditions and strong winds began on September 1 (Butts 1985). In subalpine forests, many small fires are started that burn only a few trees before going out (Kipfmueller and Baker 2000).

The Fire Season

The fire season in the Rockies is typically from April to October, but the season decreases in length with elevation. In the southern Rockies the fire season tends toward bimodality, with peaks in May to June and in September to October and a low period from about mid-July to the end of August (Cohen 1976; Ryan 1976; Floyd, Romme, and Hanna 2000). The number of fires and average fire size are typically highest in June (Ryan 1976; Floyd, Romme, and Hanna 2000). The low period after June reflects the wet period associated with the peak of the North American monsoon. In the northern Rockies, where the monsoon has less effect, the number of fires is more unimodally distributed with a peak in July (Fig. 5.3; Barrows 1951a). Larsen (1925) suggests, based on an analysis of over 13,000 fires, that the fire season in northern Idaho and Montana is bounded by the time mean air temperature is above about 10°C and monthly precipitation is <50 mm. This results in a fire season of about 150 days in the lowest forested zone and about 76 days in subalpine forests.

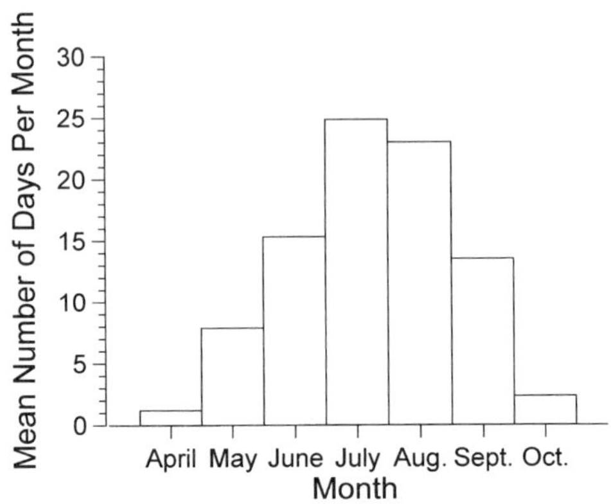

Figure 5.3. The mean number of days in each month when lightning fires have occurred in the northern Rocky Mountains (data from Barrows 1951a from 1931 to 1945 on national forests).

Fire-Spread Weather

Even if there is a climatic episode favorable to ignition, an ignited fire cannot spread unless weather conditions are also favorable. In the Rockies the major weather factors that promote fire spread, once a fire is ignited, are high temperatures and low precipitation or drought, which leads to low fuel moisture, low relative humidity, and strong winds (Table 5.2). These factors are reflected in fire-weather indexes commonly used in the region (Table 5.2). Antecedent conditions are also important. Low precipitation during the preceding winter and spring often leads to more and larger fires during the fire season (Kipfmueller and Swetnam 2000). Where there are live fine fuels (e.g., grasses, conifer needles), high precipitation one to four years previously may provide abundant fuels that lead to increased fires during dry years (Veblen et al. 1996, 2000; Kipfmueller and Swetnam 2000). Drought intensity, measured by the Palmer indexes, is strongly correlated with fire occurrence, large fires, and area burned. High temperatures and drought lead to low fuel moisture. The most significant fire spread rates during the 1988 fires were associated with 1000-hour time lag fuel moistures <13% (Renkin and Despain 1992), but the area burned was correlated with 100-hour time lag fuel moisture (Turner et al. 1994). When the moisture content of fine fuels reaches as low as 4–7%, very large fires and rapid rates of spread have occurred (Jemison 1932; Thomas 1991). Strong winds spread fires in this region, but strong winds alone are insufficient unless the relative humidity is low (Beighley and Bishop 1990). The spread of plume-dominated, rather than wind-driven, fires is promoted by atmospheric instability, reflected in steep, upper-air lapse rates and air temperature–dew point differences >6°C (Haines 1988; Werth and Ochoa 1990).

Two major synoptic climatic patterns leading to strong winds, and two synoptic patterns leading to hot, dry conditions, contribute to the occurrence of large

Table 5.2. Reported weather effects on the occurrence of fires, large fires, and the amount of area burned in the U.S. Rocky Mountains

	Fires	Large fires	Area burned	Where	Author(s)
Temperature					
Monthly mean temp.	$r = 0.79$	$r = 0.49$	$r = 0.57$	Black Hills, SD & WY	McCutchan & Main 1989
Temp. > 38°C		×		Rocky Mts.	Brown & Davis 1939
Temp. > 37°C		×		Northern ID	Jemison 1932
Temp. above avg. in July		×		Front Range, CO	Veblen et al. 1996, 2000
Temp. above avg. in summer	×			Southern CO	Baker 1992
Temp. above avg. in summer			×	Yellowstone, WY	Balling et al. 1992a
Max. temp. in summer			$r = 0.58$	Yellowstone, WY	Balling et al. 1992b
High temperatures		×		MT & SD	Potter 1996
Precipitation					
No precip. for 8 or more days		×		Rocky Mts.	Brown & Davis 1939
Precip. below avg. preceding winter/spring			×	Yellowstone, WY	Balling et al. 1992a
Precip. below avg. preceding spring	×	×		Front Range, CO	Veblen et al. 1996, 2000
Precip. below avg. in summer			×	Yellowstone, WY	Balling et al. 1992a
Precip. below avg. in summer	×			Yellowstone, WY	Renkin & Despain 1992
Precip. below avg. in summer		×		West-central ID	Steele et al. 1986
Precip. total in summer			$r = -0.52$	Yellowstone, WY	Balling et al. 1992b
Precip. days in summer			$r = -0.41$	Yellowstone, WY	Balling et al. 1992b
Precip. below avg. for year	×	×		Front Range, CO	Veblen et al. 1996, 2000
Precip. above avg. 1–3 years before	×	×		Front Range, CO	Veblen et al. 1996, 2000
Dry periods		×		Priest Range, ID	Marshall 1927
Precip. substantially below average	×			Northern Rockies	Barrett et al. 1997
Fuel Moisture					
Fuel moisture 0–7%		×		Rocky Mts.	Brown & Davis 1939
Fuel moisture (duff & branch wood) < 10%		×		Northern Rockies	Weidman 1923
Fuel moisture (duff & branch wood) < 10%		×		Northern ID	Gisborne 1927
Fuel moisture (duff & branch wood) 4–5%		×		Northern ID	Jemison 1932
100-hr time lag fuel moisture			$r = 0.52$	Yellowstone, WY	Turner et al. 1994
1000-hr time lag fuel moisture			$r = 0.36$	Yellowstone, WY	Turner et al. 1994
1000-hr time lag fuel moisture < 13%		×		Yellowstone, WY	Renkin & Despain 1992
1000-hr time lag fuel moisture			×	ID & MT	Burgan et al. 1996

Continued

Table 5.2. *Continued*

	Fires	Large fires	Area burned	Where	Author(s)
Drought					
Severe droughts		×		Glacier NP, MT	Barrett et al. 1991
Droughts	×			Northern Rockies	Barrett et al. 1997
Palmer Hydrological Index	$r = -0.86$	$r = -0.65$		Black Hills, SD & WY	McCutchan & Main 1989
Palmer Drought Index	$r = -0.84$	$r = -0.66$		Black Hills, SD & WY	McCutchan & Main 1989
Palmer Drought Index		×		Western MT	Goens 1990
Palmer Drought Index		×		State of Colorado	Cohen 1976
Palmer Drought Severity Index			$r = -0.60$	Yellowstone, WY	Balling et al. 1992b
Palmer Drought Severity Index		×		Western MT	Kipfmueller & Swetnam 2000
Relative Humidity (RH)					
RH < 20% needed before strong winds can effectively spread a fire		×		Boise, ID	Beighley & Bishop 1990
RH about 10%		×		Northern ID	Jemison 1932
Wind					
Wind > 40 km/hr		×		Rocky Mts.	Brown & Davis 1939
Winds very strong		×		Ouzel Fire, CO	Butts 1985
Wind gusts > 80 km/hr		×		Pingree Park, CO	Colo. St. Univ. 1995
Winds very strong		×		Western MT	Goens 1990
Strong afternoon winds		×		Northern ID	Gisborne 1927
Strong winds above the fire		×		Boise, ID	Small 1957
Foehn winds		×		Rocky Mts.	Heilman et al. 1994
Foehn winds		×		Western MT	Goens 1990
Atmospheric Instability					
Steep upper air lapse rates producing unstable air		×		Western U.S.	Haines 1988
Air temperature-dew point differences > 6°C		×		Western U.S.	Haines 1988
Fire Weather Indexes					
Fosberg Fire Weather Index (wind, moisture)			$r = 0.66$	Black Hills, SD & WY	McCutchan & Main 1989
Burning Index (rel. humid., temp.)			$r = 0.64$	Black Hills, SD & WY	McCutchan & Main 1989

Note: Large fires are defined using a variety of criteria. Negative reports (e.g., no effect of Palmer Drought Severity Index) are not included. The first column lists the variable used by the author; succeeding columns indicate which fire parameter(s) was found to be related. State abbreviations are: CO = Colorado, ID = Idaho, MT = Montana, SD = South Dakota, WY = Wyoming.

fires in the region. First, high-level or low-level jet streams overhead contribute to strong, gusty surface winds that lead to rapid, extensive fire spread (Schaefer 1957; Haines 1988; Goens 1990). This was what occurred during the extensive September 6–7, 1988, fire runs in western Montana (Goens 1990) and during the famous 1910 fire year in the northern Rockies (Schaefer 1957). Strong winds just above the fire can lead to blowup conditions that promote rapid fire spread (Small 1957). Second, rapidly moving dry, cold fronts that pass over a fire may produce strong, gusty surface winds that lead to extensive fire spread (Schullery 1989; Beighley and Bishop 1990; Renkin and Despain 1992). Several cold fronts passed over Yellowstone and the northern Rockies area in 1988, each leading to significant fire runs (Goens 1990; Thomas 1991). The fatalities of the South Canyon fire near Glenwood Springs, Colorado, on July 6, 1994, were in part due to a strong cold front that passed over the fire creating rapid spread (Butler et al. 1998). Third, persistent upper-level ridges or high-pressure systems over the western United States and southern Canada produce hot, dry surface conditions that are well known to contribute to fires in the region (Brotak 1983; Schullery 1989); as is the case in western Canada (Johnson and Wowchuk 1993; Nash and Johnson 1996). However, strong 500-mb zonal flow across the northern United States may also lead to dry conditions that promote large fires (Brotak 1983; Heilman, Eenigenburg, and Main 1994). Pre–Euro-American crown fires in a subalpine landscape in southeastern Wyoming spread preferentially toward the north, probably reflecting the first two cases, and toward the south, reflecting the third case (Baker and Kipfmueller 2001). Few fires spread to the east in the direction of prevailing winds, reinforcing the importance of particular synoptic conditions for significant fire spread.

Vegetation and Fuels

Vegetation Types Along Environmental Gradients

Weather alone is insufficient to lead to fire, as certain fuel conditions are needed. The quantity and quality of fuel available to a fire depend on the characteristics and successional status of the vegetation. The major pygmy-woodland, montane, and subalpine forest types in the Rockies vary primarily along elevational, topographic-moisture, and geographic gradients (Peet 1988; Table 5.3). These forest types differ in their fuel structure and in the prevailing types of fires. Pygmy conifer woodlands may have sparse herbaceous layers on rockier sites and coarser soils, but can also have dense, grassy understories, or have abundant shrubs and high fuel loads (Floyd, Romme, and Hanna 2000). Pygmy conifers are often <7 m tall and may contain multiple stems, as well as branches that reach near the ground. This ladder-fuel structure and high fuel loads commonly lead to crown fires (Hester 1952; Floyd, Romme, and Hanna 2000). However, some sites contain taller trees with grassy understories, structured more as montane forests. Montane forests, especially on more xeric sites, typically contain comparatively low-density tall conifers with straight boles and few branches near the ground, and often have a grassy or semicontinuous herbaceous understory. Such sites

Table 5.3. Major forest types and their dominant tree species in the Rocky Mountains

Forest zone	Major forest type	Dominant species	Common name
Subalpine forest	Spruce fir forest	*Picea engelmannii*	Engelmann spruce
		Abies lasiocarpa	Subalpine fir
	Xeric pine forest	*Pinus aristata*	Bristlecone pine
		Pinus flexilis	Limber pine
		Pinus albicaulis	Whitebark pine
	Lodgepole pine forest	*Pinus contorta*	Lodgepole pine
	Quaking aspen forest	*Populus tremuloides*	Quaking aspen
Montane forest	Douglas fir forest	*Pseudotsuga menziesii*	Douglas fir
	Ponderosa pine forest	*Pinus ponderosa*	Ponderosa pine
	Mixed conifer forest	*Pseudotsuga menziesii*	Douglas fir
		Pinus ponderosa	Ponderosa pine
		Abies concolor	White fir
Pygmy woodland	Pinyon juniper woodland	*Pinus edulis*	Twoneedle pinyon
		Juniperus scopulorum	Rocky Mt. juniper
		Juniperus monosperma	Oneseed juniper
		Juniperus osteosperma	Utah juniper
Riparian forest	Cottonwood forest	*Populus deltoides* ssp. *monilifera*	Plains cottonwood
		Populus deltoides ssp. *wislizeni*	Rio Grande cottonwood
		Populus angustifolia	Narrowleaf cottonwood
	Blue spruce forest	*Picea pungens*	Blue spruce

Note: Nomenclature is from the U.S. Department of Agriculture's online PLANTS database (*http://plants.usda.gov/plants/*).

typically support low-intensity surface fires and periodic crown fires (Weaver 1974; Ryan 1976; Ehle 2001). More mesic sites and mixed conifer forests are often denser, may contain abundant shrubs, and thus have more ladder fuels. While surface fires may occur, there is a higher probability of crown fire than in drier montane forests (Veblen 2000; Veblen, Kitzberger, and Donnegan 2000). Lodgepole pine forests often are dense, and have sparse, low-growing understories, although taller-shrub understories also occur. Very dense stands self-thin after a few decades, leading to high dead-fuel loads beneath dense canopies (Alexander 1979). Lodgepole pine forests are prone to crown fires, but surface fires can occur (Franklin and Laven 1991). Spruce fir forests often contain abundant ladder fuels, such as young trees, tall shrubs, and the low branches of the dominant conifers. These forests are also prone to crown fires.

The varying fuel structures of forests lead to strong differences in susceptibility to fire, even among adjoining forest types, illustrating that the structure of vegetation and fuels is important to the fire regime. Forests, relative to grasslands

Table 5.4. Ignition rates and mean fire sizes for major forest types in Colorado, ignition rates in the northern Rocky Mountains, and ignition ratios in southwestern Idaho

	Ignition rate		Mean fire size (ha)	Ignition ratio
	Colorado[a]	Northern Rockies[b]	Colorado[a]	Idaho[c]
Sagebrush-grass	3.6	—	2.22	144
Pinyon-juniper	—	—	4.01	—
Ponderosa pine	81.9	310.7	1.00	24
Douglas-fir	25.5	24.0	0.53	42
Aspen	1.9	9.0	1.43	—
Lodgepole pine	8.3	13.5	3.97	—
Spruce-fir	4.1	13.2	2.42	—

Source:
[a] Fechner and Barrows 1976.
[b] Bevins and Barney 1980.
[c] Meisner et al. 1994.
Note: The ignition rate is the number of fires per 400,000 ha (million acres) of that forest type per year. The ignition ratio is the number of lightning strikes per fire start.

or shrublands, are strongly favored locations for fire starts throughout the Rocky Mountains (Barrows 1951b, 1978). Old-growth forests in landscapes with lodgepole pine are preferential locations for ignitions (Renkin and Despain 1992), but intermediate-aged trees appear more susceptible in northern Idaho (Fowler and Asleson 1984). Forest types, such as ponderosa pine and Douglas-fir forests, that may occur within a few hundred meters of each other have quite different ignition rates (Table 5.4), an index that measures the ability of a forest type to sustain fires. Higher ignition ratios also suggest that it is more difficult to start a fire in Douglas-fir forests than in ponderosa pine forests (Table 5.4). The most important factor in ignition is fuel moisture in ponderosa pine and Engelmann spruce forests, but duff depth in lodgepole pine and Douglas-fir forests (Latham and Schlieter 1989). Mean fire sizes in the southern Rocky Mountains are highest in pinyon juniper woodlands and lodgepole pine forests and lowest in Douglas-fir, ponderosa pine, and aspen forests (Table 5.4).

Fuel Buildup with Succession

There is surprisingly little consistent difference in fuel loads among the major cover types and environments, at least in the northern Rocky Mountains. Brown and See (1981) analyzed hundreds of fuel plots, and found no trends with slope, elevation, or aspect. Brown and Bevins (1986) compared hundreds of samples of fuel loads from all the major forest types (Table 5.3) in the northern Rockies, and found insignificant variation in mean loads for each fuel component among most cover types. Fuel loads were quite varied within a cover type, presumably reflecting trends with succession and effects of disturbance.

Fuel trends appear complex and are not always consistent as succession proceeds in Rocky Mountain forests. In subalpine forests of the northern Rockies the fine fuels (litter, grass and forbs, and small branchwood), which are important in sustaining ignitions and in subsequent fire spread, are low during the first decade of postfire succession, as these fuels are typically consumed by the fire (Kessell et al. 1978). These fuels gradually increase for 150 to 200 years after a fire in the central Rockies, then slowly decline with age, but quantities of fine fuels in these forests are generally relatively small (Romme 1982). Small, dead fuels not consumed by the fire fall to the forest floor and are high immediately after the fire, but decline within the first few decades (Kessell et al. 1978) to as much as a century (Romme 1982), as do the live fuels (shrubs, grass). Clagg (1975), however, found more complex trends in medium-sized fuels with time since fire in the southern Rockies. Shrub fuels may decline in other major cover types as succession proceeds (Habeck 1976), but may also increase (Brown and See 1981). After the first few decades, most small fuels show relatively minor trends or no trend with stand age.

The largest fuel component generally increases through time (Kessell et al. 1978; Romme 1982), although Brown and See (1981) report no consistent trend with stand age, except that very old stands have high loads. Large, sound fuels are relatively unimportant to future fires, since they are typically not consumed, but large, rotten fuels are important to fire intensity, and do accumulate over time (Clagg 1975). Overall, Romme (1982) observed a decline in total dead woody fuels important to future fires through about the first century, then a gradual buildup to about 450 years after a fire. Brown and See (1981), however, question whether there are any consistent trends in fuel components with stand age in any Rocky Mountain forest.

The duff layer is important in maintaining smoldering fires during cold, wet weather and in ignition in lodgepole pine and Douglas-fir forests (Latham and Schlieter 1989). In the Rockies, duff often accumulates gradually in a nearly linear way from the time of the fire (Clagg 1975; Habeck 1976; Romme 1982). However, Clagg (1975) and Alexander (1979) found a trend toward a peak in duff/litter depth at about 125 to 200 years after a fire in lodgepole pine forests.

Interaction of Climate and Fuels

Coincidence of Fuel Buildup and Fire Weather

An intermediate view of the relative roles of climate and fuels is that fires (particularly large fires) are encouraged when fuel buildup is accompanied by suitable fire weather. Insufficient fuel buildup, in this view, may limit the size of fires even if suitable fire weather occurs. Romme and Despain (1989) emphasize that during the 250 years prior to the 1988 fires, fires in Yellowstone National Park were small because the landscape was dominated by young forests.

While fuel buildup may be important in limiting fire in some areas, this limitation can be overcome by extreme fire weather, as Bessie and Johnson (1995) suggest. When fuel moisture is very low (<10% in duff and branch wood), fires may spread with little response to fuel-load variations (Gisborne 1927). Jemison (1932) noted that when fuel moisture reaches as low as 4% or 5%, then strong winds and high temperatures can lead to exceptional rates of fire spread (>600 ha hr^{-1}) that are insensitive to fuel loads. During the Yellowstone fires of 1988, age-class boundaries that stopped past fires were ineffective, and fuel variations were apparently insignificant in shaping fire spread (Turner and Romme 1994). The mechanism behind exceptional fire spread, independent of fuel loads, is that low fuel moisture makes more of the fuel load flammable and increases fire intensity.

There are thus two present conceptual models for the interaction of weather and fuels. In the "fuel-weighted interactive model," based on Romme and Despain (1989), fire weather varies, but this variation does not control fire occurrence until fuel buildup is sufficient to lead to a combined high probability of fire. Importance is weighted toward fuel, and fire weather may just control the timing of the fire. In the "weather-weighted interactive model," based on Jemison (1932) and Bessie and Johnson (1995), fuel buildup does increase fire probability, but fire weather is often sufficiently extreme to override the importance of fuels. This model is supported by a weak tendency for area burned to increase as time-since-fire increases (Baker and Kipfmueller 2001).

Elevational Gradient of Interactions

An elevational gradient of interactions of climate and fuels can be hypothesized (Fig. 5.4a), based on trends reviewed earlier. Droughts decline with elevation, while lightning increases. Fine fuels reach a peak and ladder fuels reach a low point in montane forests. Other variables that change are length of the fire season and relative fuel moisture (Fowler and Asleson 1984).

These trends lead to different potential limitations to fire occurrence in the three zones, but all three zones are hypothesized to have fire-weather control or weather-weighted interactive control, rather than fuel control of the fire regime (Table 5.1). In pygmy conifer woodlands at low elevation, drought and ladder fuels are common, the fire season is long, and fuel moisture is relatively low. Fire may be most limited by insufficient lightning and the continuous fine fuels necessary to ignition and spread. This zone is thus hypothesized to have a fire weather controlled fire regime, in which ignition-promoting weather, rather than drought, is the primary control. In montane forests at middle elevations, droughts and lightning are common and fine fuels are abundant; fire may be most limited by the need for a combination of droughts and abundant fine fuels (Veblen, Kitzberger, and Donnegan 2000). Thus this may again be a fire weather controlled regime, but the weather control is by wet conditions promoting fine fuels followed by dry conditions leading to low fuel moisture. This regime may also be the most fuels-weather interactive fire regime, with weighting toward weather. In subalpine forests lightning and ladder fuels are common, but fine fuels are less common and droughts are uncommon, so fire occurrence is most strongly limited by the occurrence of droughts that dry out dead and live fine fuels sufficiently to carry a fire. Subalpine forests in the Rockies appear most likely to fit the model of a more fire weather controlled fire regime, with drought the primary weather control.

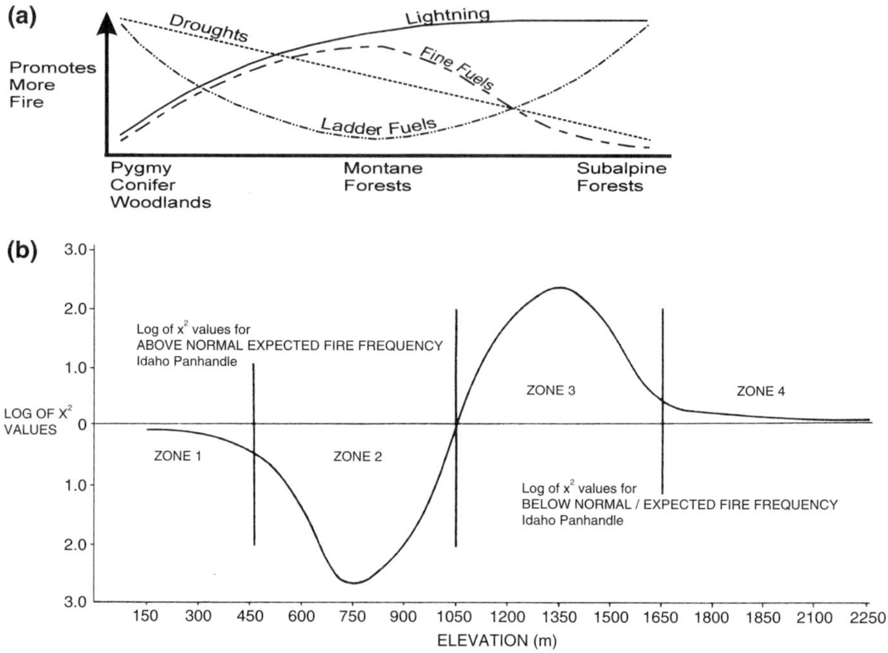

Figure 5.4. Trends in fire relationships with elevation and vegetation zone in the Rocky Mountains: (a) Hypothesized trends in droughts, lightning, fine fuels, and ladder fuels along an elevational gradient through the major forest zones in the Rocky Mountains; and (b) Empirically derived theoretical trend in chi-squared values version elevation in northern Idaho. The chi-squared value is based on the ratio of the observed/expected density of fires (number/ha), so higher values on the *y*-axis reflect more fire than expected (Fowler and Asleson 1984; reprinted with permission from *Physical Geography*, Vol. 5, No. 3, p. 243, © V.H. Winston & Son, Inc., 360 South Ocean Boulevard, Palm Beach, FL 33480; all rights reserved).

These trends suggest a predominant weather control of fire regimes, shifting from lightning occurrence, to the pattern of wet/dry episodes, to drought along the elevational gradient. A similar gradient has been suggested in the western United States as a whole, from low fire-frequency desert sites to high fire-frequency montane sites to low fire-frequency subalpine forest sites (Martin 1982). An empirical model based on 2088 lightning-caused fires between 1960 and 1971 in a 1.3 million ha area in northern Idaho (Fowler and Asleson 1984) adds an additional zone, and lacks a pygmy conifer zone (Fig. 5.4b). Fire density in zone 1 is near the mean, because the effects of a long fire season and dry fuels are offset by low lightning frequency. In zone 4, fire density is also near the mean because the effect of high lightning frequency is offset by high fuel moisture and a short fire season. However, in the middle, which I suggest to have the highest fire occurrence, they have a zone 2 with reduced fire density, where lightning is not frequent enough to offset the high fuel moisture and short fire season, and a zone 3 of elevated fire density where lightning does overcome these limits. Fire density on the western slope, but not the eastern slope of the northern Rockies, suggests a mid-elevation peak (Fig. 5.5a). The pattern is incompletely known in

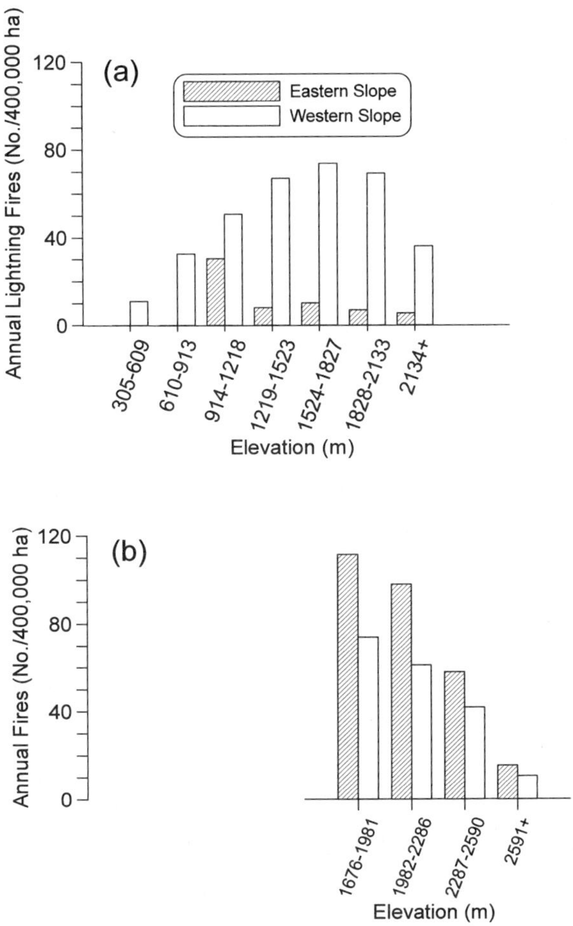

Figure 5.5. Elevational trends in mean annual fires per 400,000 ha on the eastern and western slopes of the Rocky Mountains: (a) lightning fires in the northern Rocky Mountains. Data are from Barrows (1951a). Elevation zones are in 1000 ft increments in the original data; (b) total fires in Colorado (data from Ryan (1976)).

the southern Rockies (Fig. 5.5b). Research is needed to test these ideas over a broader area.

The Contingent View

Spatial Variation, Constraints, and Dependencies

Spatial Variation in Topographic Constraints on Fire Regimes

Topography and the location of mountains control air masses that affect relative humidity and other important components of fire weather. In Colorado low relative humidity is most frequent in a broad zone at the center of the state, from the western border east to near Denver, with an outlier along the far northern foothills of the Front Range (Cohen 1976). Days characterized by dry unstable air that promotes extreme fire behavior are much more frequent in the central and southern Rockies than the northern Rockies (Werth and Werth 1998).

In mountains the wind field that affects rates and directions of fire spread is the result of synoptic-scale forcing, modified at that scale by mountain location and orientation (Barry 1992). In Colorado frequent strong winds peak in the western San Juan Mountains and on the eastern slope of the northern Colorado Front Range (Cohen 1976). In central Idaho the wind blows parallel to ridges, increasing the drying effect, while in parts of Idaho and Montana the wind blows perpendicular to ridgelines, which then successively slow the wind and decrease fire spread (Larsen and Delavan 1922). Strong downslope winds, resulting from synoptic and topographic effects, occur frequently in the lee of the Rockies, especially in the northern Colorado Front Range (Cohen 1976), and are accentuated where mountains are perpendicular to prevailing westerly winds (Goens 1990). Downslope winds are concentrated between November and March but may also occur during the beginning and end of the fire season (Cohen 1976; Barry 1992).

On a finer topographic scale, ridges and valleys may channel and accelerate synoptic-scale winds, with more exposed locations dominated by these winds, while more protected locations may develop local thermally induced winds (Sturman 1987). Terrain-forced convergence led to strong winds and large fire runs in several parts of the 1988 fires in western Montana (Goens 1990). Noctural downslope winds are most strongly developed on calm, clear nights when a surface temperature inversion and a low-level jet may develop. The low-level jet may lead to rapid spread rates on ridges in the early morning, even though adjoining valleys are calm due to the inversion, as in the 1967 Sundance fire in Idaho (Baughman 1981). The variable arrangement of hillslopes and incoming solar radiation leads to dynamic thermally induced winds (Sturman 1987). During the early part of the day, asymmetrical heating of slopes may induce a cross-valley flow and an upslope flow, while later in the afternoon the overlying flow may scour into valley bottoms through down-mixing (Sturman 1987). Rough, broken mountain terrain increases the mixing depth at night as well, and may bring strong winds aloft to the surface, as was the case during the 1988 fires in western Montana (Goens 1990).

Fire Breaks as Spatial Constraints

Fire breaks are physical and vegetational features that can stop a fire, lower its intensity, or shift its direction (Fig. 5.6). In extreme conditions fire may cross apparent barriers with ease, and fires can always spot, via airborne embers, across distances of several kilometers (e.g., Jemison 1932). Physical fire breaks (e.g., rock outcrops, lakes) do not change in their resistance to fire spread as weather changes, but vegetational fire breaks (e.g., snow avalanche tracks) may be

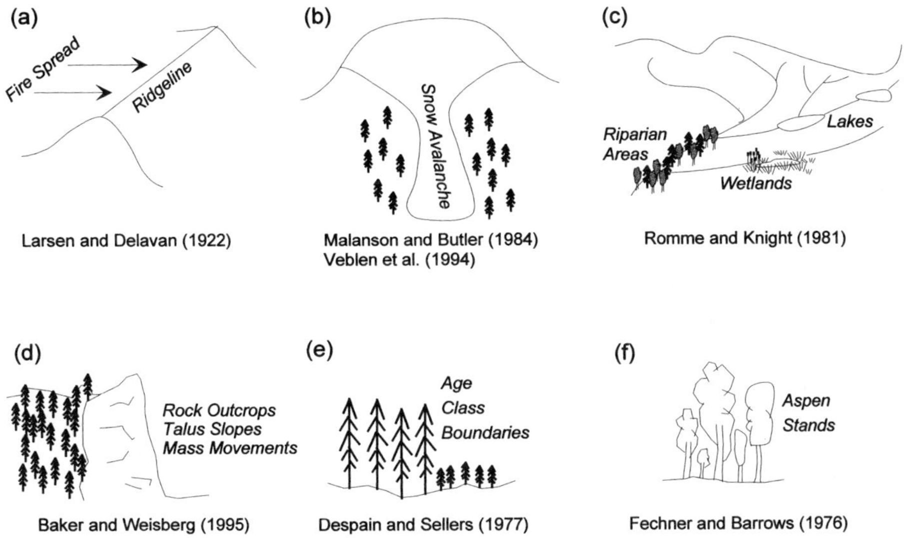

Figure 5.6. Potential fire breaks identified in the literature: (a) ridgelines, (b) avalanche tracks, (c) riparian areas and wetlands, (d) rock outcrops, talus slopes, and mass movement paths, (e) forest age class boundaries, (f) aspen stands.

effective only under less extreme conditions (e.g., Turner and Romme 1994). Even minor increases in fuel moisture can be a fire break under mild conditions (Clark 1990).

Fire breaks affect the susceptibility of the landscape to fires and, ultimately, the fire regime. First, fire breaks decrease fire sizes relative to fire sizes in similar fuels without breaks. Second, fire frequency at a point, with fire breaks in the vicinity, is decreased relative to fire frequency if the fire breaks were not nearby. This occurs because spread from adjoining areas, which contributes to fire frequency at a point, is diminished by nearby fire breaks. The fire rotation was 65% longer 6 km or more, as opposed to less than 3 km, from water breaks in a boreal forest in Canada (Larsen 1997), so this effect can be very significant. On a finer scale, patches of rock and bare soil in pinyon juniper woodlands lead to about 400-year fire rotations, while nearby shrublands, with more continuous fine fuels have fire rotations of nearly 100 years (Floyd, Romme, and Hanna 2000). Third, if fire breaks are absent, then large fires can spread, leading to synchrony and spatial homogeneity in fuel buildup, which may promote synchronous susceptibility to future fires and lead to large future fires (Baisan and Swetnam 1990). If fire breaks are present, the landscape is more likely to have asynchrony and spatial heterogeneity in fuel buildup and potential future fires. Landscapes with fire breaks are thus likely to be more continuously susceptible, than are landscapes without fire breaks, to "recording" lesser magnitude climate-related fire events, since some part of landscapes with fire breaks is likely to retain susceptible fuel loads.

Elevational Gradient in Constraints

Topographic effects and fire breaks (Fig. 5.6) likely vary along elevational gradients, although there has been no systematic study of elevational trends. Lower elevations, especially in valleys, are subject to temperature inversions and associated calm winds, but upslope and downslope thermal winds can become strong. Lower elevations probably typically have the highest density of some physical fuel breaks (i.e., rock outcrops, canyons, large rivers) but not others (lakes, ridgelines, mass movements). The highest forest elevations, in the subalpine zone up to treeline, have more exposed ridges subject to strong, synoptic-scale winds, low-level jets, and downslope winds, but are less affected by thermal winds and inversions. These high elevations have a high density of certain kinds of vegetational and physical fuel breaks (i.e., wetlands, snow avalanche paths, aspen stands, age-class boundaries, lakes, mass movements, ridgelines). In the southern Rocky Mountains a relatively flat peneplain at mid-elevations has a low density of physical fuel breaks, and perhaps has the highest fuel continuity in the mountains. These patterns of spatial constraint are only hypothetical, and additional research is warranted.

Fire as a Spatially Dependent Process

Fire is inherently a spatial spread process, which leads to spatial autocorrelation in fire regimes (Chou et al. 1990), although spotting that offsets autocorrelation also occurs, particularly in stand-replacing fire regimes. Spatial autocorrelation occurs when the probability of fire at a point is dependent on the probability of fire at adjoining points. For a forest stand with a certain structure, fuel load, and climatic setting, the long-term fire frequency or other parameters of the fire regime in the stand may differ depending on the types of nearby ecosystem, proximity to favored ignition points (e.g., ridgelines), terrain setting, nearby fuel loads, and so on. The distance over which spatial autocorrelation occurs is related to the size of the fires and the physical processes that affect fire probability (e.g., wind). The scale of significant spatial autocorrelation in the fire regime is also the scale over which spatial constraints (e.g., fuel breaks, topographic effects) have a significant effect. One study in southeastern Wyoming found significant spatial autocorrelation in mean fire intervals in a lodgepole pine forest over distances exceeding 2 km (Baker and Kipfmueller 2001), but additional research is needed.

Historical Legacies

Legacy of Past Climate and Natural Disturbance

The Little Ice Age (LIA) in the southern Rockies and the U.S. southwest, from about AD 1400 to 1850 was, on average, cold and dry relative to the present, but fluctuations occurred (Bradley and Jones 1992; Petersen 1994; Grissino-Mayer 1995), so the name is perhaps a misnomer. Droughts occurred in the late 1500s in the southern Rockies and Great Plains (D'Arrigo and Jacoby 1991; Woodhouse 2001), wet episodes in the early 1600s, and droughts around 1820 and 1860 (Woodhouse and Overpeck 1998; Cook et al. 1999), while the 1830s are identified as the wettest episode since the 1600s (Grissino-Mayer 1995). After the LIA a strong increase in pinyon pollen in southwestern Colorado suggests wetter summers from a stronger North American monsoon (Petersen 1988, 1994). Historical climatic data and tree-ring analysis for the Rockies reveal warming, but continuing significant fluctuations in climate since the end of the LIA (Stockton and Meko 1975; Cook, Meko, and Stockton 1997; Woodhouse 2001; Woodhouse and Brown 2001).

The older live trees in a forest may have originated in an earlier, and quite different environmental period, such as the LIA, leaving a legacy of forest structure that shapes present fire occurrence. Clark (1990) suggested that the present disturbance regime may not explain forest structure established earlier, but the converse is also a possibility. The structure (e.g., density, volume) of trees present in older forests may reflect a long sequence of climatic conditions and disturbances conceivably no longer present in the landscape. The composition of the forest may also reflect a previous climate, as several centuries may be required for forest composition to adjust to climate change (Campbell and McAndrews 1993).

Stand characteristics at the time of a stand-replacing burn may leave an important legacy affecting subsequent fuel loads, more important in some respects than the characteristics of the postfire stand. The attributes (e.g., density) of trees killed by a fire, compared to attributes of trees that survived or became established after the fire, are better predictors of loads of small fuels and amounts of downed rotten material in postfire *Pinus contorta* stands in Colorado (Alexander 1979). This is not true for litter, which is more related to postfire live-tree volume. In the northern Rockies there is also little relationship between fuel loads and stand age, in part because of the significant legacy from the pre-burn stand (Brown and See 1981).

The absence of a consistent trend in fuel loads during succession is also attributed in part to periodic, unpredictable disturbances and other varying sources of leaf, needle, branch, and stem input to the fuels complex (Brown and See 1981). A variety of events and influences, such as windstorms, ice storms, droughts, disease, parasites, and a host of disturbances can cause canopy trees or understory plants to contribute material to the fuel load. Insect and parasite effects on fuel loads, for example, can be quite complex, as these agents can open the canopy, leading to increased solar radiation and possibly faster fuel decomposition, as well as increased fuel contribution from the canopy (Knight 1987).

Legacy of Human Land Uses

Several human land uses influence fire regimes while they are occurring, but may also leave a legacy that affects future fires. Livestock grazing can reduce fine fuels in forests, reducing the probability of ignition and spread (e.g., Hatton 1920). However, excessive livestock grazing continued for years may also favor development of ladder fuels that increase the probability of crown fires in Douglas fir forests in the northern Rockies (Zimmerman and Neuenschwander 1984). Similar effects are suggested for ponderosa pine forests throughout the southwestern United States (Covington and Moore 1994). In the southern Rocky Mountains where the fire regime in montane forests is sensitive to fine-fuel abundance linked to the El Niño–Southern Oscillation (Veblen et al. 2000), livestock grazing, where excessive, could decrease fine fuels, reducing the sensitivity of the fire regime in these forests to the El Niño–Southern Oscillation. However, livestock grazing has been decreasing in the wildland–urban interface in some areas. A variety of other land uses, such as timber harvesting (Weatherspoon and Skinner 1995) and associated forest fragmentation (Goldammer and Price 1998), can alter the structure and composition of fuels, leaving a legacy that affects the potential response of the fire regime to climatic change.

Fire as a Temporally Dependent Process

Fire regimes do not adjust immediately to climatic change, since fire as a spatial process requires decades to centuries to fully burn through a landscape. If climate changes abruptly, it may require up to two fire rotations of the new fire regime for all the trees in a landscape to receive a fire from the new regime, in part erasing present fuel loads and forest structures that then can adjust to the new regime (Baker 1995). However, fires do not consume all the fuels.

The legacy is not fully over in two rotations, but persists until past trees killed by the new fire regime fully decompose, since large dead wood affects fire intensity. The new regime may shorten (lengthen) the rotation, leading to a lower (higher) mean age and tree size for forest stands across the landscape (Baker 1995). However, as the existing older (younger) and larger (smaller) trees are killed by fire, these large (small) stems atypical of the new fire regime will continue to affect the new regime until stems killed by the fire decompose. Downed logs in subalpine forests in the Rockies may require >150 years to fully decompose (Brown et al. 1998).

Thus the maximum time that existing stand structure may influence future fires, even if climate changes immediately, is on average about two rotations of the new fire regime (decades to centuries) to burn away existing structure, and an additional century and a half (in subalpine forests) to decompose the trees killed by the new fire regime. Baker (1995) argued that the time required for fire regimes to adjust to climatic change may often exceed the time that climate is stable, leading to perpetual temporal disequilibrium between climate, fire regimes, fuel loads, and forest structure. New climates that arise quickly may interact for decades to centuries with past fuel loads and forest structures before the new fire regime is fully adjusted. If climate changes gradually in a directional way, then the fire regime will be perpetually adjusting to the new climate, held back by an ongoing legacy of fuel loads and forest structures.

Potential Response to Climatic Change

Projected Climate Changes

General circulation models (GCMs) predict increased temperature and precipitation in the Rocky Mountains under a doubled CO_2 climate, summarized by Houghton et al. (1996) as follows: These models do not include topographic detail important in the region, but do include aerosol effects that damp projected temperature increases. Predictions are for about 1.5–3.5°C warming in winter, 0.0–0.5°C warming in summer, 0.0–0.25 mm/day more precipitation in winter, and 0.1–0.6 mm/day more precipitation in summer in central North America by about AD 2050. A net increase in soil moisture, averaging about 1 cm in both winter and summer in the central North America region, is also predicted, although a net decrease may occur in the southern Rockies. Snowpack may decrease by 25% to 100% (McCabe and Wolock 1999). Variability in climate associated with El Niño–Southern Oscillation may continue and be enhanced somewhat. GCMs do not presently simulate changes in winds very well, or some local-scale processes (e.g., topographically controlled convection and thunderstorm formation) important to fire weather in the Rocky Mountain region. More spatially precise regional-nested GCMs, which do better with local processes, have predictions congruent with the summary above but also predict a larger increase in summer precipitation (Giorgi et al. 1998; Leung and Ghan 1999).

Potential Fire Regime Changes

The Broad-Scale View

Projected climate changes may influence vegetation and fuels, ignitions, and fire spread weather (Fosberg, Stocks, and Lynham 1996). In the central and northern Rockies, not considering the effects of fire, projected warmer and wetter winters and drier summers alone may allow expansion of ranges of ponderosa pine, western larch, western red cedar, and Gambel oak and lead to significant contractions in whitebark pine and Engelmann spruce (Bartlein, Whitlock, and Shafer 1997). While simple upward or northward migration is not projected, some montane species (e.g., Douglas fir) may migrate upward and replace subalpine species (e.g., whitebark pine). With the addition of fire into the model, an increase in lodgepole pine and other fire-adapted species may occur (Keane, Arno, and Brown 1990; Bartlein, Whitlock, and Shafer 1997). If some tree species find their present ranges no longer suitable, then increased leaf senescence, stress-related mortality, and other effects may increase dead fuels (Ryan 1991). Fine dead fuels important to ignition and spread may respond most rapidly to projected climate changes, but it is difficult to predict the net outcome for fuel loads resulting from changes in fuel inputs, decomposition rates, and nutrient shifts (e.g., carbon–nitrogen ratios) (Ryan 1991). Widespread mortality of canopy trees, as has occurred during past droughts (e.g., Allen and Breshears 1998), would lead to significant increases in fire intensity. Replacement of present canopy dominants (Bartlein, Whitlock, and Shafer 1997) would have complex effects on fuels.

Lightning and the length of the fire season are expected to increase, but with uncertain consequences. Increased use of wildlands by people is also likely to increase ignitions (Ryan 1991). About a 5% to 6% increase in lightning per 1°C of global warming is the general projection (Price and Rind 1994b). However, in parts of the Rockies the fire regime may already have sufficient lightning, and additional lightning could have little effect if accompanied by increased precipitation and soil moisture, as projected, that keep fuels wetter. If the fire season increases in the Rockies, several effects might occur. More fires might burn before bud set, for example, increasing tree crown damage and mortality (Ryan 1991).

There are two GCM-based projections of effects on fire in the Rockies. In the first projection (Price and Rind 1994a), an empirical model linked to a GCM is used to predict the number of fires per month from water balance and number of lightning days. An increase from 476 to 619 (30%) lightning fires per month in May, June, and July is projected for a doubled-CO_2 climate. In the second projection (Flannigan et al. 1998), fire weather is based on mean and extreme values of the Fire Weather Index (FWI), which integrates temperature, humidity, precipitation, and wind speed to predict the intensity of spreading fire. In the more humid western part of the northern Rockies, the FWI is projected, for a doubled-CO_2 climate, to be on average 1–2 times present values. On the drier eastern slopes of the northern and central Rockies and throughout Colorado, the average FWI may be 2–5 times present values. An area centered on eastern Montana and Wyoming would have >5 times present values, the greatest increase projected for North America (Flannigan et al. 1998, Flannigan et al., Chapter 4, this volume, Fig. 5b). Extreme values of FWI are generally expected to be >1.5 times present values for most of the central and northern Rockies (Flannigan et al. 1998, Flannigan et al., Chapter 4, this volume, Fig. 5b). The Flannigan et al. projections suggest that the Rockies might be among the regions in North America most vulnerable to increases in severe fire weather.

These projections are incomplete for the Rockies, and they do not always agree with observed trends. Price and Rind (1994a) use empirical models derived for the Southwest to project Rocky Mountain changes. Flannigan et al. (1998) include only part of the Rockies, and the fire-weather index they use explains only about 44% of burned area in the Black Hills of South Dakota and Wyoming (McCutchan and Main 1989), so other factors must be important. Burned area has increased in the Rockies at a higher rate than these projections would suggest (Fig. 1a), so other factors must be having an influence. Annual burned area in Yellowstone National Park increased from 1890 to 1990, as did the Palmer Drought Severity Index, which has increased primarily due to declining winter precipitation (Balling, Meyer, and Wells 1992b). Winter precipitation in the Rockies, however, is projected by many GCMs to increase (Houghton et al. 1996; Bartlein, Whitlock, and Shafer 1997), although snowpack may decrease (McCabe and Wolock 1999). And yet more fires are being projected, which is inconsistent with the observed trend of burned area in Yellowstone. Further empirical and modeling work is clearly warranted.

The Contingent View

If the projections reviewed above were to occur, then what role might the landscape and historical legacies play in shaping the outcome? First, under the Flannigan et al. (1998) projections, the greatest effects may be in eastern Montana and Wyoming, while under the Price and Rind (1994a) projections the greatest effects may be near the Southwest, perhaps in southern Colorado. Flannigan et al. unfortunately do not include this area in their study. Second, both projections suggest an increase in fires under a doubled-CO_2 climate, but the Flannigan et al. projections suggest an increase in extreme fire weather that typically contributes most to area burned. Increases in the frequency of extreme fire weather will decrease the ability of fire breaks (Fig. 5.6) to halt or shift fire spread, so fires may spread farther.

Third, there is a significant legacy, from a century of human land uses, that may tend to diminish future fires. Human-set fires in the settlement era (Veblen and Lorenz 1991), combined with timber harvesting, have decreased the amount of old-growth forest and forest age generally, and decreased large dead wood in Rocky Mountain forests (Kaufman, Moir, and Bassett 1992). Old-growth forest is important to ignition and large dead wood to fire intensity. Large trees present in pre–Euro-American old-growth forests, but that succumbed to logging or early fires, have either become wood products or reached the late stages of decomposition on the forest floor (Brown et al. 1998), and are unlikely to affect future fires. Fine fuels have been reduced in many stands by excessive livestock grazing (Savage and Swetnam 1990), decreasing ignition potential in a way that may offset increased lightning (Price and Rind 1994a).

However, these effects are offset by a number of other trends. Fire exclusion and timber harvesting in some areas have allowed the buildup of fuels that may promote more severe fires (Covington and Moore 1994). Also in some areas dead fuels have built up from insect and disease outbreaks that appear to have been exacerbated by human land uses and fire exclusion (e.g., Hadley 1994). Sites with ladder fuels or abundant dead fuels may be epicenters from which future fires, burning under extreme weather conditions (Flannigan et al. 1998), can become significant crown fires that spread across the landscape. Air temperatures have increased in fragmented forests adjacent to roads and timber harvests, leading to increased drying of fuels inside forest interiors (Vaillancourt 1995; Goldammer and Price 1998). Increased access to and use of forests by people has also increased ignitions (Barrows, Sandberg, and Hart 1976). Thus the present landscape is generally younger, more homogeneous, and often contains less fine fuels and large dead fuels than at the time of Euro-American settlement, but also probably has drier fuels, is more likely to be ignited, and is possibly more prone to severe fires.

If increased fire is the result of future climatic change as the projections suggest, then the landscape will undergo adjustment lasting decades. Fire sizes will likely increase, since vegetational fire breaks will be less effective, and an increase in the number of fires is also predicted (Price and Rind 1994a). Increased fire size will likely decrease the fire rotation, decreasing the time needed for the landscape to adjust (Baker 1995). Present fire rotations in the Rockies are in the 60- to 300-year range generally (Baker 1995), so, at a minimum, several decades will likely be needed for the landscape to burn over under the new fire regime. Several more decades will be needed for remnant fuels to decompose. Of course, climatic change may occur gradually, in which case continued adjustment is likely.

Elevational Gradient in Potential Response

At the lowest elevations in the pygmy conifer zone and in the drier lower montane zone of the northern Rockies, fires might increase because of increased lightning (Price and Rind 1994a) in this lightning-limited zone. However, fine fuels have been depleted in some areas by excessive livestock grazing, and increased temperatures will lead to higher moisture stress that may not favor grasses and forbs needed for ignition and spread. Fires, if they do occur, may have higher intensity than pre–Euro-American fires in this zone, not because of unnatural fuel buildup (Floyd, Romme, and Hanna 2000) but because of more extreme fire weather. This zone already has the highest mean fire size (Table 5.4), but fire sizes might increase further. However, this zone has many physical fire breaks (e.g., canyons, large rivers) that may limit increases in fire spread.

In the montane zone, as in the pygmy conifer zone, fine fuels have often been depleted by livestock grazing, and grasses and forbs may not be favored by increases in moisture stress, so these factors may decrease the probability of future fires. However, dry fuels that are present will have an increased lightning-ignition source, and the buildup of larger fuels and ladder fuels, as a result of fire exclusion, may lead in some cases to more intense fires. Physical fire breaks may limit surface-fire spread, but higher-intensity crown fires may not be deterred. Surface fires will not lead to rapid adjustment of fuel loads to the new climatic regime, since canopy trees can survive and continue to affect understory fuel loads. Crown fires, if they become more common, will encourage more rapid adjustment of fuel loads, by killing overstory trees. Dead stems will still need to decompose before fuel loads will fully adjust to the new fire regime.

In the subalpine zone, lightning appears less limiting, and an increase would have little effect. However, fine fuels have not been widely depleted by livestock grazing. Fire suppression has likely had less effect, and fuels are not generally limiting in this zone. The lower part of this zone and the upper part of the montane have relatively continuous fuels, particularly in lodgepole-pine forests. Fire hazard in this zone has also been increased by forest fragmentation, and the wider availability of dry fuels in clear-cut openings. In this part of the subalpine zone, the higher frequency of extreme fire weather predicted by Flannigan et al. (1998) would likely lead to much larger fires and comparatively rapid adjustment to climatic change. In the higher subalpine zone, in contrast, there are many fire breaks that could impede the spread of fires, although extreme fire weather may overcome vegetational fire breaks. Because of the fire breaks, many individual fires will be needed before the whole zone is affected, so adjustment to climatic change will likely be longest in this area, where fire rotations are already a century or more.

Problems in Detecting Changes in Fire Regimes

Fire intervals and associated statistics (e.g., mean fire interval) are frequently used to identify change in fire regimes, but suffer from autocorrelation, lags, uncertainties, and ambiguities that make sampling difficult, and the value of these statistics uncertain (Baker and Ehle 2001). Different authors tend to use different intervals (e.g., scar-to-present) and different sampling/compositing areas, which makes comparison difficult. Purposeful sampling, while sometimes unavoidable, leads to biased estimates of fire intervals. A significant potential response to climatic change is increasing crown fires, but analysis of crown fires in montane forests has not been adequate, resulting in an insufficient baseline for understanding changes in fire regimes (Shinneman and Baker 1997; Baker and Ehle 2001). Since fire is a spatially autocorrelated process, samples may be autocorrelated, which can lead to biased estimates of fire regime parameters (e.g., mean fire interval). Samples taken over time to identify changes in fire regimes often suffer from insufficient statistical power, if not an absence of statistical analysis (Baker and Ehle 2001). Since fire is a spatial-spread process, fire intervals may represent more than one climatic regime. The fire regime requires a period of adjustment following a climatic change, and during this time the landscape is in transition, with newly burned areas adjusted to the new regime and unburned areas unadjusted (Baker 1993). If the climate changes again before adjustment, then a disequilibrium between climate and fire intervals may be maintained (Baker 1995). Many of these problems can be overcome by appropriate sampling, standardization of procedures and measures, and explicit treatment of potential errors (Baker and Ehle 2001).

Land-use changes that potentially affect fire regimes have often occurred during times when climate also changed, so potential causative agents are temporally confounded. Spatial comparisons of areas affected by a particular land use with reference areas free of the land use can potentially isolate a land-use or climatic effect (Grissino-Mayer 1995). Reference areas have included kipukas free of severe livestock grazing (Touchan, Swetnam, and Grissino-Mayer 1995), islands lacking intentional fire suppression (Bergeron and Archambault 1993), and national parks or other protected areas (Floyd, Romme, and Hanna 2000). Temporally confounded causes can also be potentially separated, where spatial control is not possible, by modeling the separate contributions of each process to the observed pattern of change. This has been used to isolate potentially competing causes of historical climatic change by quantifying the radiative forcing of each source of temperature change (Houghton et al. 1996). Clark (1988) similarly modeled the separate contributions to fires since AD 1240 from fuel buildup, the 22-year drought cycle, and the breakup of early successional stands.

Retrospective modeling of observed historical changes in fire regime parameters, such as area burned (Fig. 5.1), with climate and land-use drivers may allow the contributions of recent climatic and land-use changes in the Rockies to be isolated. This is an essential step in predicting future changes.

Summary and Conclusions

The broad-scale view links climate, fuels, and elevation. Winter snowpack in the Rockies is linked via teleconnections to the Pacific, while summer drought is also linked to the North Pacific, the El Niño–Southern Oscillation, and bi-decadal solar and lunar drought cycles. Lightning decreases from south to north, but increases with elevation, and is most limiting to fires at low elevations and in the north. Low fuel moisture and deep duff are most important to ignition of fine fuels. Fires are typically ignited during thunderstorms, and they can smolder for weeks before spreading significantly, so weather after ignition is important. Strong winds and extensive fire spread are associated with jet streams and summer cold fronts, and drought is associated with persistent high pressure or zonal flow across the northern states. Fuel loads often do not vary in a consistent way among forest types or with succession, but are influenced by the pre-burn stand and postfire disturbances, as well as time since fire.

The contingent view suggests that these general patterns and trends are shaped by spatial effects and historical legacies in the present landscape. Spatial effects include (1) geographic and topographic effects on humidity, wind, and other climate variables important to fires, (2) variation in the density and effect of physical (e.g., rivers) and vegetational (e.g., snow avalanche paths) fire breaks, and (3) spatial dependency in the fire regime related to fire sizes and the scale of topographic effects on weather affecting fires. Historical legacies include forest structures and fuel loads resulting from (1) past climates or the forest preceding the fire, (2) past human land uses, and (3) the slow adjustment of fire regimes to climatic changes.

GCM predictions for a doubled-CO_2 climate are for warming in all seasons, but concentrated in winter, accompanied by increased precipitation, with slightly more increase in summer. Significant compositional changes predicted in some forests may shift entire fire regimes, but more commonly an increase in extreme fire weather will lead to more and larger fires, although predictions are preliminary for the Rockies. If the fires do increase, it will require several decades at a minimum for the legacy of present forest structure and fuel loads to be erased. Topographically complex landscapes with many fire breaks will require longer periods to adjust.

Present methods of analyzing fire regimes suffer from problems that may hamper detection of climatically induced changes, but some problems can be overcome. Ongoing human land uses are confounded with climatic effects, requiring spatial comparisons or modeling to disentangle the contributions of each potential cause.

Acknowledgments. I appreciate the opportunity to visit South America, with support from Thomas T. Veblen and the Inter-American Institute, as this visit provided the impetus for this chapter. This chapter is based on work supported by the Cooperative State Research, Education and Extension Service, U.S. Department of Agriculture, Agreement No. 95-37106-2357, and the National Park Service, Global Change Program, Cooperative Agreement No. CA 1268-1-9009.

References

Adams, D.K., and Comrie, A.C. 1997. The North American monsoon. *Bull. Am. Meteorol. Soc.* 78:2197–2213.

Alexander, M.E. 1979. Fuels description in lodgepole pine stands of the Colorado Front Range. M.S. thesis. Colorado State University, Fort Collins.

Allen, C.D., and Breshears, D.D. 1998. Drought-induced shift of a forest-woodland ecotone: Rapid landscape response to climate variation. *Proc. Nat. Acad. Sci.* 95: 14839–14842.

Anderson, B.T., Roads, J.O., and Chen, S.-C. 2000. Large-scale forcing of summertime monsoon surges over the Gulf of California and the southwestern United States. *J. Geophys. Res.* 105:24455–24467.

Arno, S.F. 1996. The seminal importance of fire in ecosystem management-impetus for this publication. In *The Use of Fire in Forest Restoration: A General Session at the Annual Meeting of the Society for Ecological Restoration*, eds. C.C. Hardy and S.F. Arno, pp. 3–5, September 14–16, 1995 Seattle, WA. Ogden, UT: USDA Forest Service Gen. Tech. Rep. INT-GTR-341, Intermountain Research Station.

Baisan, C.H., and Swetnam, T.W. 1990. Fire history on a desert mountain range: Rincon Mountain Wilderness, Arizona, U.S.A. *Can. J. For. Res.* 20:1559–1569.

Baker, W.L. 1992. Structure, disturbance, and change in the bristlecone pine forests of Colorado, *U.S.A. Arct. Alp. Res.* 24:17–26.

Baker, W.L. 1993. Spatially heterogenous multi-scale response of landscapes to fire suppression. *Oikos* 66:66–71.

Baker, W.L. 1995. Longterm response of disturbance landscapes to human intervention and global change. *Landscape Ecol.* 10:143–159.

Baker, W.L., and Ehle, D. 2001. Uncertainty in surface-fire history: the case of ponderosa pine forests in the western United States. *Can. J. For. Res.* 31:1205–1226.

Baker, W.L., and Kipfmueller, K.F. 2001. Spatial ecology of pre–Euro-American fires in a southern Rocky Mountain subalpine forest landscape. *Prof. Geogr.* 53:248–262.

Baker, W.L., and Weisberg, P.J. 1995. Landscape analysis of the forest-tundra ecotone in Rocky Mountain National Park, Colorado. *Prof. Geogr.* 47:361–375.

Balling, R.C. Jr., Meyer, G.A., and Wells, S.G. 1992a. Climate change in Yellowstone National Park: Is the drought-related risk of wildfires increasing? *Clim. Change* 22: 35–45.

Balling, R.C. Jr., Meyer, G.A., and Wells, S.G. 1992b. Relation of surface climate and area burned in Yellowstone National Park. *Agric. For. Meteorol.* 60:285–293.

Barlow, M., Nigam, S., and Berbery, E.H. 2001. ENSO, Pacific decadal variability, and U.S. summertime precipitation, drought, and stream flow. *J. Clim.* 14:2105–2128.

Barrett, S.W., Arno, S.F., and Key, C.H. 1991. Fire regimes of western larch-lodgepole pine forests in Glacier National Park, Montana. *Can. J. For. Res.* 21:1711–1720.

Barrett, S.W., Arno, S.F., and Menakis, J.P. 1997. Fire episodes in the inland northwest (1540–1940) based on fire history data. Ogden, UT: USDA Forest Service Gen. Tech. Rep. INT-GTR-370, Intermountain Research Station.

Barrows, J.S. 1951a. Lightning fires in the northern Rocky Mountains. *USDA For. Ser. Fire Contr. Notes* 12:24–28.

Barrows, J.S. 1951b. Fire behavior in northern Rocky Mountain forests. Missoula, MT: USDA Forest Service Station Pap. 29, Northern Rocky Mountains Forest and Range Experimental Station.

Barrows, J.S. 1978. Lightning fires in southwestern forests. Report to the USDA Forest Service, Intermountain Forest and Range Experimental Station, Northern Forest Fire Lab. by Department of Forest and Wood Science, Colorado State University, Fort Collins.

Barrows, J.S., Sandberg, D.V., and Hart, J.D. 1976. Lightning fires in northern Rocky Mountain forests. Report to the USDA Forest Service, Intermountain Forest and Range Experiment Station, Northern Forest Fire Laboratory by Department of Forest and Wood Science, Colorado State University, Fort Collins.

Barry, R.G. 1992. *Mountain Weather and Climate*, 2nd ed. London: Routledge.

Bartlein, P.J., Whitlock, C., and Shafer, S.L. 1997. Future climate in the Yellowstone National Park region and its potential impact on vegetation. *Cons. Biol.* 11:782–792.

Baughman, R.G. 1981. Why windspeeds increase on high mountain slopes at night. Ogden, UT: USDA Forest Service Res. Pap. INT-276, Intermountain Forest and Range Experiment Station, 6p.

Beighley, M., and Bishop, J. 1990. Fire behavior in high-elevation timber. *Fire Manag. Notes* 51:23–28.

Bergeron, Y., and Archambault, S. 1993. Decreasing frequency of forest fires in the southern boreal zone of Quebec and its relation to global warming since the end of the "Little Ice Age." *Holocene* 3:255–259.

Bessie, W.C., and Johnson, E.A. 1995. The relative importance of fuels and weather on fire behavior in subalpine forests. Ecology 76:747–762.

Bevins, C.D., and Barney, R.J. 1980. Lightning fire densities and their management implications on northern region National Forests. *Proc. Conf. Fire For. Meteorol.* 6:127–131.

Biondi, F., Gershunov, A., and Cayan, D.R. 2001. North Pacific decadal climate variability since 1661. *J. Clim.* 14:5–10.

Bradley, R.S., and Jones, P., eds. 1992. *Climate Since A.D. 1500.* London: Routledge.

Brotak, E.A. 1983. Weather conditions associated with major wildland fires in the western United States. *Proc. Conf. Fire For. Meteorol.* 7:7–8.

Brown, A.A., and Davis, W.S. 1939. A fire danger meter for the Rocky Mountain region. *J. For.* 37:552–558.

Brown, J.K., and Bevins, C.D. 1986. Surface fuel loadings and predicted fire behavior for vegetation types in the northern Rocky Mountains. Ogden, UT: USDA Forest Service Res. Note INT-358, Intermountain Research Station.

Brown, J.K., and See, T.E. 1981. Downed dead woody fuel and biomass in the northern Rocky Mountains. Ogden, UT: USDA Forest Service Gen. Tech. Rep. INT-117, Intermountain Forest and Range Experiment Station.

Brown, P.M., Sheppard, W.D., Mata, S.A., and McClain, D.L. 1998. Longevity of windthrown logs in a subalpine forest of central Colorado. *Can. J. For. Res.* 28:932–936.

Bumstead, A.P. 1943. Sunspots and lightning fires. *J. For.* 41:69–70.

Burgan, R.E., Hartford, R.A., and Eidenshink, J.C. 1996. Using NDVI to assess departure from average greenness and its relation to fire business. Ogden, UT: USDA Forest Service Gen. Tech. Rep. INT-GTR-333, Intermountain Research Station.

Butler, B.W., Bartlette, R.A., Bradshaw, L.S., Cohen, J.D., Andrews, P.L., Putnam, T., and Mangan, R.J. 1998. Fire behavior associated with the 1994 South Canyon fire on Storm King Mountain, Colorado. Fort Collins: USDA Forest Service Res. Pap. RMRS-RP-9, Rocky Mountains Research Station.

Butts, D.B. 1985. Case study: The Ouzel fire, Rocky Mountain National Park. In *Proceedings of Symposium and Workshop on Wilderness Fire*, eds. J.E. Lotan, B.M. Kilgore, W.C. Fischer, and R.W. Mutch, pp. 248–251. Ogden, UT: USDA Forest Service Gen. Tech. Rep. INT-182, Intermountain Forest and Range Experiment Station.

Campbell, I.D., and McAndrews, J.H. 1993. Forest disequilibrium caused by rapid Little Ice Age cooling. *Nature* 366:336–338.

Carleton, A.M. 1985. Synoptic and satellite aspects of the southwestern U.S. summer "monsoon". *J. Climatol.* 5:389–402.

Carleton, A.M., Carpenter, D.A., and Weser, P.J. 1990. Mechanisms of interannual variability of the southwest United States summer rainfall maximum. *J. Clim.* 3:999–1015.

Cayan, D.R. 1996. Interannual climate variability and snowpack in the western United States. *J. Clim.* 9:928–948.

Chang, F.-C., and Smith, E.A. 2001. Hydrological and dynamical characteristics of summertime droughts over U.S. Great Plains. *J. Clim.* 14:2296–2316.

Changnon, D., McKee, T.B., and Doesken, N.J. 1993. Annual snowpack patterns across the Rockies: long-term trends and associated 500-mb synoptic patterns. *Mon. Wea. Rev.* 121:633–647.

Changnon, S.A. Jr. 1985. Secular variations in thunder-day frequencies in the twentieth century. *J. Geophys. Res.* 90:6181–6194.

Chou, Y.-H., Minnich, R.A., Salazar, L.A., Power, J.D., and Dezzani, R.J. 1990. Spatial autocorrelation of wildfire distribution in the Idyllwild quadrangle, San Jacinto Mountain, California. *Photogramm. Eng. Rem. Sens.* 56:1507–1513.

Clagg, H.B. 1975. Fire ecology in high-elevation forests in Colorado. MS thesis. Colorado State University, Fort Collins.

Clark, J.S. 1988. Effect of climate change on fire regimes in northwestern Minnesota. *Nature* 334:233–235.

Clark, J.S. 1990. Fire and climate change during the last 750 yr in northwestern Minnesota. *Ecol. Monogr.* 60:135–159.

Cohen, J.D. 1976. Analysis of Colorado mountain fire weather. M.S. Thesis. Colorado State University, Fort Collins.

Cole, J.E., and Cook, E.R. 1998. The changing relationship between ENSO variability and moisture balance in the continental United States. *Geophys. Res. Lett.* 25:4529–4532.

Colorado State University. 1995. The Hourglass fire at Pingree Park July 1, 1994. *Nature, People, and Wildfire: A Delicate Balance.* Information brochure. Colorado State University, Fort Collins, Pingree Park Campus.

Colson, D. 1957. Thunderstorm analysis in the northern Rocky Mountains. Ogden, UT: USDA Forest Service Res. Pap. No. 49, Intermountain Forest and Range Experiment Station.

Cook, E.R., Meko, D.M., Stahle, D.W., and Cleaveland, M.K. 1999. Drought reconstructions for the continental United States. *J. Clim.* 12:1145–1162.

Cook, E.R., Meko, D.M., and Stockton, C.W. 1997. A new assessment of possible solar and lunar forcing of the bidecadal drought rhythm in the western United States. *J. Clim.* 10:1343–1356.

Covington, W.W., and Moore, M.M. 1994. Southwestern ponderosa pine forest structure: changes since Euro-American settlement. *J. For.* 92(1):39–47.

Currie, R.G. 1984. Evidence for 18.6-year lunar nodal drought in western North America during the past millennium. *J. Geophys. Res.* 89:1295–1308.

D'Arrigo, R., and Jacoby, G.C. 1991. A 1000-year record of winter precipitation from northwestern New Mexico, USA: a reconstruction from tree-rings and its relation to El Niño and the Southern Oscillation. *Holocene* 1:95–101.

Diaz, H.F. 1983. Some aspects of major dry and wet periods in the contiguous United States, 1895–1981. *J. Clim. Appl. Meteorol.* 22:3–16.

Diaz, H.F., and Markgraf, V. 1992. *El Niño: Historical and Paleoclimatic Aspects of the Southern Oscillation.* Cambridge: Cambridge University Press.

Ehle, D.S. 2001. Spatial and temporal patterns of disturbance and ponderosa pine forest structure in Rocky Mountain National Park. M.A. thesis. University of Wyoming, Laramie.

Fall, P.L. 1997. Fire history and composition of the subalpine forest of western Colorado during the Holocene. *J. Biogeogr.* 24:309–325.

Fechner, G.H., and Barrows, J.S. 1976. Aspen stands as wildfire fuel breaks. Eisenhower Consortium Bulletin 4, Fort Collins, CO: Department of Forestry and Wood Science College of Forestry and Natural Resources, Colorado State University.

Flannigan, M.D., Bergeron, Y., Engelmark, O., and Wotton, B.M. 1998. Future wildfire in circumboreal forests in relation to global warming. *J. Veg. Sci.* 9:469–476.

Floyd, M.L., Romme, W.H., and Hanna, D.D. 2000. Fire history and vegetation pattern in Mesa Verde National Park, Colorado, USA. *Ecol. Appl.* 10:1666–1680.

Fosberg, M.A., Stocks, B.J., and Lynham, T.J. 1996. Risk analysis in strategic planning: fire and climate change in the boreal forest. In *Fire in Ecosystems of Boreal Eurasia*, eds. J.G. Goldammer and V.V. Furyaev, pp. 495–504. Dordrecht: Kluwer Academic.

Fowler, P.M., and Asleson, D.O. 1984. The location of lightning-caused wildland fires, northern Idaho. *Phys. Geogr.* 5:240–252.

Franklin, T.L., and Laven, R.D. 1991. Fire influences on central Rocky Mountain lodgepole pine stand structure and composition. *Proc. Tall Timbers Fire Ecol. Conf.* 17: 183–196.

Fuquay, D.M., Baughman, R.G., Taylor, A.R., and Hawe, R.G. 1967a. Characteristics of seven lightning discharges that caused forest fires. *J. Geophys. Res.* 72:6371–6373.

Fuquay, D.M., Baughman, R.G., Taylor, A.R., and Hawe, R.G. 1967b. Documentation of lightning discharges and resultant forest fires. Ogden, UT: USDA Forest Service Res. Note INT-68, Intermountain Forest and Range Experiment Station.

Gershunov, A., and Barnett, T.P. 1998. Interdecadal modulation of ENSO teleconnections. *Bull. Am. Meteorol. Soc.* 79:2715–2725.

Giorgi, F., L. Mearns, O., Shields, C., and McDaniel, L. 1998. Regional nested model simulations of present day and 2 × CO_2 climate over the central Plains of the U.S. *Clim. Change* 40:457–493.

Gisborne, H.T. 1927. Meteorological factors in the Quartz Creek forest fire. *Mon. Wea. Rev.* 55:56–60.

Gisborne, H.T. 1931. A five-year record of lightning storms and forest fires. *Mon. Wea. Rev.* 59:139–150.

Goens, D.W. 1990. Meteorological factors contributing to the Canyon Creek fire blowup September 6 and 7, 1988. In *Proceedings of the 5th Conference on Mountain Meteorology*, pp. 180–186, June 25–29, Boulder, CO. Boston: American Meteorological Society.

Goldammer, J.G., and Price, C. 1998. Potential impacts of climate change on fire regimes in the Tropics based on MAGICC and a GISS GCM-derived lightning model. *Clim. Change* 39:273–296.

Grissino-Mayer, H.D. 1995. Tree-ring reconstructions of climate and fire history at El Malpais National Monument, New Mexico. Ph.D. dissertation. University of Arizona, Tucson.

Habeck, J.R. 1976. Forests, fuels and fire in the Selway-Bitterroot Wilderness, Idaho. *Proc. Tall Timbers Fire Ecol. Conf.* 14:305–353.

Hadley, K.S. 1994. The role of disturbance, topography, and forest structure in the development of a montane forest landscape. *Bull. Torr. Bot. Club* 121:47–61.

Haines, D.A. 1988. A lower atmosphere severity index for wildland fires. *Nat. Wea. Digest* 13:23–27.

Hatton, J.H. 1920. Livestock grazing as a factor in fire protection on the national forests. *USDA Circular* 134. Washington, DC: U.S. Government Printing Office.

Heilman, W.E., Eenigenburg, J.E., and Main, W.A. 1994. Upper-air synoptic patterns associated with regional fire-weather episodes. *Proc. Conf. Fire Forest Meteorol.* 12:355–362.

Hester, D.A. 1952. The pinon-juniper fuel type can really burn. *USDA For. Ser. Fire Contr. Notes* 13:26–29.

Higgins, R.W., Mo, K.C., and Yao, Y. 1998. Interannual variability of the U.S. summer precipitation regime with emphasis on the southwestern monsoon. *J. Clim.* 11: 2582–2606.

Higgins, R.W., and Shi, W. 2001. Intercomparison of the principal modes of interannual and intraseasonal variability of the North American monsoon system. *J. Clim.* 14: 403–417.

Houghton, J.T., Meira Filho, L.G., Callander, B.A., Harris, N., Kattenberg, A., and Maskell, K., eds. 1996. *Climate Change 1995: The Science of Climate Change.* Cambridge: Cambridge University Press.

Hu, Q., Woodruff, C.M., and Mudrick, S.E. 1998. Interdecadal variations of annual precipitation in the central United States. *Bull. Am. Meteorol. Soc.* 79:221–229.

Jemison, G.M. 1932. Meteorological conditions affecting the Freeman Lake (Idaho) fire. *Mon. Wea. Rev.* 60:1–2.

Johnson, E.A., and Wowchuk, D.R. 1993. Wildfires in the southern Canadian Rocky Mountains and their relationship to mid-tropospheric anomalies. *Can. J. For. Res.* 23:1213–1222.

Kauffman, M.R., Moir, W.H., and Bassett, R.L., eds. 1992. *Old-Growth Forests in the Southwest and Rocky Mountain Regions: Proceedings of a Workshop*, March 9–13, Portal, AZ. Fort Collins, CO: USDA Forest Service Gen. Tech. Rep. RM-213, Rocky Mountains Forest and Range Experiment Station.

Keane, R.E., Arno, S.F., and Brown, J.K. 1990. Simulating cumulative fire effects in ponderosa pine/Douglas-fir forests. *Ecol.* 71:189–203.

Kessell, S.R., Potter, M.W., Bevins, C.D., Bradshaw, L., and Jeske, B.W. 1978. Analysis and application of forest fuels data. *Environ. Manag.* 2:347–363.

Kipfmueller, K.F., and Baker, W.L. 2000. A fire history of a subalpine forest in south-eastern Wyoming, USA. *J. Biogeogr.* 27:71–85.

Kipfmueller, K.F., and Swetnam, T.W. 2000. Fire-climate interactions in the Selway-Bitterroot Wilderness area. In *Wilderness Science in a Time of Change Conference.* Vol. 5: *Wilderness Ecosystems, Threats, and Management*, eds. D.N. Cole, S.F. McCool, W.T. Borrie, and J. O'Laughlin, pp. 270–275. Fort Collins, CO: USDA Forest Service Proceedings RMRS-P-15-VOL-5, Rocky Mountains Research Station.

Kitzberger, T., Swetnam, T.W., and Veblen, T.T. 2001. Inter-hemispheric synchrony of forest fires and the El Niño–Southern Oscillation. *Global Ecol. Biogeogr.* 10:315–326.

Knight, D.H. 1987. Parasites, lightning, and the vegetation mosaic in wilderness landscapes. In *Landscape Heterogeneity and Disturbance*, ed. M.G. Turner, pp. 59–83. New York: Springer-Verlag.

Kumar, A., Wang, W., Hoerling, M.P., Leetmaa, A., and Ji, M. 2001. The sustained North American warming of 1997 and 1998. *J. Clim.* 14:345–353.

Kunkel, K.E., and Angel, J.R. 1999. Relationship of ENSO to snowfall and related cyclone activity in the contiguous United States. *J. Geophys. Res.* 104:19425–19434.

Larsen, C.P.S. 1997. Spatial and temporal variations in boreal forest fire frequency in northern Alberta. *J. Biogeogr.* 24:663–673.

Larsen, J.A. 1925. The forest-fire season at different elevations in Idaho. *Mon. Wea. Rev.* 53:60–63.

Larsen, J.A., and Delavan, C.C. 1922. Climate and forest fires in Montana and northern Idaho, 1909–1919. *Mon. Wea. Rev.* 50:55–68.

Latham, D.J., and Schlieter, J.A. 1989. Ignition probabilities of wildland fuels based on simulated lightning discharges. Ogden, UT: USDA Forest Service Res. Pap. INT-411, Intermountain Research Station.

Leung, L.R., and Ghan, S.J. 1999. Pacific Northwest climate sensitivity simulated by a regional climate model driven by a GCM. Part II: $2 \times CO_2$ simulations. *J. Clim.* 12: 2031–2053.

López, R.E., and Holle, R.L. 1986. Diurnal and spatial variability of lightning activity in northeastern Colorado and central Florida during the summer. *Mon. Wea. Rev.* 114: 1288–1312.

Malanson, G.P., and Butler, D.R. 1984. Avalanche paths as fuel breaks: implications for fire management. *J. Environ. Manag.* 19:229–238.

Mantua, N.J., Hare, S.R., Zhang, Y., Wallace, J.M., and Francis, R.C. 1997. A Pacific interdecadal climate oscillation with impacts on salmon production. *Bull. Am. Meteorol. Soc.* 78:1069–1079.

Marshall, R. 1927. Influence of precipitation cycles on forestry. *J. For.* 25:415–429.

Martin, R.E. 1982. Fire history and its role in succession. In *Forest Succession and Stand Development in the Northwest*, ed. J.E. Means, pp. 92–99. Forest Science Laboratory, Oregon State University, Corvallis.

McCabe, G.J., and Wolock, D.M. 1999. General-circulation-model simulations of future snowpack in the western United States. *J. Am. Water Res. Assoc.* 35:1473–1484.

McCutchan, M.H., and Main, W.A. 1989. The relationship between mean monthly fire potential indices and monthly fire severity. *Proc. Conf. Fire Forest Meteorol.* 10: 430–435.

Meisner, B.N., Chase, R.A., McCutchan, M.H., Mees, R., Benoit, J.W., Ly, B., Albright, D., Strauss, D., and Ferryman, T. 1994. A lightning fire ignition assessment model. *Proc. Conf. Fire Forest Meteorol.* 12:172–178.

Millspaugh, S.H., Whitlock, C., and Bartlein, P.J. 2000. Variations in fire frequency and climate over the past 17,000 yr in central Yellowstone National Park. *Geology* 28: 211–214.

Minnich, R.A., Vizcaino, E.R., Sosa-Ramirez, J., and Chou, Y.-H. 1993. Lightning detection rates and wildland fire in the mountains of northern Baja California, Mexico. *Atmósfera* 6:235–253.

Mitchell, J.M. Jr., Stockton, C.W., and Meko, D.M. 1979. Evidence of a 22-year rhythm of drought in the western United States related to the Hale solar cycle since the 17th century. In *Solar-Terrestrial Influences on Weather and Climate*, eds. B.M. McCormac and T.A. Seliga, pp. 125–143. Dordrecht: Reidel.

Mitchell, V.L. 1976. The regionalization of climate in the western United States. *J. Appl. Meteorol.* 15:920–927.

Namias, J. 1982. Anatomy of Great Plains protracted heat waves (especially the 1980 U.S. summer drought). *Mon. Wea. Rev.* 110:824–838.

Nash, C.H., and Johnson, E.A. 1996. Synoptic climatology of lightning-caused forest fires in subalpine and boreal forests. *Can. J. For. Res.* 26:1859–1874.

Orville, R.E. 1994. Cloud-to-ground lightning flash characteristics in the contiguous United States: 1989–1991. *J. Geophys. Res.* 99:10833–10841.

Orville, R.E., and Huffines, G.R. 2001. Cloud-to-ground lightning in the United States: NLDN results in the first decade, 1989–98. *Mon. Wea. Rev.* 129:1179–1193.

Orville, R.E., and Silver, A.C. 1997. Lightning ground flash density in the contiguous United States: 1992–95. *Mon. Wea. Rev.* 125:631–638.

Overpeck, J.T., Rind, D., and Goldberg, R. 1990. Climate-induced changes in forest disturbance and vegetation. *Nature* 343:51–53.

Palmer, T.N., and Brankoviać, Č. 1989. The 1988 US drought linked to anomalous sea surface temperature. *Nature* 338:54–57.

Peet, R.K. 1988. Forests of the Rocky Mountains. In *North American Terrestrial Vegetation*, eds. M.G. Barbour and W.D. Billings, pp. 63–101. Cambridge: Cambridge University Press.

Petersen, K.L. 1988. *Climate and the Dolores River Anasazi*. Anthropol. Papers No. 113. Salt Lake City: University of Utah Press.

Petersen, K.L. 1994. A warm and wet Little Climatic Optimum and a cold and dry Little Ice Age in the southern Rocky Mountains, U.S.A. *Clim. Change* 26:243–269.

Potter, B.E. 1996. Atmospheric properties associated with large wildfires. Intern. *J. Wildl. Fire* 6:71–76.

Price, C., and Rind, D. 1994a. The impact of a $2 \times CO_2$ climate on lightning-caused fires. *J. Clim.* 7:1484–1494.

Price, C., and Rind, D. 1994b. Possible implications of global climate change on global lightning distributions and frequencies. *J. Geophys. Res.* 99:10823–10831.

Qu, J., and Omi, P.N. 1994. Potential impacts of global climate changes on wildfire activity in the USA. *Proc. Conf. Fire Forest Meteorol.* 12:85–92.

Reap, R.M. 1986. Evaluation of cloud-to-ground lightning data from the Western United States for the 1983–1984 summer seasons. *J. Clim. Appl. Meteorol.* 25:785–799.

Renkin, R.A., and Despain, D.G. 1992. Fuel moisture, forest type, and lightning-caused fire in Yellowstone National Park. *Can. J. For. Res.* 22:37–45.

Romme, W.H. 1982. Fire and landscape diversity in subalpine forests of Yellowstone National Park. *Ecol. Monogr.* 52:199–221.

Romme, W.H., and Despain, D.G. 1989. The long history of fire in the greater Yellowstone ecosystem. *Western Wildl.* 15(2):10–17.

Ropelewski, C.F., and Halpert, M.S. 1986. North American precipitation and temperature patterns associated with the El Niño/Southern Oscillation. *Mon. Wea. Rev.* 114: 2352–2362.

Rorig, M.L., and Ferguson, S.A. 1999. Characteristics of lightning and wildland fire ignition in the Pacific Northwest. *J. Appl. Meteorol.* 38:1565–1575.

Ryan, K.C. 1976. Forest fire hazard and risk in Colorado. M.S. thesis. Colorado State University, Fort Collins.

Ryan, K.C. 1991. Vegetation and wildland fire: Implications of global climate change. *Environ. Intern.* 17:169–178.

Savage, M., and Swetnam, T.W. 1990. Early 19th-century fire decline following sheep pasturing in a Navajo ponderosa pine forest. *Ecology* 71:2374–2378.

Schaefer, V.J. 1957. The relationship of jet streams to forest wildfires. *J. For.* 55:419–425.

Schullery, P. 1989. The fires and fire policy. *BioScience* 39:686–694.

Shindell, D., Rind, D., Balachandran, N., Lean, J., and Lonergan, P. 1999. Solar cycle variability, ozone, and climate. *Science* 284:305–308.

Shinneman, D.J., and Baker, W.L. 1997. Nonequilibrium dynamics between catastrophic disturbances and old-growth forests in ponderosa pine landscapes of the Black Hills. *Cons. Biol.* 11:1276–1288.

Simard, A.J., Haines, D.A., and Main, W.A. 1985. Relations between El Nino/Southern Oscillation anomalies and wildland fire activity in the United States. *Agric. For. Meteorol.* 36:93–104.

Small, R.T. 1957. Relationship of weather factors to rate of spread of the Robie Creek fire. *Mon. Wea. Rev.* 85:1–8.

Smith, S.R., and O'Brien, J.J. 2001. Regional snowfall distributions associated with ENSO: Implications for seasonal forecasting. *Bull. Am. Meteorol. Soc.* 82:1179–1191.

Stahle, D.W., D'Arrigo, R.D., Krusic, P.J., Cleaveland, M.K., Cook, E.R., Allan, R.J., Cole, J.E., Dunbar, R.B., Therrell, M.D., Gay, D.A., Moore, M.D., Stokes, M.A., Burns, B.T., Villanueva-Diaz, J., and Thompson, L.G. 1998. Experimental dendroclimatic reconstruction of the southern oscillation. *Bull. Am. Meteorol. Soc.* 79:2137–2152.

Steele, R., Arno, S.F., and Geier-Hayes, K. 1986. Wildfire patterns change in central Idaho's ponderosa pine-Douglas-fir forest. *W. J. Appl. For.* 1:16–18.

Stocks, B.J. 1991. The extent and impact of forest fires in northern circumpolar countries. In *Global Biomass Burning: Atmospheric, Climatic, and Biospheric Implications*, ed. J.S. Levine, pp. 199–202. Cambridge, Massachusetts: MIT Press.

Stockton, C.W., and Meko, D.M. 1975. A long-term history of drought occurrence in western United States as inferred from tree rings. *Weatherwise* (Dec):244–249.

Sturman, A.P. 1987. Thermal influences on airflow in mountainous terrain. *Progr. Phys. Geogr.* 11:183–206.

Tapia, A., Smith, J.A., and Dixon, M. 1998. Estimation of convective rainfall from lightning observations. *J. Appl. Meteorol.* 37:1497–1509.

Thomas, D.A. 1991. The Old Faithful fire run of September 7, 1988. *Proc. Conf. Fire Forest Meteorol.* 11:272–280.

Touchan, R., Swetnam, T.W., and Grissino-Mayer, H.D. 1995. Effects of livestock grazing on pre-settlement fire regimes in New Mexico. In *Proceedings: Symposium on Fire in Wilderness and Park Management*, eds. J.K. Brown, R.W. Mutch, C.W. Spoon, and R.H. Wakimoto, pp. 268–272. Ogden, UT: USDA Forest Service Gen. Tech. Rep. INT-GTR-320, Intermountain Research Station.

Trenberth, K.E., Branstator, G.W., and Arkin, P.A. 1988. Origins of the 1988 North American drought. *Science* 242:1640–1645.

Turner, M.G., Hargrove, W., Gardner, R.H., and Romme, W.H. 1994. Effects of fire on landscape heterogeneity in Yellowstone National Park, Wyoming. *J. Veg. Sci.* 5:731–742.

Turner, M.G., and Romme, W.H. 1994. Landscape dynamics in crown fire ecosystems. *Landscape Ecol.* 9:59–77.

Vaillancourt, D.A. 1995. Structural and microclimatic edge effects associated with clearcutting in a Rocky Mountain forest. M.S. thesis. University of Wyoming, Laramie.

van Loon, H., and Labitzke, K. 1988. Association between the 11-year solar cycle, the QBO, and the atmosphere. Part II: Surface and 700 mb in the northern hemisphere in winter. *J. Clim.* 1:905–920.

Veblen, T.T. 2000. Disturbance patterns in southern Rocky Mountain forests. In *Forest Fragmentation in the Southern Rocky Mountains*, eds. R.L. Knight, F.W. Smith, S.W. Buskirk, W.H. Romme, and W.L. Baker, pp. 31–54. Boulder: University Press of Colorado.

Veblen, T.T., Hadley, K.S., Nel, E.M., Kitzberger, T., Reid, M., and Villalba, R. 1994. Disturbance regime and disturbance interactions in a Rocky Mountain subalpine forest. *J. Ecol.* 82:125–135.

Veblen, T.T., and Kitzberger, T. (In press). Inter-hemispheric comparison of fire history: The Colorado Front Range, U.S.A. and the northern Patagonian Andes, Argentina. *Plant Ecol.*

Veblen, T.T., Kitzberger, T., and Donnegan, J. 1996. Fire ecology in the wildland/urban interface of Boulder County. Res. Rep. to City of Boulder Open Space by the Department of Geography, University of Colorado, Boulder.

Veblen, T.T., Kitzberger, T., and Donnegan, J. 2000. Climatic and human influences on fire regimes in ponderosa pine forests in the Colorado Front Range. *Ecol. Appl.* 10:1178–1195.

Veblen, T.T., and Lorenz, D.C. 1991. *The Colorado Front Range: A Century of Ecological Change.* Salt Lake City: University of Utah Press.

Wagner, G., Livingstone, M., Masarik, J., Muscheler, R., and Beer, J. 2001. Some results relevant to the discussion of a possible link between cosmic rays and the earth's climate. *J. Geophys. Res.* 106:3381–3387.

Watanabe, M., and Nitta, T. 1999. Decadal changes in the atmospheric circulation and associated surface climate variations in the northern hemisphere winter. *J. Clim.* 12:494–510.

Watson, A.I., Holle, R.L., and López, R.E. 1994. Cloud-to-ground lightning and upper-air patterns during bursts and breaks in the southwest monsoon. *Mon. Wea. Rev.* 122:1726–1739.

Weatherspoon, C.P., and Skinner, C.N. 1995. An assessment of factors associated with damage to tree crowns from the 1987 wildfires in northern California. *For. Sci.* 41:430–451.

Weaver, H. 1974. Effects of fire on temperate forests: Western United States. In *Fire and Ecosystems*, eds. T.T. Kozlowski and C.E. Ahlgren, pp. 279–319. New York: Academic Press.

Weidman, R.H. 1923. Relation of weather forecasts to the prediction of dangerous forest fire conditions. *Mon. Wea. Rev.* 52:563–564.

Werth, P., and Ochoa, R. 1990. The Haines index and Idaho wildfire growth. *Fire Manag. Notes* 51:9–13.

Werth, J., and Werth, P. 1998. Haines index climatology for the western United States. *Fire Manag. Notes* 58:8–17.

Woodhouse, C.A. 1993. Tree-growth response to ENSO events in the central Colorado Front Range. *Phys. Geogr.* 14:417–435.

Woodhouse, C.A. 2001. A tree-ring reconstruction of streamflow for the Colorado Front Range. *J. Am. Water Res. Assoc.* 37(3):1–9.

Woodhouse, C.A., and Brown, P.M. 2001. Tree-ring evidence for Great Plains drought. *Tree-Ring Res.* 57:89–103.

Woodhouse, C.A. and Overpeck, J.T. 1998. 2000 years of drought variability in the central United States. *Bull. Am. Meteorol. Soc.* 79:2693–2714.

Zimmerman, C.T., and Neuenschwander, L.F. 1984. Livestock grazing influences on community structure, fire intensity, and fire frequency within the Douglas-fir/ninebark habitat type. *J. Range Manag.* 37:104–110.

6. Tree-Ring Reconstructions of Fire and Climate History in the Sierra Nevada and Southwestern United States

Thomas W. Swetnam and Christopher H. Baisan

Most of the fire history research conducted in the past century has focused on case studies and local-scale assessments of pattern and process, with an emphasis on describing typical fire frequencies in forest stands and watersheds. Dominant research themes have included the characterization and analyses of fire frequencies across ranges of topographic settings and habitats. In general, these "histories" have been more about describing time-averaged processes, than elucidating the events, narratives, and contingencies of "history." Now that many crossdated fire chronologies have been developed from tree-ring analyses of fire-scarred trees, it is possible to assemble regional to global-scale networks of fire occurrence time series. These networks and time series can be used in quantitative, historical analyses that identify and separate broad-scale climate-driven patterns of fire occurrence from local, nonclimatic features of individual sites. The seasonal to annual resolution of tree rings facilitates historical fire climatology because the high temporal resolution of these data allows us to connect multiple events in space and time. The importance of climatic influence is reflected in the degree of synchrony in specific fire events and decadal to centennial trends among widely distributed sites (Swetnam and Betancourt 1990, 1998; Swetnam 1993; Veblen et al. 1999; Veblen, Kitzberger, and Donnegan 2000; Grissino-Mayer and Swetnam 1997, 2000; Heyerdahl, Brubaker, and Agee 2001, in press; Kitzberger and Veblen 1998; Kitzberger, Veblen, and Villalba 1997; Kitzberger, Swetnam, and Veblen 2001; Brown, Kaufmann, and Shepperd 1999; Brown et al. 2001; Allen 2002).

Synchrony of events across space is a fundamental principle of dendrochronology and is the basis of tree-ring dating and the identification of broad-scale environmental patterns in tree rings (Douglass 1941; Fritts and Swetnam 1989). Patterns of wide and narrow rings, for example, are highly correlated among precipitation-sensitive trees growing in arid and semi-arid regions. Significant correlations ($p < 0.05$) of standardized ring-width series extend up to 1100 km between trees and sites in the western United States (Cropper and Fritts 1982; Meko et al. 1993). The reason for these positive correlations is that broad-scale drought and wet years have acted to synchronize the relative changes in tree-ring growth of moisture-limited conifers over large geographic areas (LaMarche and Fritts 1971; Fritts 1976, 1991). Local weather and nonclimatic variations result in unique variations in tree growth at individual sites. However, by combining numerous ring-width chronologies from broad areas, the site-specific variations are averaged out, while the common climatic signals are concentrated in mean value functions, or amplitude series from principal components analysis (Fritts 1976). It is from these composite, regional tree-ring networks that climatic history is most effectively reconstructed (e.g., Fritts 1976, 1991; Meko et al. 1993; Cook et al. 1999).

The short- and long-term climatic fluctuations that have importantly affected tree growth at local to global scales have also affected fire regimes. The common link of climatic influence on tree-ring growth and forest fuels (quantity and moisture content) provides the basis for fire-climate research in dendrochronology. In this chapter we illustrate our key findings regarding climatic controls of past fire regimes in the southwestern United States and Sierra Nevada of California. Following a description of tree-ring sampling strategies and methods of fire chronology development, we illustrate with a set of examples how fire-scar networks can be used to identity fire-climate associations across a broad range of spatial scales. Of particular importance is the finding that annual resolution fire-scar networks can provide an independent indicator of changing temporal patterns of globally important climatic processes, such as of the El Niño–Southern Oscillation.

Fire-Scar Chronologies

Fire-scar chronologies were reconstructed in forest stands throughout Arizona and New Mexico, and on the west slope of the Sierra Nevada (hereafter, these regions are referred to as the "Southwest" and the "Sierras," respectively). Many of these chronologies were developed through cooperative studies with land management agencies in national forest and national park wilderness and protected areas. Presence of living or dead fire-scarred trees was obviously necessary for reconstructing fire-scar based fire history, but sample areas included a broad range of abundance of fire-scarred trees. Concerns over impacts and aesthetics, and limited access sometimes required opportunistic sampling near roads or trails. Study areas and stands to be sampled were often located in areas where prescribed fire and forest restoration efforts were underway or planned. Some collection sites were selected as areas that were judged to have vegetation and topographic characteristics that were representative of broader areas within the management units. Other collections were obtained along natural fire spread corridors, such as along coniferous canyon bottoms linking grasslands to uplands, with the explicit purpose of evaluating landscape-scale linkages and processes (e.g., Kaib et al. 1996; Kaib 1998; Barton, Swetnam, and Baisan 2001).

Given the constraints listed above, the selection of study areas and trees was necessarily nonrandom and largely subjective, so the fire frequency estimates and other aspects of the reconstructed fire regimes may not be fully representative of larger surrounding areas. Potential biases due to nonrandom sampling and problems with fire frequency analysis methods have been highlighted in recent critiques of tree-ring based fire histories (e.g., Johnson and Gutsell 1994; Baker and Ehle 2001). The scope and context of this chapter does not allow a detailed and direct response to these critiques. In general, most of the critiques involve problems in estimating fire interval distributions (i.e., fire frequency analyses) and are only indirectly relevant to our focus on the historical aspects of past fire regimes. In subsequent sections we will show that notwithstanding possible biases and limitations of the fire-scar record, well-replicated fire-scar chronologies can provide complete inventories of widespread fire events within sites, and useful indices of local to regional fire activity.

Fire-Scarred Tree Selection

Our sampling strategy was to maximize the completeness of an inventory of fire dates within study sites over as a long a time period as possible, while also collecting samples that were spatially dispersed throughout the sites. We located fire-scar specimens within sites by systematically searching throughout forest stands. Site (or forest stand) boundaries were usually delineated by cliffs, rock outcrops, scree slopes, canyon bottoms, and ridgelines. During searches we carefully examined every living tree, log, and snag with a fire scar that was observed along walking traverses throughout the site. We sampled trees with maximum numbers of well-preserved fire scars that were broadly distributed throughout the sites.

We have often collected multiple clusters of fire-scarred trees (2–5 trees) in relatively small areas (i.e., 1–5 ha) within stands. These clusters can sometimes be useful for estimating small area (point) fire frequencies by compositing the fire dates from the cluster (e.g., Kilgore and Taylor 1979; Baisan and Swetnam 1990; Brown and Swetnam 1994). Site (or stand) chronologies typically include a minimum of 10 fire-scarred trees, and encompass areas of about 10 to 100 ha. Some of our collections were from many clusters of trees along elevational transects and/or within medium to large watersheds (1000–10,000 ha). In a few cases our collections included 50 to 100 (or more) fire-scarred trees widely dispersed across entire mountain ranges or large landscapes (20,000–>50,000 ha) (e.g., see Baisan and Swetnam 1990, 1997; Caprio and Swetnam 1995; Grissino-Mayer and Swetnam 1997). More details about our collections and study sites, including summaries of fire interval statistics, can be found in Swetnam and Baisan (1996), Swetnam, Baisan, and Kaib (2001), and the many fire history papers in the Reference section.

Composite Chronologies, Filtering, and Sample Size Effects

Fire chronologies were composited (sensu Dieterich 1980, 1983) at different spatial scales to evaluate fire regime changes (e.g., Baisan and Swetnam 1990; Grissino-Mayer and Swetnam 1997; Brown and Sieg 1996, 1999; Brown, Kaufmann, and Shepperd 1999; Brown et al. 2001) (Fig. 6.1). One of the ways we have assessed fire regime variations is by "filtering" methods, whereby minimum numbers or percentages of trees scarred are used to sort and describe fire event and interval data (e.g., Swetnam and Baisan 1996; Swetnam, Baisan, and Kaib 2001). These filters helped identify fires that were probably more or less extensive within sites in a relative sense. Filtering also helped identify fire frequency estimates that were less affected by sample size (described below). Fire-scar data compilation, sorting, statistical analyses, and graphical presentation were greatly facilitated by Henri Grissino-Mayer's development of the FHX2 software (Grissino-Mayer 1995, 1999, 2001, and see *http://web.utk.edu/~grissino/fhx2.htm*). Using the FHX2 program, different minimum numbers and/or percentages of trees scarred per fire can be defined and used as a coarse filter for computing fire interval statistics for fires of different relative spatial extent within or between stands.

In using filtering approaches, our aim was to reasonably identify and classify fire events that were probably more or less widespread, while recognizing that fire-scar data analyzed in this manner provide relative (versus absolute) estimates of fire frequency and extent. In assessments of the degree and pattern of synchrony of fire events within sites, we commonly used filters of a minimum of two trees scarred per fire, and/or 10% and 25% of trees recording fires per year. Although particular fire event filters (e.g., 10% or 25%) may be arbitrary, such a priori selection of threshold quantities for testing, classifying, and sorting data is a widely accepted statistical practice (e.g., the use of specific confidence intervals, or percentile thresholds in statistical description and hypothesis testing). Use of a priori filtering thresholds also facilitates comparisons among sites because filtered fire frequencies are less affected by sample size (see below).

One of the concerns in fire history sampling is the effect of study area size, and number of fire-scarred trees sampled, on fire frequency estimates (Arno and Peterson 1983; Swetnam and Baisan 1996; Baker and Ehle 2001). As study areas increase in size the chances of encompassing additional past fire perimeters increases. Likewise, as more fire-scarred trees are sampled and included in composites, there is an increased chance of detecting additional fires that burned in previously unsampled areas, or only in small areas. The effects of changing sample size and the completeness of the inventory of fire dates within sites or study areas can be assessed in a manner that is similar to the use of species–area

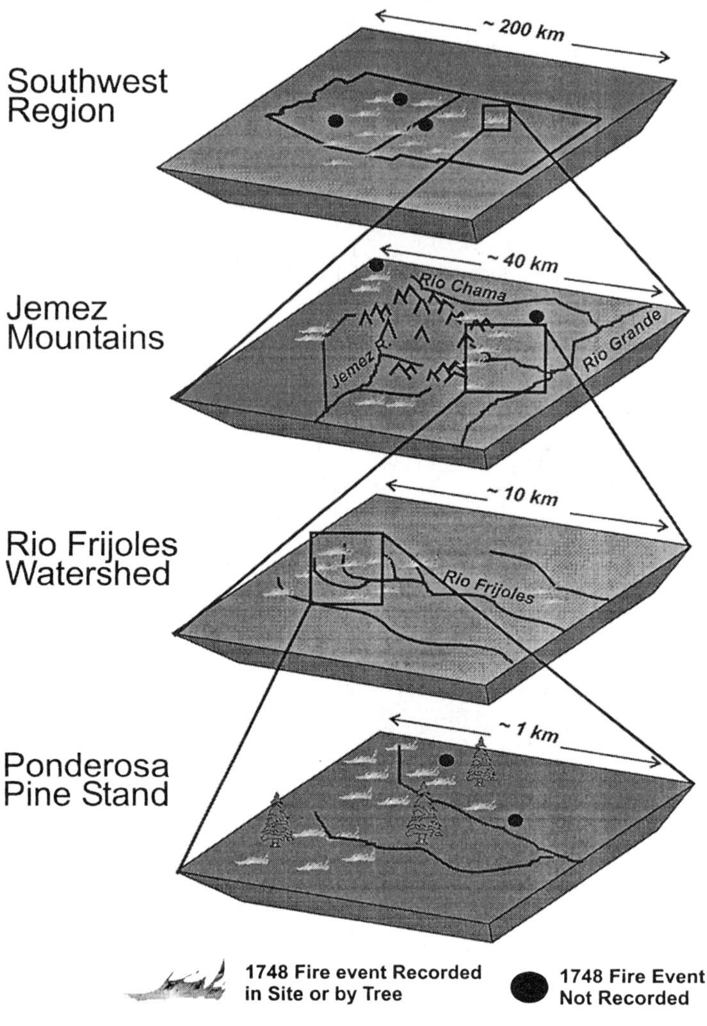

Figure 6.1. Different spatial scales of analyses in fire histories are illustrated in this hierarchical set of maps. The fire year 1748 was the most synchronous fire year in the southwestern fire-scar network, and is shown schematically as an example of cross-scale synchrony. Synchrony of fire dates between trees and nearby stands can be reasonably inferred to indicate fires that spread between sample points, although unburned areas between points, and separate fire ignitions are acknowledged possibilities. Synchrony of fire dates among stands, watersheds, and mountain ranges separated by great distances or barriers to fire spread is most probably caused by climatic entrainment of fire occurrence.

curves by botanists for assessing the completeness of inventories of plant species diversity (e.g., Colwell and Coddington 1994; Rosenzweig 1995).

We address here the issue of completeness of our fire-scar chronologies because this is relevant to our interpretations that we were able to detect widespread fires within and between sites, and that these relatively extensive fire events were associated with climatic variations. In our example, fire frequencies (fires/century) at different sample sizes were re-computed for a fixed time period and study site using randomly selected sub-sets of the sampled trees (Fig. 6.2). Re-sampling (bootstrap) methods were used to estimate the confidence intervals of the mean fire intervals recomputed at different sample sizes (Mooney and Duval 1993).

As expected, a general pattern that we commonly observed in these assessments was that fire frequencies tended to increase as more trees were added to the collection. However, when we applied the least restrictive filter of fire dates—namely the inclusion of only those fire dates recorded by two or more trees—the fire frequency estimates were typically asymptotic as a function of sample size (Fig. 6.2). This result suggests that single-tree fire-scar dates were probably representing relatively localized, small fires that occurred around those single trees. As sample size increased more of these small fires were detected, and so fire frequency continuously increased.

Presumably, with additional samples from an area of fixed size the fire frequency should eventually stabilize. If the area was large enough, as more samples were collected fire frequency would eventually reach the maximum possible frequency of one fire a year (i.e., all years with fire-scar dates). However, at the spatial scale of most of our sample areas (10–1000 ha), surface fires recorded by two or more fire-scarred trees probably represented relatively widespread fires that exposed many trees to re-scarring. Hence, when only these fire events were included, the fire frequencies tended to stabilize after a certain number of trees were sampled. In application of this kind of assessment to many of our fire-scar chronologies, we have found that in sites of less than approximately 100 ha, 10 to 15 trees were usually sufficient to reach fire frequency asymptotes using the 2-tree minimum filter. In large sample areas (1000–10 000+ ha) asymptotes were usually not achieved with the 2-tree minimum filter but often were achieved with more restrictive filters (e.g., 25% or more trees scarred per fire, unpublished data).

The main interpretation from these analyses was that most of our fire-scar chronologies were complete, or nearly complete, inventories of relatively widespread fires that occurred within the sampled areas. Frequencies of fires of any size, occurring anywhere within the study sites, however, were probably underestimated because many small fires were probably not picked up by fire-scar sample sets of these sizes.

An important point to bear in mind is that mean fire intervals (i.e., the inverse of fire frequency) estimated from composite fire-scar chronologies should not be interpreted to indicate that every square meter burned within the study area, on average, at those intervals. Even in the case of mean fire intervals computed using

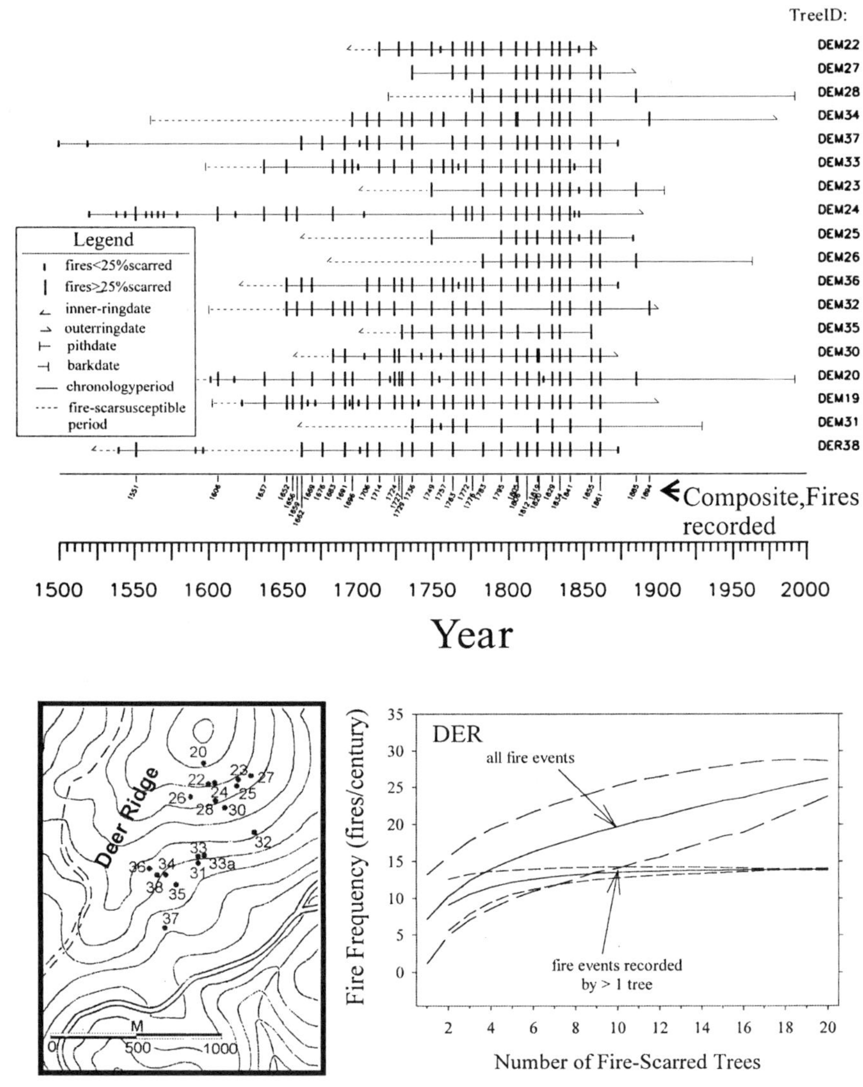

Figure 6.2. Example of a fire-scar chronology from a forest stand in the Sierra Nevada, California (Deer Ridge, Mountain Home State Forest, upper graph). Time spans of specimens from individual fire-scarred trees are shown by the horizontal lines, and the fire dates are indicated by vertical tick marks. The map (*lower left*) shows the spatial distribution and extent of this site (note that only the specimens from the central clusters of this site are included in the master fire chronology chart). The graph on the lower right illustrates the fire frequency in this stand computed as a function of sample size. The mean fire frequencies (*solid lines*) were computed from random inclusion (1000 re-samplings) of subsets of the 18 fire-scarred trees for each sample size. The time period used was 1700 to 1900 because most trees were recording fires during this period. The 95% confidence limits (*dashed lines*) of the computed fire frequencies were estimated from the mean and variance of the re-sampled sets at each sample size.

the more restrictive filters (e.g., 10% or 25%), and thereby inferring that these were intervals between relatively widespread fires, this does not imply that no areas were unburned within the sampled areas during those fire events.

In general, it is our view that fire historians have tended to overemphasize fire frequency analyses (i.e., description and testing of different fire interval distributions) as the primary goal of fire history research. Statistical descriptions and tests of fire interval distributions are inherently limited in objectivity, resolution, and reliability. One reason for this is that selection of an appropriate study area extent or time period to analyze, which very importantly affect interval distributions, will always be subjective or arbitrary at some level (Millar and Woolfenden 1999). Improved sampling methods can only go so far in estimating or correcting for biases and peculiarities in the paleorecord, which by its nature is fragmentary and preserved by only partially understood biological and physical processes (Swetnam, Allen, and Betancourt 1999).

Rather than focusing so exclusively on statistical analysis of fire interval distributions, we think that historical approaches are likely to be equally or more reliable and informative about the drivers of past fire patterns and processes, such as humans and climatic variations. Powerful explanations and understanding can be derived from the discovery of specific historical events, trends, contingencies, and patterns. These historical processes are often obscured in time-averaged summaries, statistically fitted models, and estimates of central tendency. Reasonable and convincing explanations often derive from relatively straightforward graphical assessments of the temporal-spatial patterns of event synchrony. Such patterns are often evident in fire-scar chronology composites, especially when compared with independent historical records of climate and land-use history. Statistical detection and testing of visually evident historical changes and linkages are also possible using methods such as contingency, correlation, and superposed epoch analyses. These kinds of graphical and statistical analyses emphasize the unique, historical nature of fire regimes, rather than just the time and space averaged view emphasized in fire frequency (fire interval) analyses.

Examples of Mountain Range-Scale Fire Chronologies and Historical Interpretations

Master fire chronologies from two mountain ranges in the southwestern United States illustrate the value of examining historical patterns, rather than just the time and space-averaged aspects of fire regimes (Figs. 6.3 and 6.4). The two mountain ranges are the Mogollon Mountains in the Gila Wilderness, New Mexico, and the Santa Catalina Mountains near Tucson, Arizona. Stands were sampled along elevational transects in both mountain ranges. The tree rings and fire scars in these samples were dated and composited using techniques described in detail elsewhere (Dieterich 1980; Dieterich and Swetnam 1984; Swetnam and Dieterich 1985; Baisan and Swetnam 1990; Swetnam and Baisan 1996; Abolt 1997; Swetnam, Baisan, and Kaib 2001).

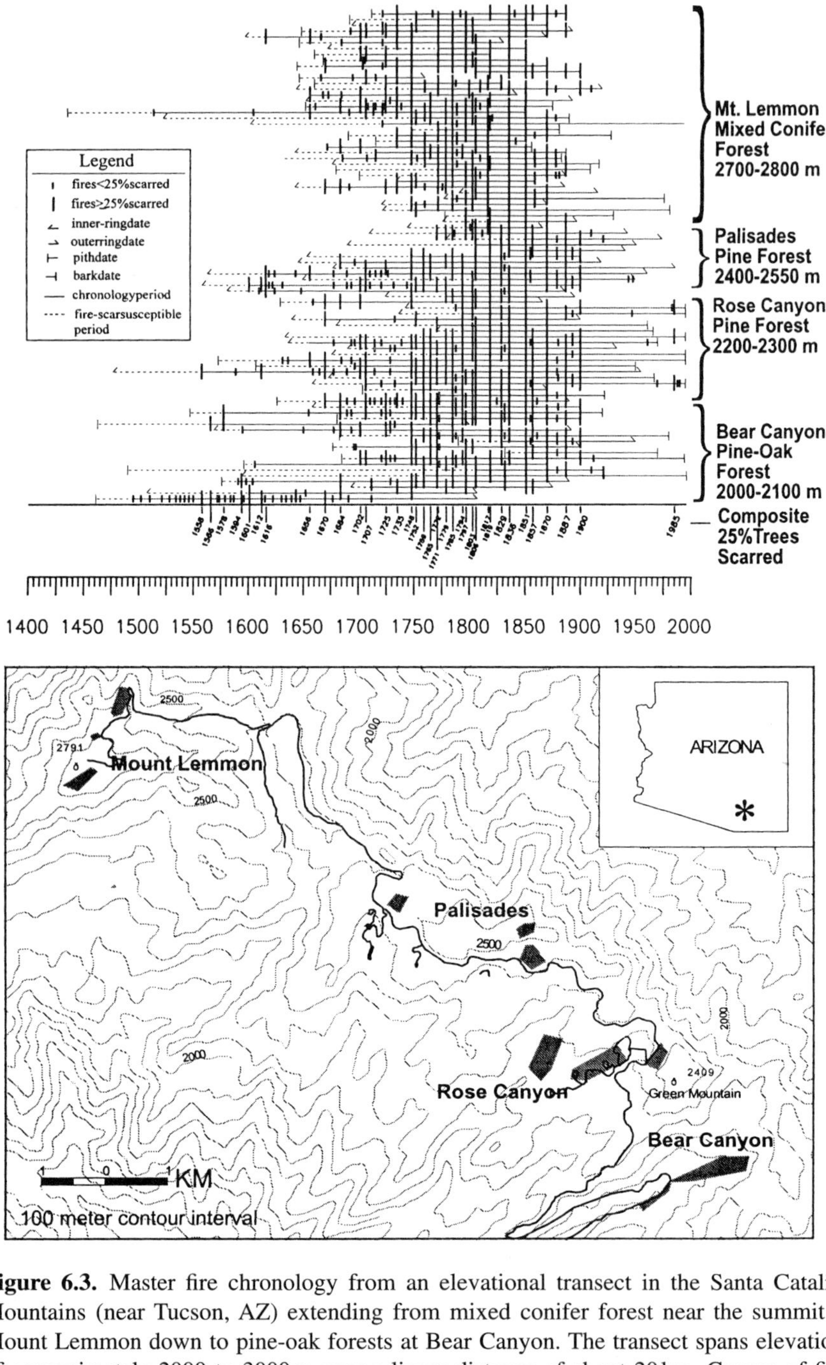

Figure 6.3. Master fire chronology from an elevational transect in the Santa Catalina Mountains (near Tucson, AZ) extending from mixed conifer forest near the summit of Mount Lemmon down to pine-oak forests at Bear Canyon. The transect spans elevations of approximately 2000 to 3000 m over a linear distance of about 20 km. Groups of fire-scarred trees sampled in sites (stands) are indicated by brackets and site names on the right. Note the high degree of synchrony of a subset of the fire dates across the elevational gradient; this is compelling evidence that widespread fires occurred during those synchronous fire years.

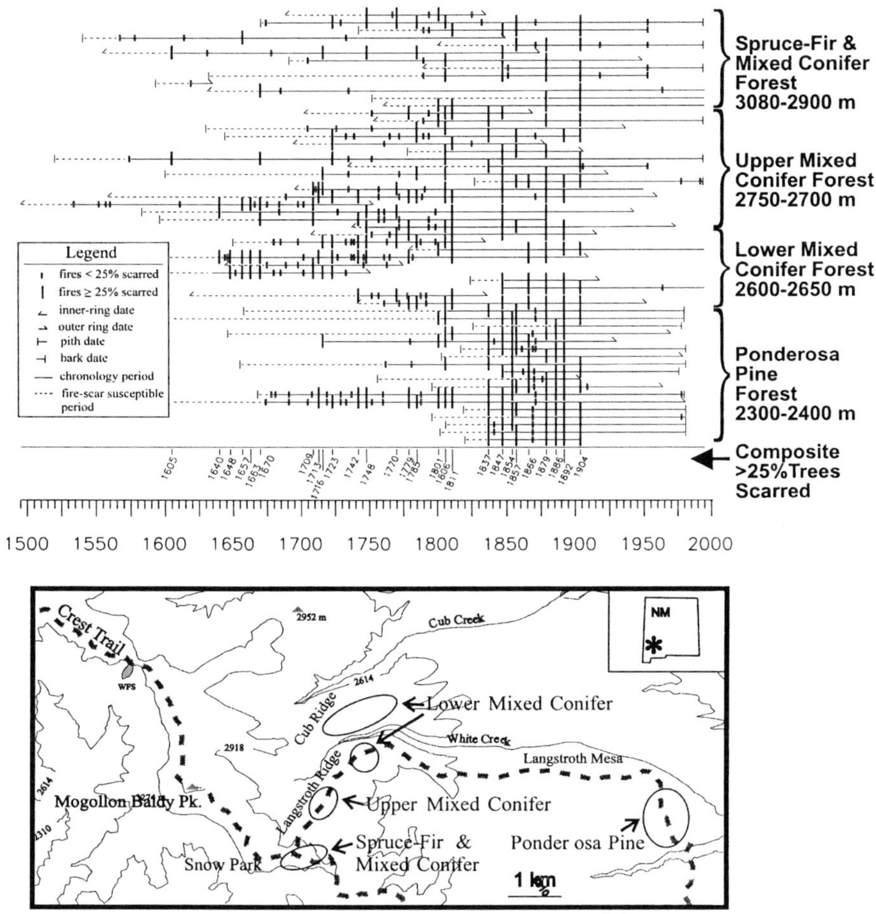

Figure 6.4. Master fire chronology from an elevational transect in the Mogollon Mountains (Gila Wilderness, NM) extending from spruce fir forest near the summit of Mogollon Baldy and down to ponderosa pine forests on Langstroth Mesa (see map at bottom). The transect spans elevations of approximately 2300 to 3080 m over a linear distance of about 15 km. Groups of fire-scarred trees sampled in sites (stands) are indicated by brackets and site names on the right. Note the apparent change in fire frequency and synchrony ca. 1800, and also in Figure 6.3.

Composite stand and transect chronologies show several common patterns in fire histories of pine and mixed-conifer forests in the Southwest and Sierras. One of the most obvious patterns is a striking change in fire frequency in the late nineteenth or early twentieth centuries (Figs. 6.3 and 6.4). This reduction in fire occurrence coincides in almost all cases to within a few years of the first documented introduction of large numbers of domestic livestock (sheep, goats, cattle, or horses). The great ranching boom of the late nineteenth century, for example, led to sheep or cattle introduction to some mountain areas as early as the 1870s, and was delayed in other more remote mountain ranges until after around 1900. The timing of the decline of frequent fires as recorded by the fire scars closely reflects these historic land use differences (see Swetnam, Baisan, and Kaib 2001 for specific examples). In general, the livestock introduction and coincident reduction in fire occurrence preceded by a decade or more the advent of organized and systematic fire suppression by government agencies. In most places limited fire fighting by a few government agents began about 1905 to 1910. Organized fire fighting was probably not very effective in many areas until increased numbers of fire fighters, lookout towers, and equipment (e.g., aircraft) became available after the 1930s or 1940s (Pyne 1982; Swetnam, Baisan, and Kaib 2001; Rollins, Swetnam, and Morgan 2001).

In the southwest, frequent fires were typically interrupted between about 1870 and 1900. Figures 3 and 4 show examples of disrupted fire regimes around 1900; see Swetnam, Baisan, and Kaib (2001) for examples of variable fire regime disruption dates from the 1870s to 1900s in southern Arizona and New Mexico. Exceptions were places where earlier introduction of livestock (especially sheep) by Hispanic or Navajo herders occurred (i.e., early nineteenth, eighteenth, or seventeenth centuries, depending on location), as documented with independent archival records (Savage and Swetnam 1990; Touchan, Allen, and Swetnam 1996; Baisan and Swetnam 1997). Other exceptions were uninterrupted fire regimes in locations where intensive livestock grazing did not occur because of topographic barriers, such as impassable lava flows (Grissino-Mayer and Swetnam 1997).

Fire regimes were not disrupted until the midtwentieth century (i.e., 1940s and 1950s) in the remote, rugged mountains of northern Mexico where permanent water or roads needed for intensive livestock and human uses were lacking. These late disruptions coincide with the "ejido reforms" of the 1940s, after which there was an increase in numbers of roads, water tank development, livestock grazing, and logging in some areas (Fulé and Covington 1997, 1999; Fulé, Covington, and Moore 1997; Kaib 1998; Swetnam, Baisan, and Kaib 2001; Heyerdahl and Alvarado, Chapter 7, this volume). These exceptions essentially prove the rule: intensive livestock grazing and associated human land uses were the initial causes of fire regime disruption in most areas of the greater Southwest. Continued absence of widespread, frequent surface fire in the mid to late twentieth century (at least on the U.S. side of the border) was probably due to a combination of livestock grazing and organized, increasingly effective fire suppression efforts by government agencies.

Climate change is an unlikely explanation for the late nineteenth- to early twentieth-century fire regime disruptions. This is because (1) the disruptions were typically asynchronous between mountain ranges that shared similar regional climate patterns, (2) droughts and wet periods during this era (i.e., 1870s–1910s) do not consistently coincide with the disruptions, whereas the dates of livestock introductions generally do coincide, (3) portions of some remote mountains in Sonora, Mexico, that were not heavily grazed continued to burn throughout the twentieth century, despite having very similar climate as nearby mountain ranges on the U.S. side where grazing occurred and frequent fire regimes were disrupted (Swetnam, Baisan, and Kaib 2001).

The frequent surface fire regimes of mid-elevation forests (2000 to 3000 m) in the Sierras were typically disrupted earlier than in most southwestern sites. The last widespread fire in our sites on the west slope of the Sierras occurred between about 1850 and 1870 (Fig. 6.2, and see Caprio and Swetnam 1995). This corresponds with movement of large sheep herds into the Sierras during and following a severe drought in the early 1860s, which forced sheepherders in the Central Valley to seek forage in the high mountain meadows (Vankat 1977). This intensive grazing led to denudation of large tracts of formerly grassy areas in the high Sierras by the 1870s, as decried by John Muir; he called these sheep herds "hooved locusts" (Muir 1911).

Native Americans and High-Frequency Fire Regimes

The decline of frequent fire regimes in the Southwest and elsewhere has sometimes been attributed to the forced removal of Native Americans from these landscapes during the nineteenth century and earlier (Pyne 1982, 1985). Drawing primarily from written historical documents, and interviews of Native Americans during the twentieth century, some cultural and environmental historians argue that human manipulation of vegetation with fire was ubiquitous for many millennia before the arrival of Europeans (e.g., Dobyns 1978; Pyne 1982, 1985; Denevan 1992; Anderson 1996). A general conclusion is that humans were the dominant and overriding influence on fire regimes. "Natural" (nonhuman) factors, such as climate and lightning variability, are also acknowledged as important drivers of past fire regimes but are typically considered to be of secondary importance, or as merely complementary to the human drivers.

Although the written histories that the cultural historians depend on is extensive, alternative views on the universality of human dominance of past fire regimes, particularly for the western United States, have been presented (e.g., Vale 1998; Vale 2002). One of the chief points made in recent papers is that lightning was a more frequent and dominant cause of fires in western U.S. landscapes than was appreciated by almost all nineteenth- and early twentieth-century observers (e.g., Allen 2002; Baker 2002). It is only in the past couple of decades that with the new lightning detection technologies, comprehensive maps have become available showing millions of lightning strikes per year over regions the size of individual western states (e.g., Gosz et al. 1995). In a recent study of detected lightning fires during the twentieth century, we have found rates of ignition in southern Arizona mountains as high as two fires per km^2/y (unpublished data). A lack of knowledge of the very high rates of fire ignitions by lightning in some western forests, combined with anti-Indian biases in the nineteenth century and earlier, probably led to erroneous attribution of some fires to Native Americans, while under estimates of the importance of lightning as causes of forest fires (Allen 2002; Baker 2002).

Based on our research in the Southwest and Sierras, we conclude that Native American control of past fire regimes was very time and place specific, and cannot be broadly generalized as ubiquitous or dominant in all places and times. Fire regimes in large portions of these regions would probably have had similar characteristics (fire frequency, seasonality, extent, etc.) if people had never entered the Americas. It is clear, however, that people profoundly affected fire regimes in particular places and times. For example, in a study of more than 200 fire-related quotations in Spanish, Mexican, and American archival documents (relevant to the Southwest) extending back to the seventeenth century, Kaib (1998) found that more than 70% were in the context of warfare with the Apache people of southern Arizona and New Mexico. Intentional burning of large areas was very rare, except during times of warfare. The use of fire against enemies was a common practice used by all sides—Apache, Spaniard, Mexican, and American soldiers. Combatants burned particular places (campsites, livestock watering and grazing areas, etc.) during conflicts, but intentional burning of broader areas was only rarely mentioned in the documentary sources. The general picture was one of great temporal and spatial variability and specificity in the firing of landscapes during warfare.

This emphasis on the time and place specific influence of Native Americans on past regimes in the Southwest is supported by tree-ring studies. For example, a tree-ring study of eighteenth- and nineteenth-century fire history in several mountain ranges of southern Arizona and northern Mexico revealed that fire frequency generally tracked the occurrences of peacetime and wartime (Kaib et al. 1996; Kaib 1998). Based on place name references in archival documents, it was evident that some of the sampled stands were located near historic campsites or travel routes. Highest fire frequencies occurred during periods of maximal conflict among all sides, while reduced fire frequencies occurred when truces with Apaches were in effect. Other fire-scar studies in the Chiricahua Mountains of Arizona (Seklecki et al. 1996) and the Organ Mountains of New Mexico (Morino 1996) also found evidence of changing fire frequencies and seasonal timing that were speculated to be related to presence or absence of Apaches. Again, these study sites were located in specific areas where independent documentary sources indicate historical usage by Apaches.

In a detailed case study in the Sacramento Mountains, Kaye and Swetnam (1999) used independent documentary records and tree-ring dates of "culturally modified trees" to pinpoint the presence of Apaches in both time and place. In this study the culturally modified trees were "peeled" ponderosa pines that the Apaches had used as a food source by peeling the bark and cambium layer from a section of the lower bole (Swetnam 1984). The soft cambium provided carbohydrate and other nutrients (Martorano 1981) and was probably used primarily as an emergency food source (Swetnam 1984). Tree-ring dates from the peelings, and documented dates of skirmishes between Apaches and soldiers within and near the study area, were used to assess frequency and season of fires during known occupation periods versus other times. We also assessed regional climatic associations with fire dates and fire frequency trends.

We found that Apaches may have increased fire frequencies during some periods, and altered the seasonal timing of a few fires. Overall, however, the results were equivocal. Even in this unique case study, where detailed independent sources of temporal and spatial evidence were available to assess possible Native American influence on past fire regimes, it was not possible to strongly conclude that they significantly altered the character of fire regimes from what would have prevailed with lightning alone as an ignition source.

A broader-scale study of fire histories within the Sacramento Mountains, including the chronologies used by Kaye and Swetnam (1999), confirmed that climatic variations (drought/wet years) were dominant controls of past fire regime variations at the landscape scale (Brown et al. 2001). Again, the most significant and demonstrable effect of humans on past fire regimes was the disruption of frequent, widespread surface fires in the late nineteenth and early twentieth centuries when large numbers of livestock were introduced, and organized fire suppression began.

Twentieth-Century Verification of Fire Events

A common observation in fire chronologies from the Southwest and Sierras are a few scattered fire-scar dates in the twentieth century (Figs. 6.3, 6.4, and 6.5). There is usually a good correspondence of these dates with known twentieth-century fires in these areas. For example, almost all fires greater than 10 acres (4 ha) documented in fire atlases maintained by the U.S. Forest Service for the portion of our elevation transect in Gila Wilderness (Rollins, Swetnam, and Morgan 2001) were confirmed by the fire-scar dates from these areas (Fig. 6.4). In fact the particular trees that recorded fires corresponded well with the mapped perimeters of these fires. For example, a 1953 wildfire is know to have burned only within the area in the uppermost site, whereas a 1978 "prescribed natural fire" burned only with the areas of the lowermost site (Fig. 6.4) (Abolt 1997). The widespread 1904 fire in this chronology was referred to in both old Forest Service records and the local newspaper, with very specific place names that locates this fire within our study sites (Abolt 1997).

In the Santa Catalina Mountains of Arizona the last widespread fire in 1900 along our sampled transect was described and photographed by government surveyors who fought this low-intensity surface fire (Swetnam, Baisan, and Kaib 2001). This fire was clearly recorded as an extensive fire-scar event along the 20-km transect (Fig. 6.3). The 1985 fire was also documented in this network of site chronologies as occurring only within the Rose Canyon site (Fig. 6.3). Verification of dozens of other fire-scar dates, through references in documents or mapped fire perimeters in fire atlases, provides a high degree of confidence to our interpretation that fire-scar collections were generally complete and accurate recorders of past fires (for additional examples, see Dieterich and Swetnam 1984; Swetnam and Dieterich 1985; Baisan and Swetnam 1990; Caprio and Swetnam 1995; Swetnam, Baisan, and Kaib 2001).

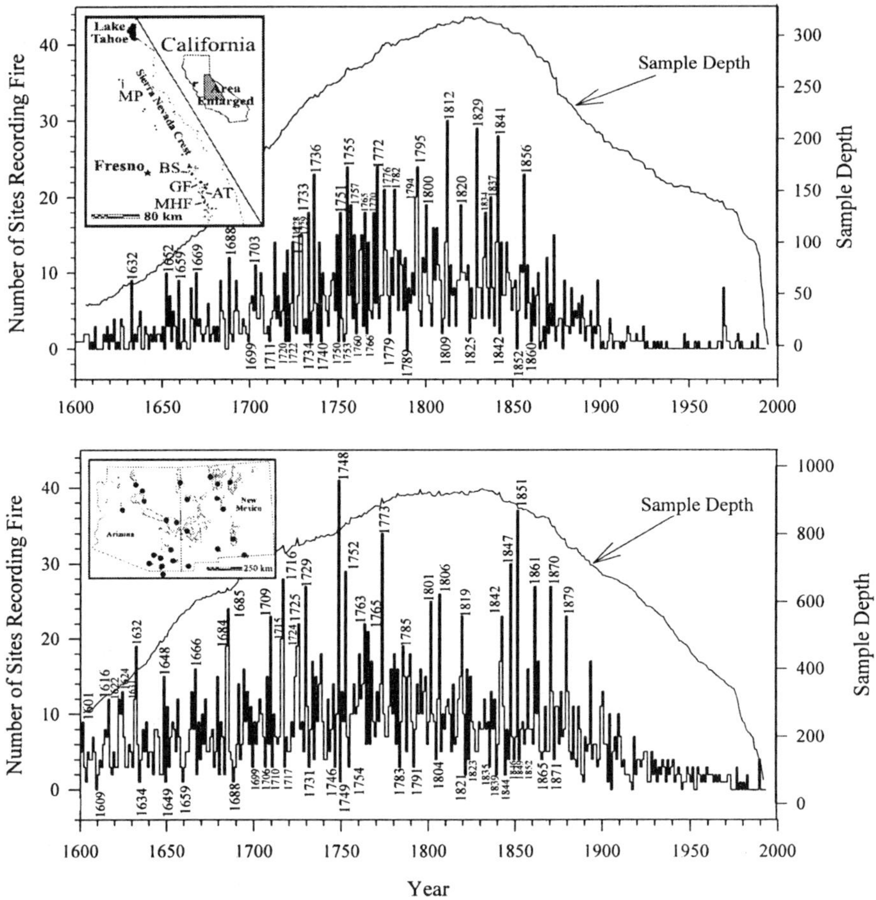

Figure 6.5. Composite time series of fire events in the Sierra Nevada (upper graph) and Southwest (lower graph) from regional networks of fire-scar chronologies. Number of sites recording fire each year are shown (AD 1600–1995). The number of fire-scarred trees included in the data sets during each year (sample depth) are also shown. The map insert of the Sierras shows locations of the five giant sequoia groves (letter codes). Small irregular dots show approximate range of sequoia groves. The 49 sites from the Sierras included in the composite are from four elevational transects adjacent to the Mariposa Grove (MP), the Big Stump Grove (BS), Giant Forest (GF), and Mountain Home State Forest (MHF). The map insert of the Southwest shows 26 mountain ranges (as dots) where the 63 sites included in the composite are located. The irregular outline on this map is the approximate range of ponderosa pine in Arizona and New Mexico.

Synchrony Within Stands, Watersheds, and Mountain Ranges

An outstanding feature of many fire-scar chronologies in the Southwest and Sierras is a high degree of synchrony of fire-scar dates among trees across a broad range of spatial scales, from stands to regions. The high degree of synchrony of some fires over linear distances of more than 10km and elevation gradients of 1000 to 2000m (Figs. 6.3 and 6.4) leads to a simple and logical interpretation: relatively large areas burned within these study areas during these synchronous years.

It is likely that some of these synchronous events represent separately ignited fires that did not coalesce into contiguous burned areas. It is also very likely that some unburned areas existed between sampled trees and sites along these transects and within the surrounding areas. Despite these considerations our basic interpretation is still reasonable, that relatively greater areas probably burned during the highly synchronous years than during less synchronous years (i.e., fire years recorded by a single tree or a few trees; Figs. 6.2, 6.3, and 6.4). It is also very likely that many pre-1900 fires burned over very large areas because lightning ignitions occur as early as April in some years in the Southwest, and fires are known to have burned for weeks to months. Nineteenth-century newspapers, for example, reported that wildfires burned for long periods of time and achieved enormous sizes; some fires exceeded 500,000ha (Bahre 1985).

The synchrony of multiple tree and site fire events is often statistically significant ($p < 0.05$) across a range of spatial scales. For example, contingency analysis of the fire dates common to 3, 4, or 5 sampled giant sequoia groves over the past 1300 years showed that the odds of obtaining this observed degree of synchrony of events by chance was less than 1 in 1000 (Swetnam 1993). In general, we have interpreted significant synchrony of fire dates among trees within stands to be indicative of widespread fire at this scale. Synchrony among widely scattered sites—especially where effective fire barriers or distance separate the sites (as in the giant sequoia example)—is indicative of regional climatic influence on fire occurrence (e.g., Swetnam and Betancourt 1990, 1992, 1998; Grissino-Mayer and Swetnam 2000; Swetnam and Baisan 1996; Kaib et al. 1996).

Fire Drought Patterns in the Southwest and Sierras

Regional Composites and Synchronous Fire Years

The regional networks of fire-scar chronolgies we have assembled are from 63 sites in 26 mountain ranges in the Southwest, and 49 sites from four elevational transects on the west slope of the Sierras. Our Sierran collections include five giant sequoia fire-scar chronologies, which will be described separately. The influence of interannual climatic variation is evident as years when many sites (and trees) have recorded fires during particular years, and as years when no, or few sites (and trees) have recorded fire events (Fig. 6.5). The interpretation of climate as the primary driver of this synchrony is reasonable because there is no other known factor that operates at these spatial and temporal scales that could result in such a high degree of year-to-year synchrony. Also, as will be demonstrated below, these synchronous dates are statistically associated with independent records of interannual wet and dry conditions.

The synchrony is visually obvious (Figs. 6.3, 6.4, and 6.5), but it is reasonable to ask: Is the degree of observed synchrony statistically significant? Specifically, if this number of independent, random time series were combined could the observed synchrony among the series have occurred purely by chance? The statistical strength of the observed synchrony is illustrated by a contingency calculation. Fire frequency within 63 individual sites in the Southwest averaged about one fire per 7.5 years from 1700 to 1900. Using this average fire frequency and simple binomial joint probability calculations, strictly by chance we would expect about one coincidence of the same fire date in 21 of the 63 sites (one-third) in about a 35,000-year period. Yet 15 different years met or exceeded this criterion in the 201-year period (Fig. 6.5). The probability of 41 of 63 sites recording the same fire date by chance, as in 1748, is vanishingly small. These probability calculations oversimplify the contingency of fire events among multiple sites because the fire interval distributions and probabilities are not necessarily binomial; they are different from site to site, and they change through time. Nevertheless, these probability estimates indicate that it is highly likely that our general conclusion is robust: the degree of synchrony observed is much greater than one would expect to occur by chance.

The relative, year-to-year strength of the synchrony is difficult to assess directly because the regional time series contains trends that are in part due to the sample depth (number of fire-scarred trees that were alive and recording fire-scar dates each year). Some of these trends, however, are probably related to climatic variability. An example of a decade-scale variation in regional fire occurrence and climate will be described in the next section, but first we focus on the extreme year-to-year (interannual) variations and their associations with climate variability.

The years of highest synchrony are labeled in Fig. 5 and were identified as years that exceeded the 95th percentile of smallest or largest values in a ranking of the fire years based on the number of sites recording fires per year in 20-year moving periods. By using a moving period for the percentile rankings we adjusted for the changing sampling depth. The year-by-year values of the 95th percentile threshold were variable (i.e., the values produced a somewhat jagged curve, not shown) because the moving period included or excluded the particularly large or small values as it was shifted along the time series. The result was that some "extreme" years exceeding the 95th percentile were included or excluded in a somewhat arbitrary fashion. Therefore we used the 95th percentile curves (upper and lower) as a general guide for selecting the years to include or exclude in the analyses. Overall, this approach led to the inclusion or exclusion of only a few additional years (either large or small), and in a separate analyses we found that the basic results were not changed relative to use of only years strictly defined by the moving period.

Although the ranking in moving periods provided some adjustment for sample size, we decided it was best to exclude the pre-1700 and post-1860 periods of the Sierra regional chronology, and the post-1880 period of the Southwest regional chronology. The sample depth in the earliest period (before ca. 1700) in the

Sierras drops below approximately 100 trees and 10 sites, and therefore it is doubtful that we are accurately identifying all regional extremes with this reduced sample size, especially small fire years. Some regional large fire events were evident in the 1600s (Fig. 6.5, upper graph) and these were included in the analyses. The many apparent low fire activity years during the 1600s, however, were probably due to the small sample size, and so the extreme small events in this century were not included in the analysis. In general, as more sites and trees enter the data sets in later years, the number of zero value years decline, and the regional small years tend to become more apparent (Fig. 6.5). The Southwest network included more than 200 trees and 20 sites back to 1600, so regional large and small events were included in the analysis back through the 1600s.

The post-livestock-grazing eras were evident in both regional chronologies as declines in numbers of sites recording fires in the late 1800s. Several large event years (e.g., 1871, 1898, and 1970 in the Sierras, and 1891 and 1899 in the Southwest) and many small event years appear after the onset of intensive grazing in the two regions. We chose to exclude these post-livestock-grazing periods in the fire-climate analysis because of the known change in fuels in these periods relative to the preceding periods, and the obvious change in the nature of the fire-scar record at these times (e.g., Figs. 6.2, 6.3, 6.4, and 6.5; see also discussion and literature cited in previous sections). Interestingly the 1970 large event in the Sierras is traceable to extensive prescribed burning along one of the four elevation transects—in Sequoia National Park. These fires were set by the National Park Service in an ambitious prescribed burning effort during this particular year (unpublished Sequoia and Kings Canyon national parks fire history database).

The decline in sample depth through the twentieth century was due to our selective sampling of primarily dead fire-scarred trees (i.e., stumps, snags, and logs) to maximize chronology length and minimize impacts on living trees. The outer ring dates of these dead specimens were often in the early or midtwentieth century (e.g., Figs. 6.2, 6.3, and 6.4). Although this decline in sample depth probably affected our ability to detect some fires during the late twentieth century, we doubt that this effect was very pronounced. Support for this interpretation is the fact that the twentieth-century fire-scar records were commonly confirmed by the independent documentary record (e.g., 1970 example, and other examples mentioned previously). Also in most sites, where it was permissible and possible, we also sampled a few living trees with fire scars for the purpose of obtaining the full record of twentieth-century fire dates. Most of the time, these living fire-scarred trees had frequent fire scars extending up to the disruption period near the turn of the century, then no fire scars, or only one or two fire scars recorded during the twentieth century (e.g., Figs. 6.2, 6.3, and 6.4).

Interannual Fire Associations with Dry/Wet Patterns

We used superposed epoch analyses (SEA) to evaluate the interannual relations between extreme fire years (large and small) as identified in the two regional fire chronologies (Fig. 6.5). This method involved computing the average (or departure from average) climate condition during, before and after the extreme years. Monte Carlo techniques were used to estimate the confidence intervals of the observed averages (or departures) (Mooney and Duvall 1993). A similar technique was first used in studies of the potential effect of volcanic eruptions on global climate patterns, and was adapted by Baisan and Swetnam (1990), Swetnam and Betancourt (1992), Swetnam (1993), and Grissino-Mayer (1995) for use in fire history studies.

The FHX2 software includes a subroutine written by Richard Holmes to carry out the SEA computations (Grissino-Mayer 2001). The program requires the input of a list of key dates and a continuous time series of an environmental variable, such as a precipitation or drought index. In the present case, for the key years we used the extreme large and small fire years in the regional chronologies (years labeled in Fig. 6.5). The environmental time series we used were two recently developed tree-ring reconstructions of summer (June–August) Palmer Drought Severity Index (PDSI) from the Southwest and the Sierras (Meko et al. 1993; Cook et al. 1999). These PDSI reconstructions are based on large networks of drought-sensitive tree-ring-width chronologies, and they were derived via calibration and validation using linear regression techniques. Details of the calibration and validation statistics of these reconstructions are described on the worldwide web (at *http://www.ngdc.noaa.gov/paleo/pdsi.html*; see also Meko et al. 1993 and Cook et al. 1994, 1999). The reconstructed values were summer (June–August) PDSI, but in general also reflect persistent moisture conditions during the preceding month (i.e., May) because the PDSI algorithm includes lagging water balance effects of preceding periods.

The SEA results (Fig. 6.6) were similar to patterns observed in the other SEA studies of fire associations with interannual precipitation or drought variables (e.g., Veblen et al. 1999; Veblen, Kitzberger, and Donnegan 2000; Donnegan, Veblen, and Sibold 2001). In particular, large fire years (on average) tended to be significantly dry ($p < 0.001$, Fig. 6.6, upper and lower left graphs). Small fire years tended to be significantly wet in the Southwest ($p < 0.05$). The association of fire and drought was not surprising, but more interesting results were the findings of lagging relationships in fire–PDSI comparisons. For example, summer PDSI in the year before small fire years was consistently low (dry) in both the Southwest and Sierras ($p < 0.001$, Fig. 6.6, upper and lower right graphs). Summer conditions in years preceding large fire years tended to be wet, but this was consistent and statistically significant only in the Southwest regional composite. We interpret the importance of lagging patterns in the Southwest to be due to a high importance of fine fuel accumulation during wet years in these relatively dry sites. The widespread fires within and among sites throughout the region were largely a function of the accumulation of a continuous fuel layer of grass and tree needles. A series of one to three years of wet conditions was often important for the development of a continuous fuel layer that carried the spreading surface fires. Understory fuel accumulation and dynamics were also important because the frequently occurring fires consumed these fine fuel layers. In semi-arid conditions, it probably required one to several years of relatively wet

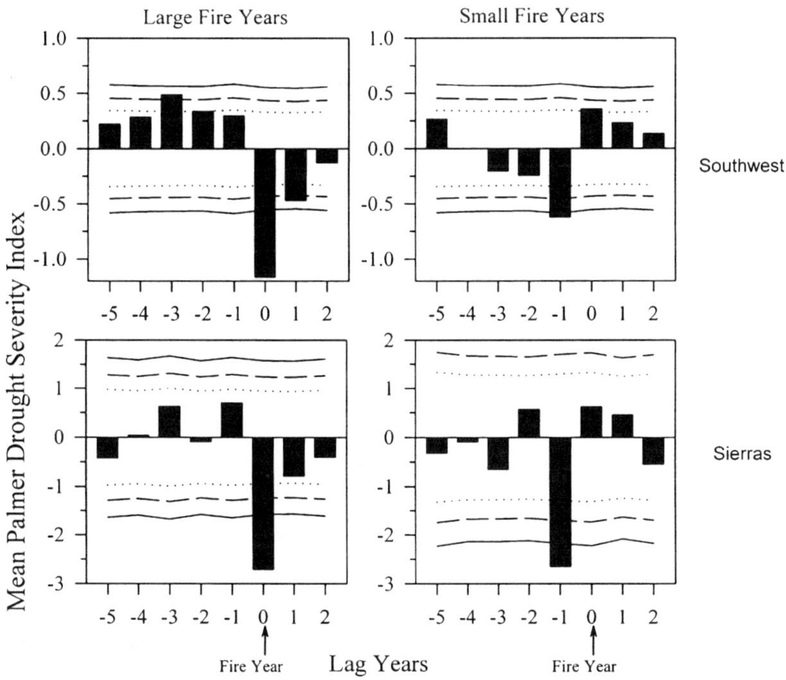

Figure 6.6. Results of superposed epoch analysis (SEA) comparing summer Palmer Drought Severity Indexes (PDSI) during relatively large (extensive) and small (less extensive) fire years in the Southwest (*top row*) and Sierras (*bottom row*). (See text for explanation of how "extensive" and "less extensive" were defined and time periods analyzed.) Horizontal dotted, dashed, and solid lines are 99.9, 99.0, and 95.0 confidence intervals, respectively, computed using a resampling procedure (Swetnam and Betancourt 1992).

conditions (and lack of fire) to rebuild continuous surface fuels. The importance of dry years preceding the smallest regional fire years was probably due, in part, to the occurrence of extensive fires during these preceding dry years, thus limiting the ignition and spread of fires during the next year. Dry preceding years also limited fuel production necessary for fire ignition and spread in the subsequent year, especially if the subsequent year was wet (i.e., in the Southwest comparison, Fig. 6.6 upper right).

The different fire–PDSI lagging patterns in the Southwest and Sierras were probably due to the different mixtures of tree species and understory conditions in the two regions. In other studies we have sorted study sites into those with significant ponderosa pine or Jeffrey pine components, versus somewhat higher elevation, mixed conifer sites where these pine species were relatively minor components or were absent (Swetnam and Baisan 1996; Caprio and Swetnam 1995; Swetnam and Betancourt 1998). We found that the lagged wet conditions preceding large fire years were restricted to the pine-dominant sites. Mixed conifer sites tended to show no significant previous years wet patterns, but drier conditions occurred during large fire years than in the ponderosa pine sites. As just described, we think this difference was due to the high importance of understory fuel amounts in the relatively xeric, pine-dominated forests. In contrast, low fuel moisture was probably more important for successful fire ignition and spread in the relatively mesic, and productive mixed conifer forests (i.e., fuels were generally not limiting). Hence the lack of significantly wet years preceding regional large fire years in the Sierras could be because most of these sites were in relatively productive mixed conifer stands, whereas the majority of the southwestern sites were in dry ponderosa pine stands.

This interpretation is supported by similar SEA results in Oregon and Washington (Heyerdahl, Brubaker, and Agee, in press) where precipitation is greater and mixed conifer forests are more productive than in the southwestern pine-dominant stands. Also the relatively dry pine forests sampled in Colorado (Veblen, Kitzberger, and Donnegan 2000; Donnegan, Veblen, and Sibold 2001), Mexico (Heyerdahl and Alvarado, Chapter 7, this volume), and *Austrocedrus chilensis* woodlands in Argentina (Kitzberger, Veblen, and Villalba 1997; Kitzberger and Veblen 1998; Veblen et al. 1999) had similar wet years preceding large fire years.

El Niño–Southern Oscillation and Fire Relationships

The importance of wet/dry sequences to synchronized fire activity in some regions is at least partly explainable by El Niño–Southern Oscillation (ENSO) teleconnections to regional rainfall patterns. ENSO events are known to affect seasonal rainfall amounts through changes in atmospheric circulation (e.g., position, strength, and sinuosity of the jet stream) and frequency of tropical and subtropical storms (Aceituno 1988; Andrade and Sellers 1988; Nicholls 1992; Diaz and Markgraf 2000; Harrington, Cerveny, and Balling 1992). Weak to moderate correlations have been identified between modern fire occurrence and fire-scar records and various indexes of the Southern Oscillation in the Southwest, Colorado Front Range, Oregon, Washington, Mexico, and in Patagonia (Swetnam and Betancourt 1990, 1992; Kitzberger, Veblen, and Villalba 1997; Kitzberger and Veblen 1998; Fulé and Covington 1999; Veblen, Kitzberger, and Donnegan 2000; Donnegan, Veblen, and Sibold 2001; Heyerdahl, Brubaker, and Agee, in press; Heyerdahl and Alvarado, Chapter 7, this volume).

A key finding of these studies was that synchronized, regional fire events tended to occur during dry years that were often associated with La Niña events (in the Southwest, Colorado, and Patagonia). These dry, regional fire years tended to follow one to several wet years that were often associated with El Niño events. Wet/dry patterns and regionally synchronized fire events were not entirely consistent within regions or through time, but were sufficiently strong as to be detectable in both twentieth-century and paleo-fire and climate comparisons (Fig. 6.7). Moreover, as expected, reverse correlations were noted in the Pacific Northwest, where El Niños tended to produce drier conditions and increased fire activity (Morgan et al. 2001; Heyerdahl, Brubaker, and Agee, in press). As

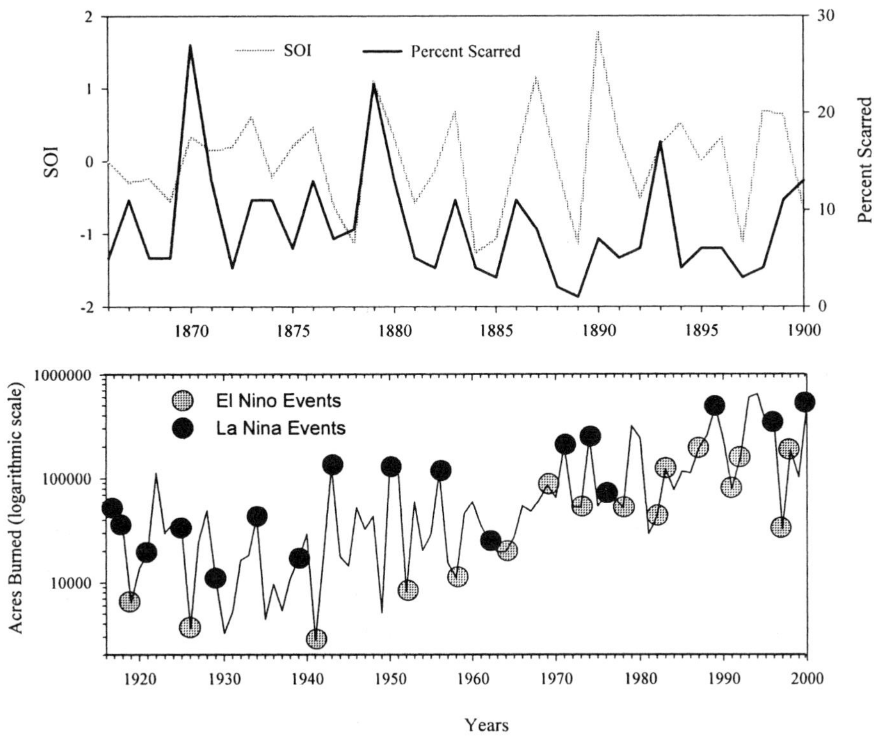

Figure 6.7. Time series of the percentage of trees scarred per year in a network of 15 sites in Arizona and New Mexico compared with the estimated Darwin-Tahiti Southern Oscillation Index (upper graph). The Spearman rank correlation from 1866 to 1905 is 0.46, $p = 0.002$ (Swetnam and Betancourt 1990). In the lower graph the annual area burned in all federal, state, and private lands in the Arizona and New Mexico (1905–1994) is compared with El Niño and La Niña events.

remarkable as these regional fire-climate relationships were, an even more interesting pattern recently emerged at the global scale. We discovered that fire occurrence time series from the Southwest and Patagonia shared similar interannual to decadal scale variations (discussed below) (Kitzberger, Swetnam, and Veblen 2001). Given that ENSO climate teleconnections are similar in the two regions, perhaps it should not be surprising that ENSO might act as a pacemaker, synchronizing fire activity at interhemispheric (i.e., global) scales.

Decadal-Scale Changes in Fire Frequency and Climate

In addition to interannual fire-climate variations and correlations we have also detected decadal-scale fire-climate patterns. One of the most interesting decadal-scale changes occurred in the Southwest from about 1780 to 1840. (Other examples of decadal-scale fire-climate changes will be described in the next section on giant sequoia fire history.) In recent years the evidence for this change in the

Southwest and other regions, and its association with global-scale climate patterns, has continued to build (Swetnam and Betancourt 1998; Grissino-Mayer and Swetnam 2000; Kitzberger, Swetnam, and Veblen 2001; Heyerdahl, Brubaker, and Agee, in press). At present, there are five lines of evidence pointing to a major climate-driven fire regime change in the late eighteenth and early nineteenth centuries: (1) unusually long intervals between fires during this period, (2) a shift from higher to lower fire frequency (and a related shift from less synchronous to more synchronous fire events), (3) a shift in seasonality of fires, (4) a striking decrease in the interannual correlation of fire events and climate indexes, and (5) the existence of a similar secular change in northern Patagonia, Argentina.

The first indication of a late eighteenth to early nineteenth century fire regime shift that we noticed was an unusually long interval between surface fires in the Gila Wilderness, New Mexico (Swetnam and Dieterich 1985). Since then, we have identified unusually long fire-free intervals around this time in many other (but not all) chronologies in the Southwest (Fig. 6.8). In some areas a long interval begins as early as the 1780s, and in others the interval does not begin until the early 1800s (e.g., Figs. 6.4 and 6.8). In some sites a few small fires (i.e., recorded by one or a few trees) occurred during the long interval, but there was a notable lack of widespread (highly synchronous) fires (Figs. 6.4, 6.8, and note also the slight dip in the number of sites recording fire in the Southwest during the early 1800s in Fig. 6.5).

The second indication of an important fire regime shift was a decrease in fire frequency after ca. 1800, and a notable increase in synchrony of fire events between trees (Figs. 6.3 and 6.4). This kind of change in frequency and synchrony was also noted in our giant sequoia studies during another time period (i.e., a change around AD 1300). Such frequency/synchrony (extent) shifts may reflect the natural feedbacks between fire frequency, fuel amounts, types, and spatial arrangements (Swetnam 1993). During relatively high frequency periods, fuels become more of a limiting factor to fire ignition and spread because the lags between fire events are too short for fuel continuity (amounts and spatial connectedness) to build to the point where fires will spread extensively through stands. This feedback between fires and fuels leads to spatially heterogeneous fuel layers and fire extent patterns. During relatively low fire frequency periods, fuels are less limiting because the longer lags enable fuel continuity to increase. When fires do occur, they tend to spread through the relatively abundant, spatially continuous fuels. Recent dynamic simulation models, incorporating climate and fuels components, generally support these interpretations with direct comparisons between simulated spatial and temporal patterns of fire frequency and extent and actual fire history data (Miller and Urban 1999, 2000).

A third line of evidence pointing to fire-regime and climate changes at the turn of eighteenth to nineteenth centuries is an apparent shift in seasonality of fire in a set of fire-scar chronologies from west central New Mexico (Grissino-Mayer and Swetnam 2000). Allen (1989) noted a similar seasonality change in a fire-scar data set from the Jemez Mountains in northern New Mexico. By examining the intraannual position of fire scars, we were able to infer the relative timing of past

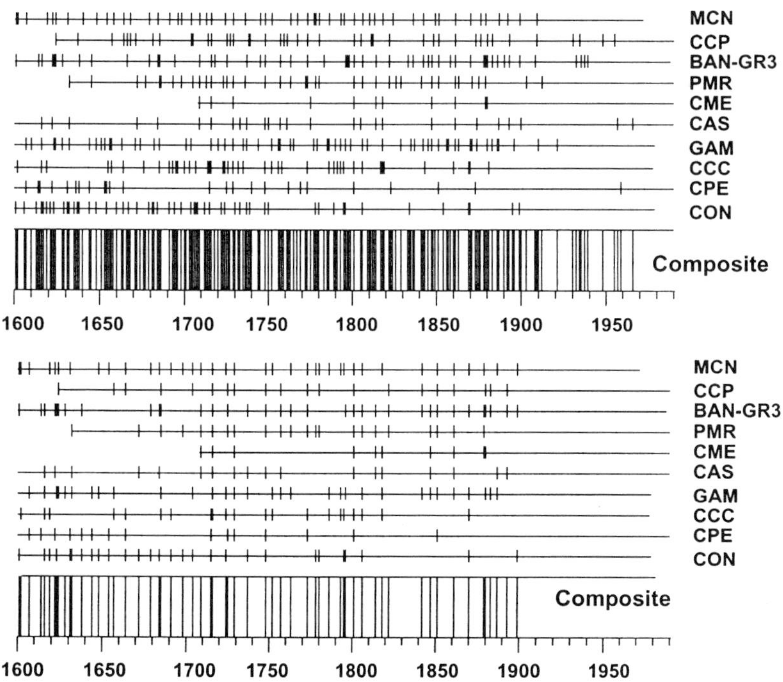

Figure 6.8. Composite fire-scar chronologies from the Jemez Mountains, New Mexico. These 10 stands are very broadly distributed around the mountain range, over an area of about 50,000 ha (see schematic map in Fig. 1). The horizontal lines and tick marks in the upper graph show time spans and fire dates, respectively, of fires recorded by any sampled fire-scarred tree within the stand. The bottom graph shows the same chronologies, but only fire dates recorded by 25% or more of the trees within each of the stands. The long vertical lines at the bottom show the composite of all dates for each graph. Note that the 25% filter emphasizes fires that were probably relatively widespread, both within and among stands. The fire regime disruption at around 1900 is evident in both graphs. Early and persistent fire regime disruption is evident in the three lowermost stands (CCC, CPE, and CON), and this has been attributed to early livestock grazing by Hispanic ranchers in these specific sites (Touchan, Allen, and Swetnam 1996). An early 1800s gap in fire occurrence in all chronologies is most apparent in the 25% filtered chronologies (bottom graph).

fires in relation to the cambial growth and dormant seasons (Dieterich and Swetnam 1984; Ortloff 1996). In a compilation of several hundred intraannual ring position observations, it was apparent that a secular change in fire seasonality began in the early 1800s (Fig. 6.9). Moreover the composite chronologies from this subregion of the Southwest show a pattern of reduced fire frequency ca. 1780, and more synchronous fire events after this time (Grissino-Mayer and Swetnam 2000).

SEA analysis of the periods before and after the shift reveals changes in the responses of fire occurrence to interannual climate patterns (Fig. 6.10). Our

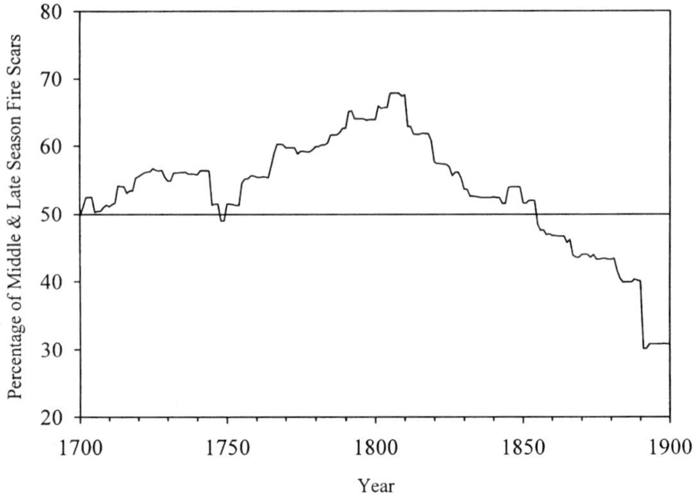

Figure 6.9. The relative position of fire scars within tree rings at El Malpais, New Mexico, changed through time, with a decreasing percentage of middle to late season scars (probably July–September) after ca. 1800 (from Grissino-Mayer and Swetnam 2000; reprinted with permission from *The Holocene*, © Arnold Publishers).

general conclusions from these analyses were that fire seasonality changes were probably related to a shift in seasonality of rainfall patterns (Grissino-Mayer and Swetnam 2000). In particular, the shift from a late-season dominant fire regime prior to 1800 to more early season fires after 1800 (Fig. 6.9) could have been a

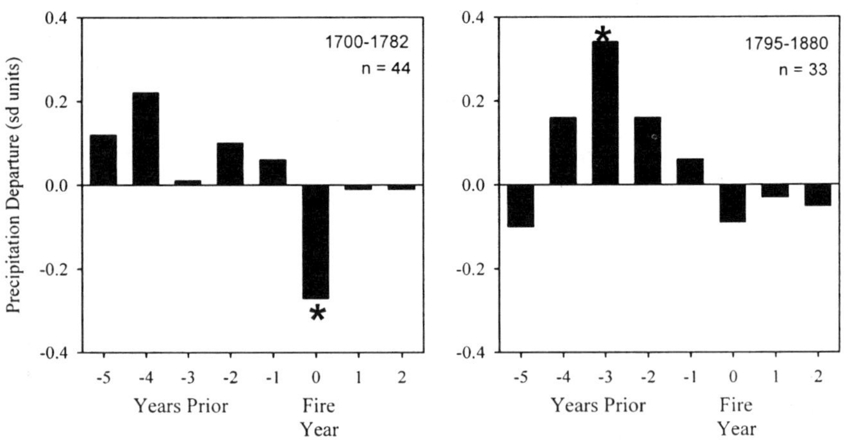

Figure 6.10. Superposed epoch analysis of the fire events before (*left*) and after (*right*) ca. 1800 at El Malpais, NM, suggests a change in the lagging relations between fires and climate in the two periods (from Grissino-Mayer and Swetnam 2000; reprinted with permission from *The Holocene*, © Arnold Publishers). Asterisks indicate significant values at the 95% confidence level.

consequence of fewer El Niño events after circa 1800 than before. El Niño events tend to result in relatively wet winters and early springs (Andrade and Sellers 1988), and a reduction in summer monsoonal rainfall (Harrington, Cerveny, and Balling 1992; Gutzler and Preston 1997). Hence, with more frequent El Niño events before circa 1800, the peak dry conditions and fire season in the Southwest would have often been relatively late, that is, from July to September. Higher fire frequencies in the pre-1800 period than in the post-1800 period could also have been partly related to more frequent El Niños, which would lead to increased fuel production in the relatively dry Southwest forests.

The SEA (Fig. 6.10) suggests that moist conditions in prior years were generally important both before and after 1800 (but this pattern was not statistically significant before 1800). Drought conditions were strongly associated with extensive fire events before but not after 1800. This pattern may have developed because the post-1800 period had an increasing frequency of dry, late springs/early summers (and increasing numbers of fire events occurring during this season, e.g., Fig. 6.9). In a climatic situation when dry springs were the norm, drier than average conditions (relative to the whole period) could not have been very important for fire ignitions and extensive fire spread.

The fourth line of evidence for a change in fire-climate relations ca. 1780 to 1840 was a large drop in correlation between regional drought indexes and fire occurrence over the entire Southwest during this period (Fig. 6.11). For this analysis, first differences were computed (see equation in the caption to Fig. 6.11) for both the regional drought and fire-scar series, so only the year-to-year variations were retained in the series and all long-term variations (e.g., decadal to centennial) were removed. Remarkably high interannual correlations were evident in the periods preceding and following ca. 1780 to 1840, with Pearson r-values exceeding 0.8 during the 1730s to 1780s and 1840s. Again, the importance of extreme switching between relatively wet and dry years (e.g., see especially the mid-1700s in Fig. 6.11) appears to be a key to regional fire and climate synchrony. Decreased climate and fire variance and correlation during the 1780s to 1840s period points to a weakening of the interannual switching of wet to dry conditions.

The fifth line of evidence offers a plausible climatic explanation for the decadal-scale change. A very similar reduction in fire occurrence during ca. 1780 to 1840 occurred in Patagonia, Argentina (Kitzberger, Swetnam, and Veblen 2001). Cross-spectral analyses of the Southwest and Patagonia regional fire time series showed moderate coherence in the 2- to 10-year portion of the spectrum, with clear changes in coherence during the 1780 to 1840 period. We also noted that this period had the lowest frequency of El Niño and La Niña events in the past two to three hundred years, as determined from a broad range of paleoclimatic reconstructions (ice cores, tree-rings, coral layers, and archival documents) (Kitzberger, Swetnam, and Veblen 2001). The early 1800s (i.e., ca. 1810s–1830s) was notable as a pronounced cold period throughout the Northern Hemisphere (Mann, Bradley, and Hughes 1998), and some extremely cold years occurred during these decades that were probably related to major volcanic eruptions (e.g., the cold year of 1816 which followed the eruption of Tambora in 1815). Finally,

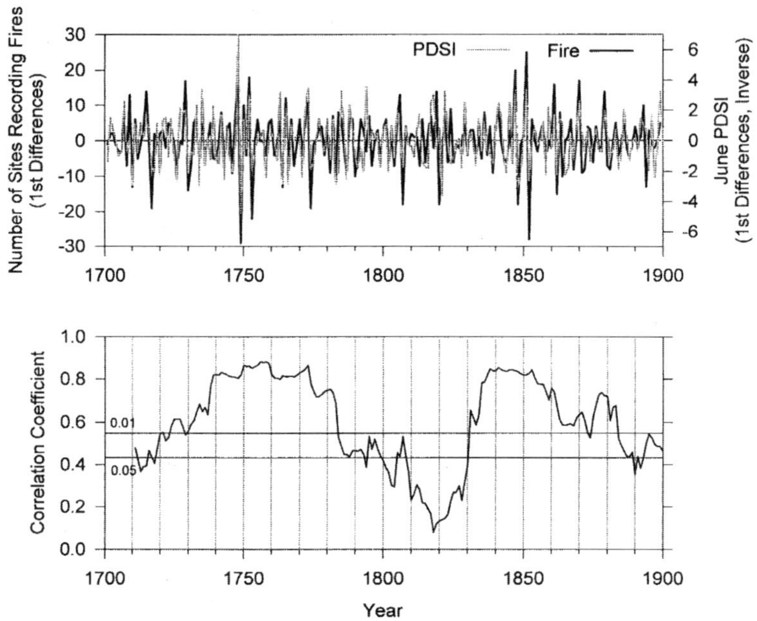

Figure 6.11. A composite time series of fire events in the Southwest (number of sites recording fires each year) is compared with a composite of Palmer Drought severity grid point reconstructions for June to August (from Cook et al. 1999) (upper graph). The interannual variations in the two time series are emphasized in this comparison by transforming (filtering) them by computing the first differences (i.e., first difference = value (year t) – value (year $t-1$)). Note that PDSI values were multiplied by –1 so that dry years (negative values) would be positive and correspond with large fire years (positive values). The lower graph shows a 20-year running correlation (plotted on the eleventh year of the period) between the two time series (from Swetnam and Baisan 1996; and Swetnam and Betancourt 1998, reprinted from *Journal of Climate*, © American Meteorological Society).

Heyerdahl's study in the Pacific Northwest shows a very similar decline in fire frequency during the early 1800s (Heyerdahl, Brubaker, and Agee, in press). Although the precise climatic mechanisms for reduced fire activity in such broadly scattered regions as the Pacific Northwest, the Southwest, and Patagonia are unclear, the evidence would suggest that wet/dry oscillations associated with ENSO, and/or anomalous global-scale cold conditions were probably involved.

Giant Sequoia Fire History and Climate

Giant sequoias are remarkable recorders of past surface fires. By sampling dozens of fire-scarred sequoia stumps, logs, and snags in five sequoia groves on the western slope of the Sierras, we reconstructed a network of fire histories that span the past 2000 to 3000 years (Stephenson, Parsons, and Swetnam 1989; Swetnam et al. 1991, 1992; Swetnam 1993). The composite record of fire dates from five groves shows that fire regimes varied across a range of temporal scales, from interannual to decadal, to centennial (Fig. 6.12).

The fire history work in giant sequoia groves provides an example of extreme sampling constraints and difficulties that fire historians face in reconstructing long and well-replicated fire-scar chronologies. Fire-scar cavities are common on ancient sequoias, but there are aesthetic, ethical, and regulatory constraints in obtaining cross-sectional samples from these magnificent living trees. These constraints required that we obtain our specimens entirely from dead trees. The sampling involved very arduous cutting with large chain saws (1–2-m length bars).

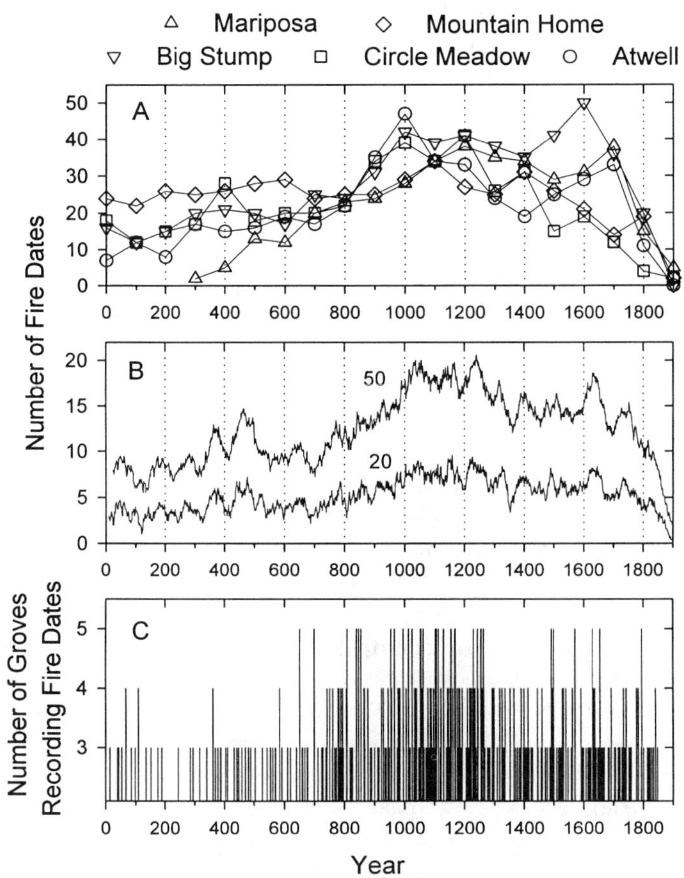

Figure 6.12. Fire occurrence in 5 sequoia groves since 1 BC. The upper graph shows centennial fire frequencies (number of fires/century) computed in each of the 5 groves, plotted on the first year of the century. The middle graph shows moving-period fire frequencies among all groves (sum of all years with fires in any of the 5 groves) for 50- and 20-year periods, plotted on the 25th and 10th years, respectively. The lower graph shows synchronous fire years in 3, 4, or 5 groves for each year (reprinted with permission from Swetnam, T.W. 1993, Fire history and climate change in giant sequoia groves, *Science* 262:885–889, Copyright 1993 American Association for the Advancement of Science).

Each sampled tree had several deep fire-scar cavities and a dozen or more cross sections were typically removed per tree. Careful judgment and selection of the "best" trees for sampling (and the best locations of those trees) was imperative because most dead trees with fire-scar cavities clearly did not have well-preserved, long records of past fires. In addition to loss of fire-scar evidence because of decay, and burning off of old fire scars, there were practical limitations in obtaining specimens from some trees because the fire-scar cavities were too deep to use conventional chain saws, or were at angles and heights that were unsafe for cutting. In sum, random or rigidly systematic sampling designs (e.g., grids) were thoroughly impractical in this forest type.

Despite the sampling difficulties and potential biases in selection of particular sequoia trees, we were able to obtain very long, well-replicated records and to detect substantial common variation in fire events and trends among the groves (Fig. 6.12). A variety of evidence indicate that these temporal and spatial changes in fire regimes were largely associated with past climatic variability. As previously mentioned, contingency analyses confirmed that synchrony of fires among the five groves (and synchrony of years without fires) was much greater than would be expected to occur by chance ($p < 0.01$) during most centuries. A SEA, using independent tree-ring chronologies and precipitation reconstructions from drought sensitive trees (Hughes and Graumlich 1996; Graybill and Funkhouser 1999), also confirmed that fire event synchrony was associated with drought, and lack of fire events was associated with wet years (Fig. 6.13). The drought-fire association was strongest during the most extensive fire event years (i.e., the more groves recording a fire event per year, the drier the average conditions) (Fig. 6.13).

A composite time series of fire occurrence in all groves showed substantial decadal to century-scale variability, and this series was significantly correlated ($p < 0.02$) with growing season temperatures estimated from independent foxtail pine (Graumlich 1993) and bristlecone pine tree-ring chronologies from the region (LaMarche 1974) (Fig. 6.14). An interesting result of this analysis was that at these time scales of decades and centuries, no significant correlations with the precipitation time series were identified ($p > 0.05$). But, as noted in the SEA, precipitation was associated with the occurrence of synchronous (widespread) fire events (Fig. 6.13). In contrast, SEA revealed no association between synchronous fire events and the growing season temperature estimates from foxtail and bristlecone pine tree-ring widths (results not shown).

Hence there appears to be a frequency-dependent response of giant sequoia fire regimes to precipitation and temperature. High-frequency (interannual) variations in precipitation, but not temperature, were associated with regionally synchronous fire events (Fig. 6.13). Low-frequency (decadal to centennial) variations in temperature, but not precipitation, were associated with variations and trends in fire frequency (Fig. 6.14). A plausible interpretation of these results is that interannual variations in fire activity were largely driven by moisture content of fuels. The interannual variance of growing season temperature is typically lower than the interannual variance of precipitation. Conversely, there is typically more

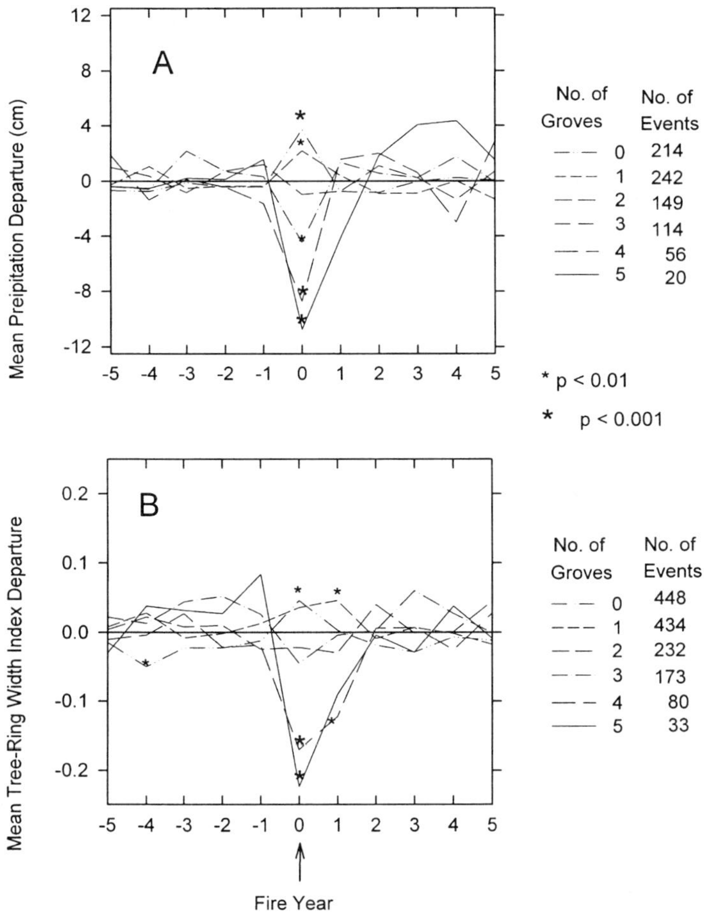

Figure 6.13. Superposed epoch analysis (SEA) of sequoia fire events versus precipitation time series. The upper graph shows the SEA using a reconstruction of winter precipitation in the Sierra from AD 1060 to 1850 (Graybill and Funkhouser 1999), and the lower graph shows the SEA using a drought-sensitive bristlecone pine chronology from the lower forest border in the White Mountains, CA, from AD 500 to 1850 (LaMarche 1974; Hughes and Graumlich 1996). Note that the more extensive fire events (i.e., synchronous fire events in 4 or 5 groves) had the strongest drought-fire signal (reprinted with permission from Swetnam, T.W. 1993, Fire history and climate change in giant sequoia groves, *Science* 262:885–889, Copyright 1993 American Association for the Advancement of Science).

decadal- to centennial-scale variance in reconstructed temperatures than in reconstructed precipitation time series (Graumlich 1993; Hughes and Graumlich 1996). It may be that the decadal- to centennial-scale responses of fire regimes to similar time-scale temperature regimes (Fig. 6.14) are a natural consequence of the concentration of climatic variability in this part of the spectrum. Moreover we suspect that the highest fire frequencies in sequoia groves occurred when decadal-scale warm temperatures coincided with high interannual variability in precipitation.

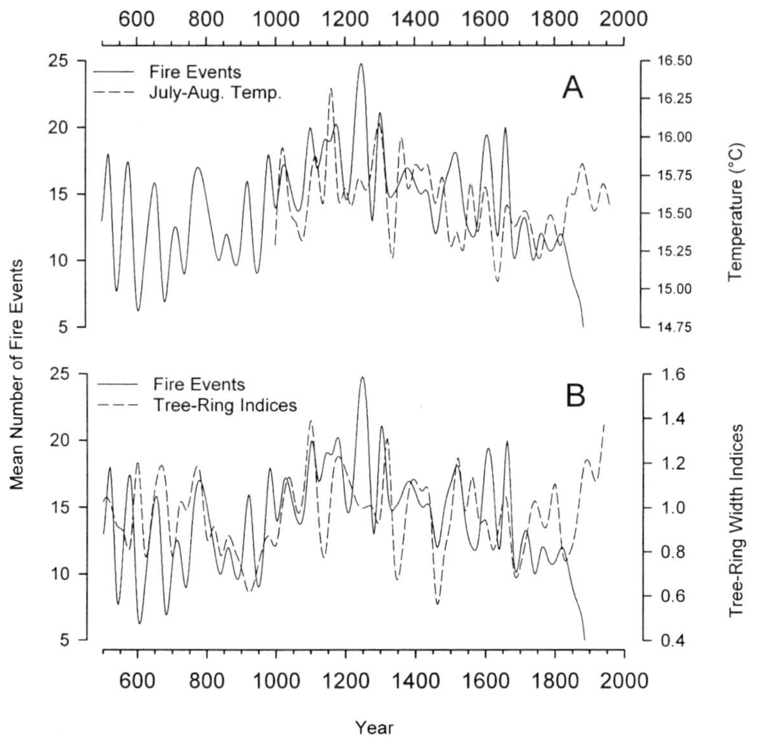

Figure 6.14. Decadal and centennial variations in estimated temperatures and fire occurrence in the Sierras are compared. The fire occurrence time series was computed from a weighted sum of fire events in the five sequoia groves in 20-year nonoverlapping periods (i.e., each year had a value of 0 to 5 depending on number of groves recording fire). The temperature series were 20-year, nonoverlapping means, and both the temperature and fire occurrence series were slightly smoothed with a cubic spline (for graphical purposes, but not for the statistical analyses). The upper graph shows a comparison of fire activity with reconstructed summer temperature from foxtail pine in the Sierras (Graumlich 1993), and the lower graph shows a comparison with a temperature responsive, upper treeline bristlecone pine chronology from the White Mountains, CA (LaMarche 1974). The Pearson correlation between the foxtail reconstructed temperature and fire series was $r = 0.41$, $p = 0.006$, and the correlation between bristlecone ring-width chronology and fire series was 0.30, $p = 0.012$ (unsmoothed values used in correlation analysis; reprinted with permission from Swetnam, T.W. 1993, Fire history and climate change in giant sequoia groves, *Science* 262:885–889, Copyright 1993 American Association for the Advancement of Science).

These conditions would be conducive to production of copious fuels during warm and wet years, and abundant fire ignitions and extensive fire spread during the warm and dry years.

Examples of such situations may have occurred during some decades of the so-called Medieval Warm Period, which appears to have been strongly expressed in the Sierra Nevada region from ca. AD 900 to 1300 (LaMarche 1974;

Graumlich 1993; Stine 1994). The highest fire frequencies in the past 2000 years occurred during this period (Fig. 6.14). This period, and the subsequent Little Ice Age (ca. AD 1400–1840, Grove 1988) have often been overextrapolated by various researchers, with unwarranted assumptions that these were monolithic periods of temporally consistent climate in virtually all regions of the Northern Hemisphere (see a critique of these assumptions regarding the Medieval Warm Period by Hughes and Diaz 1994). We agree that there is a high degree of regional variability in climate, and a lack of strong evidence for anything like a Medieval Warm Period or Little Ice Age in many parts of the world. Nevertheless, the climate history of the Sierras apparently coincided with the approximate timing and climatic conditions usually ascribed to these two periods (warm and cold, respectively).

In addition to the tree-ring width evidence (LaMarche 1974; Graumlich 1993) and lake level evidence (Stine 1994), now fire history may be added as another line of independent evidence in support of the occurrence of a generally warm period ca. 900 to 1300 and a subsequent cool period in the Sierra Nevada (regardless of whether they are given the appellation "Medieval Warm Period" or "Little Ice Age"). As noted above, fire frequencies were highest during the late Middle Ages (especially ca. 1100–1300) and decreased fire frequencies occurred after 1300s, especially during the major cold episodes of the mid 1400s and late 1600s. Although fire can only be considered an indirect proxy for past climatic variations, it is arguably not any less directly related to climate than, for example, lake levels.

Conclusion

Regional synchrony of ecological process is the hallmark of climatic influence and is an emergent property evident in fire occurrence time series aggregated over regions to continents (e.g., Swetnam and Betancourt 1998; Kitzberger, Swetnam, and Veblen 2001). Although fire history is often a function of site-specific environmental and cultural variables, it is clear that with network approaches, involving massive replication of high-resolution fire-scar time series across multiple points in space, it is possible to reconstruct very useful proxies of ecologically effective climatic change. The synchrony of fire regime variations in different regions can be compared and contrasted to elucidate historical climatic and cultural events and variations.

Disentangling climatic and human effects on past fire regimes is very challenging but not impossible. Multiple case studies and comparisons across networks of fire history sites is a key to identifying and distinguishing the effects of humans and climate on past forest fire regimes. More comparisons are needed of fire-scar chronologies with independent reconstructions and records of both climate and human history (e.g., from documentary sources or culturally modified trees). So far we have identified a few cases in the Southwest where Native American effects on fire frequency and seasonality before 1900 may be discernable. The most striking and clearly identified effect of humans on nineteenth- and early twentieth-century fire regimes in the Southwest and Sierras was the disruption of fire regimes by the introduction of intensive livestock grazing.

Interesting time periods showing coherent and significant fire and climate changes, such as the early 1800s and transition from Medieval Warm Period to Little Ice Age (1300–1400), offer unique opportunities for fire historians and paleoclimatologists to target specific regions and mechanisms for testing. For example, as we learn more about regionally consistent and specific terrestrial teleconnections to ocean-atmosphere patterns (El Niño–Southern Oscillation, Pacific Decadal–Oscillation, North Atlantic Oscillation, etc.), we could target key "sensitive" regions for new fire history collections and reconstructions. We have learned that climatic teleconnections in some regions are opposite in response relative to other regions. The Pacific Northwest, and northern U.S. Rockies, for example, tend to have opposite drought and fire responses to ENSO relative to the Southwest. The changing and variable nature of these inverse patterns should be thoroughly assessed using combinations of twentieth-century climate and fire occurrence data (fire atlases) and tree-ring based fire histories (Morgan et al. 2001). Direct comparisons between existing fire atlases and broad-scale networks of fire histories will be one way to do this, but development of more extensive networks is needed, especially in regions where relatively few crossdated fire-scar chronologies have been developed, such as in southwest Canada and the Pacific Northwest, northern Rockies, Great Basin, and northern Mexico.

References

Abolt, R.A.P. 1997. Fire histories of upper elevation forests in the Gila Wilderness, New Mexico via fire scar and stand age structure analyses. M.S. thesis, School of Renewable Natural Resources, University of Arizona, Tucson. 120p.

Aceituno, P. 1988. On the functioning of the Southern Oscillation in the South American sector. Part 1. Surface climate. *Mon. Wea. Rev.* 116:505–524.

Allen, C.D. 1989. Changes in the landscape of the Jemez Mountains, New Mexico. Ph.D. dissertation. University of California, Berkeley.

Allen, C.D. 2002. Lots of lightning and plenty of people: An ecological history of fire in the upland Southwest. In *Fire, Native Peoples, and the Natural Landscape*, ed. T.R. Vale, pp. 173–194. Covelo, CA: Island Press.

Anderson, M.K. 1996. Tending the wilderness. *Restor. Manag. Notes* 14(2):154–166.

Andrade, E.R. Jr., and Sellers, W.D. 1988. El Niño and its effect on precipitation in Arizona and western New Mexico. *J. Clim.* 8:403–410.

Arno, S.F., and Petersen, T.D. 1983. Variation in estimates of fire intervals: A closer look at fire history on the Bitterroot National Forest. USDA Forest Service Res. Pap. INT-301.

Bahre, C.J. 1985. Wildfire in southeastern Arizona between 1859 and 1890. *Desert Plants* 7(4):190–194.

Baisan, C.H., and Swetnam, T.W. 1990. Fire history on a desert mountain range: Rincon Mountain Wilderness, Arizona, U.S.A. *Can. J. For. Res.* 20:1559–1569.

Baisan, C.H., and Swetnam, T.W. 1997. Interactions of fire regime and land-use history in the central Rio Grande Valley. USDA Forest Service Res. Pap. RM-RP-330. 20p.

Baker, W.L. 2002. Indians and fire in the U.S. Rocky Mountains: the wilderness hypothesis renewed. In *Fire, Native Peoples, and the Natural Landscape*, ed. T.R. Vale. pp. 41–76. Covelo, CA: Island Press.

Baker, W.L., and Ehle, D. 2001. Uncertainty in surface-fire history: the case of ponderosa pine forests in the western United States. *Can. J. For. Res.* 31(7):1205–1226.

Barton, A.M., Swetnam, T.W., and Baisan, C.H. 2001. Arizona pine (*Pinus arizonica*) stand dynamics: Local and regional factors in a fire-prone madrean gallery forest of Southeast, Arizona, USA. *Landscape Ecol.* 16(4):351–369.

Brown, P.M., and Sieg, C.H. 1996. Fire history in interior ponderosa pine communities of the Black Hills, South Dakota, USA. *Int. J. Wildl. Fire* 6(3):97–105.

Brown, P.M., and Sieg, C.H. 1999. Historical variability in fire at the ponderosa pine–Northern Great Plains prairie ecotone, southeastern Black Hills, South Dakota. *Ecoscience* 6(4):539–547.

Brown, P.M., and Swetnam, T.W. 1994. A crossdated fire history in a coast redwood forest near Redwood National Park, California. *Can. J. For. Res.* 24:21–31.

Brown P.M., Kaufmann, M.R., and Shepperd W.D. 1999. Long-term, landscape patterns of past fire events in a montane ponderosa pine forest of central Colorado. *Landscape Ecol.* 14(6):513–532.

Brown, P.M., Kaye M.W., Huckaby, L.S., and Baisan, C.H. 2001. Fire history along environmental gradients in the Sacramento Mountains, New Mexico: Influences of local and regional processes. *Ecoscience* 8(1):115–126.

Caprio, A.C., and Swetnam, T.W. 1995. Historic fire regimes along an elevational gradient on the west slope of the Sierra Nevada, California. In *Proceedings of Symposium on Fire in Wilderness and Park Management*, tech. coords. J.K. Brown, R.W. Mutch, C.W. Spoon, and R.H. Wakimoto, pp. 173–199. USDA Forest Service Gen. Tech. Rep. INT-GTR-320.

Colwell, R.K., and Coddington, J.A. 1994. Estimating terrestrial biodiversity through extrapolation. *Philos. Trans. Roy. Soc. London, (ser. B) Biolog. Sci.* 345(1311):101–118.

Cook, E.R., Meko, D.M., Stahle, D.W., and Cleaveland, M.K. 1994. Tree-ring reconstructions of past drought across the coterminous United States: Tests of a regression method and calibration/verification results. In *Tree-Rings, Environment, and Humanity*, eds. J.S. Dean, D.M. Meko, and T.W. Swetnam, pp. 155–169. Tucson, AZ: Radiocarbon.

Cook, E.R., Meko, D.M., Stahle, D.W., and Cleaveland, M.K. 1999. Drought reconstructions for the continental United States. *J. Clim.* 12(4):1145–1162.

Cropper, J.P., and Fritts, H.C. 1982. Density of tree-ring grids in western North America. *Tree-Ring Bull.* 42:3–10.

Denevan, W.M. 1992. The pristine myth—The landscape of the Americas in 1492. *Ann. Assoc. Am. Geogr.* 82(3):369–385.

Diaz, H.F., and Markgraf, V. 2000. *El Niño and the Southern Oscillation: Multiscale Variability and Global and Regional Impacts*. Cambridge: Cambridge University Press.

Dieterich, J.H. 1980. The composite fire interval—A tool for more accurate interpretations of fire history. In *Proceedings of the Fire History Workshop*, tech. coords. M.A. Stokes, and J.H. Dieterich, pp. 8–14, October 20–24, Tucson, AZ. USDA Forest Service Gen. Tech. Rep. RM-81.

Dieterich, J.H. 1983. Fire history of southwestern mixed conifer: A case study. *For. Ecol. Manag.* 6:13–31.

Dieterich, J.H., and Swetnam, T.W. 1984. Dendrochronology of a fire-scarred ponderosa pine. *For. Sci.* 30(1):238–247.

Dobyns, H.F. 1978. From fire to flood: Historic human destruction of Sonoran desert riverine oases. Ballena Press Anthropological Papers No. 20. Soccoro, NM.

Donnegan, J.A., Veblen, T.T., and Sibold, J.S. 2001. Climatic and human influences on fire history in Pike National Forest, central Colorado. *Can. J. For. Res.* 31(9): 1526–1539.

Douglass, A.E. 1941. Crossdating in dendrochronology. *J. For.* 39(10):825–831.
Frittts, H.C. 1976. *Tree Rings and Climate*. London: Academic Press.
Fritts, H.C. 1991. *Reconstructing Large-Scale Climatic Patterns from Tree-Ring Data*. Tucson: University of Arizona Press.
Fritts, H.C., and Swetnam, T.W. 1989. Dendroecology: A tool for evaluating variations in past and present forest environments. *Adv. Ecol. Res.* 19:111–189.
Fulé, P.Z., and Covington, W.W. 1997. Changing fire regimes in Mexican pine forests: Ecological and management implications. *J. For.* 94(10):33–38.
Fulé, P.Z., and Covington, W.W. 1999. Fire regime changes in La Michilia Biosphere Reserve, Durango, Mexico. *Conserv. Biol.* 13(3):640–652.
Fulé, P.Z., Covington, W.W., and Moore, M.M. 1997. Determining reference conditions for ecosystem management of southwestern ponderosa pine. *Ecol. Appl.* 7(3):895–908.
Graumlich, L.J. 1993. A 1000-year record of temperature and precipitation in the Sierra Nevada. *Quat. Res.* 39:249–255.
Graybill, D.A., and Funkhouser, G.S. 1999. Dendroclimatic reconstructions during the past millennium in the Southern Sierra Nevada and Owens Valley, California. In *Proceedings of Southern California Climate Symposium on Trends and Extremes of the Past 2,000 Years*, eds. M.R. Rose and P.E. Wigand, pp. 239–269, October 25, 1991. Natural History Museum of Los Angeles County, Tech. Rep., Number 11.
Grissino-Mayer, H.D. 1995. Tree-ring reconstructions of climate and fire history at El Malpais National Monument, New Mexico. Ph.D. dissertation, Department of Geosciences, University of Arizona, Tucson. 407p.
Grissino-Mayer, H.D. 1999. Modeling fire interval data from the American Southwest with the Weibull distribution. *Int. J. Wildl. Fire* 9(1):37–50.
Grissino-Mayer, H.D. 2001. FHX2—Software for analyzing temporal and spatial patterns in fire regimes from tree rings. *Tree-Ring Res.* 57(1):113–122.
Grissino-Mayer, H.D., and Swetnam, T.W. 1997. Multi-century history of wildfire in the ponderosa pine forests of El Malpais National Monument. *New Mexico Bur. Mines Mineral Resources Bull.* 156:163–171.
Grissino-Mayer, H.D., and Swetnam, T.W. 2000. Century-scale climate forcing of fire regimes in the American Southwest. *Holocene* 10(2):213–220.
Grove, J.M. 1988. *The Little Ice Age*. London: Methuen.
Gutzler, D.S., and Preston, J.W. 1997. Evidence for a relationship between spring snow cover in North America and summer rainfall in New Mexico. *Geophys. Res. Lett.* 24(17):2207–2210.
Harrington, J.A. Jr., Cerveny, R.S., and Balling, R.C. Jr. 1992. Impact of the Southern Oscillation on the North American Southwest monsoon. *Phys. Geogr.* 13:318–330.
Heyerdahl, E.K., Brubaker, L.B., and Agee, J.K. 2001. Spatial controls of historical fire regimes: A multiscale example from the interior west, USA. *Ecology* 82(3):660–678.
Heyerdahl, E.K., Brubaker, L.B., and Agee, J.K. (In press). Annual and decadal influence of climate on fire regimes (1687–1994) of the Blue Mountains, USA. *Holocene*.
Hughes, M.K., and Diaz, H.R. 1994. Was there a Medieval Warm Period, and if so, where and when? *Clim. Change* 26:109–142.
Hughes, M.K., and Graumlich, L.J. 1996. Multimillennial dendroclimatic records from the western United States. In *Climatic Variations and Forcing Mechanisms of the last 2000 Years*, eds. R.S. Bradley, P.D. Jones, and J. Jouzel, pp. 109–124. NATO Advanced Studies Workshop Series. New York: Springer-Verlag.
Johnson, E.A., and Gutsell, S.L. 1994. Fire frequency models, methods and interpretations. *Adv. Ecol. Res.* 25:239–287.
Kaib, M. 1998. Fire history in riparian canyon pine-oak forests and the intervening desert grasslands of the Southwest borderlands: A dendroecological, historical, and cultural inquiry. M.S. thesis, University of Arizona, Tucson. 234p.
Kaib, M., Baisan, C.H., Grissino-Mayer, H.D., and Swetnam, T.W. 1996. Fire history in the gallery pine-oak forests and adjacent grasslands of the Chiricahua Mountains of

Arizona. In *Effects of Fire on Madrean Province Ecosystems: A Symposium Proceedings*, tech. coord. P.F. Ffolliott, L.F. DeBano, M.B. Maker Jr., G.J. Gottfried, G. Solis-Garza, C.B. Edminster, D.G. Neary, L.S. Allen, and R.H. Hamre, pp. 253–264. USDA Forest Service Gen. Tech. Rep. RM-GTR-289.

Kaye, M.W., and Swetnam, T.W. 1999. An assessment of fire, climate, and Apache history in the Sacramento Mountains, New Mexico, USA. *Phys. Geogr.* 20(4):305–330.

Kilgore, B.M., and Taylor, D. 1979. Fire history of a sequoia-mixed conifer forest. *Ecology* 60(1):129–142.

Kitzberger, T., and Veblen, T.T. 1998. Influences of humans and ENSO on fire history of Austrocedrus chilensis woodlands in northern Patagonia, Argentina. *Ecoscience* 4(4): 508–520.

Kitzberger, T., Veblen, T.T., and Villalba, R. 1997. Climatic influences on fire regimes along a rain forest to xeric woodland gradient in northern Patagonia, Argentina. *J. Biogeogr.* 24(1):35–47.

Kitzberger, T., Swetnam, T.W., and Veblen, T.T. 2001. Inter-hemispheric synchrony of forest fires and the El Nino-Southern Oscillation. *Global Ecol. Biogeogr.* 10(3): 315–326.

LaMarche, V.C. Jr. 1974. Paleoclimatic inferences from long tree-ring records. *Science* 183:1043–1048.

LaMarche, V.C. Jr., and Fritts, H.C. 1971. Anomaly patterns of climate over the western United States, 1700–1930, derived from principal components analysis of tree-ring data. *Mon. Wea. Rev.* 99(2):138–142.

Mann, M.E., Bradley, R.S., and Hughes, M.K. 1998. Global-scale temperature patterns and climate forcing over the past six centuries. *Nature* 392:779–787.

Martorano, M.A. 1981. Scarred Ponderosa Pine Trees Reflecting Cultural Utilization of Bark. M.S. thesis, Department of Anthropology, Colorado State University, Fort Collins. 127p.

Meko, D., Cook, E.R., Stahle, D.W., Stockton, C.W., and Hughes, M.K. 1993. Spatial patterns of tree-growth anomalies in the United States and Southeastern Canada. *J. Clim.* 6(9):1773–1786.

Miller, C., and Urban, D.L. 1999. A model of surface fire, climate and forest pattern in the Sierra Nevada, California. *Ecol. Model.* 114(2–3):113–135.

Miller, C., and Urban, D.L. 2000. Connectivity of forest fuels and surface fire regimes. *Landscape Ecol.* 15(2):145–154.

Millar, C.I., and Woolfenden, W.B. 1999. The role of climate change in interpreting historical variability. *Ecol. Appl.* 9(4):1207–1216.

Mooney, C.Z., and Duvall, R.D. 1993. Bootstrapping: A non-parametric approach to statistical inference. Sage University Paper Series on Quantitative Applications in the Social Sciences 07-095, Newbury Park, CA. 73p.

Morgan, P., Hardy, C., Swetnam, T.W., Rollins, M.G., and Long, D.G. 2001. Mapping fire regimes across time and space: Understanding coarse and fine-scale patterns. *Int. J. Wildl. Fire* 10(3–4):329–342.

Morino, K.A. 1996. Reconstruction and interpretation of historical patterns of fire occurrence in the Organ Mountains, New Mexico. M.S. thesis. University of Arizona, Tucson. 144p.

Muir, J. 1911. *My First Summer in the Sierra*. Boston: Houghton Mifflin.

Nicholls, N. 1992. Historical El Niño/Southern Oscillation variability in the Australasian region. In *El Niño: Historical and Paleoclimatic Aspects of the Southern Oscillation*, eds. H.F. Diaz and V. Markgraf, pp. 151–173. Cambridge: Cambridge University Press.

Ortloff, W. 1996. Wood anatomical evidence of fire seasonality. In *Tree-Rings, Environment, and Humanity*, eds. J.S. Dean, D.M. Meko, and T.W. Swetnam, pp. 89–93. Tucson, AZ: Radiocarbon.

Pyne, S.J. 1982. *Fire in America: A Cultural History of Wildland and Rural Fire*. Princeton: Princeton University Press.

Pyne, S.J. 1985. Vestal fires and virgin lands: a historical perspective on fire and wilderness. In *Proceedings of Symposium and Workshop on Wilderness Fire*, tech. coord. J.E. Lotan, B.M. Kilgore, W.C. Fischer, and R.W. Mutch, pp. 254–262. November 15–18, 1983, Missoula, MT. USDA Forest Service Gen. Tech. Rep. INT-182.

Rollins, M., Swetnam, T.W., and Morgan, P. 2001. Evaluating a century of fire patterns in two Rocky Mountain wilderness areas using digital fire atlases. *Can. J. For. Res.* 31(12):2107–2123.

Rosenzweig, M.L. 1995. *Species diversity in space and time*. Cambridge: Cambridge University Press.

Savage, M., and Swetnam, T.W. 1990. Early and persistent fire decline in a Navajo ponderosa pine forest. *Ecol.* 70(6):2374–2378.

Seklecki, M.T., Grissino-Mayer, H.D., and Swetnam, T.W. 1996. Fire history and the possible role of Apache-set fires in the Chiricahua Mountains of southeastern Arizona. In *Effects of Fire on Madrean Province Ecosystems: A Symposium Proceedings*, tech. coord. P.F. Ffolliott, L.F. DeBano, M.B. Maker Jr., G.J. Gottfried, G. Solis-Garza, C.B. Edminster, D.G. Neary, L.S. Allen, and R.H. Hamre, pp. 238–246. USDA Forest Service General Tech. Rep. RM-GTR-289.

Stephenson, N.L., Parsons, D.J., and Swetnam, T.W. 1989. Restoring natural fire to the sequoia-mixed conifer forest: should intense fire play a role? *Proceedings 17th Tall Timbers Fire Ecology Conference. High Intensity Fire in Wildlands: Management Challenges and Options*, pp. 321–337, May 18–21, Tallahassee, FL.

Stine, S. 1994. Extreme and persistent droughts in California and Patagonia during mediaeval times. *Nature* 369:546–549.

Swetnam, T.W. 1984. Peeled ponderosa pine trees: A record of inner bark utilization by Native Americans. *J. Ethnobiol.* 4(2):177–190.

Swetnam, T.W. 1993. Fire history and climate change in giant sequoia groves. *Science* 262:885–889.

Swetnam, T.W., and Baisan, C.H. 1996. Fire effects in southwestern forests. *Proceedings of the Second La Mesa Fire Symposium,* March 29–31, 1994, Los Alamos, NM. USDA Forest Service Gen. Tech. Rep. RM-GTR-286.

Swetnam, T.W., and Betancourt, J.L. 1990. Fire-southern oscillation relations in the southwestern United States. *Science* 249:1017–1020.

Swetnam, T.W., and Betancourt, J.L. 1992. Temporal patterns of El Nino/Southern Oscillation—Wildfire patterns in the southwestern United States. In *El Nino: Historical and Paleoclimatic Aspects of the Southern Oscillation*, eds. H.F. Diaz and V.M. Markgraf, pp. 259–270. Cambridge: Cambridge University Press.

Swetnam, T.W., and Betancourt, J.L. 1998. Mesoscale disturbance and ecological response to decadal climatic variability in the American Southwest. *J. Clim.* 11(12):3128–3147.

Swetnam, T.W., and Dieterich, J.H. 1985. Fire history of ponderosa pine forests in the Gila Wilderness, New Mexico. In *Proceedings-Symposium and Workshop on Wilderness Fire*, tech. coords. J.E. Lotan, B.M. Kilgore, W.C. Fischer, and R.W. Mutch, pp. 390–397. November 15–18, 1983, Missoula, MT. USDA Forest Service Gen. Tech. Rep. INT-182.

Swetnam, T.W., Allen, C.D., and Betancourt, J.L. 1999. Applied historical ecology: Using the past to manage for the future. *Ecol Appl.* 9(4):1189–1206.

Swetnam, T.W., Baisan, C.H., Caprio, A.C., Touchan, R., and Brown, P.M. 1992. Tree-ring reconstruction of giant sequoia fire regimes. Report on Cooperative Agreement DOI 8018-1-0002 to National Park Service. University of Arizona. 90p.

Swetnam, T.W., Baisan, C.H., and Kaib, J.M. 2001. Forest fire histories in the sky islands of La Frontera. In *Changing Plant Life of La Frontera: Observations on Vegetation in the United States/Mexico Borderlands*, eds. G.L. Webster and C.J. Bahre, pp. 95–119. Albuquerque: University of New Mexico Press.

Swetnam, T.W., Touchan, R., Baisan, C.H., Caprio, A.C., and Brown, P.M. 1991. Giant sequoia fire history in Mariposa Grove, Yosemite National Park. In *Proceedings of the Yosemite Centennial Symposium*, pp. 249–255. El Portal, CA: Yosemite Association.

Touchan, R., Allen, C.D., and Swetnam, T.W. 1996. Fire history and climatic patterns in ponderosa pine and mixed-conifer forests of the Jemez Mountains, northern New Mexico. In *Fire Effects in Southwestern Forests: Proceedings of the Second La Mesa Fire Symposium*, ed. C.D. Allen, pp. 33–46. USDA Forest Service Gen. Tech. Rep. RM-GTR-286.

Vale, T.R. 1998. The myth of the humanized landscape: An example from Yosemite National Park. *Natural Areas J.* 18(3):231–236.

Vale, T.R., editor. 2002. *Fire, Native Peoples, and the Natural Landscape*. Covelo, CA: Island Press.

Vankat, J.L. 1977. Fire and man in Sequoia National Park. *Ann. Assoc. Am. Geogr.* 67(1): 17–27.

Veblen, T.T., Kitzberger, T., and Donnegan, J. 2000. Climatic and human influences on fire regimes in ponderosa pine forests in the Colorado Front Range. *Ecol. Appl.* 10(4): 1178–1195.

Veblen, T.T., Kitzberger, T., Villalba, R., and Donnegan, J. 1999. Fire history in northern Patagonia: The roles of humans and climatic variation. *Ecol. Monogr.* 69(1):47–67.

7. Influence of Climate and Land Use on Historical Surface Fires in Pine-Oak Forests, Sierra Madre Occidental, Mexico

Emily K. Heyerdahl and Ernesto Alvarado

The rugged mountains of the Sierra Madre Occidental, in north-central Mexico, support a mosaic of diverse ecosystems. Of these, the high-elevation, temperate pine-oak forests are ecologically significant for their extensiveness and biodiversity. They cover nearly half the land area in the states of Durango and Chihuahua (42%), and comprise a similar percentage of the temperate coniferous forest in Mexico as a whole (45%; World Forest Institute 1994; SARH 1994). These forests are globally significant centers of vascular plant diversity, and of endemism in both plant and animal species (Bye 1993; Manuel-Toledo and Jesús-Ordóñez 1993). For example, they have the highest number of pine and oak species in the world (Rzedowski 1991) and contain many of Mexico's *Pinus*, *Quercus*, and *Arbutus* species (33%, 30%, and 66%, respectively; Bye 1995). Surface fires were historically frequent in these forests, and variations in their frequency may have contributed to the maintenance of this biodiversity (Dieterich 1983; Fulé and Covington 1997, 1999; Park 2001). However, we know little about the drivers of variation in historical fire regimes.

Forest fires are controlled by processes acting across a broad range of spatial scales (Tande 1979; Payette et al. 1989; Swetnam and Baisan 1996; Taylor and Skinner 1998; Heyerdahl, Brubaker, and Agee 2001). At coarse spatial scales, annual extremes in regional climate can synchronize the occurrence of fires across broad areas (Swetnam and Betancourt 1998; Swetnam and Baisan, Chapter 6, this volume). For example, fires were widespread during years of regionally low precipitation at sites in North and South America (Veblen et al. 1999; Veblen,

Kitzberger, and Donnegan 2000; Kitzberger, Swetnam, and Veblen 2001; Heyerdahl, Brubaker, and Agee in press; Swetnam and Baisan, Chapter 6, this volume). Climate varies at annual scales in Mexico, partly in response to the El Niño–Southern Oscillation (ENSO), which significantly affects precipitation in the Sierra Madre Occidental (Ropelewski and Halpert 1986, 1987, 1989; Kiladis and Diaz 1989; Cavazos and Hastenrath 1990; Stahle et al. 1998, 1999). We would expect such temporal variations in climate to synchronize the occurrence fire across this region by affecting the amount and moisture content of the fine fuels that carry surface fires. Assessing the annual relationship between climate and fire requires long accurate records. Unfortunately, detailed archival records of fire occurrence and climate are rare for much of the Sierra Madre Occidental. However, multicentury records of both can be reconstructed from annually dated tree-ring series for the region (Fulé and Covington 1997, 1999; Stahle et al. 1998, 1999).

Climate is not the only factor that drives variation in fire regimes through time. In the western United States, for example, fire regimes were dramatically affected by late nineteenth- and early twentieth-century changes in land use, such as grazing, road building, and timber harvesting (e.g., Leopold 1937; Savage and Swetnam 1990; Baisan and Swetnam 1997; Fulé and Covington 1997, 1999; Kaib 1998; Veblen et al. 1999; Veblen, Kitzberger, and Donnegan 2000; Heyerdahl, Brubaker, and Agee, in press). These land-use activities also intensified in the Sierra Madre Occidental in the mid-1900s with changes in the *ejido* system of land tenure in Mexico and may have affected fire regimes there.

Our objective was to infer the role of annual variation in regional climate and changes in land use in driving the occurrence of widely synchronous surface fires in pine-oak forests of the Sierra Madre Occidental of Mexico. Specifically, we reconstructed a multicentury history of fire from tree rings and fire scars at eight sites in the states of Durango and Chihuahua. We compared this history to existing tree-ring reconstructions of precipitation and ENSO activity (Stahle et al. 1998, 1999) and to archival records of land use.

Study Area

Sampling Sites

We relied on the knowledge of local foresters and researchers to judgmentally locate eight largely unlogged sites (2–6 ha each) containing relatively old, fire-scarred trees. The sites are distributed over nearly 700 km on the dry east side of the crest of the Sierra Madre Occidental in north-central Mexico (Fig. 7.1). All the sites are high in elevation (2440–2950 m, Table 7.1), but vary in slope (16–65%), aspect (3–343°) and topographic position (hill slopes: SSP, AJT, FCT, CHI, LBA; mesas: CAR, MLC; rocky ridge: ALF). The shallow, coarse-textured volcanic soils at most of our sites are typical of the region in general (Challenger 1998; Ferrusquía Villafranca 1998).

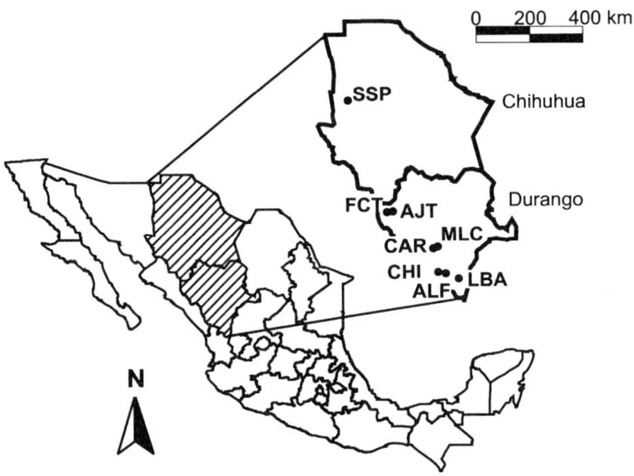

Figure 7.1. Mexico and the states of Durango and Chihuahua, showing the location of the eight sites at which we reconstructed fire history.

Forest composition at these sites, typical of this portion of the Sierra Madre Occidental (Bye 1995), was dominated by four pine species (*Pinus durangensis* Mart., *P. teocote* Schl. & Cham., *P. ayacahuite* Ehren., or *P. engelmannii* Carr.), but other species also occurred (*P. arizonica* Engelm., *P. herrerai* Mart., *P. lumholtzii* Robins. & Fern. and *Pseudotsuga menziesii* Mirb. Franco). Several species of *Quercus* were common at all sites, and a few species of *Arbutus* and *Juniperus* occurred at some southern sites. The understory was dominated by grasses and herbs.

Table 7.1. Location and topographic position of the sampling sites

Site name	Site code	Ownership	Nearby town	Elevation (m)	Aspect (degrees)	Slope (%)	Area sampled (ha)
Salsipuedes	SSP	El Largo	Madera	2620	314	47	2
Alto del Jiguital	AJT	El Tecuan	Tamazula	2440	18	34	3
Falda de la Cañada	FCT	Santa Ana	Tamazula	2660	212	26	4
El Carpintero	CAR	La Victoria–Miravalles	San Miguel	2790	295	18	4
Mesa de los Ladrónes	MLC	La Victoria–Miravalles	San Miguel	2830	179	16	3
Las Chivas	CHI	La Victoria	El Salto	2950	224	42	6
Arroyo de las Flores	ALF	La Campana	El Salto	2800	3	65	5
Las Bayas	LBA	UJED Research Forest	La Flor	2900	343	38	3

Note: Sites are ordered from north to south (top to bottom). All sites but LBA are owned by the *ejidos* indicated. UJED is the Universidad Juárez del Estado de Durango. All sites except SSP are in Durango.

Instrumental Climate

The Sierra Madre Occidental has a monsoonal climate with warm, wet summers, a long dry period in the spring and a shorter one in the fall (Fig. 7.2; Mosiño Alemán and García 1974). Most annual precipitation (70–80%) falls during the summer (June–September) as a result of the monsoon that develops over southern Mexico in May and spreads north along the Sierra Madre Occidental to reach Arizona and New Mexico by July (Mosiño Alemán and García 1974; Hales 1974; Douglas et al. 1993). Annual precipitation, derived from low-elevation stations for the states of Durango and Chihuahua, averages 40 and 56 cm, respectively (1945–1993; Douglas and Englehart 1995). While the seasonal distribution of precipitation at our high-elevation sampling sites is probably similar to these statewide averages, total precipitation is likely higher. For example, El Salto (elevation ca.2500 m), near the southern end of our sampling area, annually receives 92 cm of rain (1940–1993; Fig. 7.2). Winter precipitation can fall as snow at high elevations in the Sierra Madre Occidental, but persistent snow packs are rare (Mosiño Alemán and García 1974; Challenger 1998). Precipitation in the Sierra Madre Occidental varies through time, partly in response to global processes like ENSO. Winters are wetter than average during El Niño years and drier than average during La Niña years (Ropelewski and Halpert 1986, 1987, 1989; Kiladis and Diaz 1989; Cavazos and Hastenrath 1990). Temperatures are generally mild in this region, with an annual maximum in June (e.g., 16°C at El Salto, Fig. 7.2; Mosiño Alemán and García 1974).

Most modern fires in our study area burn in the spring (January–May, SEMARNAP 2000) as temperatures warm and fine fuels dry, but before monsoon rains increase fine-fuel moisture and encourage new growth of grasses and herbs. Lightning is most common from April to October and has been inferred

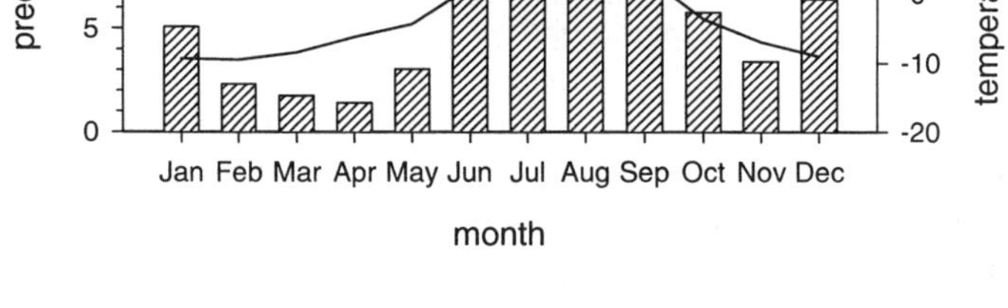

Figure 7.2. Climate of El Salto, Durango (1940–1993; elevation ca.2500 m). Total monthly precipitation is shown as bars, average monthly minimum, mean, and maximum temperatures are shown as lines.

as an ignition source for fire elsewhere in the Sierra Madre Occidental (Turman and Edgar 1982; Fulé and Covington 1999).

Historical Climate from Tree Rings

Precipitation has been reconstructed from tree rings for Durango (1386–1995; Stahle et al. 1999). These reconstructions explain 56% of the variance in the instrumental record of winter precipitation (previous November–March) and 53% of that in early summer (May–June, 1942–1983). For each of these seasons, modern precipitation varies similarly in Durango and Chihuahua ($r = 0.57$ and 0.59, for winter and early summer, respectively, $p < 0.01$, 1945–1994; Douglas and Englehart 1995). Consequently the reconstruction for Durango probably captures variation in precipitation at our sites in both states.

Variation in the strength and phase of ENSO is captured by an index of the Southern Oscillation, computed as the normalized difference in monthly surface pressure between Tahiti and Darwin, Australia, two measurement stations near the oscillating centers of high and low pressure (Enfield 1992; Allan, Lindesay, and Parker 1996). Years of low (high) values of the Southern Oscillation Index (SOI) are typically El Niño (La Niña) years (Deser and Wallace 1987). Winter SOI (December–February) has been reconstructed from tree rings and explains 53% of the variance in instrumental SOI (1706–1977, Cook 1985; Allan, Lindesay, and Parker 1996; Stahle et al. 1998).

Methods

Fire Regimes

Over an area of 2 to 6 ha per site, we used a chain saw to remove scarred sections from 19 to 32 of those trees that we judged to have the greatest number of visible, well-preserved scars (Arno and Sneck 1977). More than half of these trees (56%) were alive when sampled. We sanded the scarred sections until the cell structure was visible with a binocular microscope and assigned calendar years to tree rings using a combination of visual crossdating of ring widths and crosscorrelation of measured ring-width series (Holmes 1983). The crossdating was confirmed by another dendrochronologist for nearly half the dated sections (47%). We excluded 13% of the sampled trees from further analyses because they could not be crossdated.

We used fire scars as evidence of surface fires and identified them as discontinuities between cells, within a ring or along a ring boundary, where the cambium had been killed but not mechanically damaged, followed by overlapping, curled rings (Dieterich and Swetnam 1984). Additionally we obtained a small amount of supporting evidence of surface fires (5% of fire-scar dates) from abrupt changes in the width of annual rings (e.g., Landsberg et al. 1984; Sutherland, Covington, and Andariese 1991). However, because factors other than surface fires can cause abrupt changes in cambial growth (e.g., Brubaker 1978), we used such a change in a given sample as evidence of a surface fire only when it coincided with a fire scar in other samples at the same site.

We identified the calendar year in which each scar formed to determine the year of fire occurrence, and the position of each scar within the ring (ring boundary, earlywood, latewood, or unknown) as an indication of the season of fire occurrence (Dieterich and Swetnam 1984; Baisan and Swetnam 1990). In the Northern Hemisphere the season of cambial dormancy (i.e., the period corresponding to the ring boundary) spans two calendar years: from the time the cambium stops growing in the fall of one year until it resumes in the spring of the following year. For this study we assigned ring-boundary scars to the *following* calendar year because modern fires in the Sierra Madre Occidental generally burn in the spring, as they do under monsoonal climates elsewhere (Baisan and Swetnam 1990; Fulé and Covington 1997, 1999; SEMARNAP 2000). Scar position could not always be determined where it was obscured by rot or insect galleries or where rings were narrow.

For each site we composited the dates from all trees into a single record of fire occurrence (Dieterich 1980) and computed the intervals between years in which a fire scarred at least one tree at that site. We analyzed fire intervals for the period after which at least five trees (17–29% of trees) per site had scarred at least once and before any major recent shifts in fire regimes (Table 7.2). Two-parameter Weibull distributions fit the fire-interval density distribution at seven of the sites ($p > 0.05$, one-sample Kolmogorov-Smirnov goodness-of-fit test) and marginally fit the distribution at the remaining site (AJT, $p = 0.03$). Consequently we used percentiles of the fitted Weibull distribution to characterize the distribution of fire intervals at each site (Grissino-Mayer 1999, 2001).

Table 7.2. Size of sampling areas and amount of fire evidence collected

Site	Number of trees crossdated	Number of fire scars	Abrupt changes in ring width	Earliest year sampled	Analysis start year	Analysis end year
SSP	18	212	9	1629	1785	1951
AJT	22	191	18	1669	1772	1893
FCT	25	86	6	1754	1857	1994
CAR	24	234	8	1700	1795	1951
MLC	29	222	13	1729	1797	1951
CHI	23	165	11	1791	1898	1994
ALF	22	236	0	1779	1841	1994
LBA	17	123	4	1687	1817	1951
Total	180	1469	69			

Note: Earliest years are dates of first rings found at each site, while analysis start year is the first year for which at least five trees at the site had scarred at least once. Analysis end year is either the last year of record or the approximate year of an abrupt decrease in fire frequency at each site. Number of scars are for the entire period of record.

Drivers of Temporal Variation in Historical Fire Regimes

To identify climate drivers of fire at annual scales, we determined whether variation in regional climate was associated with variation in the occurrence of widespread surface fires in our study area. Specifically, we assessed whether climate during widespread- and non-fire years was significantly different from climate during the preceding and following years (±5 years), using superposed epoch analysis (SEA; Baisan and Swetnam 1990; Swetnam and Betancourt 1992; Grissino-Mayer 1995). We used this analysis to test for departures in climate during two sets of years at our eight sites: *widespread fire years*, namely those with at least 50% of sites recording a fire (~1 standard deviation above the mean; 31 years); and *non-fire years*, namely those with no sites recording fire (68 years). For both sets of years, we computed departures in three climate parameters: *winter precipitation* (previous November–March; Stahle et al. 1999), *early summer precipitation* (May–June; Stahle et al. 1999), and *winter SOI* (December–February; Stahle et al. 1998). We identified significant departures as those with $p < 0.05$, determined by bootstrapping (1000 trials; Swetnam and Betancourt 1992; Mooney and Duvall 1993; Grissino-Mayer 1995). We conducted this analysis from 1772 to 1977, the period after which at least five trees per site had scarred at least once (17–29% of trees per site; Table 7.2) to the end of the record of reconstructed SOI (1977). However, the tree-ring record started after 1772 for some sites, so we computed the percentage of sites burning during a given year as a percentage of those sites that had a record for that year. Finally, we repeated these SEA analyses but included existing fire history reconstructions from an additional four sites in Durango (Fulé and Covington 1997).

To identify nonclimatic drivers of surface fire, we determined whether changes in land use were synchronous with variation in surface fire occurrence in our study area. We used regional trends in land use to make inferences about the effects of land use on the history of fire at our sites because we lack site-specific land-use histories. Specifically, we used a national record of the amount of land redistributed via the *ejido* system (Sanderson 1984), as an indication of likely settlement in the forests of the Sierra Madre Occidental. We compared this time series to that of percentage of sites recording fire per year. To emphasize decadal variation, we smoothed the time series of fire occurrence using a cubic spline that retained 50% of the variance present in the original series at periods of 20 years (Diggle 1990).

Results

Fire Regimes

We removed fire-scarred sections from 206 trees, most of which were *Pinus durangensis* (40%), *P. teocote* (14%), *P. ayacahuite* (10%), *P. engelmannii* (6%) or unknown species (26%; Table 7.2). The remaining samples came from

P. arizonica (1%), *P. herrerai* (1%), *P. lumholtzii* (1%) or *Pseudotsuga menziesii* (1%). We were able to crossdate 180 of these trees, yielding 1469 fire scars, and 69 abrupt changes in ring width (Fig. 7.3; Dieterich 1980; Grissino-Mayer 2001).

We were able to assign an intra-ring position to most scars (73% of 1341 scars during the analysis periods; Table 7.2). The distribution of scars by intra-ring position was similar among sites. Of the scars to which we could assign an intra-

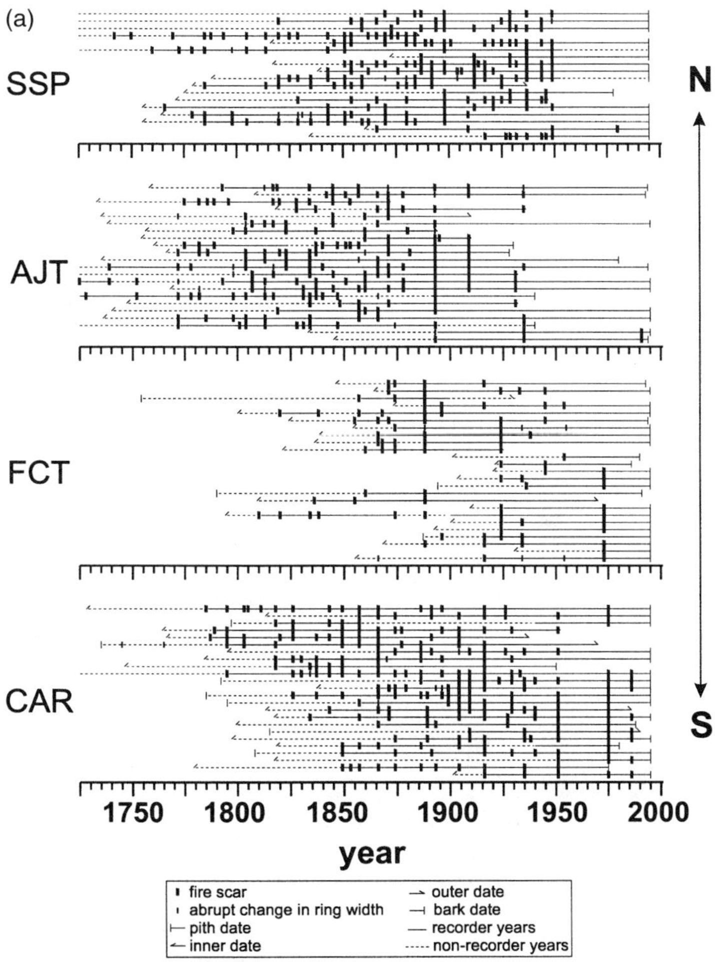

Figure 7.3. Fire charts. Each horizontal line shows the fires recorded by a single tree through time. Recorder years generally follow the first scar on each tree. Nonrecorder years precede the formation of the first scar on each tree but also occur when tree rings are consumed by subsequent fires or rot. Inner and outer dates are the dates of the earliest or latest rings sampled for trees where pith or bark were not sampled.

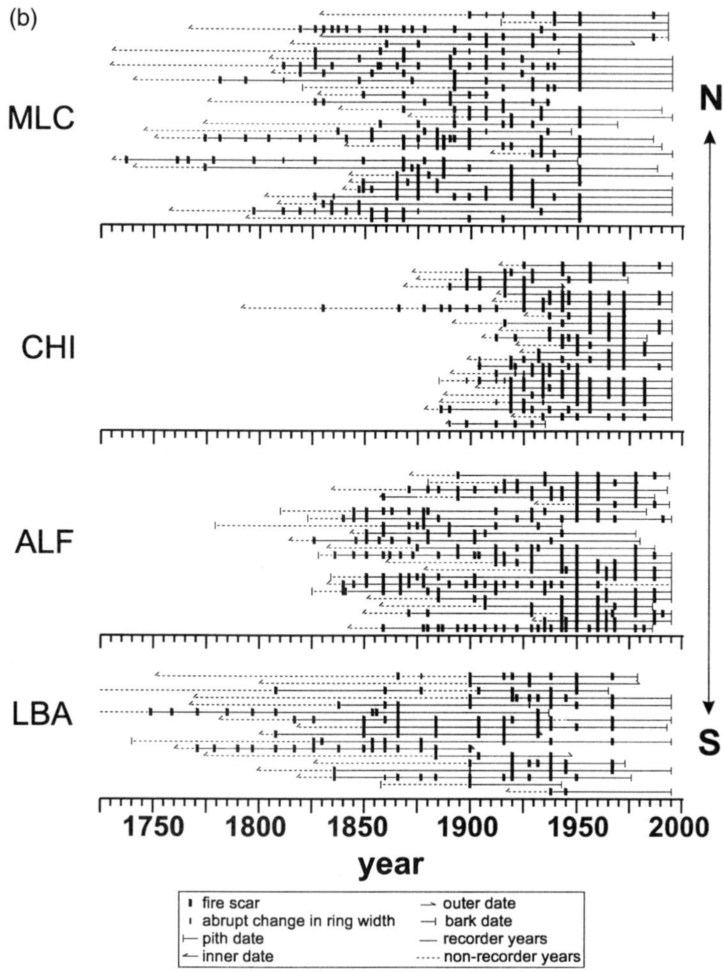

Figure 7.3. *Continued*

ring position, most were created by fires burning when the cambium was dormant (63% ring-boundary scars; Fig. 7.4). Most of the rest of the scars were created during the growing season (35% earlywood scars), and of these, most were formed early in that season (51% in the first third of the earlywood, 35% in the middle third). Only a few scars were created by fires burning late in the cambial growing season (2% latewood scars).

The distribution of intervals was similar for the composite surface fires from our sample of trees at most sites, although intervals were slightly longer and more variable, at FCT and LBA than at the other sites (sampled areas 2–6 ha; Fig. 7.5). Weibull median intervals were 3 to 6 years, minimum intervals 1 to 2 years and maximum intervals 9 to 20 years. Most fires (76–100% per site), were recorded

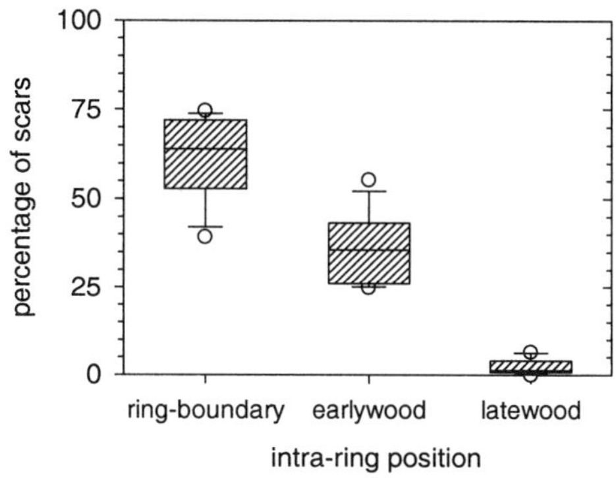

Figure 7.4. Distribution among sites of intra-ring position of fire scars, as a percentage of scars per site for which position could be determined (974 scars or 56–83% per site). Ring-boundary scars were formed by fires that burned between growing seasons, when the cambium was dormant, whereas earlywood and latewood scars were formed by fires that burned during the growing season. The boxes enclose the 25th to 75th percentiles of the distribution. The whiskers enclose the 10th to 90th percentiles and the horizontal line across each box indicates the 50th percentile. Circles mark all values lying outside the 10th to 90th percentiles.

by more than one tree, with an average of 5 trees recording a fire per site (range: 1–23).

At some sites surface fire regimes changed abruptly in the late nineteenth to midtwentieth century, with sites near one another generally experiencing syn-

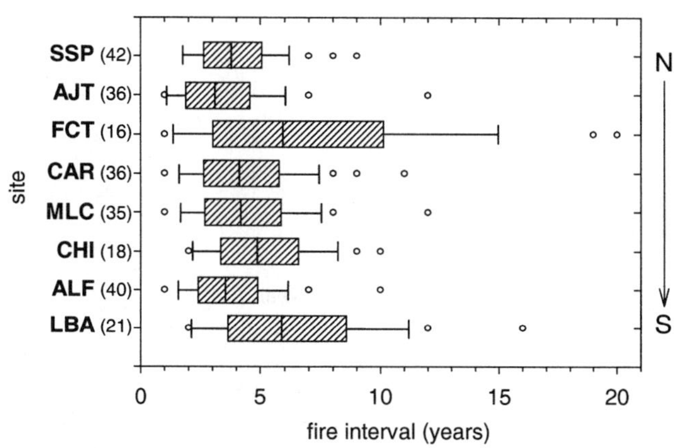

Figure 7.5. Composite fire intervals by site, with the number of intervals in parentheses. The box-and-whisker sets are as defined for Figure 7.4, but mark the percentiles of Weibull distributions fit to the composite fire intervals at each site, for the analysis periods indicated in Table 7.2. Trees were sampled over 2 to 6 ha per site (Table 7.1).

chronous changes. Specifically, fires nearly ceased after the late 1800s at AJT and after about 1950 at SSP, CAR, MLC, and LBA (Fig. 7.3). In contrast, surface fires remained frequent until the time of sampling at CHI and ALF. The abrupt cessation of fire at some sites is not likely an artifact of sampling dead trees, and hence low twentieth-century sample size, because an average of 14 trees (range: 7–20) had a record extending into the late twentieth century at each site. Fires may have been frequent at FCT before about 1950, as they were at nearby AJT. However, the record at FCT is less than 150 years for most trees, which is too short to determine if the fire regime changed at this site around 100 years ago, as it did at AJT.

Drivers of Temporal Variation in Historical Fire Regimes

Annual variation in climate was a strong driver of surface fires in the Sierra Madre Occidental. Not surprisingly, fires were widespread in years with significantly dry winters and early summers, but did not burn during significantly wet years (Fig. 7.6a, b). Consistent with these results, fires were widespread during years of significantly high SOI (Fig. 7.6c), which tend to be La Niña years and have dry winters. In contrast, variation in SOI was not significantly associated with non-fire years.

Climate in preceding years was also an important driver of surface fires in our study area. Specifically, fires were widespread following several years with wet

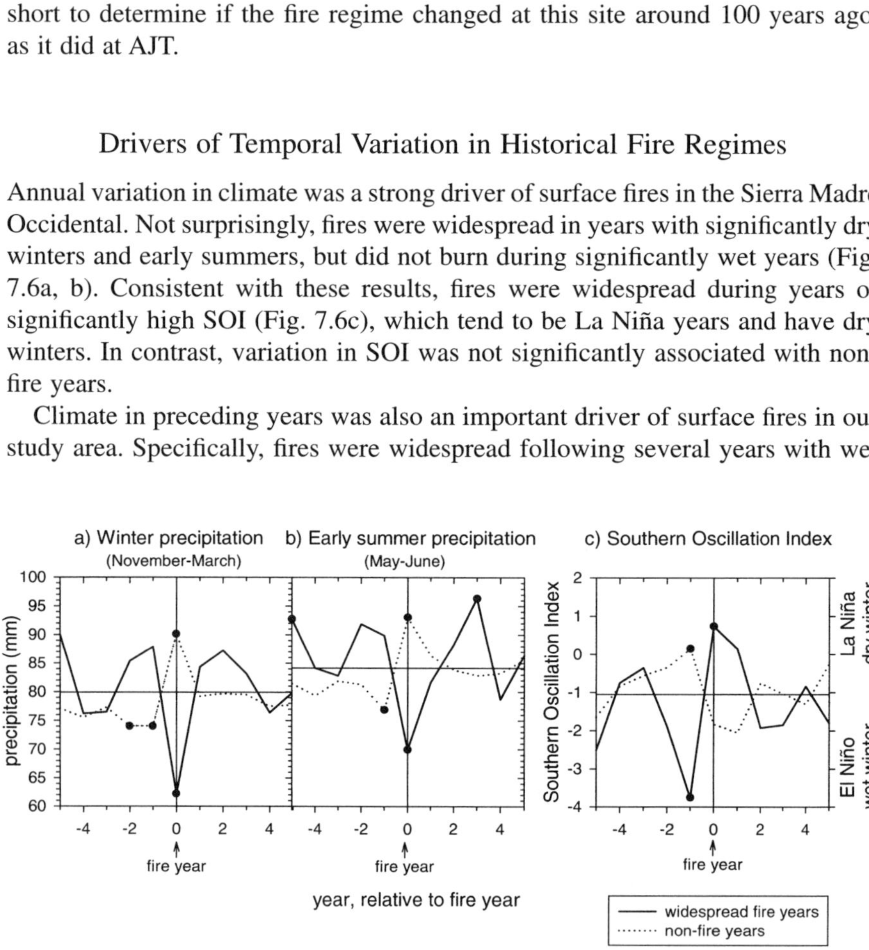

Figure 7.6. Annual association of fire and climate. Average departure from climate during widespread fire years (31 years, >50% of sites recording fire) and non-fire years (68 years, no sites recording fire), and for years immediately before and after these years. Solid dots mark departures that fall outside the 95% confidence interval, determined by bootstrapping. The horizontal lines indicate average precipitation or SOI for the analysis period (1772–1977).

winters and early summers (although not significantly wet), while fires did not burn following 1 to 2 significantly dry years (Fig. 7.6a, b). Consistent with these results, fires were widespread following a year with significantly low SOI (Fig. 7.6c), which tend to be El Niño years and have wet winters. This association is reversed for non-fire years, which followed a year of significantly high SOI (La Niña years).

Surface fires over a broader area were similarly driven by climate. When we repeated the SEA analyses including four additional existing fire history reconstructions from Durango (Fulé and Covington 1997), we found nearly identical patterns of significant climate departures for both widespread and non-fire years.

In addition to varying at annual time scales, the occurrence of widespread fires varied at decadal time scales, sometimes due to variation in the synchrony of fires among sites but sometimes to a lack of fire (Fig. 7.7). Compared to the period from the late 1700s to about 1930, fires were somewhat less synchronous among sites for brief periods around 1810 and 1910. However, the decrease in synchrony around 1810 could be due to low sample size because few of the sampled trees have a record before this time. The occurrence of widespread fires declined sharply beginning around 1930, due to an abrupt cessation of fires at some sites (Fig. 7.3). This abrupt decline was synchronous with the beginning of extensive distribution of *ejido* lands in Mexico.

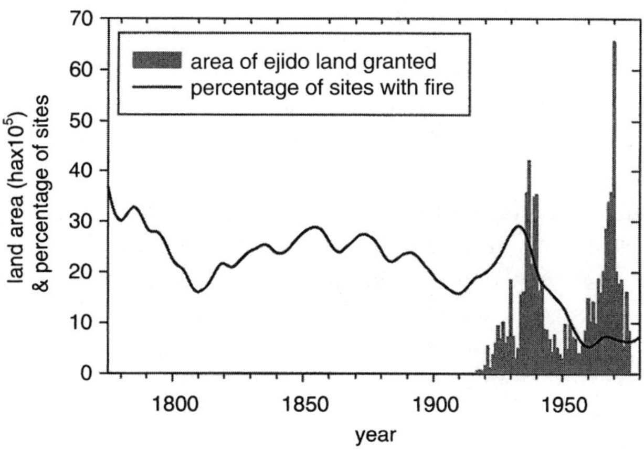

Figure 7.7. Decadal variation in the occurrence of synchronous fires, compared to changes in land tenure in Mexico. The percentage of sites recording fire per year was determined from the combined composite records of fire occurrence for the analysis periods identified for each site in Table 7.2, smoothed using cubic splines with a 50% frequency cutoff at 20 years. Land tenure is the amount of land distributed to *ejidos* (Sanderson 1984).

Discussion

Fire Regimes

Based on our sample of trees, composite surface fire intervals were remarkably similar across the study area, despite topographic variation among the sites (Fig. 7.5, Table 7.1). Topographically driven variation in solar insolation was an important driver of spatial variation in historical surface fire regimes farther north (e.g., Taylor and Skinner 1998; Heyerdahl, Brubaker, and Agee 2001). However, differences in solar energy input to steep slopes of different aspect are not as great in Mexico as they are at higher latitudes (Holland and Steyn 1975) and so may not drive differences in fire frequency as they do farther north. Furthermore the frequency of fire at these sites may not be driven only by the topographic characteristics of the sampled area but may also depend on the frequency of fire in surrounding areas because our sites are not surrounded by fire breaks (Agee, Finney, and de Gouvenain 1990; Bergeron 1991; Heyerdahl, Brubaker, and Agee 2001). We do not have a clear explanation for the long and variable intervals that we found at FCT and LBA, relative to the other sites. The record at FCT may entirely postdate a change in fire intervals because this site is near AJT. Fires at AJT nearly ceased in the late 1800s and major changes in fire regimes are generally synchronous among sites that are near one another. However, we compare fire intervals among our sites cautiously because these sites were not selected to capture spatial variation in fire frequency. Rather, we selected sites and trees that we expected to yield relatively long records of surface fires in order to explore the role of climate in driving widespread fires. Consequently we may not have captured the full range of variability in fire frequency across the landscape (Baker and Ehle 2001; Lertzman, Fall, and Dorner 1998). Furthermore the fire intervals we report may be affected by the small differences in area over which they were composited (2–6 ha; Table 7.1; Arno and Petersen 1983; Baker and Ehle 2001).

The intervals we report probably include fires of different sizes, although we did not reconstruct this parameter of fire regimes. The number of scarred trees per fire at our sites yields little information about the size of those fires because we sampled trees over relatively small areas (2–6 ha). Most fires were recorded by at least several trees at a site (average of 73% of fire years per site recorded by ≥3 trees). However, even fires recorded by a single tree may be extensive because our sites are not surrounded by fire breaks so that fires may have spread into them from surrounding areas.

Most of the fires we reconstructed probably burned in the spring, before the onset of the monsoon rains that wet litter fuel and encourage new growth of grasses and herbs. This is consistent with the seasonality of most modern fires in the Sierra Madre Occidental which burn during the dry spring when lightning is most common (Mosiño Alemán and García 1974; Hales 1974; Turman and Edgar 1982; Douglas et al. 1993; SEMARNAP 2000), and with written reports of spring burning by indigenous people (Sheridan and Naylor 1978; Graham 1994). Most fire years with ring-boundary scars on some trees also had scars in the first third of the earlywood on other trees (62%), consistent our assumption that most fires burned early in the year, when some of the trees had begun growing. Likewise, no fire years had ring-boundary scars on some trees and latewood scars on others, suggesting that few fires burned late in the year. However, some fall or winter fires may have burned in our study area because some fire years (24%) had only ring-boundary scars. Consequently we cannot determine whether these fires burned during the fall, after growth ceased, or during the following spring, before growth began again. Although lightning is not as common in the fall and winter as in the summer, humans could have ignited fires in these forests during the brief fall dry season.

Historically surface fires in our study area, and at sites elsewhere in Durango (Fulé and Covington 1997, 1999) probably burned earlier in the year than surfaces fires in the Mexico/U.S. borderlands. Most historical fires in our study area burned during the season of cambial dormancy whereas in the borderlands, they burned during the cambial growing season (Swetnam, Baisan, and Kaib 2001). Based on the few existing studies of cambial phenology, fires in the borderlands burned during the warm spring dry period (April–June) consistent with the seasonality of lightning and modern fires in that region (Baisan and Swetnam 1990; Swetnam, Baisan, and Kaib 2001). We know of no studies of cambial phenology in the pine-oak forests of the Sierra Madre Occidental, but the early spring seasonality we inferred from fire scars for this region is also consistent with the seasonality of modern precipitation, lightning, and fires. However, these differences in the intra-ring position of fire scars could result from differences in cambial phenology between the two regions, rather than from a difference in the season of burning.

Climate Was a Strong Driver of Surface Fire Regimes

Current year's climate synchronized the occurrence of widespread surface fires among our sites in the Sierra Madre Occidental, probably by affecting fuel moisture and perhaps by affecting fuel amount (Fig. 7.6). In this region, where winters are relatively dry and cold, fires burn primarily in the spring, before the flush of live surface fuels and the onset of monsoon rains in early summer which wet surface fuels and inhibit fire ignition and spread. Winter precipitation probably affects fire by influencing soil moisture and hence the growth of live surface fuels. Consequently, after dry winters, the spring flush of grasses and herbs may be delayed, lengthening the fire season and increasing the likelihood of widespread fires in this region. The opposite may occur after wet winters, when high soil moisture leads to an early spring flush and a relatively short fire season. Winter precipitation probably does not affect the moisture content of fine fuels during the subsequent fire season because any increased moisture will evaporate quickly with warm, dry weather. However, the onset of monsoon rain in early summer can affect fine fuel moisture at the beginning of the fire season. During years when the onset of the monsoon rains was delayed (i.e., years with low early summer precipitation), fine fuels remained dry. As a result the fire season was relatively long and the probability of synchronous fires was greater than during years when the monsoon rains began early. These relationships are consistent with the effect of precipitation on fire regimes in monsoonal climates elsewhere (Swetnam and Baisan, Chapter 6, this volume).

The current year associations we found between surface fire and ENSO are generally consistent with those we found between fire and precipitation, because these two measures of climate are strongly associated in the study area. In the Sierra Madre Occidental, dry La Niña winters may have resulted in a delay in the spring flush of grasses and herbs and hence a relatively long fire season, increasing the probability of widespread fires, as described above. Historical ENSO activity also affected the length of the fire season elsewhere in North America (Heyerdahl, Brubaker, and Agee, in press). We would expect wet El Niño winters to have the opposite effect, suppressing widespread fire activity, as they do in the American Southwest (Swetnam and Betancourt 1990). However, El Niño years were not significantly associated with non-fire years, perhaps because the effect of ENSO on weather, and hence fire, varies from one event to the next (Enfield 1992; Allen 2000; Kitzberger, Swetnam, and Veblen 2001). Specifically, in the Sierra Madre Occidental, ENSO activity sometimes affects winter temperature as well as winter precipitation. For example, the El Niño winters of 1982–1983 and 1997–1998 were very cold as well as wet in northern Mexico (SEMARNAP 2000). Consequently heavy snow broke tree limbs and tops, increasing fuel loads so that extensive areas burned when these fuels dried in the spring (Alvarado 1984). We do not know how common these cold El Niño winters were historically, because there are no reconstructions of winter temperature for this region. However, the occurrence of some cold El Niño winters would explain why fire activity is not strongly suppressed during El Niño years when viewed over several centuries in our study area.

Prior year's climate also strongly synchronized the occurrence of widespread surface fires among our sites in the Sierra Madre Occidental, probably by affecting fuel amount, rather then fuel moisture. The growth of grasses and herbs was probably enhanced during wet years, increasing the amount of fine-fuel available to carry surface fires in subsequent dry years. This enhanced growth may also have increased fuel continuity so that fires spread more effectively, similar to the effect of wet years on fine-fuel production inferred for dry pine forests elsewhere (Swetnam and Baisan 1996; Baisan and Swetnam 1997; Swetnam and Betancourt 1998; Veblen, Kitzberger, and Donnegan 2000). In contrast, these fine live fuels were probably reduced during prior dry years. Specifically, dry winters may have delayed or inhibited the spring flush of grasses and herbs, especially given the poor moisture retention of the coarse soils at our sites. Fires during dry prior years probably also consumed these fuels, further limiting the amount of fine fuel available to carry fire in subsequent years (Swetnam and Betancourt 1998).

The prior year associations we found between surface fire and ENSO are generally consistent with those we found between fire and precipitation. At our sites, fires were widespread in years following wet El Niño years but did not burn in years following dry La Niña years, consistent with the effect of precipitation on the growth and consumption of fine live fuels, discussed above. ENSO varies with a period of two to five years (Enfield 1992; Stahle 1998), so the association we found between widespread fire and prior year's ENSO activity is probably not an artifact of the intrinsic scale of variation in ENSO. Last, fires were widespread during La Niña years and following prior El Niño years. This switching from one atmospheric state to another is characteristic of the ENSO system (Kiladis and Diaz 1989), and it drives widely synchronous fires elsewhere in North and South America (Swetnam and Betancourt 1998; Kitzberger, Swetnam, and Veblen 2001).

Land-Use Change Caused Recent Cessation of Surface Fires

The recent abrupt cessation of surface fires at some of our sites likely resulted from a complex mix of local changes in land use rather than from regional variation in climate, since fires did not cease synchronously at all sites (Fig. 7.3). Fire at individual sites can be dramatically impacted by grazing, fire use or suppression, timber harvesting, and the construction of roads and railways (e.g., Leopold 1937; Dieterich 1983; Savage and Swetnam 1990; Baisan and Swetnam 1997; Fulé and Covington 1997, 1999; Kaib 1998; Veblen et al. 1999; Veblen, Kitzberger, and Donnegan 2000; Heyerdahl, Brubaker, and Agee, in press). However, local variation in the intensity of these activities can impact fire regimes differently among sites, particularly for small, widely dispersed sites such as those we sampled. We lack local land-use histories for our sites but speculate that the differences in timing of fire exclusion among them probably resulted from differences in the type and timing of changes in land use.

Mid-twentieth-century changes in Mexican land tenure probably resulted in local increases in human occupation of the high-elevation pine-oak forests at some of our sites (Fulé and Covington 1997). There is little quantitative information on human use of the remote and rugged Sierra Madre Occidental before the twentieth century. However, before 1900 these mountains were sparsely populated by indigenous people, such as the Tarahumara, Tepehuano, Mayo, and Yaqui, who occupied the lower valleys and deep canyons in winter and the upper mountains in summer. They practiced slash-and-burn agriculture and used fire for hunting and religious purposes (Bye 1976; Sheridan and Naylor 1978; Graham 1994). There is little evidence that the high-elevation pine-oak forests of this region were densely occupied until the mid-twentieth century, in response to reform in the land tenure system in Mexico (Sanderson 1984; Thompson and Wilson 1994). In the early 1900s, shortly after the Mexican Revolution, new legislation (Agrarian Law 1915; Mexican Constitution 1917) legalized the *ejido* system, the reallocation of land to small communities of landless people. Despite this legalization not much land was actually distributed until the administration of Lazaro Cárdenas (1934–1940) when nearly 800,000 people in Mexico received land grants of about 20 million hectares (Sanderson 1984; Thompson and Wilson 1994). The distribution of *ejido* lands brought a wave of people from low-elevation agricultural areas to settle the forested mountains, resulting in a change from traditional land use. Today all but one of our sites are owned by *ejidos* (Table 7.1).

The movement of people to forest *ejidos* in the Sierra Madre Occidental in the mid-1900s may have affected fire regimes by introducing, or intensifying, cattle grazing, road building, or logging (Fulé and Covington 1997; Kaib 1998), and perhaps by changing traditional uses of fire. We speculate that some or all of these changes in land use may have caused the mid-1900s cessation of fire at four of our sites (SSP, CAR, MLC, and LBA). Cattle were introduced to southern Mexico in the early 1500s and rapidly spread north (Rouse 1977; Jordan 1993). However, while cattle grazed on the lower slopes of the Sierra Madre Occidental (Leopold 1937), they were probably not grazed in great numbers in the high elevations of our sampling sites until the major distribution of *ejido* land in the mid-1900s. The introduction of livestock grazing may have resulted in fire exclusion at some of our sites at this time, as it has elsewhere in Mexico and the American Southwest, by reducing both the amount and continuity of the fine fuel that carries surface fires in these forests (Baisan and Swetnam 1997; Leopold 1924, 1937; Madany and West 1983; Savage and Swetnam 1990; Grissino-Mayer and Swetnam 1997; Kaib 1998; Mast, Veblen, and Linhart 1998; Fulé and Covington 1999; Swetnam, Baisan, and Kaib 2001). Grazing may not be the only cause of change in fire regimes at this time. Roads and trails built to access *ejido* lands, and harvest timber can interrupt fuel continuity and may have reduced the number of fires that spread into our sites. Changes in human use of fire may also have contributed to the exclusion of fire in the mid-1900s at some of our sites. We have little quantitative information on the use of fire by indigenous people, but the occupation of *ejido* lands probably curtailed their ignition of fire. This may have contributed to the decline in fire if these ignitions were an important cause of the fires we reconstructed at our sites. Twentieth-century fire suppression is not a likely cause of the changes we reconstructed in fire regimes because fire-fighting resources were limited during this time (Leopold 1937; Dieterich 1983; González-Cabán and Sandberg 1989; Fulé and Covington 1999; Kaib 1998).

We speculate that the abrupt cessation of fire at some of our sites (AJT and perhaps FCT) in the late 1800s could have been caused by a dramatic increase in travel routes, decades before the major distribution of *ejido* lands. The Sierra Madre Occidental is a high and rugged mountain range (200–3000 m) over which few easy travel routes exist (Jordon 1993). Consequently few roads crossed it in the early twentieth century (Leopold 1937). In Mexico, a few kilometers of railroad were constructed in the nineteenth century, but the major construction of rail lines, including those from southern Mexico northward into the central highlands, began in 1880 (Coatsworth 1981). In that year, there were 770 km of railroad but this had expanded to 24,700 km by 1911 (Powell 1921). These roads may have allowed access to parts of the Sierra Madre Occidental, resulting in changes in land use that affected fire regimes. For example, silver mines near AJT and FCT may have been established at this time and resulted in timber harvesting, leading to a decrease in surface fires.

We speculate that frequent surface fires continued to burn into the late 1990s at two sites (CHI and ALF) because they were relatively inaccessible and ignition of fires remained frequent. ALF is a rocky ridge that may have been a poor site for grazing or a difficult area for road building and timber harvesting. At CHI, most of the trees are young. Perhaps this forest regenerated after logging or a stand-replacing fire in the mid-1800s and may not have been suitable for harvesting or grazing during the time of major *ejido* land distribution in the mid-1900s.

Conclusion

Our objective was to infer the drivers of temporal variation in fire regimes in pine-oak forests of the Sierra Madre Occidental in north-central Mexico. We reconstructed a multicentury history (1772–1994) of the occurrence of surface fires from 1469 fire scars on 180 trees sampled at 8 sites over nearly 700 km in the states of Durango and Chihuahua. We compared our fire histories to existing tree-ring reconstructions of winter and early summer precipitation and the Southern Oscillation Index. Fire intervals were similar among our sites, with Weibull median fire intervals of 3 to 6 years. Most fires probably burned in the warm, dry spring, based on the intra-ring position of fire scars (98% formed during the season of radial dormancy or early in the growing season) and the seasonality of precipitation, lightning, and modern fires in this region. However, some fall or winter fires may have occurred. Annual variation in precipitation and El Niño–Southern Oscillation were strong drivers of current year's fire, probably through their effects on fuel moisture. Extensive fires generally burned during dry years but not during wet ones. Extensive fires also typically burned during La Niña years, which tend to have dry winters in this region. Climate in prior years was also a strong driver of fire, through its effect on fuel amount. Widespread fires often burned following one to two wet years and also following El Niño years, which tend to have wet winters in this region. Likewise fires were not widespread following dry years and following La Niña years. Prior year's climate probably affected the growth of grass and herbaceous fuel. Changes in land use, rather than climate, probably caused the near cessation of fire that we reconstructed at some sites because these shifts did not occur synchronously (some ca.1900, some ca.1950). Frequent surface fires continued to burn until the time of sampling at two of our sites.

Acknowledgments. For help with field sampling, we thank Jeffrey R. Bacon, Jorge Bretado Velazquez, Jose Coria Quiñonez, Jon Datillo, Stacy Drury, Kat Maruoka, A. Enrique Merlin Bermudez, Fernando Najera, Humberto Ortéga, Gonzalo Rodrigez Lara, Octaviano Rosales, Santiago Guadalupe Salazar Hernandez, Rosalba Salazar, Francisco Soto Rodriguez, Godofredo Soto Rodrigez, Jesús Soto Rodriguez, Miguel Soto, and Bob Vihnanek. For help with sample preparation, we thank Jon Datillo and Travis Kern. We thank Steven J. McKay for assisting with laboratory and data analysis, Stacy Drury for providing vegetation data for the Las Bayas site, and Tom Thompson for drafting Figure 7.1. For reviews of the manuscript, we thank J. K. Agee, W. L. Baker, S. Drury, P. Z. Fulé, M. Harrington, S. J. McKay, D. L. Peterson, E. K. Sutherland, S. Sutherland, T. W. Swetnam, and one anonymous reviewer. Partial funding for this project came from the USDA Forest Service, Pacific Northwest Research Station.

References

Agee, J.K., Finney, M., and de Gouvenain, R. 1990. Forest fire history of Desolation Peak, Washington. *Can. J. For. Res.* 20:350–356.

Allan, R.J. 2000. ENSO and climatic variability in the past 150 years. In *El Niño and the Southern Oscillation: Multiscale Variability and Global and Regional Impacts*, eds. H.F. Diaz, and V. Markgraf, pp. 3–55. Cambridge: Cambridge University Press.

Allan, R.J., Lindesay, J., and Parker, D. 1996. *El Nino/Southern Oscillation and Climatic Variability*. Victoria, Australia: CSIRO Publishing.

Alvarado, C.E. 1984. Health diagnostics of wind-blown and snow-damaged trees in the Forest Management Unit No. 2 PROFORMEX, Durango. B.S. thesis. Chapingo, Mexico: University of Chapingo.

Arno, S.F., and Petersen, T.D. 1983. *Variation in Estimates of Fire Intervals: A Closer Look at Fire History on the Bitterroot National Forest*. Res. Pap. INT-301. Ogden, UT: USDA Forest Service, Intermountain Forest and Range Experiment Station.

Arno, S.F., and Sneck, K.M. 1977. *A Method for Determining Fire History in Coniferous Forests of the Mountain West*. Gen. Tech. Rep. GTR-INT-42. Ogden, UT: USDA Forest Service, Intermountain Forest and Range Experiment Station.

Baisan, C.H., and Swetnam, T.W. 1990. Fire history on a desert mountain range: Rincon Mountain Wilderness, Arizona, USA. *Can. J. For. Res.* 20:1559–1569.

Baisan, C.H., and Swetnam, T.W. 1997. *Interactions of Fire Regimes and Land Use in the Central Rio Grande Valley*. Res. Pap. RM-RP-330. USDA Forest Service, Rocky Mountain Forest and Range Experiment Station, Fort Collins, CO.

Baker, W.L., and Ehle, D. 2001. Uncertainty in surface-fire history: The case of ponderosa pine in the western United States. *Can. J. For. Res.* 31:1205–1226.

Bergeron, Y. 1991. The influence of island and mainland lakeshore landscapes on boreal forest fire regimes. *Ecology* 72:1980–1992.

Brubaker, L.B. 1978. Effects of defoliation by Douglas-fir tussock moth on ring sequences of Douglas-fir and grand fir. *Tree-Ring Bull.* 38:49–60.

Bye, R. 1993. The role of humans in the diversification of plants in Mexico. In *Biological diversity of Mexico: Origins and distribution*, eds. T.P. Ramamoorthy, R. Bye, A. Lot, J. Fa, pp. 707–731. New York: Oxford University Press.

Bye, R. 1995. Prominence of the Sierra Madre Occidental in the biological diversity of Mexico. In *Biodiversity and Management of the Madrean Archipelago: The Sky Islands of Southwestern United States and Northwestern Mexico*, tech. coord. L.F. DeBano, and P.F. Ffolliot, pp. 19–27. General Technical Report RM-GTR-264, Fort Collins, CO: USDA Forest Service, Rocky Mountain Forest and Range Experiment Station.

Cavazos, T., and Hastenrath, S. 1990. Convection and rainfall over Mexico and their modulation by the Southern Oscillation *Int. J. Climatol.* 10:377–386.

Challenger, A. 1998. Utilizacion y conservacion de los ecosistemas terrestres de Mexico. Pasado, presente y futuro. Comision Nacional para el Conocimiento de la Biodiversidad. Mexico, D.F.

Coatsworth, J.H. 1981. *Growth against Development: The Economic Impact of Railroads in Porfirian Mexico*. DeKalb: Northern Illinois University Press.

Cook, E.R. 1985. A time series approach to tree-ring standardization. Ph.D. dissertation. Tucson: University of Arizona.

Deser, C., and Wallace, J.M. 1987. El Niño events and their relation to the Southern Oscillation. *J. Geophys. Res.* 92:14189–14196.

Dieterich, J.H. 1980. The composite fire interval: a tool for more accurate interpretation of fire history. In *Proceedings of the Fire History Workshop*, tech. coord. M.A. Stokes, and J.H. Dieterich, pp. 8–14. October 20–24, 1980, *Tucson. Gen.* Tech. Rep. RM-81., Fort Collins, CO: USDA Forest Service, Rocky Mountain Forest and Range Experiment Station.

Dieterich, J.H. 1983. *Historia de los incendios forestales en la Sierra de los Ajos, Sonora.* Instituto Nacional de Investigaciones Forestales, Centro de Investigaciones Forestales del Norte. Nota Tecnica no. 8, PR-04.

Dieterich, J.H., and Swetnam, T.W. 1984. Dendrochronology of a fire-scarred ponderosa pine. *For. Sci.* 30:238–247.

Diggle, P.J. 1990. *Time Series: A Biostatistical Introduction.* Oxford Statistical Science Series 5. New York: Oxford University Press.

Douglas, A.V., and Englehart, P.J. 1995. Diagnostic studies of the Mexican monsoon. In *Proceedings of the Nineteenth Annual Climate Diagnostics Workshop*, pp. 202–206. U.S. Department of Commerce, National Oceanic and Atmospheric Administration, Divisional data computed from the Global Historical Climatology Network (available from National Climatic Data Center, Asheville, NC).

Douglas, M.W., Maddox, R.A., Howard, K.W., and Reyes, S. 1993. The Mexican monsoon. *J. Clim.* 6:1665–1677.

Enfield, D.B. 1992. Historical and prehistorical overview of El Niño/Southern Oscillation. In *El Niño: historical and paleoclimatic aspects of the Southern Oscillation*, eds. H.F. Diaz, and V. Markgraf, pp. 95–117. New York: Cambridge University Press.

Ferrusquía Villafranca, I. 1998. Geologia de Mexico: Una sinopsis. In *Diversidad biologica de Mexico*, eds. T.P. Ramamoorthy, R. Bye, A. Lot, and J. Fa, pp. 3–108. Mexico City: Instituto de Biologia, Universidad Autonoma de Mexico.

Fulé, P.Z., and Covington, W.W. 1997. Fire regimes and forest structure in the Sierra Madre Occidental, Durango, Mexico. *Acta Botánica Mexicana* 41:43–79.

Fulé, P.Z., and Covington, W.W. 1999. Fire regime changes in La Michilía Biosphere Reserve, Durango, Mexico. *Conserv. Biol.* 13:640–652.

González-Cabán, A., and Sandberg, D.V. 1989. Fire management and research needs in Mexico. *J. For.* 87:20–26.

Graham, M. 1994. *Mobile Farmers: An Ethnoarchaeological Approach to Settlement Organization among the Rarámuri of Northwestern Mexico.* International Monographs in Prehistory. Ann Arbor, MI.

Grissino-Mayer, H.D. 1995. Tree-ring reconstructions of climate and fire history at El Malpais National Monument, New Mexico. Ph.D. dissertation. University of Arizona, Tucson.

Grissino-Mayer, H.D. 1999. Modeling fire interval data from the American Southwest with the Weibull distribution. *Int. J. Wildl. Fire* 9:37–50.

Grissino-Mayer, H.D. 2001. FHX2—Software for analyzing temporal and spatial patterns in fire regimes from tree rings. *Tree-Ring Res.* 57:115–124.

Grissino-Mayer, H.D., and Swetnam, T.W. 1997. Multi-century history of wildfire in the ponderosa pine forests of El Malpais National Monument. *New Mexico Bur. Mines Mineral Resources, Bull.* 156:163–171.

Hales, J.E. 1974. Southwestern United States summer monsoon source—Gulf of Mexico or Pacific Ocean? *J. Appl. Meteorol.* 13:331–342.

Heyerdahl, E.K., Brubaker, L.B., and Agee, J.K. 2001. Spatial controls of historical fire regimes: A multiscale example from the Interior West, USA. *Ecology* 82:660–678.

Heyerdahl, E.K., Brubaker, L.B., and Agee, J.K. (In press). Annual and decadal climate forcing of historical fire regimes in the Interior Pacific Northwest, USA. *Holocene.*

Holland, P.G., and Steyn, D.G. 1975. Vegetational responses to latitudinal variations in slope angle and aspect. *J. Biogeogr.* 2:179–183.

Holmes, R.L. 1983. Computer-assisted quality control in tree-ring dating and measurement. *Tree-Ring Bull.* 43:69–78.

Jordan, T.G. 1993. *North American Cattle-Ranching Frontiers: Origins, Diffusion, and Differentiation*. Albuquerque: University of New Mexico Press.

Kaib, J.M. 1998. Fire history in riparian canyon pine-oak forests and the intervening desert grasslands of the southwest borderlands: A dendroecological, historical, and cultural inquiry. M.S. thesis. Tucson: University of Arizona.

Kiladis, G.N., and Diaz, H.F. 1989. Global climatic anomalies associated with extremes in the Southern Oscillation. *J. Clim.* 2:1069–1090.

Kitzberger T., Swetnam T.W., and Veblen T.T. 2001. Inter-hemispheric synchrony of forest fires and the El Niño-Southern Oscillation. *Global Ecol. Biogeogr.* 10:315–326.

Landsberg, J.D., Cochran, P.H., Finck, M.M., and Martin, R.E. 1984. *Foliar Nitrogen Content and Tree Growth after Prescribed Fire in Ponderosa Pine*. Res. Note PNW-412. Portland, OR: USDA Forest Service, Pacific Northwest Forest and Range Experiment Station.

Leopold, A. 1924. Grass, brush, timber and fire in southern Arizona. *J. For.* 22:1–10.

Leopold, A. 1937. Conservationist in Mexico. *Am. For.* 43:118–120, 146.

Lertzman, K., Fall, J., and Dorner, B. 1998. Three kinds of heterogeneity in fire regimes: At the crossroads of fire history and landscape ecology. *Northwest Sci.* 72:4–23.

Madany, M.H., and West, N.E. 1983. Livestock grazing—Fire regime interactions within montane forests of Zion National Park, Utah. *Ecology* 64:661–667.

Manuel-Toledo, V., and Jesús-Ordóñez, M.de. 1993. The biodiversity scenario of Mexico: A review of terrestrial habitats. In *Biological Diversity of Mexico: Origins and Distribution*, eds. T.P. Ramamoorthy, R. Bye, A. Lot, and J. Fa, pp. 757–777. New York: Oxford University Press.

Mast, J.N., Veblen, T.T., and Linhart, Y.B. 1998. Disturbance and climatic influences on age structure of ponderosa pine at the pine/grassland ecotone, Colorado Front Range. *J. Biogeogr.* 25:743–755.

Mooney, C.Z., and Duvall, R.D. 1993. Bootstrapping: A nonparametric approach to statistical inference. Newbury Park, CA: Sage University Paper Series on Quantitative Applications in the Social Sciences 07-095.

Mosiño Alemán, P.A., and García, E. 1974. The climate of Mexico. In *World Survey of Climatology. Vol. 11: Climates of North America*, ed. R.A. Bryson, and F.K. Hare, pp. 345–404. New York: Elsevier.

Park, A.D. 2001. Environmental influences on post-harvest natural regeneration in Mexican pine-oak forests. *For. Ecol. Manag.* 144:213–228.

Payette, S., Morneau, C., Sirois, L., and Desponts, M. 1989. Recent fire history of the Northern Quebec biomes. *Ecology* 70:656–673.

Powell, F.W. 1921. *The Railroads of Mexico*. Boston: Stratford, 1921.

Rzedowski, J. 1978. *Vegetation de Mexico*. Mexico: Limusa.

Rzedowski, J. 1993. Diversity and origins of the phanerogamic flora of Mexico. In *Biological Diversity of Mexico: Origins and Distribution*, eds. T.P. Ramamoorthy, R. Bye, A. Lot, and J. Fa, pp. 129–144. New York: Oxford University Press.

Ropelewski, C.F., and Halpert, M.S. 1986. North American precipitation and temperature patterns associated with the El Niño/Southern Oscillation (ENSO). *Mon. Wea. Rev.* 114: 2352–2362.

Ropelewski, C.F., and Halpert, M.S. 1987. Global and regional scale precipitation patterns associated with the El Niño/Southern Oscillation. *Mon. Wea. Rev.* 115:1606–1626.

Ropelewski, C.F., and Halpert, M.S. 1989. Precipitation patterns associated with the high index phase of the Southern Oscillation. *J. Clim.* 2:268–284.

Rouse, J.E. 1977. *The Criollo: Spanish Cattle in the Americas*. Norman: University of Oklahoma Press.

Sanderson, S.R.W. 1984. *Land Reform in Mexico: 1910–1980*. New York: Academic Press.

Savage, M., and Swetnam, T.W. 1990. Early 19th-century fire decline following sheep pasturing in a Navajo ponderosa pine forest. *Ecology* 71:2374–2378.
Secretaría de Agricultura y Recursos Hidráulicos (SARH). 1994. *Memoria nacional del inventario nacional forestal periodico 1992–1994*. Subsecretaria Forestal y de Fauna Silvestre. Secretaria de Agricultura y Recursos Hidraulicos. Mexico, D.F.
Secretaría de Medio Ambiente y Recursos Naturales (SEMARNAP). 2000. Programa nacional de proteccion contra los incendios forestales. Resultados 1995–2000. Secretaria de Medio Ambiente Recursos Naturales y Pesca. Mexico, D.F.
Sheridan, T.E., and Naylor, T.H. 1979. *Raramuri: A Tarahumara Colonial Chronicle 1607–1791*. Flagstaff, AZ: Northland Press.
Stahle, D.W., D'Arrigo, R.D., Krusic, P.J., Cleaveland, M.K., Cook, E.R., Allan, R.J., Cole, J.E., Dunbar, R.B., Therrell, M.D., Gay, D.A., Moore, M.D., Stokes, M.A., Burns, B.T., Villanueva-Diaz, J., and Thompson, L.G. 1998. Experimental dendroclimatic reconstruction of the Southern Oscillation. *Bull. Am. Meteorol. Soc.* 79:2137–2152. (Data archived at the World Data Center for Paleoclimatology, Boulder, Co.)
Stahle, D.W., Cleaveland, M.K., Therrell, M.D., and Villanueva-Diaz, J. 1999. Tree-ring reconstruction of winter and summer precipitation in Durango, Mexico, for the past 600 years. In *10th Symposium on Global Change Studies*, ed. T.R. Karl, pp. 317–318, January 10–15, 1999, Dallas, Tex. Boston: American Meteorological Society.
Sutherland, E.K., Covington, W.W., and Andariese, S. 1991. A model of ponderosa pine growth response to prescribed burning. *For. Ecol. Manag.* 44:161–173.
Swetnam, T.W., and Baisan, C.H. 1996. Historical fire regime patterns in the southwestern United States since AD 1700. In *Fire Effects in Southwestern Forests, Proceedings of the Second La Mesa Fire Symposium*, tech. coord. C.D. Allen, pp. 11–32. Gen. Tech. Rep. RM-GTR-286, Fort Collins, CO: USDA Forest Service, Rocky Mountain Forest and Range Experiment Station.
Swetnam, T.W., and Betancourt, J.L. 1990. Fire–Southern Oscillation relations in the southwestern United States. *Science* 249:1017–1020.
Swetnam, T.W., and Betancourt, J.L. 1992. Temporal patterns of El Niño/Southern Oscillation-wildfire teleconnections in the southwestern United States. In *El Nino: Historical and Paleoclimatic Aspects of the Southern Oscillation*, eds. H.F. Diaz, and V. Markgraf, pp. 259–269. New York: Cambridge University Press.
Swetnam, T.W., and Betancourt, J.L. 1998. Mesoscale disturbance and ecological response to decadal climatic variability in the American Southwest. *J. Clim.* 11:3128–3147.
Swetnam, T.W., Baisan, C.H., and Kaib, J.M. 2001. Forest fire histories of the Sky Islands of *La Frontera*. In *Changing Plant Life of La Frontera: Observations on Vegetation in the United States/Mexico Borderlands*. eds. G.L. Webster, and C.J. Bahre, pp. 95–119. Albuquerque: University of New Mexico Press.
Tande, G.F. 1979. Fire history and vegetation pattern of coniferous forests in Jasper National Park, Alberta. *Can. J. Bot.* 57:1912–1931.
Taylor, A.H., and Skinner, C.N. 1998. Fire history and landscape dynamics in a late-sucessional reserve, Klamath Mountains, California, USA. *For. Ecol. Manag.* 111:285–301.
Thompson, G.D., and Wilson, P.N. 1994. Ejido reforms in Mexico: Conceptual issues and potential outcomes. *Land Economics* 70:448–465.
Turman, B.N., and Edgar, B.C. 1982. Global lightning distributions at dawn and dusk. *J. Geophys. Res.* 87:1191–1206.
Veblen, T.T., Kitzberger, T., and Donnegan, J. 2000. Climatic and human influences on fire regimes in ponderosa pine forests in the Colorado Front Range. *Ecol. Appl.* 10: 1178–1195.
Veblen, T.T., Kitzberger, T., Villalba, R., and Donnegan, J. 1999. Fire history in northern Patagonia: The roles of humans and climatic variation. *Ecol. Monog.* 69:47–67.
World Forest Institute. 1994. *Mexico: Forestry and the Wood Products Industry*, 2nd ed. Portland, OR: World Forest Institute.

8. Impact of Past, Present, and Future Fire Regimes on North American Mediterranean Shrublands

Jon E. Keeley and C.J. Fotheringham

Mediterranean shrublands occur in five regions of the world, under a climate of mild wet winters and hot summer–fall droughts lasting six months or more. In California they dominate landscapes below 2000 m in the central and southern coastal ranges and foothills of the Sierra Nevada. One consequence of this distribution is that these shrublands, more than any other vegetation type, interface with urban areas (Fig. 8.1). These shrublands are subject to periodic massive wildfires (Fig. 8.2) that account for 40% of all wildland acreage burned in the United States (Lillard 1961), creating a particularly hazardous urban–wildland interface. Contributing to this fire hazard are the moderate temperatures during the rainy winter and spring, which prolong the growing season and generate broad bands of dense contiguous fuels. The long drought makes these fuels readily ignitable and the autumn foëhn winds that come each year at the end of the dry season produce the worst fire climate conditions in the country (Schroeder et al. 1964).

This chapter examines the past, present, and future fire regimes in California shrublands, particularly chaparral and coastal sage scrub. Although shrublands are recorded from nearly all counties in the state (Callaham 1985), this review will focus on those in the central and southern coastal ranges with the largest expanses of contiguous shrubland (Fig. 8.3). Of particular concern are the extent to which humans have altered this regime in the past and the extent to which future global change will affect fire regimes and vegetation patterns.

Humans directly influence fire regimes in two ways: they ignite fires and they suppress fires. Evaluating the net effect of these impacts is not simple because

Figure 8.1. Interface between urban environments and evergreen chaparral (*right*) and semi-deciduous coastal sage scrub (*left*) in southern California (by J.E. Keeley).

their relative importance varies across the landscape. For example, in the montane coniferous forests of the Southwest, lightning-ignited fires are abundant and human ignitions are far less important than in lower-elevation shrublands of southern California where lightning is uncommon and humans cause the majority of fires (Fig. 8.4). Also fire suppression has been far more effective in western coniferous U.S. forests, often achieving nearly complete fire exclusion (Skinner and Chang 1996; Agee 1993), but this "fire-suppression = fire-exclusion" equation does not apply to shrublands of southern and central coastal California (Keeley and Fotheringham 2001b).

Determinants of Brushland Fire Regimes

Fire regimes are determined by the temporal and spatial pattern of ignitions, fuels, weather, and topography (Pyne, Andrews, and Laven 1996), and with regard to Californian shrublands there are two schools of thought on their relative importance. One is based on deductions from Rothermel's fire behavior model (Rothermel 1972) and argues that fire regime is a highly deterministic process driven by fuel load (Rothermel and Philpot 1973; Philpot 1974a,b, 1977). Under this model fire occurrence is unaffected by external drivers such as ignitions or weather, rather it is viewed as entirely dependent on community patterns of fuel accumulation (Minnich 1989, 1995,1998, 2001; Minnich and Cho 1997). The

Figure 8.2. Crown fire in chaparral (photo by USFS, Riverside Fire Lab).

alternative model argues that the fire regime is controlled by the coincidence of ignitions occurring under severe current and antecedent weather conditions that influence fuel flammability (Phillips 1971; Keeley et al. 1989; Davis and Michaelson 1995; Keeley and Fotheringham 2001a,b). Under this model any of these factors may be limiting, and the importance of each varies spatially and temporally with external drivers such as severe fire weather being of paramount importance in coastal California. These models have very different implications for fire management and affect our perception of anthropogenic impacts on fire regime and our ability to sort out future climatic signals.

Patterns of Ignition

In order to appreciate fully the role humans play in shrubland fire regimes, we need to first examine how ignitions, fuels, and weather interact to determine fire behavior. In California humans have been a source of ignitions for more than

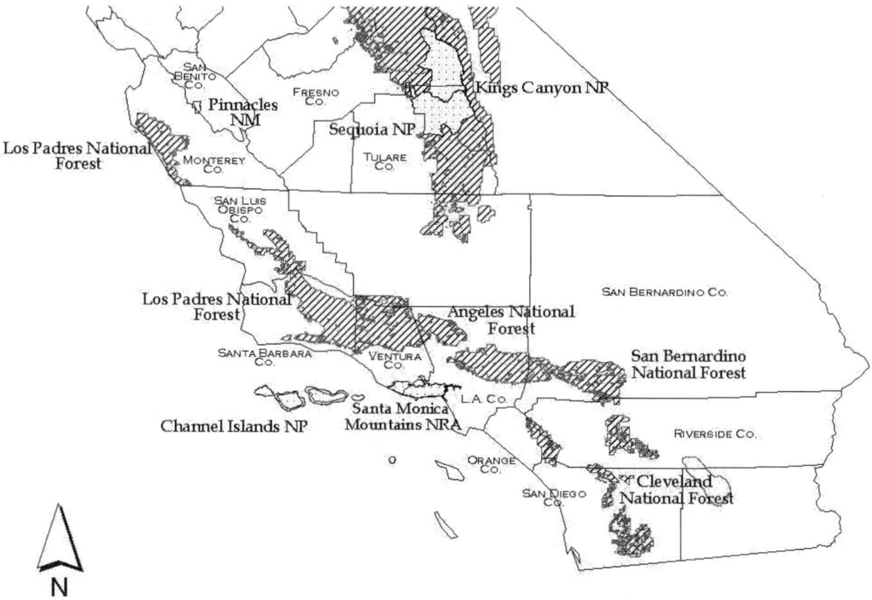

Figure 8.3. Central and southern California regions considered in this chapter. Central coastal California includes Monterey, San Luis Obispo, Santa Barbara, and Ventura counties, and southern California includes Los Angeles, San Bernardino, Riverside, Orange, and San Diego counties. Collectively these nine counties comprise nearly two million hectares of shrubland (Table 8.1).

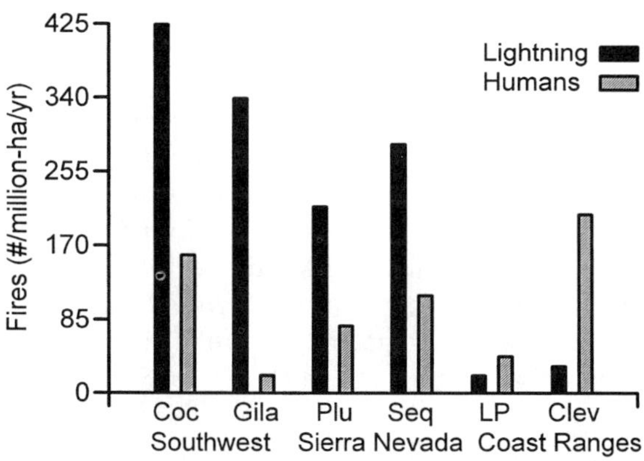

Figure 8.4. Regional comparison of lightning- and human-caused fires on USFS national forests. The Southwest includes the Coconino (Coc) in Arizona and Gilia in New Mexico. In California the Sierra Nevada forests are the Plumas (Plu) and Sequoia (Seq), and the California coastal ranges national forests are the Los Padres (LP) and Cleveland (Clev). Fire occurrence data from the published U.S. Forest Service, National Forest Fire Reports, 1970–1979, and forest area from (*http://www.fs.fed.us/land/*).

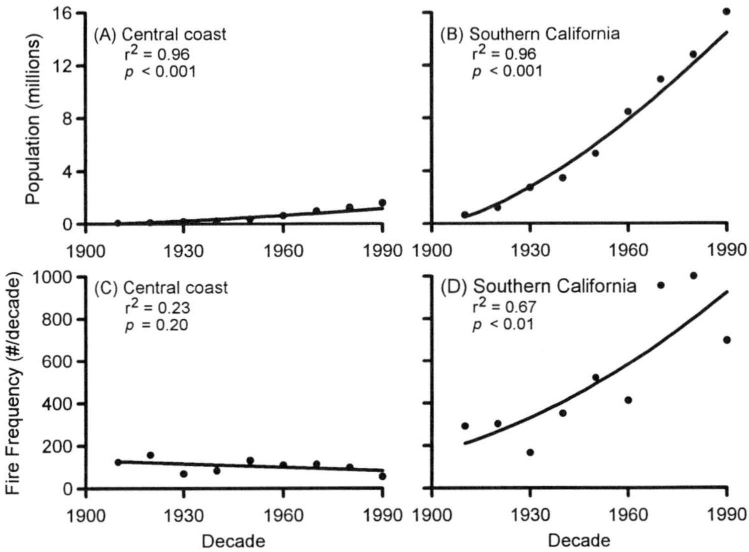

Figure 8.5. Decadal variation in population density (A–B) and fire frequency (C–D) for central coastal and southern California. Population data from the U.S. Department of Commerce, *http://www.census.gov/populations/cencounts/ca190090.txt*. (Fire data from the Statewide Fire History Data Base, California Department of Forestry, Fire and Resource Assessment Program (FRAP), Sacramento, CA, which includes historical fire records from the U.S. Forest Service national forests, California Division of Forestry ranger units and other protected areas, plus city and county records; minimum fire size recorded varied between 16 and 40 ha, depending on the agency).

10,000 years, but they likely have had a greater influence in the twentieth century due to the near exponential rise in population density and fire frequency in the southern part of the state (Fig. 8.5).

Under natural conditions lightning is a source of ignition but far less predictable than in other parts of the Southwest (Fig. 8.4). Within the state, lightning-ignited fires vary spatially because thunderstorms are rare near the coast and most frequent at higher elevations in the interior (Radtke, Atndt, and Wakimoto 1982; Keeley 1982; Greenlee and Moldenke 1982; Knipper 1998). Lightning is the dominant ignition source in the Sierra Nevada, but it is a far less common ignition source in the coastal ranges. Within the coastal ranges lightning varies with elevation; for example, in San Diego County lightning strikes are 10 times more abundant above 1800 m than below 500 m, and they vary temporally with 85% occurring between July and September (Wells and McKinsey 1994, 1995). Similar patterns are evident further south in Baja California (Minnich et al. 1993). The annual density of lightning discharges in this region is roughly 1 per 100 ha (Michael L. Wells, personal communication; Minnich et al. 1993). Based on the frequency of fires ignited by lightning in this region (Keeley 1982; Minnich et al. 1993), it would appear that only 2% to 5% of all lightning discharges ignite a wildfire. In other words, 95% of all lightning discharges strike inadequate fuels, or are extinguished by rain, before they reach a detectable size.

Lightning ignitions in coastal and southern California shrublands account for a highly variable amount of burning, ranging from less than 1% to more than 50% of the landscape per decade (Table 8.1). Both spatial and temporal factors are involved. Considering all of California, lightning ignitions account for an increasing fraction of burning from the coast to the interior and from south to north (Keeley 1982). Occasionally lightning may coincide with severe weather and fuel conditions and result in massive fires such as the Marble Cone Fire in 1977 on the Los Padres National Forest (Table 8.1). Longer-term data sets for the Los Padres show this to be an infrequent event (Davis and Michaelsen 1995), suggesting that lightning fires in these coastal ranges are capable of reaching extraordinary size but the temporal variance is high.

Lightning is more predictable in the higher interior Sierra Nevada Range (Fig. 8.4), and it varies inversely with elevation (van Wagtendonk 1992). In Sequoia National Park (located in the southern Sierra Nevada, Fig. 8.3) lightning-ignited fires reach a peak at elevations between 2000 and 3000 m and are considerably less frequent in the lower-elevation shrubland-dominated foothills (Parsons 1981; Vankat 1985). Within the park the lower-elevation shrublands experience fewer lightning-ignited fires than would be expected based on shrubland area ($p < 0.001$ with χ^2 test), and the opposite is true for higher-elevation mixed-coniferous forests. This pattern is repeated throughout the Sierra Nevada;

Table 8.1. Total number of fires and hectares burned and percentage due to lightning during the 1970s decade for lower-elevation foothills (California Division of Forestry Jurisdiction) and higher-elevation interior mountains (U.S. Forest Service national forests) in southern and central–coastal California

CDF Ranger Unit/USFS National Forest	Total fires (10^6 ha/ decade)	Total area burned (ha)	Fires due to lightning (%)	Area due to lightning (%)
Foothills (CDF)				
Monterey/San Benito	3,140	53,570	2	<1
San Luis Obispo	3,310	44,130	2	<1
San Bernardino	9,680	12,240	4	11
Riverside	17,620	332,950	1	5
Orange	42,900	120,830	<1	<1
San Diego	9,450	20,930	3	6
Mountains (USFS)				
Los Padres	2,340	49,720	9	56[a]
Angeles	4,980	214,460	15	4
San Bernardino	4,400	41,030	24	6
Cleveland	4,870	121,370	11	<1

Source: Keeley 1982.

Note: Sites are arranged from north to south, and national forest locations are shown in Figure 3. All of these ranger units or forests are dominated by chaparral, but they also include mixtures of grassland, sage scrub, woodlands, and forests.

[a] Much of this is due to a single lightning-ignited fire (Marble Cone Fire) in 1977.

foothill shrublands average about 10 lightning-ignited fires per year per million hectares and the higher-elevation montane forests experience 100 to 200 per year per million ha (Keeley 1982). Of course, making predictions about the elevational patterns of burning by lightning alone (i.e., in the absence of anthropogenic interference) is complicated by the likelihood that along this elevational gradient, conditions conducive to fire spread are inversely related to lightning fire frequency. Modeling is perhaps the only means of understanding the natural fire regimes in these ecosystems (e.g., Greenlee and Langenheim 1980; Davis and Burrows 1993; Davis and Michaelsen 1995; Zedler and Seiger 2000).

In general, rain or high humidity accompanies lightning fires, and there is often a time lag between ignition and changes in weather conducive to rapid fire spread. Thus, in forested ecosystems where lightning is the dominant ignition source (e.g., the Southwest, Fig. 8.4), fire suppression has been extraordinarily effective. Fire detection has become increasingly more reliable (Chandler 1960), and there is reason to believe that many suppressed lightning-ignited fires, in both forests and shrublands, would have burned out if never detected. This is supported by a greater number of reports of lightning-ignited wildfires in the latter half of the twentieth century (Keeley 1977, 1982; Greenlee and Moldenke 1982; Vankat 1985); however, it could reflect changes in fuel structure as well (Weatherspoon and Skinner 1996).

In the coniferous forests of the Sierra Nevada, lightning-ignited fires peak in the summer months of July and August and match closely the monthly distribution of human-ignited fires (Parsons 1981; Vankat 1985; van Wagtendonk 1992). Throughout the chaparral-dominated coastal ranges lightning-ignited fires are also concentrated in the summer months of July and August (Keeley 1982). However, humans are the dominant source of ignition (Fig. 8.4), and their impact on fire season varies from apparently very little effect in the central coast, as illustrated by a summer peak in burning to a much greater impact in the south, where anthropogenic fires result in a longer fire season and greater autumn burning (Fig. 8.6). Thus, in contrast to the situation in forests throughout the western United States where lightning is the dominant source of ignition and humans have successfully suppressed most fires, the vast majority of fires in chaparral and coastal sage scrub in the coastal ranges are ignited by humans (Keeley, Fotheringham, and Morais 1999; Keeley and Fotheringham 2001b). In short, fire suppression has not eliminated burning on this shrubland landscape. Human impact is most pronounced at lower elevations and in proximity to metropolitan areas. On shrubland landscapes under natural conditions, lightning is a predictable source of ignition but variably distributed in time and space.

Fuels and Weather

The spatial and temporal arrangement of fuels is a critical determinant of fire behavior, and fuel loading is determined largely by differences in site productivity and vegetation age. The extent to which fire will propagate across a landscape

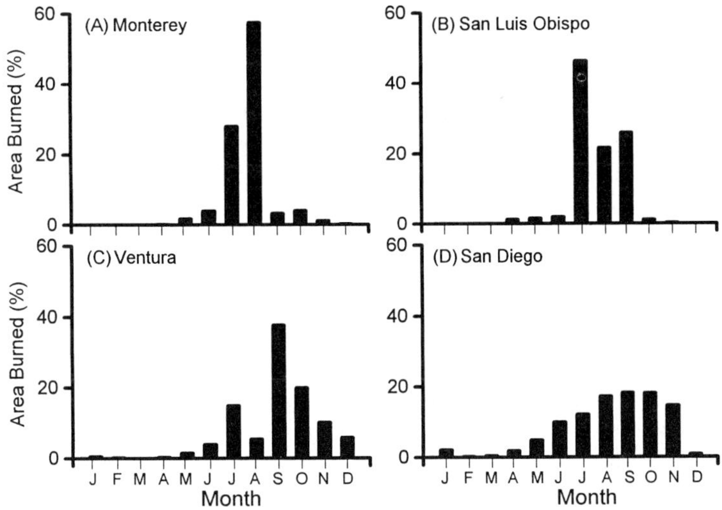

Figure 8.6. Seasonal distribution of burning reported for 1970 to 1999 for selected counties (data from the California Statewide Fire History Database; see Fig. 5).

is determined by the spatial arrangement of fuels and weather conditions prior to and during the fire. Fuel structure needs to be considered at different scales. In a stand of vegetation on a single slope face, the important fuel characteristics are the vertical and horizontal placement of fuels, fuel surface-area/volume ratio, and the moisture status of leaves and stems. At this scale shrubs are of uniform age and may be rather coarse grained in monotypic stands, becoming finer grained as the mixture of species increases. At the landscape level fuels are fine grained, and large expanses of homogeneous fuels are the exception. Barriers of reduced fuel loading, which could include rocks, rivers, alluvial fans, young age classes, or less flammable vegetation types, may inhibit fire spread. As seasonal drought progresses, different portions of the landscape are added as potential fuels, further contributing to inherent landscape heterogeneity of fuels. This interaction among landscape structure, fuels, and moisture limits the ability of models to predict fire spread, and the fine-grain nature of fuels leads to potentially large errors (Kessell and Cattelino 1978).

Fuel Structure and Fuel Moisture

In mature shrublands, surface fuels are insufficient to carry fire, and thus fires propagate through the canopy as crown fires. Recently burned sites have sufficient herbaceous growth to carry surface fires (Haidinger and Keeley 1993), and this may be exacerbated by artificial seeding of nonnative grasses (e.g., Zedler,

Gautier, and McMaster 1983). However, sufficient herb biomass to carry surface fires is unlikely following dry winters or on highly infertile coarse-textured soils, such as occur in certain coastal sites (e.g., Lompoc, CA) or the interior ranges of Baja California (Franco-Vizcaino and J. Sosa-Ramirez 1997).

Normally, following a wet winter, high fuel moisture in chaparral shrubs makes them relatively resistant to fire in spring and early summer. However, as the amount of herbaceous matter in the stand increases, the seasonal window of burning increases. Dead herbaceous fuels dry rapidly and are capable of carrying fire within days of a rainfall event (Chandler 1963), and species composition plays a role as nonnative grasses typically die many weeks earlier than native herbs (Keeley, personal observations). As a result certain herbaceous fuels greatly extend the length of the fire season.

Shrublands that have been partially or fully type-converted to grasslands (e.g., Bentley 1967) have a greater probability of igniting but do not represent an extreme fire hazard as fire intensities are low and the fine herbaceous fuels fail to sustain embers or create the vortexes that carry the fire ahead of the moving front (Regelbrugge 2000). Even so, fires in a dense growth of non-native herbs, such as mustards (*Brassica nigra* and *Hirschfeldia incana*) on steep slopes, have been known to generate fire intensities sufficient to destroy homes (J. Keeley, personal observation).

For intact shrublands, two factors affect woody fuel moisture: the physiological activity (water potential) of live foliage and the quantity of dead fuels (Green 1981). Shrub species differ markedly in moisture status of foliage due in part to differences in rooting depth (Davis, Kolb, and Barton 1998)—shallow-rooted shrubs, such as chamise (*Adenostoma fasciculatum*) and *Ceanothus* spp., typically experience water potentials two to three times lower than more deeply rooted shrubs such as scrub oak (*Quercus berberidifolia*). Under prescription weather conditions fires may readily spread through *Adenostoma*-dominated chaparral but extinguish when they encounter patches of scrub oak (Chandler 1957; Green 1981). However, under extended drought, foliage moisture in scrub oak may drop to levels conducive to rapid-fire spread (Olsen 1960; Pirsko and Green 1967; Green 1981).

Dead fuels lack an internal water source and respond rapidly to changes in humidity; small diameter stems can dry completely within hours and larger fuels within days of experiencing low humidity (Chandler 1963; McCutchan 1977). Dead fuels not only combust readily, but as the proportion of dead/live material increases, there is an elevated potential for dead fuel combustion to cause drying of living foliage to a level sufficient for combustion. Because dead fuel carries fire and live fuel absorbs energy, the ratio of dead/live fuel is critical. This increase in combustibility of live fuels is enhanced by the common position of dead fuels beneath the living foliage. Topography plays a similar role. On steep terrain, head fires burning upslope enhance the combustion of fuels ahead of the front and may spread two to three times faster than on level ground—fire spread will roughly double for each 13 degree rise in slope (Green 1981).

Fuel Structure and Wind

At low wind speed, fuel structure and arrangement plays a critical role in fire spread. For example, fine-textured, low, compact fuels—particularly subshrubs with extremely high levels of volatiles, for example *Salvia* spp. (sage)—may readily combust and spread fire rapidly. However, under the same weather conditions, fire might naturally extinguish in a taller chaparral stand in which fuels are more widely scattered in the canopy, and there is little continuity with ground-level fuels (Green 1981). Under low to moderate wind conditions species-specific fuel characteristics in chaparral can promote fire spread. Many characteristics of *Adenostoma fasciculatum* (chamise) make it far more flammable than associated shrub species. About two-thirds of the plant is composed of twigs <25 mm diameter and thus has a stem surface area–volume ratio greater than that of other species (Conard and Regelbrugge 1994). Individual chamise leaves have a relatively low surface area/volume ratio, but they have an extremely high content of volatile compounds that vaporize and increase combustibility (Philpot 1969). On a whole plant basis, *Adenostoma* leaves have a very high surface area/volume ratio; they comprise 67% of surface area but only 16% of plant volume, reflecting the loose packing of foliage (Countryman and Philpot 1970; Barro and Conard 1991).

One of the key factors affecting flammability of *Adenostoma* is the fact that it does not self-prune dead twigs and branches; instead, they are held aloft in the canopy and increase canopy porosity (shrub canopy volume/leaf and stem volume), which often exceeds 99% (Rundel, Parsons, and Baker 1980). High canopy porosity increases flammability and extends the seasonal window of flammability. Also experimental studies demonstrate that this natural retention of dead branches substantially increases fire intensity over an artificial treatment of clipping and leaving as surface fuels (Schwilk 2000). Species with more densely packed fuels, and that self-prune dead branches and have thicker twigs and stems (e.g., scrub oak, *Quercus*, or chaparral holly *Heteromeles arbutifolia*), often will not burn under conditions suitable for fire spread in *Adenostoma*-dominated chaparral. It has been hypothesized that characteristics enhancing flammability have adaptive value (Mutch 1970) and shrubs with seedling recruitment restricted to postfire environments (*Adenostoma, Ceanothus, Arctostaphylos*) have significantly higher flammability than species that recruit independently of fire (*Prunus, Rhamnus, Quercus*) (Bond, unpublished data).

While high-canopy porosity increases flammability, it leads to lower bulk density (mass/volume) and fuel loading (mass/area), reducing the total energy available for combustion. Thus the *Adenostoma* fuel structure increases flammability under a wide range of conditions, whereas the *Quercus* fuel structure is limited in the range of conditions suitable for burning, but under the severest conditions *Quercus* fuels should be expected to generate the highest intensities.

Fuel structure appears to play a less deterministic role under windy conditions, but there is a complex interaction of wind, humidity, fuels, temperature, and topography. Cool moist marine air will extinguish fires (Coffin 1959), whereas warm dry air will lead to fire spread in fuels that otherwise would not burn. Wind accelerates oxygen supply and thus combustion (Green 1981) and is the primary mode of heat transfer. It carries heated air to adjacent fuels on the downwind side, raising the fuel temperature and driving off moisture. Wind also carries away water vapor as well as firebrands, which often occur when gusts are greater than 16 km/hr (Green 1981). Topographic features frequently cause unstable and erratic changes in velocity and direction as winds adapt to the topography. On coastal-facing slopes onshore winds are channeled up-canyon and produce eddies at ridgelines that may become turbulent and erratic. The typical pattern is for local daytime up-canyon wind and nighttime down-canyon winds, and on coastal slopes in the central coastal region extraordinarily strong down-canyon winds known as Sundowners are occasionally experienced (Ryan 1996).

Overriding synoptic-scale winds can upset these local wind patterns, e.g., foëhn winds known as "north winds" or "mono winds" in central California (Greenlee and Langenheim 1980) and Santa Ana winds further south (Lessard 1988) (Fig. 8.7). These winds are controlled by regional synoptic patterns that include a Great Basin high-pressure cell and Pacific Coast trough of low pressure, but their ultimate manifestation is a result of local topography (Schroeder and Buck 1970; Fosberg et al. 1966). For example, in the southern Sierra Nevada, the steep eastern escarpment and lack of low passes keeps these winds aloft (Mitchell 1969), and thus foëhn winds are not experienced on the lower western slopes. In Ventura and Los Angeles counties these winds are funneled through passes in the east west trending Transverse Ranges and thus are predominantly northern or northeastern winds (Weide 1968; Schroeder et al. 1964). In San Diego County they are strictly eastern due to the north–south orientation of the Peninsular Ranges winds (Campbell 1906; Sommers 1978). These ranges extend southward into Baja California where their sharp eastern escarpment, coupled with the Gulf of California to the east, limit the formation of foëhn winds on the west slopes of the Sierra San Pedro Mártir (Keeley and Fotheringham 2001a,b).

In southern California these hot, dry Santa Ana winds often have less than 10% relative humidity and may exceed 100 km per hour (Fosberg et al. 1966; Ryan 1969). Although referred to as "desert winds," the high temperatures and low humidity are the result of compression as air descends to form the "basin air mass" (Mitchell 1969), and on a local scale as it descends through coastal passes (Krick 1933). Santa Ana winds are most common in the autumn (Fig. 8.8). They have a mean life of about three days but may last two (Fosberg 1965) or three weeks (Campbell 1906), a critical factor since fire size is often determined by the duration of high wind conditions (McCutchan 1977). Under Santa Ana wind conditions fire spread is rapid. For example, the Kanan fire in the Santa Monica Mountains of southern California consumed 10,121 ha in 3 hours (Franklin 1987), and such fires may exceed 30,000 ha in a single day (Phillips 1971). Such fires are unimpeded by many potential barriers, since firebrands may be carried as much as 8 km beyond the front, igniting numerous new spot fires (Countryman

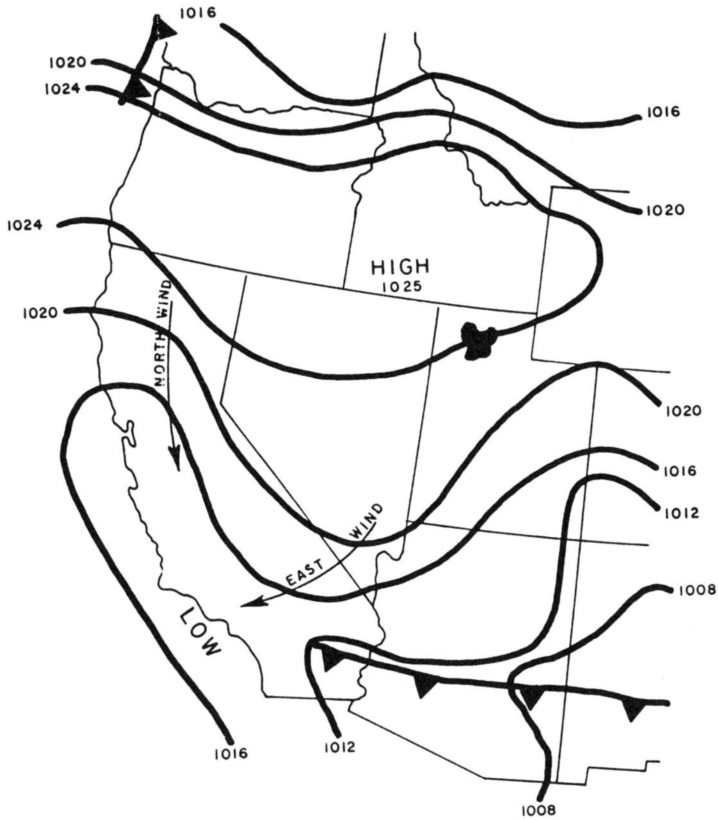

Figure 8.7. Surface weather map during the Great Basin high-pressure air mass that generates foëhn winds in central and southern California (from Phillips 1971).

1974). Under these conditions stands may burn regardless of stand age or species composition (Keeley, Fotheringham, and Morais 1999).

Fuel Mass and Stand Age

It has long been held that fuel mass increases with stand age (Philpot 1977), but this has been criticized as oversimplistic because it ignores tremendous species-specific variability in rates of biomass accumulation (Fig. 8.9). For example, some *Ceanothus* species may accumulate many times more biomass in less than 20 years than *Adenostoma fasciculatum* does in 60 years (Riggan et al. 1994; Regelbrugge 2000). Also at 10 years of age north-facing aspects may have greater biomass accumulation than drier south-facing slopes do at 80 years of age (Black 1987). This fact alone makes landscape-scale predictions of flammability based on stand age extremely difficult. Complicating the prediction of flammability with stand age is the increasing proportion of biomass in large diameter stems that

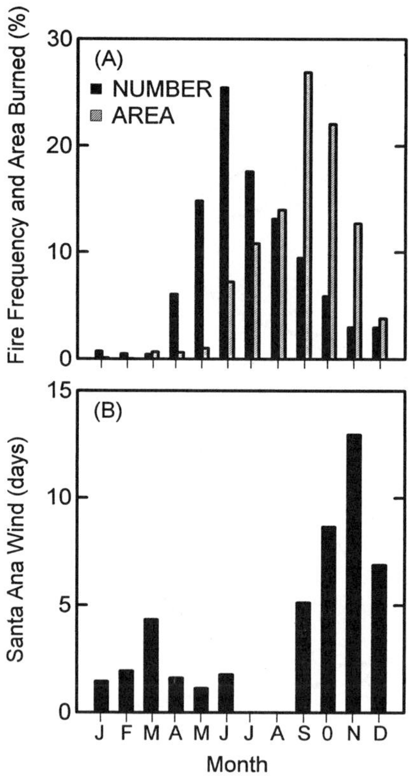

Figure 8.8. Seasonal distribution of fire occurrence and area burned during the twentieth century in Los Angeles County (data from Statewide Fire History Database; see Fig. 8.5) and seasonal distribution of Santa Ana winds (from Weide 1968).

combust only under the most extreme burning conditions. Also highly productive stands are often more mesic sites, and this, plus greater fuel density and higher fuel moisture, may reduce flammability. However, under extreme conditions, once ignited, productive sites sustain greater energy release than less productive stands (Riggan et al. 1988).

Successional changes in biomass (live and dead kg/ha) range from 1200 to 9000 in the first postfire year to 8000 to 13,000 after a decade, and 30,000 to 66,000

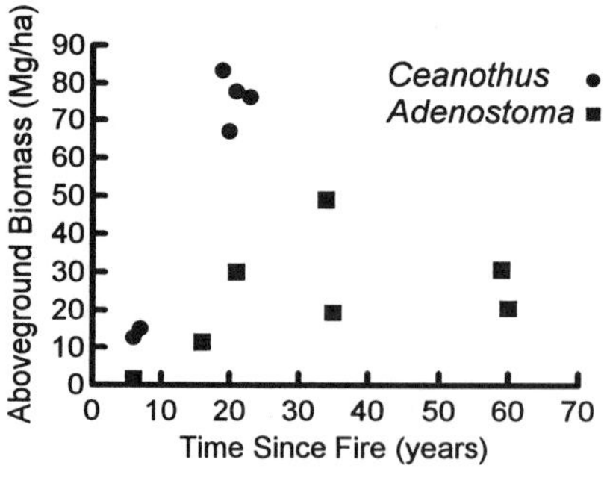

Figure 8.9. Live and dead aboveground biomass for chaparral shrubs at different times since fire based on several studies (data from Regelbrugge 2000).

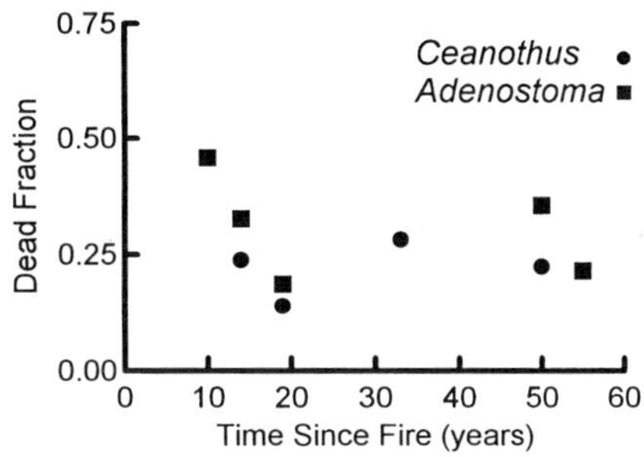

Figure 8.10. Fraction of total biomass comprising dead material for chaparral shrubs at different times since fire based on several studies (data from Paysen and Cohen 1990; Regelbrugge 2000).

(sometimes 100,000) in mature stands (Specht 1969, 1981; Green 1970; Keeley and Keeley 1984). Specht (1969) reported that the proportion of dead biomass exceeded 50% in mature chamise chaparral, and Green (1970) found 66% dead in mature *Cercocarpus betuloides*. These early reports led to the generalization of 1% dead for each year after canopy closure (Green 1981). However, more extensive studies (Fig. 8.10) report 30% dead/live ratios across the span from 20 to 60 years and no significant relationship with age (Paysen and Cohen 1990; Conard and Regelbrugge 1994; Regelbrugge 2000). It is apparent that dead/live ratios are a complicated function of many aspects of site composition and history. For example, unusually severe soil droughts may dramatically increase mortality, particularly of shallow-rooted *Ceanothus* shrubs, and this can occur in young or old stands (Keeley 2000; Davis et al. 2002). Also prior fire history may play a role; for example, chaparral stands burned by light fires leave large volumes of standing dead biomass that can produce very high dead/live ratios in young successional stands where high volume of dead fuels is not expected (e.g., Fig. 8.10).

In general, chaparral less than 25 years old has less than 20% dead, and this is insufficient to carry fire under "prescribed fire weather conditions" (Green 1981). Under severe weather conditions stand age (and total biomass and proportion dead) is less important in determining fire spread (Dunn 1989; Keeley, Fotheringham, and Morais 1999; Zedler and Seiger 2000).

The conclusion that older stands of chaparral generate fires of greater intensity needs to be viewed with caution. Fire intensity, which is often measured as fireline intensity or energy released per meter of fire front (Borchert and Odion 1995), can vary greatly depending on the interaction between weather and fuels. Sometimes intensity is equated with fire severity, which is defined as the ecological impact of the fire, and is often measured by mortality or the amount of plant biomass consumed, or alteration of nutrient cycles. However, a fast-moving fire that consumes little fuel and a slow-moving fire that consumes more fuel can achieve the same fireline intensity, and thus intensity and severity can not always be equated. In general, fire intensity is important to understanding options for fire suppression (Countryman 1974), whereas fire severity is most relevant to postfire ecosystem recovery (Keeley 1998b). Lastly, large fires often are equated with fires of high intensity, but they need not be. Large fires or mass fires are often described as catastrophic fires, but this latter term best refers to the impact of fire upon property and lives.

Past and Present Shrubland Fire Regimes

Understanding the extent of human impact on chaparral ecosystems requires that we reconstruct historical fire regimes. Stand-replacing crown fires typical of shrublands (Fig. 8.2) are not conducive to the formation of a tree-ring record of fires, as with surface fire regimes in montane coniferous forests. Thus reconstructing historical burning patterns for chaparral requires alternative approaches such as interpretation of sedimentary charcoal records.

Charcoal deposits in varved sediment cores from the Santa Barbara Channel have generated estimates of prehistoric fire frequency. Byrne, Michaelsen, and Soutar (1977) calibrated this procedure by comparison of annual varves from modern cores with U.S. Forest Service fire records. They found a significant correlation between large charcoal deposition events and incidence of large fires (>20,000 ha) in the adjacent mountain range less than 50 km from the core site. Using a core for the period from AD 730 to 1505, they were able to detect significant charcoal deposition but less than in the modern period, suggesting a lack of frequent small fires, unlike the contemporary pattern (Moritz 1997). They did, however, find two major peaks approximately 100 years apart with smaller peaks at 20- to 60-year intervals, and suggested this period had few fires, widely spaced, which became large conflagrations capable of generating large pulses of charcoal. Mensing, Michaelsen, and Byrne (1999) analyzed similar cores at a finer resolution and concluded that large fires were a feature of this region long before modern fire suppression.

Native American Impacts

Tree-ring records of fire scars from the coastal ranges and the Sierra Nevada have been interpreted to suggest that during the few hundred years prior to Euro-American colonization fire frequencies exceeded the level expected from lightning alone (Reynolds 1959; Greenlee and Langenheim 1990). From historical records and ethnographic accounts there can be no doubt that California Indians regularly utilized fire to manage their environment (e.g., Lewis 1973; Timbrook, Johnson, and Earle 1982; Wickstrom 1987; Anderson and Moratto 1996). The extent to which this management practice altered landscapes is a matter of debate. Due to the naturally high fire frequency of lightning fires in the coniferous forests of the Sierra Nevada, Vale (1998) has argued that the additional burning by Indians did not alter landscapes except in localized areas (but cf. Anderson, Barbour, and Whitworth 1998). On the other hand, it has been hypothesized that direct use of fire by Native Americans greatly altered landscape patterns in the lower elevation coastal range foothills, primarily through type conversion of shrublands and woodlands to grasslands and other herbaceous associations (Cooper 1922; Wells 1962; Huenneke 1989; Keeley 1990, 2002; Hamilton 1997). This hypothesis is supported by the low lightning activity, high Indian populations, shrub-dominated landscapes, limited resources for Native Americans in undisturbed shrublands, and weak resilience of shrublands to high fire frequency (Keeley, in review).

Euro-American Settlement Impacts

Euro-American settlers further increased fire frequency during the nineteenth century, primarily for the purpose of expanding rangeland into chaparral and coastal sage scrub dominated landscapes. The economy of the Spanish and later Mexican period was primarily based on pastoralism, and most historical sources indicate extensive grasslands at the time of colonization and limited need for immediate rangeland expansion (Keeley, 2002). Nonetheless, there are historical reports of these early pastoralists using fire to open up shrublands and increase forage (Kinney 1887), and this is reflected in increases in grass pollen from sediment cores (Russell 1983).

By the middle of the nineteenth century there was increasing pressure for rangeland expansion, and this was felt most severely in the coastal ranges south of San Francisco where 80% of livestock production was confined (Ewing et al. 1988). Following the Gold Rush of 1849, with an influx of American settlers, brush burning for the improvement of grazing became extensive throughout California. Ranchers in the foothill regions regularly burned large areas of brushland, and it became the practice of itinerant sheepherders, after leaving a grazing area, to set fires (Brown 1945; Bauer 1974; Nichols, Adams, and Menke 1984). Burcham (1957) contends that all rangelands in the state were fully occupied by 1880. A similar perspective is that of Brown and Show (1944) who stated, "It is generally conceded that what is known as the 'pastoral era' of California ended in 1870. In that year, good pasture land, which was also agricultural in character, rose to a price of from 75 cents to $6.00 per acre." In the succeeding decades there was extensive pressure to utilize fire for the purpose of opening up shrublands and increasing forage (Lee and Bonnicksen 1978). The burning by these stockmen in mountain watersheds of southern California were thought to be responsible for damaging floods on both the coastal and interior sides of the San Gabriel Mountains, leading to its designation as the first forest reserve in California (Lockmann 1981).

One factor contributing to the use of fire in the opening up of shrublands was apparently the homestead laws that allowed acquisition of 65-ha parcels from public domain land (Lee and Bonnicksen 1978). Such parcel sizes were generally sufficient to maintain a homestead based on stock production, but this plan did not work in the rugged hills of southern California, where homesteads were centered in small valleys known as *potreros*, surrounded by impenetrable chaparral. "Since the potreros were too small to support an economically sound cattle operation, stockmen supplemented meadow grazing with forage produced by periodically burning the adjacent chaparral" (Lee and Bonnicksen 1978). Since brush burning was an essential resource use practice for stockmen, they burned extensive areas of chaparral (Barrett 1935; Brown and Show 1944; Brown 1945). For instance, in 1887 it was reported, that in the southern portion of San Diego County that "at least one third of the land covered with brush, grass and oak timber has been burnt off by settlers in the past eighteen months" (Lee and Bonnicksen 1978). As a consequence of early settler burning, fire control laws were enacted soon after statehood in 1850 (Clar 1959). Not surprisingly, in the early part of the twentieth century, ranchers were often the primary opponents to fire exclusion policies, which in southern California was prompted by the need for watershed protection in the coastal plain (Lee and Bonnicksen 1978).

In summary, it is apparent that during this settlement period the primary alteration in fire regime was to increase the frequency of fires on shrubland landscapes. This was an era of very limited fire suppression, and yet fires were much as they are today in that large crown fires covering tens of thousands of hectares were not uncommon (Kinney 1900; Barrett 1935; Brown and Show 1944; Brown 1945; Minnich 1987). For example, one of the largest fires in Los Angeles County (24,000 ha) occurred in 1878 (Keeley, Fotheringham, and Morais 1999), and the largest fire in Orange County's history was over a quarter million hectares and occurred in 1889 (Lee and Bonnicksen 1978).

Twentieth-Century Patterns of Burning

Burning patterns during the twentieth century are shown for the nine counties in central and southern coastal California (Fig. 8.11). Most counties exhibited little or no change in area burned except for Los Angeles and Riverside counties in southern California, which exhibited highly significant increases in area burned during the twentieth century. In contrast to the situation in western U.S. coniferous forests, fire suppression clearly has not excluded fire from these shrubland landscapes. Collectively the 1920s, 1940s, and 1970s were high decades, and the 1930s and 1960s were low. Possible explanations for these patterns are that they result from (1) decadal-scale variation in climate, (2) natural cycles resulting from fuel buildup, and/or (3) human demographic patterns.

Role of Climate/Weather

There are numerous suggestions in the literature of extended droughts contributing to extraordinarily severe fire seasons, but with a few notable exceptions, most lack statistical rigor. Minnich (1983) reported that there was a significant positive relationship between precipitation and area burned in coastal sage scrub of southern California and adjacent Baja California, but he presented no statistics to support this contention. He also inspected patterns of chaparral burning over this time period and concluded no such relationship existed with chaparral. However, others have reported a relationship between precipitation and burning

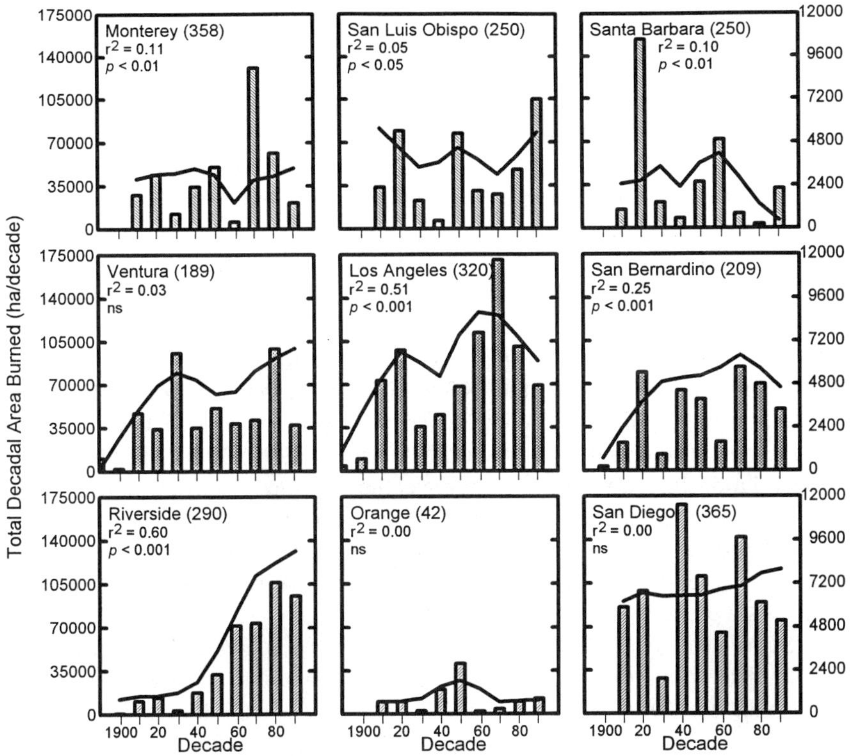

Figure 8.11. Area burned per decade and 10-year running annual average during the twentieth century for nine counties in central and southern California (data from the Statewide Fire History Database; see Fig. 8.5). Shrubland area in thousands of hectares shown in parentheses following the county name (from Callaham 1985).

in chaparral. One line of evidence is the spatial relationship between average precipitation and fire occurrence within the chaparral zone of San Diego County (Krausmann 1981). Another line of evidence is the demonstration that chaparral burning varies temporally with changes in precipitation; little area is burned following rainfall years where spring precipitation is >200 mm (Davis and Michaelsen 1995). These observations have been interpreted to mean that more rain translates into more biomass and thus greater fuels for burning in the subsequent fire season.

Using the FRAP data set (Fig. 8.11), we found few statistically significant correlations between patterns of rainfall and burning for chaparral and coastal sage shrublands combined. For each county separately, or all counties collectively, there was no significant relationship between total acreage burned per year and the nearest station with long-term records for:

1. total annual (January–December) precipitation,
2. growing season (November–June) precipitation,

3. spring (January–May) precipitation,
4. summer (June–August) precipitation, or
5. previous growing season's precipitation.

There was, however, a weak, but significant negative correlation ($p < 0.05$, $r^2 = 0.05$–0.06, $n \geq 88$) between October precipitation and area burned in each of the southern California counties, indicating that early autumn rains cut short the fire season at its peak.

Weather conditions affecting autumn foëhn wind-driven fires are most critical in determining area burned. Santa Ana wind conditions are largely responsible for fires becoming large and is reflected by the strong correlation between fire size and high temperatures. On the Los Padres National Forest fires generally ignite on days when the temperature is 3 to 5°C greater than the monthly average, and large fires never originate on days where temperatures are <25°C at the Santa Barbara airport (Davis and Michaelsen 1995; Moritz 1997). Moritz (1999) examined this relationship between severe fire weather (defined as days with maximum temperatures at the Santa Barbara airport ≥32°C) and extreme fire events in the central portion of the Los Padres National Forest. He found that large fires (>4000 ha) were strongly associated with severe fire weather. In this part of California severe fire weather is often, but not always, associated with foëhn winds (Schroeder et al. 1964; Dunn and Piierto 1987; Ryan 1996). However, farther south, for example, in the Santa Monica Mountains, all large fires appear to be driven by Santa Ana winds (NPS, Santa Monica Mountains National Recreation Area, unpublished data). In general, the largest wildfires in the central and southern coastal region are during severe fire weather conditions that include high temperatures, coupled with low humidity and high winds (Coffin 1959; Pirsko 1960; Schroeder et al. 1964; Weide 1968; Countryman, McCutchan, and Ryan 1969; Phillips 1971; Countryman 1974; Dunn and Piierto 1987; Gomes et al. 1993; Davis and Michaelson 1995; Minnich and Chou 1997).

Role of Fuel Cycles

Fuel accumulation was implicated in burning patterns in California shrublands by modeling studies published in the 1970s (Rothermel and Philpot 1973; Philpot 1974a,b). Based on untested assumptions about rates of fuel accumulation and effectiveness of fire suppression, it was concluded that large fires were increasingly more common because of an accumulation of older age classes of vegetation (Fig. 8.12). In support of this idea are many anecdotal references that fire fighters and fire researchers often relate about the tendency of fires to stop upon encountering young age classes of fuels (e.g., Philpot 1974a,b; Minnich 1998). An example of how these anecdotes are often used is the story about the 1970 Laguna Fire (one of the largest in California's history), in which it is claimed the fire died out when it encountered young age classes of vegetation (Rich Minnich, public communication, National Public Radio's "All Things Considered" radio broadcast, June 10, 1999). While that observation may be true, the deduction that

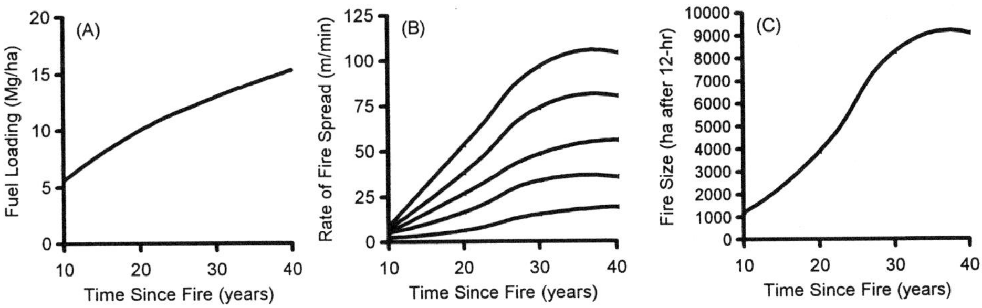

Figure 8.12. Modeling studies by Philpot (1974a, 1974b). (A) Assumed successional changes in fuel loads, (B) predicted rate of fire spread at increasing wind speeds from 10 to 50 kph, and (C) predicted fire size after 12 hours burning under sustained 50 kph wind speed. From these models it was concluded that as chaparral stands increase in age due to fire exclusion, there is a resultant increase in fuels, fire spread rate, and fire size. Following suggestions by Countryman (1974), these models were interpreted to support a fire management policy that relied heavily on prescription burning to produce a landscape comprising a mosaic of age classes.

there is a causal relationship is doubtful because the Laguna Fire burned over 10,000 ha of young vegetation (5–20 years) prior to its being extinguished (Dunn 1989), and the fire was contained only after a week of very severe Santa Ana winds subsided (Keeley, personal observations). In this fire, as well as other catastrophic fires, changes in fire behavior leading to containment often have had more to do with temporal changes in weather than spatial changes in fuels (Dunn and Piirto 1987). Although one can point to various fire perimeters that suggest fuel age is a barrier to fire spread (e.g., Philpot 1974), there are others that indicate it is not, such as half of the 5900 ha Romero Fire that burned above Santa Barbara in 1971 consumed seven-year old fuels from the 1964 Coyote Fire (Gomes et al. 1993). In short, there is no statistical evidence to support the notion that southern California landscapes supporting young vegetation are effective barriers to the spread of catastrophic fires. This of course is not meant to suggest that stand age has no effect on fire spread, only that its effectiveness is strongly controlled by weather (see the section below on Future Fire Management Strategies).

Fire history data also have been used to support the idea of fuel-driven fire behavior. Radtke, Arndt, and Wakimoto (1982) observed that peak decades of burning were followed by decades of very little burning in the Santa Monica Mountains of Ventura and Los Angeles counties. It was suggested that these decadal variations in burning represented a cyclical pattern driven by fuel loading. Their confidence in this model is illustrated by their future prediction that for the Santa Monica Mountains, the 1980s decade would be a peak and would be followed by a decline in burning during the 1990s. In retrospect we now know that, although the 1980s were high, the 1990s were even higher (Santa Monica Mountains Recreation Area, unpublished data). The primary weakness in explaining

Table 8.2. Shrubland area,[a] population density,[b] and estimated fire rotation intervals[c] for the shrub-dominated counties in California, arranged north to south

County	Brush (10^3 ha)	People/ 10^6 ha brush	Fire rotation interval (yr) pre-1951	Fire rotation interval (yr) post-1950
Monterey	358	0.99	115	64
San Luis Obispo	250	0.87	60	48
Santa Barbara	250	1.48	47	81
Ventura	189	3.54	121	34
Los Angeles	320	27.69	44	30
San Bernardino	209	6.79	46	37
Riverside	290	4.04	225	38
Orange	42	57.39	36	29
San Diego	365	6.84	35	41

[a] Area as of 1985, from Callaham 1985.
[b] Population density for 1990, from *http://www.census.gov/population/cencounts/ca190090.txt.*
[c] From Keeley et al. 1999.

decadal variations in burning by changes in fuel loads is the fact that the total burning during a decade comprises only a fraction of the fuels on the landscape and substantial fuel loads are available for burning every decade. For example, fire rotation intervals (Table 8.2) indicate that in most counties only 20% to 30% of the landscape burned in any given decade; thus decades of peak burning should not automatically be assumed to alter the future course of burning by leaving the landscape with limited fuels.

For southern and central coastal California, fire history data refute the contention made by Minnich (1989, 1998, 2001; Minnich and Cho 1997) that chaparral fire occurrence is constrained by the rate of fuel accumulation. Fire hazard estimates are either independent of age (Moritz 1999) or only weakly dependent up to 20 years of age (Schoenberg et al. 2001; Peng and Schoenberg 2001). In addition stand-age classes burned in the eight largest wildfires in the Santa Monica Mountains illustrate that these extreme events are not dependent on accumulations of older fuels (Keeley, Fotheringham, and Morais 1999). Indeed, in this range the greatest proportion of burned vegetation is in the younger aged stands, for both coastal sage scrub and chaparral (Fig. 8.13). Also vegetation type, which has a profound influence on fuel distribution (e.g., Fig. 8.9), has been shown to have little influence on fire history in the Los Padres National Forest (Moritz 1999).

Alterations in the landscape distribution of fuels have also been implicated in changes in fire size. It has been proposed that due to fire suppression, there has been an increase in the age and homogeneity of fuel distribution leading to larger and higher-intensity fires (e.g., Minnich 1989, 1995, 1998; Minnich and Cho 1997). The only data in support of this model are the high frequencies of small fires south of the U.S. border, which are interpreted as solely the result of natural burning cycles in the absence of fire suppression. However, north of the border

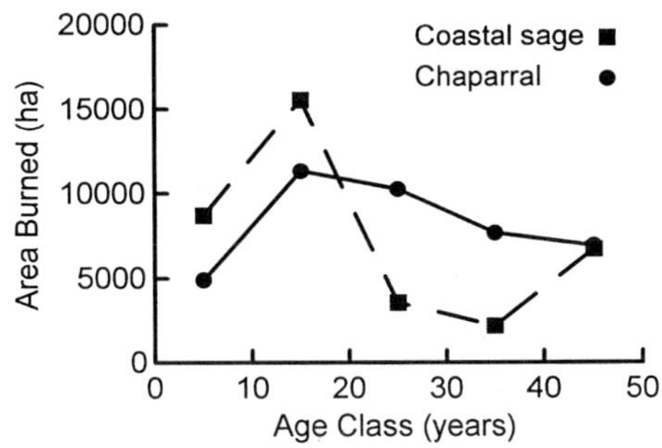

Figure 8.13. Age classes of chaparral and coastal sage scrub stands burned by all fires over 5000 ha from 1967 to 1996 in the Santa Monica Mountains (data from the U.S. National Park Service, Santa Monica Mountains National Recreation Area, Thousand Oaks, CA). Greater burning of young age classes of coastal sage scrub is likely due to more flammable fuels, longer fire season, and the concentration of this vegetation adjacent to urban centers, which are major sources of ignition.

fire suppression activities have not resulted in fire exclusion (Moritz 1997; Conard and Weise 1998; Keeley, Fotheringham, and Morais 1999; Weise et al., in press). Thus the patterns north and south of the border, while interesting, cannot be held up as an example of what happens to landscapes subjected to a fire suppression policy. Such fire management policies can not be held responsible for large destructive wildfires (Keeley and Fotheringham 2001a,b) as large fires have been a common feature of the southern California landscape throughout the ninteenth and twentieth centuries (Keeley, Fotheringham, and Morais 1999; Keeley and Fotheringham 2001a). Additionally sediment cores show the frequency of large fires has not changed during the past 450 years (Mensing, Michaelsen, and Byrne 1999), and colorful, but less authoritative, is the Digueño Indian legend of a large catastrophic fire sufficient to lead to migrations of tribes in San Diego County at about the time of Columbus (Odens 1971, p. 8). All of these observations suggest large fires are not a modern artifact of fire suppression as proposed elsewhere (Minnich 1989, 1995, 1998; Minnich and Dezzani 1991; Minnich and Chou 1997).

We now know that although the models developed by Philpot and others may be sound, their conclusions were flawed by incorrect assumptions. The assumption of a steady increase in fuels was inaccurate (Figs. 8.9 and 8.10), and the assumption that fire suppression was effectively excluding fire from shrubland landscapes was wrong (Fig. 8.11). In short, patterns of burning on shrubland landscapes cannot be explained solely by changes in accumulation of fuels (Moritz 1997, 1999; Conard and Weise 1998; Keeley, Fotheringham, and Morais 1999; Peng and Schoenberg 2001). Indeed, modeling studies that consider landscape patterns of fire spread conclude that stand age alone cannot constrain fire size (Zedler and Seiger 2000). If that were true, then just a single large Santa Ana wind-driven fire would reset the landscape to the same age class, which would forever be doomed to burn as a single large unit. Zedler and Seiger's model shows that even in the absence of Santa Ana fires, if stand age were the only controlling factor, over time, burn units would coalesce and become larger and larger with each fire cycle.

Role of Human Demography

Clearly, humans have perturbed shrubland fire regimes, but unlike the situation in many western U.S. forests, the primary impact has been through increased fire frequency (Table 8.1, Figs. 8.4 and 8.5c–d) and not through fire exclusion (Fig. 8.11). Collectively, across all counties considered in Figure 11, there was a significant correlation between fire frequency and population density and between fire frequency ($r^2 = 0.51$, $p < 0.05$, $n = 9$) and area burned ($r^2 = 0.71$, $p < 0.01$, $n = 9$). Southern California (defined in Fig. 8.3 legend), with the highest rate of population growth (Fig. 8.5b), also has had the greatest increase in wildfire ignitions (Fig. 8.5d). In contrast, the central coastal region has far fewer human ignitions (Figs. 8.4 and 8.5c), which is in line with the much lower population (Fig. 8.5a).

Indirectly the public infrastructure of roads contributes to patterns of burning. The central coastal region has substantial portions of its landscape lacking public roads, which is in stark contrast to the vast highway network connecting most parts of southern California. Fully one-third of all human-caused fires on Forest Service and CDF protected lands in southern California occur along roads (Gee 1974; Conard and Weise 1998). On shrubland landscapes near metropolitan areas, such as the Santa Monica Mountains, the vast majority of fires originate along roadways (Los Angeles County Fire Department, unpublished data).

In light of these considerations it seems probable that some portion of the decadal variation in burning during the twentieth century (Fig. 8.11) may have a human dimension. In the early part of the twentieth century populations in many parts of California were increasing rapidly (Fig. 8.5a–b). With this influx of people, came increased anthropogenic impact on the natural fire regime, driven largely by the increased mobility the automobile afforded; car registrations in California rose from 191,000 in 1915 to 1,500,000 in 1925 (Davis 1967). During the period 1908 to 1920, every county in southern California voted large bonds for road building (Davis 1967). Roads provided increased access to wildland areas. For example, a doubling in wildland use between 1916 and 1920 (Show and Kotok 1923) coincided with a marked increase in wildfire incidence in southern California (Fig. 8.5d). This rapid population growth and increased mobility strained the ability of fire protection in California during the early part of the twentieth century (Clar 1959). With the expanding population came an expansion of development at the urban–wildland interface, which then increased public susceptibility to wildfire impacts. As a consequence the decade of the 1920s witnessed some particularly destructive wildfires that increased public pressure for fire protection and prevention (Clar 1959).

In response, during the 1930s fire management agencies stepped up their attack on wildland fires by the introduction of lookout towers and aircraft for better reconnaissance, which decreased the size of some fires due to early detection (Clar 1969; Pyne 1982). During this period various innovations were introduced to suppress fires, although effective suppression was elusive due to the inaccessibility of remote wilderness areas (Brown and Show 1944; Pyne 1982). A system of fuel breaks was one early answer to this problem, and creation of 200 CCC (Civilian Conservation Corp) camps throughout the state during the Depression contributed significantly to this network. Increased fire suppression activities due to an excess of man power from federal relief programs (Clar 1969 described it as a "forced feed" of the California Division of Forestry), coupled with reduced "motor touring" (e.g., AAA memberships dropped 40% in the five years following 1929; Davis 1967) perhaps contributed to the drop in area burned during the Great Depression in many counties.

Diversion of resources to the "war effort" during the first half of the 1940s contributed to diminished fire suppression capacity (Brown 1945; Clar 1969) and may account for the peak burning that occurred in some counties during that decade (Fig. 8.11). San Diego County stands out because its worst decade for wildfires was the 1940s. Zahn (1944) suggests the extraordinary fires of this era were the result of the aircraft industry, which had concentrated a great deal of the war effort in San Diego County. He described the situation at the time as follows: "Bootleg fuel, high payrolls and a yen for the open spaces have resulted in hundreds of aircraft workers motoring to the hills—night or day—between work shifts. Most of these workers are newcomers to California, unfamiliar with the tinder-box potentialities of local brush."

The modern era of effective fire suppression was introduced in the 1950s with the development of air tankers for fire fighting (Pyne 1982), and this impact was evident in a 10-fold drop in burning across the country (Dombeck 2001). However, in California these techniques have not proved very successful in halting fires during extreme Santa Ana wind conditions (Countryman 1974).

In short, despite innovations, fire suppression has not diminished the wildland fire problem in California. Indeed, since the 1950s, there has been an increase in the allocation of funds to the California brushfire problem (Bonnicksen and Lee 1979; Kinney 1984), and an increase in the loss of property and lives (Rogers 1982; Martin and Sapsis 1995). Additionally, due to television, there has been increased public awareness of large-scale wildfires. Over this period there have been a number of workshops, conferences, and proceedings volumes published on this wildland fire problem—roughly one every 5 to 10 years since 1950—and these offer a diversity of opinions on the role of fire in the California landscape. Although not a popular view, it has been frequently suggested that the problem stemmed in large part to the burgeoning population and poor zoning regulations attendant with urban sprawl into the foothills. The problem was evident 50 years ago. For example, Zivnuska, Arnold, and Arment (1950) warned of this "potentially explosive situation," and noted "it is known that one of the significant trends in recent population changes has been the increase in number of residences in the flash-fuel types adjacent to primary watersheds." Under these conditions catastrophic fires are not necessarily the largest fires, as witnessed by the rather small

Oakland Hills Tunnel Fire in October 1991 (725 ha) that burned nearly 3000 structures and killed 25 people (Booker, Dietrich, and Collins 1995).

In summary, severe fire weather occurring each autumn coupled with human demographic patterns would seem to explain patterns of burning (Fig. 8.11) far better than changes in available fuels. During the twentieth century any changes in the fire regime have been dwarfed by the changes in land development patterns, which have increasingly placed more people at risk to the natural forces long present on the landscape (Davis 1965; Bradshaw 1987). This pattern continues—for example, in the 25 years prior to 1980, 2408 homes and other structures were destroyed by wildfires in California but in the subsequent 14 years the number tripled (*http://www.prefire.ucfpl.ucop.edu/wildfire.htm*). Preference for a rural lifestyle and the skyrocketing cost of suburban housing in large metropolitan areas has progressively increased the urban–wildland interface. Of particular concern is the prediction that rural population will exceed urban growth in the foreseeable future (Bradshaw 1987). For both economic and political reasons the notion that urban sprawl is responsible for natural wildfires becoming catastrophic fires is unpopular, in part, because it seems to defy the inherent belief that it is possible to engineer solutions to all environmental problems.

The Contemporary Versus Natural Fire Regime

There is reason to believe that the contemporary fire regime in these shrublands mirrors the natural crown fire regime far more than is generally accepted (cf. Bonnicksen and Lee 1979; Minnich 1983; Pyne 1982). Today in southern California, fire incidence peaks in the summer, but most area burned is from autumn fires (Fig. 8.8a). Likewise the natural fire regime was probably characterized by many small summer lightning-ignited fires and a few large autumn fires driven by Santa Ana or Mono winds that burned large areas (Keeley and Fotheringham 2001a). This model would seem to be contradicted by the fact that Santa Ana or Mono winds are northeast winds, whereas summer thunderstorms are associated with south winds, and the two do not commonly coincide (Coffin 1959). Consequently today it is rare for Santa Ana wind-driven fires to be other than anthropogenic in origin. However, under natural conditions the fact that lightning fires burned for months (Minnich 1987), coupled with the relatively close temporal juxtaposition of the July–August lightning fire season (Keeley 1982) with the September–November Santa Ana winds (Fig. 8.8b), makes it inevitable that lightning ignitions would occasionally have been spread by these foëhn winds (Keeley and Fotheringham 2001a). While such events could not have been frequent, we know from historical documents that summer lightning-ignited fires can burn for more than a month and consume on the order of 10^3 ha (Minnich 1987). This pales in comparison to the 10^4 ha that are often covered in a single day by a Santa Ana wind-driven fire (Phillips 1971).

Davis and Burrows (1993, 1994) modeled the long-term fire regime in chaparral by linking physical models based on fire spread equations to fuel models of stand senescence. Their simulations predicted a prehistoric fire regime of variable sized fires that produced a landscape mosaic of different age classes. With one ignition every 10 years (typical lightning-ignited fire frequency for coastal California; Keeley 1982) their model predicted that most fires would be large and over 80% of the landscape would burn at ages greater than 95 years. These conclusions are supported by other evidence that points to a natural fire regime of large fires and long fire return intervals for these coastal range landscapes (Greenlee and Langenheim 1990; Byrne, Michaelsen, and Soutar 1977; Mensing, Michaelsen, and Byrne 1999; Keeley and Fotheringham 2001a).

Alternatively, it has been argued that prior to the current fire suppression policy, landscapes were immune to Santa Ana wind-driven fires because lightning fires kept the shrublands in a fine-scale mosaic of young age classes (Minnich 1989, 1995, 1998). It is presumed that this mosaic was quickly erased by highly effective fire suppression during the first couple decades of the twentieth century (Minnich 1990). However, historical records do not support this notion. For example, 90% of the 214,000 ha of shrublands on the San Jacinto Forest Reserve were estimated to be 30 years or older when surveyed at the end of the nineteenth century, which represent far older age classes then present today (Keeley and Fotheringham 2001a). Clearly, this landscape was not a fine-scale mosaic immune to large fires. In addition the early history of forest protection does not support the idea that highly effective fire suppression was present in the opening decades of the twentieth century (Clar 1959; Lockmann 1981). Minnich and Chou's (1997) suggestion that fire suppression activities "culminated in extensive fire outbreaks as early as 1919" is contradicted by historical documentation that reports large fires in the region long before this date, and before any fire suppression activities (e.g., Kinney 1900; Barrett 1935; Brown and Show 1944; Brown 1945; Lee and Bonnicksen 1978; Radtke, Arndt, and Wakimoto 1982). Nationwide there is no evidence of substantive reductions in area burned due to fire suppression until midway through the twentieth century (Dombeck 2001).

Historically fire intensity was variable, and there is no credible evidence that it has increased during the era of fire suppression (Keeley, Fotheringham, and Morais 1999). The primary changes in the fire regime are that humans have replaced lightning as the primary source of ignition and fire frequency has increased, particularly in areas of high population density such as southern California (Figs. 8.4 and 8.5). Because fire prevention has been ineffective at eliminating human fires, presently and for the foreseeable future, fire suppression is required just to maintain some semblance of the natural fire regime.

Impacts on Vegetation

In contrast to the paradigm suggested for many western U.S. forests, ecosystem health of shrublands is threatened not by a lack of fire but by high fire frequencies that exceed the resilience of many species. Examples of high fire frequency induced extirpations are numerous (e.g., Gause 1966; Zedler et al. 1983;

Haidinger and Keeley 1993; Zedler 1995; Keeley 2000). Generally, the threat of high fire frequency is lessened on very low nutrient soils where postfire annual biomass is limited and less likely to carry a repeat fire. Where fires occur more than once in a decade, nonsprouting chaparral shrubs are entirely lost from the system. Commonly exotic grasses and forbs will take their place, and as these increase in importance, they appear to competitively displace the native annuals. A similar course is evident in coastal sage scrub under higher fire frequency. In both of these crown fire ecosystems, high fire frequency favors annuals over woody plants, and this advantage increases with increasing soil aridity (Wells 1962).

As fire frequency increases, fuel structure changes and subsequent fire behavior changes. With increasing exotic herbaceous cover, the seasonal window of flammability increases (Radtke, Arndt, and Wakimoto 1982), and fire behavior becomes a mixture of crown and surface fires. This has two very important consequences. Surface fires connect the woody fuels where otherwise they might be too widely spaced to carry a crown fire, and thus exotic herbs shorten the fire return interval. Another important consequence is that fire intensity is lower where surface fires occur and this contributes to increased survivorship of exotic annual seed banks (Fig. 8.14). With continued disturbance these nonnative invasives may replace the entire ecosystem (Keeley, 2001), and type conversions of shrublands to exotic grasslands are well documented (e.g., Cooper 1922; Bentley 1967; CDF 1978; Biswell 1989; Minnich and Dezzani 1998). As a consequence exotic grasslands tend to replace shrublands in the proximity to urban environments, where the higher ignition sources in the company of flashy fuels have the potential for even greater fire frequency. Evidence of this is seen in the substantially shorter fire return interval in grassland vegetation at the urban–wildland interface than observed for shrublands at the interface (J. Spero, California Division of Forestry, personal communication, 1999).

The extent of such type conversion is unknown because of past disturbances, which includes Indian burning throughout the Holocene and burning coupled with intensive livestock grazing in the past 200 years. In the coastal counties from Monterey southward (Fig. 8.3) exotic annual grasslands cover nearly two million hectares or 25% of the wildland landscape, and less than 1% has significant patches of native perennial bunchgrass (Huenneke 1989). Although it is often taken as a matter of faith that these landscapes have always been grassland (Heady 1977), there is evidence that many exotic grasslands were formerly dominated by woody associations (Cooper 1922; Wells 1962; Oberbauer 1978; Huenneke 1989; Keeley 1990, Hamilton 1997). Today these landscapes comprise a mosaic of vegetation patterns (Fig. 8.15) that appear to be disturbance induced (Wells 1962). Grasslands on this modern landscape comprise a new quasi-equilibrium of nonnative annuals that are somewhat resistant to recolonization by native shrubs. It is a dynamic process whereby as disturbances increase or wane, vegetation physiognomy shifts between exotic grassland and shrubland/woodland (Hobbs 1983; Freudenberger, Fish, and Keeley 1987; Callaway and Davis 1993).

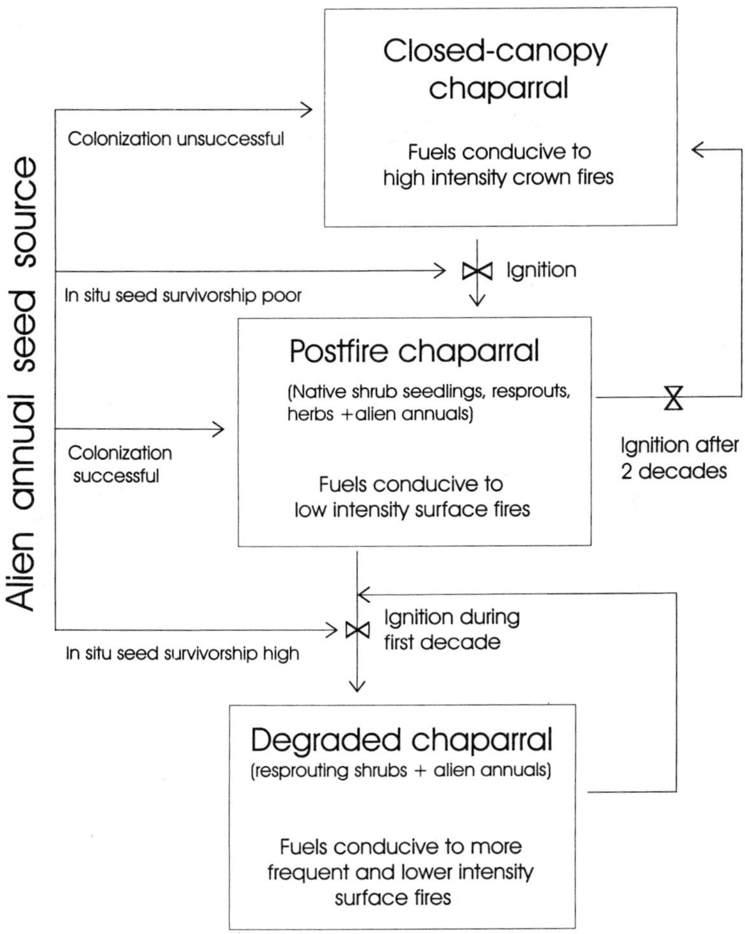

Figure 8.14. Schematic diagram of how rate of fire ignitions in chaparral affects alien plant invasion and how alien invasions affect fuel loads, which in turn alter fire frequencies, making sites more conducive to further invasion (from Keeley, 2001).

Future Fire Management Strategies

It has been suggested that "after nearly a century of suppression" there is a need for a reintroduction of fire into chaparral through prescribed burning (Minnich and Dezzani 1991; Minnich and Franco-Vizcaíno 1999). However, fire history data do not support this management strategy. On most shrubland landscapes there is an abundance of fire, and 60- to 70-year-old stands, considered to be the normal age for burning (Minnich 1989), are rare at the present time (Keeley 1992). Indeed, the current fire rotation interval of 30 to 40 years is shorter than

Figure 8.15. Vegetation mosaic of nonnative annual grassland and shrublands in the central coastal ranges of California (photo by J. Keeley).

that calculated for the early part of the twentieth century (Keeley, Fotheringham, and Morais 1999). In light of the expected trends in population growth in California, and the close association between population density and fire incidence (Fig. 8.5), increased fire prevention is far more important to protecting natural resources than prescription burning or other methods of "fire restoration."

Consequently there is a need to reevaluate prescribed burning strategies for California shrubland landscapes. There are two common motivations for prescription burning: (1) for the benefit of natural resources and (2) as a fuel manipulation technique, primarily to reduce fire hazard but also to reduce the threat of soil erosion or air quality hazards, which may be worse under wildfire conditions. In many western U.S. forests, prescription burning provides both resource benefits and a reduction in fire hazard However, the reality for some ecosystems is that prescriptions reducing fire hazard, may not always enhance resource values and sometimes may detract (Johnson and Miyanishi 1995; Keeley and Fotheringham 2001b).

Prescription Burning for Resource Benefit

There may be little justification for using fire for resource benefit, since vast portions of shrubland landscape currently experience a higher than normal fire frequency. Lack of fire does not appear to pose a risk because postfire studies demonstrate that both chaparral and coastal sage scrub regeneration are highly resilient to even the most extreme fire events occurring after a long hiatus of burning (Keeley 1998, 2000).

One proposed benefit of prescribed burns is that they are done under more moderate weather conditions than are typical for wildfires, leading to less intense fires and less severe impacts on plant and soil resources (Green 1981; Moreno and Oechel 1991; Riggan et al. 1994; Wohlgemuth, Beyers, and Conard 1999). However, some of the experimental work demonstrating fire intensity effects on seed banks and soils have been done on piles of cut fuels, which do not accurately represent the fuel structure under natural conditions and are likely to generate unnaturally high soil temperatures. More importantly, however, even the most extreme fire wildfire events today probably do not fall outside the natural range of variation for these ecosystems.

Other resource benefits from prescription burning include invasive plant control, but the primary invasive problems involve herbaceous species, which invade shrublands when fire frequency increases (Fig. 8.14). It is not likely that prescription burning would displace these invasive species, unless the target is vulnerable to a particular seasonal window of burning. In shrublands there are no such windows of opportunity that are not equally damaging to some native species. Nonnative legume shrubs known as brooms (*Cytisus scoparius* and *Genista monspesulanus*) are sometimes targeted for removal with prescription burning, but these are inevitably replaced by exotic grasses (D'Antonio 2000). However, prescription burning for restoration of shrubland communities may be useful if accompanied by vigorous revegetation with native shrubs and herbs.

In general, there are few places where fire-dependent shrublands are threatened by the lack of fire and few instances where prescription burning is needed for natural resource benefits. The primary justification for prescription burning is for fire hazard reduction. However, in these ecosystems any additional fire carries with it the potential for negative impacts on resources. Negative impacts may arise not just from burning but can be associated with other fuel manipulations. For example, fuel breaks are possible corridors for bringing nonnative invasive species into wildland areas (Keeley, 2001).

Prescription Burning for Fire Hazard Reduction

Prescription burning carries with it a risk of fires escaping, and escaped fires are quite hazardous in crown fire ecosystems, most particularly chaparral landscapes with a complex urban–wildland interface. In order to ensure successful containment of a prescribed burn, there are strict limitations on the acceptable wind speed, air temperature, relative humidity, and fuel moisture—typically wind speeds below 17 kph (10 mph), relative humidities above 30%, air temperature below 32°C (95°F), and fuel moisture above 75% (Fenner, Arnold, and Buck 1955; Green 1981). This, of course, varies with the fuel load and landscape, and various combinations will produce acceptable prescriptions (Paysen,

Narog, and Cohen 1998). One approach to reducing the risk of escaped fires is to burn in the spring, assisted by pretreatment of mechanical crushing and drying. Thus the target fuels are surrounded by less flammable living vegetation (Wolfram 1962). This procedure is expensive, and it has the potential for producing resource damage. For example, unseasonable application of fire inhibits postfire vegetation recovery (Florence 1985; Rundel, Parsons, and Baker 1987; Parker 1990) and is correlated with increased soil erosion (Turner and Lampinen 1983).

Because prescriptions are designed for safety, they are often marginal for burning. Under prescription weather conditions, fire spread is markedly influenced by fuel structure, and fire spread is often inhibited in stands less than 20 years of age (Green 1981; Paysen and Cohen 1990; Conard and Regelbrugge 1994). This is largely due to the lack of sufficient dead fuels required to spread fire to live foliage, and to the lack of fuel continuity between the ground and the shrub canopy and between adjacent canopies, factors that are extremely critical to fire spread under low wind and high humidity. Controlled burning of younger stands requires either prescriptions with risky weather conditions or pretreatment with biodegradable herbicides (which increase the dead fuels) coupled with seeding of exotic grasses that increase flashy (readily ignitable) fuels and increase surface fire spread.

Evaluating the effectiveness of prescribed burning at reducing fire hazard is complicated by the fact that such fuel management practices are never going to be fully effective against all fires. Wildfires are often more readily contained when they encounter young stands of vegetation, largely because lower fire intensities allow for safer access by fire suppression forces (Countryman 1974). However, landscape age mosaics created by rotational burning will not pose a barrier to wildfires ignited under severe fire weather, since the high winds readily push fires through young age classes (e.g., Fig. 8.13). Under these conditions young vegetation is of minimal value in halting the forward spread, and also firebrands are capable of spreading the fire kilometers beyond the front. Containment of shrubland fires burning under severe weather conditions usually requires a change to more favorable weather (Rogers 1982; Dunn and Piirto 1987; Gomes et al. 1993).

Thus prescription burning presents a catch-22 situation. It can only be done safely under weather conditions that require mature chaparral, 20 years of age or more, but stands of vegetation this age and younger will not form effective barriers to fire spread under severe weather conditions. Modeling studies indicate that to be effective even under moderate weather conditions requires a substantial portion of the landscape be treated (Mark Finney, public communication, 2001). Thus, while landscapes managed by rotational burning may contribute to easier containment of fires burning under moderate weather conditions, they are of limited value during severe weather. However, these latter fires are the ones that become truly catastrophic and are responsible for the greatest losses. Consequently National Forest Service policy of landscape-scale rotational burning to produce a mosaic of age classes needs to be reconsidered (Conard and Weise

1998). This type of fuel management is extremely expensive, unlikely to prevent catastrophic wildfires, and has little resource benefit.

Future fire management policy needs to steer away from extensive landscape-scale prescription burning and focus on intensive and strategic use of fire hazard reduction techniques, both to minimize negative impacts of high fire frequency on natural resources and to maximize fire hazard reduction. The marked differences observed between the central coastal ranges and southern California (Fig. 8.5) suggests that regions may require different fire management strategies. Greater focus needs to be given to transportation corridors as roadways are primary sites of ignitions, and since roadways are required to connect developments, as the urban/wildland interface increases, these fire hazards increase. Roads could also play a role in minimizing the negative impacts of fire hazard reduction programs, since many of the negative impacts of fuel reduction techniques (e.g., aesthetic impacts, promoting invasive plants and animals) are also shared by roadways. Thus greater attention needs to be given to co-locating roads and fuel manipulations such as fuel breaks.

Considering the psychology of many who inhabit the urban–wildland interface, it is questionable whether or not education can play a substantive role in reducing future losses from wildfires (Gardiner, Cortner, and Widaman 1987). Regulations requiring fire "safe" construction have been implicated in reducing property losses in the past and will possibly reduce the degree of future losses (Cohen 2000). It seems inevitable that fire management policy will increasingly require involvement of city and county planners in order to solve the primary fire hazard problem of how to constrain the ever-expanding urban–wildland interface. Fire managers can play a key role in providing accurate analytical models of causal factors driving extreme fire events and educating planners on the limitations to fire hazard reduction (e.g., Sapsis 2001).

Global Change Impacts on Future Fire Regimes

Fire regime is an emergent property of landscapes arising from the interaction of vegetation, weather, topography, and land management (Davis and Michaelsen 1995). Fire regime is influenced directly by vegetation through flammability characteristics and the structural distribution of fuels. Weather affects fire regimes through timing of ignitions, and through frequency and severity of burning conditions as well as direct effects on vegetation distribution. Topography affects rates of natural lightning ignitions and wind patterns that ultimately control fire behavior. Land management affects the distribution of vegetation types and thus landscape patterns of fuels. Land management, in the broad sense, also controls the extent and pattern of the urban–wildland interface, which acts as a porous boundary where fires diffuse across in both directions. Wildland fires may diffuse out from the urban–wildland interface, but the most catastrophic fires result from wildfires burning into the urban environment. Global changes, including direct effects of increased atmospheric CO_2 levels, climate changes, and changing land use, all have the potential for changing fire regimes by altering vegetation, weather, and land management.

Future increases in atmospheric CO_2 levels may directly affect plant growth and potentially alter patterns of fuel distribution. In chaparral the effects are predicted to be variable and strongly dependent on levels of other resources (Oechel et al. 1995). Along gradients of increasing soil fertility we might expect increased biomass production, but the increased leaf area may place greater demands on the limited soil water resources in this semi-arid region, dampening potential increases in primary production. Further exacerbating this dampening effect is the expected increase in summer temperature. However, this could be offset by increased water use efficiency expected with elevated CO_2.

Climate change in California shrubland landscapes over the next half-century is predicted to increase winter and summer temperatures by 3°C and 1°C, respectively, and to increase winter precipitation by 25% (Field et al. 1999). Warmer and much wetter winter conditions will almost certainly contribute to higher primary production, although the magnitude is likely to decline with decreasing soil nutrients. It is assumed that this increased production will lead to higher fuel accumulation and more intense fires. However, these climate changes may also accelerate decomposition of dead fuels, which are critical to fire spread, and the importance of this dampening effect on fuel accumulation has not been evaluated. Expected increases in C:N ratios of dead fuels imply variations in rates of decomposition along soil fertility gradients, paralleling expected increases in primary production. Thus sites with the greatest increases in fuels may also experience the greatest increases in decomposition. Even if the net effect is an increase in rate of fuel accumulation, this should not automatically be assumed to lead to major changes in the fire regime. This is based on the fact that currently rates of fuel accumulation do not play a highly deterministic role in shrubland fire regimes (Moritz 1999; Schoenberg et al. 2001; Peng and Schoenberg 2001).

Expected changes in climate will affect vegetation structure through changes in energy balance as well as nutrient cycling, but this involves such complexity that presently one can only speculate what the future holds (Oechel et al. 1995). Attempts to understand how changes in precipitation and temperature will affect vegetation composition include documentation of contemporary climatic responses (Westman 1991) and growth simulations (Malanson and Westman 1991a, b; Westman and Malanson 1992; Malanson and O'Leary 1995). Realistic parameterization of these models is one limitation to their current usefulness, and thus the primary conclusion one can draw at this point is that changes in the relative abundance of species are to be expected. Another possibility is that changes in fire intensity due to greater fuel loads may affect changes in postfire recovery, although shrublands currently exhibit extraordinary resilience to a wide range of fire intensities (Keeley 1998). Ecotones are expected to be sites of greatest sensitivity to climate change (Peteet 2000), and the complex vegetation mosaic of California landscapes (e.g., Fig. 8.15) presents many opportunities for shifts in vegetation distribution. Considering the large role played by human interference, it seems likely that the greatest alteration in fire regimes will occur at the urban–wildland ecotone.

In general, GCM predictions for twenty-first-century climates in California are of limited value in understanding future fire regimes. Patterns of burning are driven by extreme events (Moritz 1997), and these are not well modeled (Field et al. 1999). One of the primary determinants of area burned is the coincidence of ignition with severe weather, and future changes in patterns of ignition might be expected to play a determining role in fire regimes. Climate-based models predict the California region will have a few percent increase in lightning fires (Price and Rind 1994), but this may not affect these shrubland landscapes where humans are the primary ignition source (Table 8.1, Figs. 8.4 and 8.5).

Future changes in land use are likely to have a more profound impact on shrubland fire regimes than other types of global change. Land use may also be the primary driver behind losses in biodiversity in California as well as in other Mediterranean-climate regions (Sala et al. 2000). Diversity loss is expected to result from increased population growth contributing to habitat loss, habitat fragmentation, and loss of corridors. Some of these factors will affect fire regimes, but, we expect that increased fire ignitions predicted from increased population growth will have a far more profound impact on these landscapes. As the fire return interval shortens, the native shrublands are degraded to mixtures of exotic grasses and forbs, and these invasives contribute to further decreases in fire return interval and loss of native plant diversity (Fig. 14). However, dampening this potential impact of shortened fire return intervals is the stepped-up rate of post-fire shrub recovery expected from predicted increases in winter temperature and precipitation. The impact of land-use changes on these landscapes makes it likely that it will far outweigh other global change impacts on fire regimes.

Conclusion

Throughout much of the shrubland landscape humans play a dominant role in promoting fires beyond what was likely the natural fire cycle. Future climate change is expected to have a minor role in altering fire regimes relative to other global changes such as population growth and habitat fragmentation. Future fire management needs to take a strategic approach to fuel manipulations and move beyond evaluating effectiveness strictly in terms of area treated. Fire management should consider designing strategies tailored to different regions as there are marked differences between the central coastal region and southern California in source of ignition (e.g., Table 8.1, Fig. 8.4), season of burning (Fig. 8.6), and historical patterns of population growth (Fig. 8.5a–b) and burning (Figs. 8.5c–d and 8.11). Presently we know relatively little about fire regimes in shrublands in the foothills of the Sierra Nevada and interior foothills of the northern coastal ranges, and thus it would be prudent to not transfer the conclusions drawn here too broadly until we have a clearer understanding of the extent of regional variation in shrubland fire regimes. One of the primary threats that all regions share is the increasing number of people being placed at risk to the natural wildfire threat because of the rapidly expanding urban–wildland interface. Fire management will need to play an increasingly active role in the planning process through critical analysis of causal factors driving fire regimes and the limitations to hazard reduction.

Acknowledgments. We thank Jim Agee, Max Moritz, Carl Skinner, Nate Stephenson, and Paul Zedler for helpful comments on an earlier version of this ms. CJF acknowledges funding from EPA S.T.A.R. Graduate Fellowship #U-915606. We thank Karen Folger and Denise Krieger for assistance with data acquisition.

References

Agee, J.K. 1993. *Fire Ecology of Pacific Northwest Forests*. Covelo, CA: Island Press.

Anderson, M.K., Barbour, M.G., and Whitworth, V. 1998. A world of balance and plenty. In *Contested Eden. California before the Gold Rush*, eds. R.A., Gutierrez and R.J., Orsi, pp. 12–47. Los Angeles: University of California Press.

Anderson, M.K., and Moratto, M.J. 1996. Native American land-use practices and ecological impacts. In *Sierra Nevada Ecosystem Project: Final Report to Congress: Status of the Sierra Nevada*, vol. 2, eds. SNEP Team, pp. 187–206. Davis: Centers for Water and Wildland Resources, University of California.

Barrett, L.A. 1935. *A Record of Forest and Field Fires in California from the Days of the Early Explorers to the Creation of the Forest Reserves*. San Francisco: USDA Forest Service.

Barro, S.C., and Conard, S.G. 1991. Fire effects on California chaparral systems: An overview. *Environ. Int.* 17:135–149.

Bauer, D.R. 1974. A history of forest-fire control in southern California. In *Symposium on Living with the Chaparral, Proceedings*, ed. M. Rosenthal, pp. 121–129. San Francisco: Sierra Club.

Bentley, J.R. 1967. *Conversion of Chaparral to Grassland: Techniques Used in California*. Washington, DC: USDA Forest Service, Agriculture Handbook 328.

Biswell, H.H. 1989. *Prescribed Burning in California Wildlands Vegetation Management.* Los Angeles: University of California Press.

Black, C.H. 1987. Biomass, nitrogen and phosphorus accumulation over a southern California fire cycle chronosequence. *In Plant Response to Stress: Functional Analysis in Mediterranean Ecosystems*, eds. J.D. Tenhunen, F.M. Catarino, O.L. Lange, and W.C. Oechel, pp. 445–458. Berlin: Springer.

Bonnicksen, T.M., and Lee, R.G. 1979. Persistence of a fire exclusion policy in southern California: A biosocial interpretation. *J. Environ. Manag.* 8:277–293.

Booker, F.A., Dietrich, W.M., and Collins, L.M. 1995. The Oakland Hills fire of October 20, 1991, an evaluation of post-fire response. In *Brushfires in California Wildlands: Ecology and Resource Management*, eds. J.E. Keeley, and T. Scott, pp. 163–170. Fairfield, WA: International Association of Wildland Fire.

Borchert, M.I., and Odion, D.C. 1995. Fire intensity and vegetation recovery in chaparral: A review. In *Brushfires in California Wildlands: Ecology and Resource Management*, eds. J.E. Keeley, and T. Scott, pp. 91–100. Fairfield, WA: International Association of Wildland Fire.

Bradshaw, T.D. 1987. The intrusion of human population into forest and range lands of California. In *Proceedings of the Symposium on Wildland Fire 2000*, April 27–30,

South Lake Tahoe, CA, eds. J.B. Davis, and R.E. Martin, pp. 15–21. Berkeley: USDA Forest Service, Pacific Southwest Forest and Range Experiment Station, Gen. Tech. Rep. PSW-101.

Brown, W.S. 1945. History of Los Padres National Forest. Goleta, CA. USDA Forest Service, Unpublished rep. on file.

Brown, W.S., and Show, S.B. 1944. *California Rural Land Use and Management: A History of the Use and Occupancy of Rural Lands in California*. Berkeley: USDA Forest Service, California Region.

Burcham, L.T. 1957. *California Range Land: an Historic-Ecological Study of the Range Resources of California*. Sacramento: State of California, Department of Natural Resources, Division of Forestry.

Byrne, R., Michaelsen, J., and Soutar, S. 1977. Fossil charcoal as a measure of wildfire frequency in southern California: A preliminary analysis. In *Proceedings of the Symposium on Environmental Consequences of Fire and Fuel Management in Mediterranean Ecosystems*, eds. H. A. Mooney, and C. E. Conrad, pp. 361–367. Washington, DC: USDA Forest Service, Gen. Tech. Rep. WO-3.

Callaham, R.Z. 1985. *California's Shrublands: A Vast Area in Transition and Need.* Berkeley: University of California, Wildland Resources Center.

Callaway, R.M., and Davis, F.W. 1993. Vegetation dynamics, fire, and the physical environment in coastal central California. *Ecol.* 74:1567–1578.

Campbell, A. 1906. Sonora storms and Sonora clouds of California. *Mon. Wea. Re.* 34: 464–465.

CDF. 1978. *Brushland Range Improvement*. Annual report 1974–1977 inclusive. Sacramento: California Department of Forestry.

Chandler, C.C. 1957. *"Light burning" in Southern California fuels*. Berkeley: USDA Forest Service, California Forest and Range Experiment Station, Forest Res. Notes 119.

Chandler, C.C. 1960. How good are statistics on fire causes? *J. For.* 58:515–517.

Chandler, C.C. 1963. *A Study of Mass Fires and Conflagrations*. Berkeley: USDA Forest Service, Pacific Southwest Forest and Range Experiment Station, Res. Note PSW-22.

Clar, C.R. 1959. *California Government and Forestry from Spanish Days until the Creation of the Department of Natural Resources in 1927*. Sacramento: State of California, Department of Natural Resources, Division of Forestry.

Clar, C.R. 1969. *California Government and Forestry—II. During the Young and Rolph Administrations*. Sacramento: State of California, Department of Natural Resources, Division of Forestry.

Coffin, H. 1959. Effect of marine air on the fireclimate in the mountains of southern California. Berkeley: USDA Forest Service, Pacific Southwest Forest and Range Experiment Station, Tech. Pap. 39.

Cohen, J.D. 2000. Preventing disaster: Home ignitability in the wildland–urban interface. *J. For.* 98:15–21.

Conard, S.G., and Regelbrugge, J.C. 1994. On estimating fuel characteristics in California chaparral. In *12th Conference on Fire and Forest Meteorology*, pp. 120–129. Boston: Society of American Foresters.

Conard, S.G., and Weise, D.R. 1998. Management of fire regime, fuels, and fire effects in southern California chaparral: Lessons from the past and thoughts for the future. *Tall Timbers Ecol. Conf. Proc.* 20:342–350.

Cooper, W.S. 1922. *The Broad-Sclerophyll Vegetation of California: An Ecological Study of the Chaparral and Its Related Communities.* Washington, DC: Carnegie Institution of Washington, Pub. 319.

Countryman, C.M. 1974. Can southern California wildland conflagrations be stopped? Berkeley: USDA Forest Service, Pacific Southwest Forest and Range Experiment Station, Gen. Tech. Note PSW-7.

Countryman, C.M., McCutchan, M.H., and Ryan, B.C. 1969. Fire weather and fire behavior at the 1968 Canyon Fire. Berkeley: USDA Forest Service, Pacific Southwest Forest and Range Experiment Station, Res. Pap. PSW-55.

Countryman, C.M., and Philpot, C.W. 1970. Physical characteristics of chamise as wildland fuel. Berkeley: USDA Forest Service, Pacific Southwest Forest and Range Experiment Station, Res. Pap. PSW-66.

D'Antonio, C.M. 2000. Fire, plant invasions, and global changes. In *Invasive Species in a Changing World*, eds. H.A. Mooney, and R.J. Hobbs, pp. 65–93. Covelo, CA: Island Press.

Davis, F.W., and Burrows, D.A. 1993. Modeling fire regime in Mediterranean landscapes. In *Patch Dynamics*, eds. S.A. Levin, T.M. Powell, and J.H. Steele, pp. 247–259. New York: Springer-Verlag.

Davis, F.W., and Burrows, D.A. 1994. Spatial simulation of fire regime in Mediterranean-climate landscapes. In *The Role of Fire in Mediterranean-Type Ecosystems*, eds. J.M. Moreno, and W.C. Oechel, pp. 117–139. New York: Springer-Verlag.

Davis, F.W., and Michaelsen, J. 1995. Sensitivity of fire regime in chaparral ecosystems to climate change. In *Global Change and Mediterranean-Type Ecosystems*, eds. J.M. Moreno, and W.C. Oechel, pp. 435–456. New York: Springer-Verlag.

Davis, J.A. 1967. *The Friend to All Motorists: The Story of the Automobile Club of Southern California through 65 Years, 1900–1965*. Los Angeles: Automobile Club of Southern California.

Davis, L.S. 1965. *The Economics of Wildfire Protection with Emphasis on Fuel Break Systems.* Sacramento: State of California, Resources Agency, Division of Forestry.

Davis, S.D., Ewers, F.W., Sperry, J.S., Portwood, K.A., Crocker, M.C., and Adams, G.C. 2002. Shoot dieback during prolonged drought in *Ceanothus* (Rhamnaceae) chaparra: a possible case of hydraulic failure. *Amer. J. Bot.* 89:820–828.

Davis, S.D., Kolb, K.J., and Barton, K.P. 1998. Ecophysiological processes and demographic patterns in the structuring of California chaparral. In *Landscape Diversity and Biodiversity in Mediterranean-Type Ecosystems*, eds. P.W. Rundel, G. Montenegro, and F.M. Jaksic, pp. 297–310. New York: Springer-Verlag.

Dombeck, M. 2001. How can we reduce the fire danger in the interior West? *Fire Management Today* 61(1):5–13.

Dunn, A.T. 1989. The effects of prescribed burning on fire hazard in the chaparral: Toward a new conceptual synthesis. In *Proceedings of the Symposium on Fire and Watershed Management*, ed. N.H. Berg, pp. 23–29. Berkeley: USDA Forest Service, Pacific Southwest Forest and Range Experiment Station, Gen. Tech. Rep. PSW-109.

Dunn, A.T., and Piirto, D. 1987. The Wheeler fire in retrospect: factors affecting fire spread and perimeter formation. Riverside: USDA Forest Service, Pacific Southwest Research Station, unpublished report on file.

Ewing, R.A., Tosta, N., Tuaszon, R., Huntsinger, L., Marose, R., Nielson, K., Motroni, R., and Turan, S. 1988. *California's Forests and Rangelands: Growing Conflict Over Changing Uses.* Sacramento: State of California, Department of Forestry and Fire Protection.

Fenner, R.L., Arnold, R.K., and Buck, C.C. 1955. Area ignition for brush burning. Berkeley: USDA Forest Service, California Forest and Range Experiment Station, Tech. Pap. 10.

Field, C.B., Daily, G.C., Davis, F.W., Gaines, S., Matson, P.A., Melack, J., and Miller, N.L. 1999. *Confronting Climate Change in California. Ecological Impacts on the Golden State.* Cambridge, MA, and Washington, DC: Union of Concerned Scientists and Ecological Society of America.

Florence, M.A. 1985. Successional trends in plant species composition following fall, winter and spring prescribed burns of chamise chaparral in the central coast range of California. M.S. thesis: California State University, Sacramento.

Fosberg, M.A. 1965. A case study of the Santa Ana winds in the San Gabriel Mountains. Berkeley: USDA Forest Service, Pacific Southwest Forest and Range Experiment Station, Res. Note PSW-78.

Fosberg, M.A., O'Dell, C.A., and Schroeder, M.J. 1966. Some characteristics of the three-dimensional structure of Santa Ana winds. Berkeley: USDA Forest Service, Pacific Southwest Forest and Range Experiment Station, Res. Pap. PSW-30.

Franco-Vizcaíno, E., and. Sosa-Ramirez, J., 1997. Soil properties and nutrient relations in burned and unburned mediterranean-climate shrublands of Baja California, Mexico. *Acta Oecol.* 18:503–517.

Franklin, S.E. 1987. Urban-wildland fire defense strategy, precision prescribed fire: The Los Angeles County approach. In *Proceedings of the Symposium on Wildland Fire 2000*, April 27–30, 1987, South Lake Tahoe, CA, eds. J.B. Davis, and R.E. Martin, pp. 22–25. Berkeley: USDA Forest Service, Pacific Southwest Forest and Range Experiment Station, Gen. Tech. Rep. PSW-101.

FRAP. 1999. Fire management for California ecosystems. Sacramento: State of California, Resources Agency, California Department of Forestry, Fire and Resource Assessment Program, *http://frap.cdf.ca.gov/projects/fire_mgmt/ftp_main.html.*

Freudenberger, D.O., Fish, B.E., and Keeley, J.E. 1987. Distribution and stability of grasslands in the Los Angeles Basin. *Bull. Southern California Acad. Sci.* 86:13–26.

Gardner, P.D., Cortner, H.J., and Widaman, K. 1987. The risk perceptions and policy response toward wildland fire hazards by urban home-owners. *Landscape Urban Plan.* 14:163–172.

Gause, G.W. 1966. Silvical characteristics of bigcone Douglas-fir. Berkeley: USDA Forest Service, PSW-39.

Gee, P.J. 1974. Roadside fire hazard in California. M.S., thesis. University of California, Berkeley.

Gomes, D., Graham, O.L., Jr., Marshall, E.H., and Schmidt, A.J. 1993. Sifting through the ashes: Lessons learned from the Painted Cave Fire. Graduate Program for Public Historical Studies, University of California, Santa Barbara.

Green, L.R. 1970. An experimental prescribed burn to reduce fuel hazard in chaparral. Berkeley: USDA Forest Service, Pacific Southwest Forest and Range Experiment Station, Res. Note PSW-216.

Green, L.R. 1981. Burning by prescription in chaparral. Berkeley: USDA Forest Service, Pacific Southwest Forest and Range Experiment Station, Gen. Tech. Rep. PSW-51.

Greenlee, J.M., and Langenheim, J.H. 1980. The history of wildfires in the region of Monterey Bay. Sacramento: California Department of Parks and Recreation, unpublished rep.

Greenlee, J.M., and Langenheim, J.H. 1990. Historic fire regimes and their relation to vegetation patterns in the Monterey Bay area of California. *Am. Midland Natural.* 124:239–253.

Greenlee, J.M., and Moldenke, A. 1982. History of wildland fires in the Gabilan Mountains region of central coastal California. San Francisco: USDI National Park Service, Unpublished rep.

Haidinger, T.L., and Keeley, J.E. 1993. Role of high fire frequency in destruction of mixed chaparral. *Madroño* 40:141–147.

Hamilton, J.G. 1997. Changing perceptions of pre-European grasslands in California. *Madroño* 44:311–333.

Heady, H.F. 1977. Valley grasslands. In *Terrestrial Vegetation of North America*, eds. M.G. Barbour, and J. Major, pp. 491–514. New York: Wiley.

Hobbs, E.R. 1983. Factors controlling the form and location of the boundary between coastal sage scrub and grassland in southern California. Ph.D. dissertation. University of California, Los Angeles.

Huenneke, L.F. 1989. Distribution and regional patterns of Californian grasslands. In *Grassland Structure and Function: California Annual Grassland*, eds. L.F. Huenneke, and H.A. Mooney, pp. 1–12. Dordrecht: Kluwer Academic.

Johnson, E.A., and K., Miyanishi. 1995. The need for consideration of fire behavior and effects in prescribed burning. *Restor. Ecol.* 3:271–278.

Keeley, J.E. 1977. Fire dependent reproductive strategies in *Arctostaphylos* and *Ceanothus*. In *Proceedings of The Symposium on Environmental Consequences of Fire and Fuel Management in Mediterranean Ecosystems*, eds. H.A. Mooney, and C.E. Conrad, pp. 371–376. Washington, DC: USDA Forest Service, Gen. Tech. Rep. WO-3.

Keeley, J.E. 1982. Distribution of lightning and man-caused wildfires in California. In *Proceedings of the Symposium on Dynamics and Management of Mediterranean-Type Ecosystems*, eds. C.E. Conrad, and W.C. Oechel, pp. 431–437. Berkeley: USDA Forest Service, Pacific Southwest Forest and Range Experiment Station, Gen. Tech. Rep. PSW-58.

Keeley, J.E. 1990. The California valley grassland. In *Endangered Plant Communities of Southern California*, ed. A.A. Schoenherr, pp. 2–23. Fullerton: Southern California Botanists, Special Publication 3.

Keeley, J.E. 1992. Demographic structure of California chaparral in the long-term absence of fire. *J. Veg. Sci.* 3:79–90.

Keeley, J.E. 1998a. Coupling demography, physiology and evolution in chaparral shrubs. In *Landscape Diversity and Biodiversity in Mediterranean-Type Ecosystems*, eds. P.W. Rundel, G. Montenegro, and F.M. Jaksic, pp. 257–264. New York: Springer-Verlag.

Keeley, J.E. 1998b. Postfire ecosystem recovery and management: the October 1993 large fire episode in California. In *Large Forest Fires*, ed. J.M. Moreno, pp. 69–90. Leiden, The Netherlands: Backhuys.

Keeley, J.E. 2000. Chaparral. In *North American Terrestrial Vegetation*, eds. M.G. Barbour, and W.D. Billings, pp. 201–251. Cambridge: Cambridge University Press.

Keeley, J.E. (in press). Fire and invasives in Mediterranean-climate ecosystems of California. Tall Timbers Research Station Miscellaneous Publication 11:81–94.

Keeley, J.E. 2002. Native American impacts on fire regimes of the California coastal ranges. *J. Biogeogr.* 29:303–320.

Keeley, J.E., and Fotheringham, C.J. 2001a. Historic fire regime in California shrublands. *Conserv. Biol.* 15:1534–1548.

Keeley, J.E., and Fotheringham, C.J. 2001b. History and management of crown-fire ecosystems: A summary and response. *Conserv. Biol.* 15:1561–1567

Keeley, J.E., Fotheringham, C.J., and Morais, M. 1999. Reexamining fire suppression impacts on brushland fire regimes. *Science* 284:1829–1832.

Keeley, J.E., and Keeley, S.C. 1984. Postfire recovery of California coastal sage scrub. *Am. Midland Natural.* 111:105–117.

Keeley, J.E., Zedler, P.H., Zammit, C.A., and Stohlgren, T.J. 1989. Fire and demography. In *The California Chaparral: Paradigms Reexamined*, ed. S.C. Keeley, pp. 151–153. Los Angeles: Natural History Museum of Los Angeles County, Science Series 34.

Kessell, S.R., and Cattelino, P.J. 1978. Evaluation of a fire behaviour information integration system for southern California chaparral wildlands. *Environ. Manag.* 2:135–159.

Kinney, A., 1887. Report on the forests of the counties of Los Angeles, San Bernardino, and San Diego, California. Sacramento: First Biennial Report, California State Board of Forestry.

Kinney, A. 1900. *Forest and Water*. Los Angeles: Post.

Kinney, W. 1984. Economics and policy of shrubland management. In *Proceedings of the Chaparral Ecosystems Research Conference*, ed. J.J. DeVries, pp. 129–136. Davis: University of California, Water Resources Center, Rep. 62.

Knipper, C. 1998. Fire: The rejuvenating force. *Explorer* 5(8):8.

Krausman, W.J. 1981. An analysis of several variables affecting fire occurrence and size in San Diego County, California. M.A., thesis. San Diego State University.

Krick, I.P. 1933. Foehn winds of southern California. *Beitr. Geophys.* 39:399–407.

Lee, R.G., and Bonnicksen, T.M. 1978. Brushland watershed fire management policy in southern California: biosocial considerations. Davis. University of California, California Water Resources Center, Contribution 172.

Lessard, A.G. 1988. The Santa Ana wind of southern California. *Weatherwise* 41:100–104.

Lewis, H.T. 1973. *Patterns of Indian Burning in California: Ecology and Ethnohistory.* Ramona, CA: Ballena Press.

Lillard, R.G. 1961. Black horizons. *Westways* 62(10):17–19, 64–65.

Lockmann, R.F. 1981. *Guarding the Forest of Southern California.* Glendale, CA: Clark.

Malanson, G.P. 1985. Fire management in coastal sage-scrub, southern California, USA. *Biolog. Conserv.* 12:141–146.

Malanson, G.P., and O'Leary, J.F. 1995. The coastal sage scrub—Chaparral boundary and response to global climatic change. In *Global Climate Change in Mediterranean-Type Ecosystems*, eds. J.M. Moreno, and W.C. Oechel, pp. 203–224. Berlin: Springer-Verlag.

Malanson, G.P., and Westman, W.E. 1991a. Climatic change and the modeling of fire effects in coastal sage scrub and chaparral. In *Fire and the Environment: Ecological and Cultural Perspectives, Proceedings of an International Symposium*, eds. S.C. Nodvin, and T.A. Waldrop, pp. 91–96. USDA Forest Service Station, Southeastern Forest and Experiment Station, Gen. Tech. Rep. SE-69.

Malanson, G.P., and Westman, W.E. 1991b. Modeling interactive effects of climate change, air pollution, and fire on a California shrubland. *Clim. Change* 18:363–376.

Martin, R.E., and Sapsis, D.B. 1995. A synopsis of large or disastrous wildland fires. In *The Biswell Symposium: Fire Issues and Solutions in Urban Interface and Wildland Ecosystems*, eds. D.R. Weise, and R.E. Martin, pp. 35–38. Berkeley: USDA Forest Service, Gen. Tech. Rep. PSW-GTR-158.

McCutchan, M.H. 1977. Climatic features as a fire determinant. In *Proceedings of the Symposium on Environmental Consequences of Fire and Fuel Management in Mediterranean Ecosystems*, eds. H.A. Mooney, and C.E. Conrad, pp. 1–11. Washington, DC: USDA Forest Service, Gen. Tech. Rep. WO-3.

Mensing, S.A., Michaelsen, J., and Byrne, R. 1999. A 560-year record of Santa Ana fires reconstructed from charcoal deposited in the Santa Barbara Basin, California. *Quat. Res.* 51:295–305.

Minnich, R.A. 1983. Fire mosaics in southern California and northern Baja California. *Science* 219:1287–1294.

Minnich, R.A. 1987. Fire behavior in southern California chaparral before fire control: the Mount Wilson burns at the turn of the century. *Ann. Assoc. Am. Geogr.* 77:599–618.

Minnich, R.A. 1989. Chaparral fire history in San Diego County and adjacent northern Baja California: An evaluation of natural fire regimes and the effects of suppression management. In *The California Chaparral: Paradigms Reexamined*, ed. S.C. Keeley, pp. 37–47. Los Angeles: Natural History Museum of Los Angeles County, Science Series 34.

Minnich, R.A. 1990. Fire suppression in chaparral: what the United States can learn from Mexico. In *Environmental Hazards and Bioresource Management in the United States-Mexico Borderlands*, eds. P. Ganster, and H. Walter, pp. 329–342. Los Angeles: UCLA Latin American Center Publications, University of California.

Minnich, R.A. 1995. Fuel-driven fire regimes of the California chaparral. In *Brushfires in California: Ecology and Resource Management*. eds. J.E. Keeley, and T. Scott, pp. 21–27. Fairfield, WA: International Association of Wildland Fire.

Minnich, R.A. 1998. Landscapes, land-use and fire policy: where do large fires come from? In *Large Forest Fires*, ed. J.M. Moreno, pp. 133–158. Leiden, The Netherlands: Backhuys.

Minnich, R.A. 2001. An integrated model of two fire regimes. *Conservation Biology* 15:1549–1553.

Minnich, R.A., and Chou, Y. H. 1997. Wildland fire patch dynamics in the chaparral of southern California and northern Baja California. *Int. J. Wild. Fire* 7:221–248.

Minnich, R.A., and Dezzani, R.J. 1991. Suppression, fire behavior, and fire magnitudes in Californian chaparral at the urban/wildland interface. In *California Watersheds at the Urban Interface, Proceedings of the Third Biennial Watershed Conference*, ed. J. J. DeVries, pp. 67–83. Davis: University of California, Water Resources Center, Report 75.

Minnich, R.A., and Dezzani, R.J. 1998. Historical decline of coastal sage scrub in the Riverside-Perris Plain, California. *Western Birds* 29:366–391.

Minnich, R.A., and Franco-Vizcaíno, E. 1999. Prescribed mosaic burning in California chaparral. In *Proceedings of the Symposium on Fire Economics, Planning, and Policy: Bottom Lines*, eds. A. González-Cabán, and P.N. Omi, pp. 243–246. Berkeley: USDA Forest Service, Pacific Southwest Research Station, Gen. Tech. Rep. PSW-GTR-173.

Minnich, R.A., Franco-Vizcaíno, E., Sosa-Ramirez, J., and Chou, Y., 1993. Lightning detection rates and wildland fire in the mountains of northern Baja California, Mexico. *Atmósfera* 6:235–253.

Mitchell, V.L. 1969. The regionalization of climate in montane areas. Ph.D. dissertation. University of Wisconsin, Madison.

Moreno, J.M., and Oechel, W.C. 1991. Fire intensity effects on germination of shrubs and herbs in southern California chaparral. *Ecology* 72:1993–2004.

Moritz, M.A. 1997. Analyzing extreme disturbance events: fire in the Los Padres National Forest. *Ecol. Appl.* 7:1252–1262.

Moritz, M.A. 1999. Controls on disturbance regime dynamics: fire in Los Padres National Forest. Ph.D. dissertation. University of California, Santa Barbara.

Mutch, R.W. 1970. Wildland fires and ecosystems: a hypothesis. *Ecology* 51:1046–1051.

Nichols, R., Adams, T., and Menke, J. 1984. Shrubland management for livestock forage. In *Shrublands in California: Literature Review and Research Needed for Management*, ed. J.J. DeVries, pp. 104–121. Davis: University of California, Water Resources Center, Contribution 191.

Oberbauer, A.T. 1978. Distribution dynamics of San Diego County grasslands. M.S. thesis. San Diego State University.

Odens, P. 1971. *The Indians and I. Visits with Dieguenos, Quechans, Fort Mojaves, Zumis, Hopis, Navajos and Piutes*. El Centro, CA: Imperial Printers.

Oechel, W.C., Hastings, S.J., Vourlitis, G.L., Jenkins, M.A., and Hinkson, C.L. 1995. Direct effects of elevated CO_2 in chaparral and Mediterranean-type ecosystems. In *Global Change and Mediterranean-Type Ecosystems*, eds. J.M. Moreno, and W.C. Oechel, pp. 58–75. New York: Springer-Verlag.

Olsen, J.M. 1960. 1959 green-fuel moisture and soil moisture trends in southern California. Berkeley: USDA Forest Service, Pacific Southwest Forest and Range Experiment Station, Res. Note 161.

Parker, V.T. 1990. Problems encountered while mimicking nature in vegetation management: An example from a fire-prone vegetation. In *Ecosystem Management: Rare Species and Significant Habitats. Proceedings of the 15th Annual Natural Areas Conference*, eds. R.S. Mitchell, C.J. Sheviak, and D.J. Leopold, pp. 231–234. Albany New York State Museum, Bulletin 471.

Parsons, D.J. 1981. The historical role of fire in the foothill communities of Sequoia National Park. *Madroño* 28:111–120.

Payson, T.E., and Cohen, J.D. 1990. Chamise chaparral dead fuel fraction is not reliably predicted by age. *Western J. For.* 5:127–131.

Paysen, T.E., Narog, M.G., and Cohen, J.D. 1998. The science of prescribed fire: to enable a different kind of control. *Tall Timbers Ecol. Conf. Proc.* 20:31–36.

Peng, R., and Schoenberg, F. 2001. Estimation of wildfire hazard using spatial-temporal fire history data. *J. Am. Stat. Assoc.*, in press.

Peteet, D. 2000. Sensitivity and rapidity of vegetational response to abrupt climate change. *Proc. Nat. Acad. Sci.* 97:1359–1361.

Phillips, C.B. 1971. *California Aflame! September 22–October 4, 1970.* Sacramento: State of California, Department of Conservation, Division of Forestry.

Philpot, C.W. 1969. Seasonal changes in heat content and ether extractive content of chamise. Berkeley: USDA Forest Service, Intermountain Forest and Range Experiment Station, Res. Pap. INT-61.

Philpot, C.W. 1974a. The changing role of fire on chaparral lands. In *Symposium on Living with the chaparral, Proceedings*, ed. M. Rosenthal, pp. 131–150. San Francisco: Sierra Club.

Philpot, C.W. 1974b. New fire control strategy developed for chaparral. *Fire Manag.* 37: 3–7.

Philpot, C.W. 1977. Vegetative features as determinants of fire frequency and intensity. In *Proceedings of the Symposium on Environmental Consequences of Fire and Fuel Management in Mediterranean Ecosystems*, eds. H.A. Mooney, and C.E. Conrad, pp. 12–16. Washington, DC: USDA Forest Service, Gen. Tech. Rep. WO-3.

Pirsko, A.R. 1960. 1960 fire weather severity in California. Berkeley: USDA Forest Service, Pacific Southwest Forest and Range Experiment Station, Miscellaneous Pap. 54.

Pirsko, A.R., and Green, L.R. 1967. Record low fuel moisture follows drought in southern California. *J. For.* 65:642–643.

Price, C., and Rind, D. 1994. Lightning fires in a $2 \times CO_2$ world. In *12th Conference on Fire and Forest Meteorology*, October 26–28, Jekyll Island, GA, pp. 77–84. Washington, DC: Society of American Foresters.

Pyne, S.J. 1982. *Fire In America: A Cultural History of Wildland and Rural Fire.* Princeton, NY: Princeton University Press.

Pyne, S.J., Andrews, P.L., and Laven, R.D. 1996. *Introduction to Wildland Fire*. New York: Wiley.

Radtke, K.W.H., Arndt, A.M., and Wakimoto, R.H. 1982. Fire history of the Santa Monica Mountains. In *Proceedings of the Symposium on Dynamics and Management of Mediterranean-Type Ecosystems*, eds. C.E. Conrad, and W.C. Oechel, pp. 438–443. Berkeley: USDA Forest Service, Pacific Southwest Forest and Range Experiment Station, Gen. Tech. Rep. PSW-58.

Regelbrugge, J.C. 2000. Role of prescribed burning in the management of chaparral ecosystems in southern California. In *2nd Interface between Ecology and Land Development in California*, eds. J.E. Keeley, M.B. Keeley, and C.J. Fotheringham, pp. 19–26. Sacramento: U.S. Geological Survey Open-File Rep. 00–62.

Reynolds, R.D. 1959. Effect of natural fires and aboriginal burning upon the forest of the central Sierra Nevada. M.A., thesis. University of California, Berkeley

Riggan, P.J., Franklin, S.E., Brass, J.A., and Brooks, F.E. 1994. Perspectives on fire management in Mediterranean ecosystems of southern California. In *The Role of Fire in Mediterranean-Type Ecosystems*, eds. J.M. Moreno, and W.C. Oechel, pp. 140–162. New York: Springer-Verlag.

Riggan, P.J., Goode, S., Jacks, P.M., and Lockwood, R.W. 1988. Interaction of fire and community development in chaparral of southern California. *Ecol. Monogr.* 58: 155–175.

Rogers, M.J. 1982. Fire management in southern California. In *Proceedings of the Symposium on Dynamics and Management of Mediterranean-Type Ecosystems*, eds. C.E. Conrad and W.C. Oechel, pp. 496–497. Berkeley: USDA Forest Service, Pacific Southwest Forest and Range Experiment Station, Gen. Tech. Rep. PSW-58.

Rothermel, R.C. 1972. *A Mathematical Model for Predicting Fire Spread in Wildland Fuels.* Ogden, UT: USDA Forest Service, INT-115.

Rothermel, R.C., and Philpot, C.W. 1973. Predicting changes in chaparral flammability. *J. For.* 71:640–643.
Rundel, P.W., Baker, G.A., Parsons, D.J., and Stohlgren, T.J. 1987. Postfire demography of resprouting and seedling establishment by *Adenostoma fasciculatum* in the California chaparral. In *Plant Response to Stress: Functional Analysis in Mediterranean Ecosystems*, eds. J.D. Tenhunen, F.M. Catarino, O.L. Lange, and W.C. Oechel, pp. 575–596. Berlin: Springer-Verlag.
Rundel, P.W., Parsons, D.J., and Baker, G.A. 1980. The role of shrub structure and chemistry in the flammability of chaparral shrubs. In Fire Ecology: *Proceedings of the Second Conference on Scientific Research in National Parks*, vol. 10, pp. 248–260. Washington, DC: USDI National Park Service.
Russell, E.W.B. 1983. Pollen analysis of past vegetation at Point Reyes National Seashore, California. *Madroño* 30:1–11.
Ryan, B.C. 1969. A vertical perspective of Santa Ana winds in a canyon. Berkeley: USDA Forest Service, Pacific Southwest Forest and Range Experiment Station, Res. Pap. PSW-52.
Ryan, G. 1996. Downslope winds of Santa Barbara, California. Washington, DC: U.S. Department of Commerce, National Oceanic and Atmospheric Administration, National Weather Service, NOAA Tech. Memo. NWS WR-240.
Sala, O.E., et al. 2000. Global biodiversity scenarios for the year 2100. *Science* 287:1770–1774.
Sampson, A.W. 1944. Plant succession and burned chaparral lands in northern California. Berkeley: University of California, Agricultural Experiment Station, Bull. 685.
Sapsis, D. 2001. Development patterns and fire suppression. Sacramento: State of California, Resources Agency, California Department of Forestry, Fire and Resource Assessment Program, *http://frap.cdf.ca.gov/publications/development_patterns/toc.html.*
Schoenberg, F., Peng, R., Huang, Z., and Rundel, P. 2001. Exploratory analysis of wildfire data in Los Angeles County, California. *http://www.stat.ucla.edu/~frederic/papers/fire1.pdf.*
Schroeder, M.J., et al. 1964. Synoptic weather types associated with critical fire weather. Washington, DC: U.S. Department of Commerce, National Bureau of Standards, Institute for Applied Technology, AD 449–630.
Schroeder, M.J., and Buck, C.C. 1970. *Fire Weather . . . A Guide for Application of Meteorological Information to Forest Fire Control Operations.* Washington, DC: USDA Forest Service, Agricultural Handbook 360.
Schwilk, D.W. 2000. Flammability as niche construction: Canopy architecture's effect on the flammability of a chaparral species. In *Mediterranean-Type Ecosystems: Past, Present and Future*, pp. 68–69. Stellenbosch, South Africa: MEDECOS 2000, Stellenbosh University.
Show, S.B., and Kotok, E.I. 1923. *Forest Fires in California 1911–1920: An Analytical Study.* Washington, D.C.: U.S. Department of Agriculture, Circular 243.
Skinner, C.N., and Chang, C.-R. 1996. Fire regimes, past and present. In *Sierra Nevada Ecosystem Project: Final Report to Congress. Status of the Sierra Nevada*, eds. SNEP Team, pp. 1041–1069. Davis: Centers for Water and Wildland Resources, University of California.
Sommers, W.T. 1978. LFM forecast variables related to Santa Ana wind occurrences. Mon. *Wea. Rev.* 106:1307–1316.
Specht, R.L. 1969. A comparison of the sclerophyllous vegetation characteristics of Mediterranean type climate in France, California and Southern Australia. I. Structure, morphology, and succession. *Austral. J. Bot.* 17:277–292.
Specht, R.L. 1981. Primary production in Mediterranean-climate ecosystems regenerating after fire. In *Ecosystems of the World: Mediterranean-Type Shrublands*, vol. 2, eds. F. di Castri, D.W. Goodall, and R.L. Specht, pp. 257–268. New York: Elsevier Scientific.

Timbrook, J., Johnson, J.R., and Earle, D.D. 1982. Vegetation burning by the Chumash. *J. Cal. Great Basin Anthropol.* 4:163–186.

Turner, K.M., and Lampinen, B.D. 1983. Prescribed burning of chaparral: some effects on soil movement. Sacramento: State of California, Resources Agency, Department of Water Resources.

Vale, T.T. 1998. The myth of the humanized landscape: an example from Yosemite National Park. *Natural Areas J.* 18:231–236.

van Wagtendonk, J.W. 1992. Spatial analysis of lightning strikes in Yosemite National Park. In *Proceedings of the 11th Conference on Fire and Forest Meteorology*, eds. P.L. Andrews, and D.F. Potts, pp, 605–611. Bethesda, MD: Society of American Foresters.

Vankat, J.L. 1985. General patterns of lightning ignitions in Sequoia National Park, California. *Proceedings—Symposium and Workshop on Wilderness Fire*, eds. J.E. Lotan, B.M. Kilgore, W.C. Fischer, and R.W. Mutch, pp. 408–411. Fort Collins, CO: USDA Forest Service, Intermountain Forest and Range Experiment Station, Gen. Tech. Rep. INT-182.

Weatherspoon, C.P., and C.N., Skinner. 1996. Landscape-level strategies for forest fuel management. In *Sierra Nevada Ecosystem Project: Final report to Congress. Status of the Sierra Nevada*, eds. SNEP Team, pp. 1471–1492. Davis: Centers for Water and Wildland Resources, University of California.

Weide, D.L. 1968. The geography of fire in the Santa Monica Mountains. M.S. thesis. California State University, Los Angeles.

Weise, D.R., Regelbrugge, J.C., Paysen, T.E., and Conard, S.G., (in press). Fire occurrence on southern Californian national forests—Has it changed recently? In *Proceedings of Fire in California Ecosystems: Integrating Ecology, Prevention, and Management*, eds. N.G. Sugihara, and M.I. Borchert. Davis: University of California.

Wells, M.L., and McKinsey, D.E. 1994. The spatial and temporal distribution of lightning strikes in San Diego County, California. *GIS/LIS Proc.* 2:768–777.

Wells, M.L., and McKinsey, D.E. 1995. Lightning strikes and natural fire regimes in San Diego County, California. In *Biswell Symposium: Fire Issues and Solutions in Urban Interface and Wildland Ecosystems*, eds. D.R. Weise, and R.E. Martin, pp. 193–194. Berkeley: USDA Forest Service, Gen. Tech. Rep. PSW-GTR-158.

Wells, P.V. 1962. Vegetation in relation to geological substratum and fire in the San Luis Obispo quadrangle, California. *Ecol. Monogr.* 32:79–103.

Westman, W.E. 1991. Measuring realized niche spaces: Climatic response of chaparral and coastal sage scrub. *Ecology* 72:1678–1684.

Westman, W.E., and Malanson, G.P. 1992. Effects of climate change on Mediterranean-type ecosystems in California and Baja California. In *Global Warming and Biological Diversity*, eds. R.L. Peters, and T.E. Lovejoy, pp. 258–276. New Haven: Yale University Press.

Wickstrom, C.K.R. 1987. Issues concerning Native American use of fire: a literature review. Yosemite National Park, CA: Yosemite Research Center, Publ. Anthropol. 6.

Wohlgemuth, P.M., Beyers, J.L., and Conard, S.G. 1999. Postfire hillslope erosion in southern California chaparral: A case study of prescribed fire as a sediment management tool. In *Proceedings of the Symposium on Fire Economics, Planning, and Policy: Bottom Lines*, eds. A. González-Cabán, and P.N. Omi, pp. 269–276. Berkeley: USDA Forest Service, Pacific Southwest Research Station, Gen. Tech. Rep. PSW-GTR-173.

Wolfram, H. 1962. Brush can be burned in the early spring. Sacramento: State of California, Department of Natural Resources, California Division of Forestry, Range Improvement Studies 6.

Zahn, C. 1944. The San Diego fires . . . an inquest. *Am. For.* 50:161–164.

Zedler, P.H. 1995. Fire frequency in southern California shrublands: Biological effects and mana-gement options. In *Brushfires in California: Ecology and Resource Management*. eds. J.E. Keeley, and T. Scott, pp. 101–112. Fairfield, WA: International Association of Wildland Fire.

Zedler, P.H., Gautier, C.R., and McMaster, G.S. 1983. Vegetation change in response to extreme events: The effect of a short interval between fires in California chaparral and coastal scrub. *Ecology* 64:809–818.

Zedler, P.H., and Seiger, L.A. 2000. Age mosaics and fire size in chaparral: A simulation study. In *2nd Interface between Ecology and Land Development in California*, eds. J.E. Keeley, M.B. Keeley, and C.J. Fotheringham, pp. 9–18. Sacramento: U.S. Geological Survey Open-File Rep. 00–62.

Zivnuska, J.A., Arnold, K., and Arment, C. 1950. Wildfire damage and cost far-reaching. *Cal. Agric.* 4(9):8–10.

3. South America

9. Fire History and Vegetation Changes in Northern Patagonia, Argentina

Thomas T. Veblen, Thomas Kitzberger, Estela Raffaele, and Diane C. Lorenz

In recent decades a new understanding of forest dynamics has helped both scientists and resource managers appreciate the role of natural disturbances in the development of stand-level and landscape-level forest patterns in many parts of the world (Attiwell 1994; Rogers 1996). An emphasis on the role of disturbance, and especially of fire, is a consequence of the replacement of traditional equilibrium paradigms by nonequilibrium paradigms of vegetation dynamics (Glenn-Lewin, Peet, and Veblen 1992; Wu and Loucks 1995), and this dynamics is integral to effective landscape management. The widespread acceptance of nonequilibrium paradigms of vegetation dynamics has profound implications for ecosystem management.

In this chapter we examine the history of fire and its ecological consequences along one of the world's most striking vegetation gradients—the west-to-east replacement of Andean rain forests by the northern Patagonian steppe at ca. 40°S latitude, Argentina (Fig. 9.1). Fire has played a major role in the structural pattern of this landscape. In this chapter we emphasize the roles of humans in altering fire regimes, and the interaction between landscape patterns and fire behavior. We stress the profound and long-lasting impacts on the landscape of short periods of exceptional rates of burning associated with human activities, droughts, and fuel changes related to the life cycle of dominant understory plants (bamboos). The role of interannual and longer-term climatic variability, as a conditioning factor permitting years of widespread fire, of both natural and anthropogenic origin, is discussed in detail in Kitzberger and Veblen (Chapter 10, this volume).

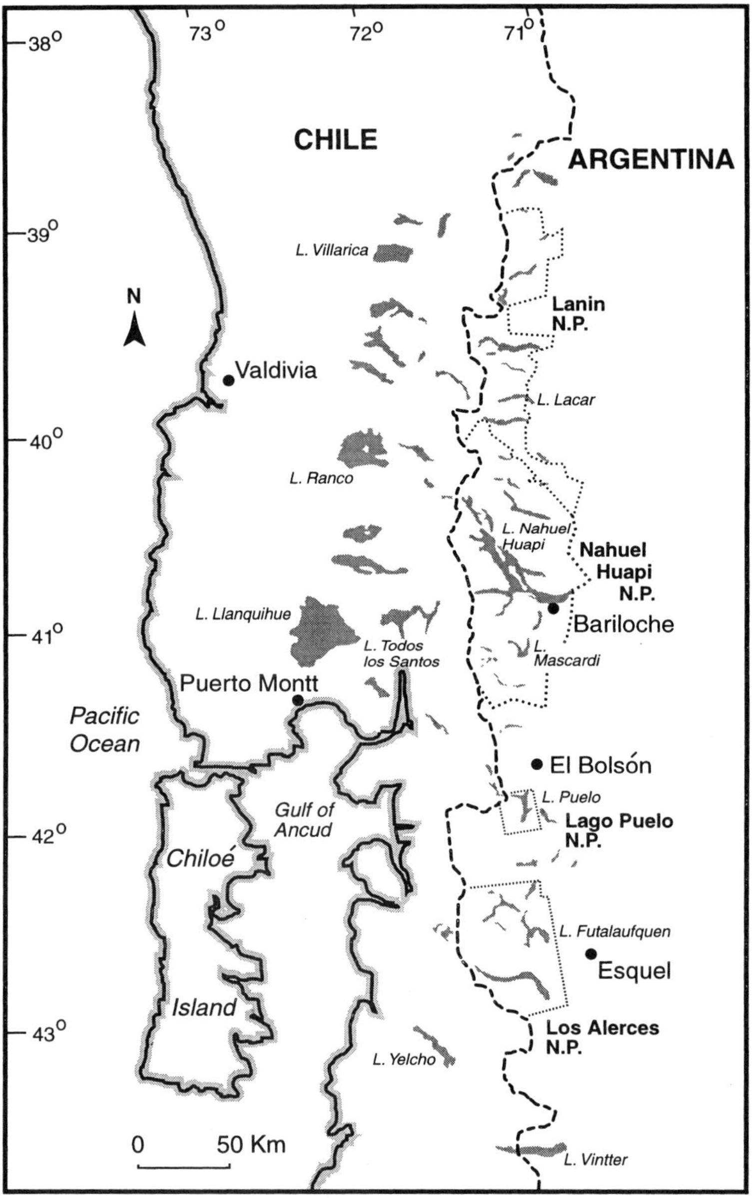

Figure 9.1. Location map of south-central Chile and northern Patagonia, Argentina showing Lanín, Nahuel Huapi, Lago Puelo, and Los Alerces national parks in Argentina.

Early scientific observers made astute observations about fire behavior and its ecological role in northern Patagonia, and they provide an important foundation for the modern study of fire ecology in the region (Rothkugel 1916; Willis 1914; Tortorelli 1947). Ironically this foundation was largely ignored by scientists and vegetation managers until the 1980s. Since about 1985, fire has become a focal point of research on vegetation dynamics in northern Patagonia (Veblen and Lorenz 1987, 1988; Gobbi and Sancholuz 1992; Gobbi 1994; Veblen, Kitzberger, and Lara 1992; Veblen et al. 1999; Dezzotti 1996; Kitzberger and Veblen 1997, 1999; Raffaele and Veblen 1998). Most of the Andean area between ca. 39° and 43°S is in one of four large Argentine national parks: Lanín, Nahuel Huapi, Lago Puelo, and Los Alerces (Fig. 9.1). These national parks are divided into western sections of strict reserves and eastern sections where controlled extractive resource use is permitted (mainly livestock raising). Since at least the 1930s the parks have executed a policy of fire suppression.

Variation in Fire Along the Rain Forest-to-Steppe Gradient

The Environment of Northern Patagonia

From west to east, northern Patagonia includes the Andean cordillera (>2000 m elevation), the lower foothills intersected by glacial lakes and valleys, and the Patagonian plains at ca. 700 m. Throughout the region, soils are derived from Quaternary volcanic ash. Because of the rainshadow effect of the Andes on the westerlies, mean annual precipitation declines from ca. 3000 mm at the continental divide to less than 500 mm only 80 km to the east in the steppe (Barros et al. 1983). Approximately 60% of the annual precipitation falls from May through August, and more than 90% of fires occur during the warm and dry season from October through March (Kitzberger, Veblen, and Villalba 1997). Regional climatic variation and its relationship to broad-scale synoptic climatic controls is described in Kitzberger and Veblen (Chapter 10, this volume).

The strong west-to-east decline in precipitation is paralleled by a dramatic vegetation gradient of: rain forest, mesic forest, xeric forest, open woodland and tall shrubland, to grass- and low shrub-steppe. At ca. 40°S, western montane rain forests (ca. 800–1100 m elevation) are dominated by 40-m tall evergreen *Nothofagus dombeyi*, and in the wettest areas they also include shade-tolerant trees such as *Laureliopsis philippiana*, *Saxegothaea conspicua* and *Dasyphyllum diacanthoides*. Typically these forests have dense understories of 3- to 6-m-tall bamboo (*Chusquea culeou*). At latitudes 41° to 43°S relatively small areas in the highest precipitation zone (>3000 mm annual precipitation) are dominated by the giant conifer *Fitzroya cupressoides* (Veblen et al. 1995), which can attain an age of at least 3000 years in northern Patagonia (Tortorelli 1956). Eastward, as precipitation declines, there is an extensive zone of pure *N. dombeyi* with understories of the 3- to 6-m-tall bamboo *Chusquea culeou*. With increasing aridity, the conifer *Austrocedrus chilensis* forms mixed stands with *N. dombeyi* and then pure conifer stands; understories become less dense as *Chusquea* is replaced by xeric shrubs and small trees such as *Aristotelia chilensis* and *Lomatia hirsuta*. Near the ecotone with the steppe, *Austrocedrus* stands form open woodlands with bunch grasses and low shrubs such as *Discaria articulata* and *Mulinum spinosum*.

The small deciduous tree *Nothofagus antarctica* often dominates sites that are unfavorable for development of tall forest including (1) relatively xeric sites that are transitional to the steppe; (2) bottoms of broad valleys that have more finely textured soils and a high probability of cold-air drainage resulting in temperature inversions; (3) sites along streams and bogs with elevated water tables; (4) mid-slope sites of shallow soils, often on north-facing slopes that become extremely dry during the summer; and (5) high-elevation sites exposed to strong winds that prevent a protective snow cover from forming (McQueen 1976; Seibert 1982). Open woodlands of 4- to 6-m-tall *N. antarctica* are most common in broad valleys in the transition toward steppe. On midslopes from 900 to 1200 m, at intermediate positions along the precipitation gradient, *N. antarctica* forms dense 2- to 4-m-tall shrublands with xerophyllous tree or shrub species such as *Schinus patagonicus*, *Lomatia hirsuta*, *Embothrium coccineum*, *Diostea juncea*, *Maytenus boaria*, and/or the bamboo *Chusquea culeou* (Rodríguez et al. 1978; Seibert 1982). Subalpine forests of the deciduous *Nothofagus pumilio* occur at elevations of 1100 to 1200 m, above either midslope shrublands or mesic montane forests. North of ca. 40°30′, the deciduous *Nothofagus nervosa* (syn. *alpina*, *procera*) and *N. obliqua* occur at mid-elevations below subalpine *N. pumilio* forests and above the more xeric habitat of *Austrocedrus* (Veblen et al. 1996). *Nothofagus obliqua–N. nervosa* forests are mainly restricted to elevations of 600 to 900 m in southwestern Lanín National Park. North of 40°20′S, *Araucaria araucana* occurs along the west-to-east precipitation gradient from mesic forests with *N. dombeyi* or *N. pumilio* to tall shrublands of *N. antarctica* and open woodlands with steppe shrubs and grasses (Veblen et al. 1995). Its main distribution is centered around 39°S in northern Lanín National Park.

Fire Behavior and Its Consequences in the Major Ecosystem Types of Northern Patagonia

Rain Forest Dominated by Fitzroya Cupressoides *and* Nothofagus Dombeyi

Despite the high precipitation characteristic of *Fitzroya*-dominated forests, fire is a major source of disturbance. Fires in these rain forests occur naturally or are set by humans during infrequent dry years (Kitzberger, Veblen, and Villalba 1997; Veblen et al. 1999). Because the bark is typically greater than 20-cm thick on large (>1.5 m diameter) trees, *Fitzroya* often survive intense fires that kill all associated thin-barked tree species (e.g., *Laureliopsis philippiana*, *Nothofagus dombeyi*, and *Saxegothaea conspicua*; Table 9.1). Age structures in these forests indicate that scattered old (>1000 years) *Fitzroya* may continue to dominate a site through several cycles of fire-induced mortality and regeneration of *N. dombeyi* and *S. conspicua* (Veblen et al., unpublished data). Stand-replacing fires also create open conditions suitable for the seedling establishment of the highly

Table 9.1. Traits of common trees and shrubs of northern Patagonia relevant to their resistance to and recovery from fire

Species	Think-barked fire and resistant	Resprouting capacity	New recruitment from: Prolific postfire seeding	New recruitment from: Buried viable seed	Notes
Trees					
Araucaria araucana	×	×			Irregularly root suckers and basal sprouts
Austrocedrus chilensis					Large trees resist fire
Fitzroya cupressoides	×	×			Large trees resist fire; irregularly root suckers
Nothofagus antarctica		×			Vigorously resprouts from lignotubers and basal buds
N. dombeyi			×		Large trees weakly resist fire
N. nervosa	×	×	×		Large trees resist fire; resprouting is irregular
N. obliqua	×	×	×		Large trees resist fire; irregularly resprouts
N. pumilio					Small trees are thin-barked and easily killed by fire
Saxegothaea conspicua				×	Even large trees are thin-barked and easily killed by fire
Shrubs					
Aristotelia chilensis		×	×		Resprouts from basal buds
Berberis spp.		×			Vigorously resprouts from basal buds
Chusquea culeou		×			Vigorously resprouts from extensive rhizomes
Diostea juncea		×		×	Vigorously resprouts from roots and stem
Discaria articulata		×			Resprouts from rhizomes; highly flammable
Embothrium coccineum		×			Highly flammable; resprouts from basal buds
Fabiana imbricata		×	×		Resprouts from large tap roots
Lomatia hirsuta		×			Highly flammable; vigorously resprouts from basal buds
Maytenus boaria		×			Resprouts from basal buds and lateral roots
Schinus patagonicus		×			Vigorously resprouts from basal roots

Sources: Tortorelli 1947, 1956; McQueen 1976; Seibert 1982; Veblen and Lorenz 1987; Ghermandi 1992.

shade-intolerant *Fitzroya*, which tends to be replaced by other tree species in the absence of coarse-scale disturbance. New cohort development plus resistance of large trees to burning, ensures the persistence of *Fitzroya* when rain forests are burned. The dependence of *Fitzroya* on coarse-scale disturbance (fire, landslides, and floods) over most of its range in northern Patagonia results in stands of old cohorts with scant regeneration, which formerly was interpreted incorrectly as evidence of a species in decline due to long-term climatic change (Kalela 1941; Tortorelli 1956; Rodríguez et al. 1978).

Nothofagus Dombeyi *Dominated Mesic Forests*

Throughout the zone of mesic *Nothofagus dombeyi* forests, postfire age structures and charcoal in the soil indicate the widespread importance of fire (Eskuche 1968; Singer 1971; Seibert 1982; Veblen and Lorenz 1987). *N. dombeyi* is thin-barked and does not regenerate vegetatively. Small-diameter stems, such as those in young postfire cohorts, are easily killed by fire, but sporadic large-diameter trees can survive and provide seed sources (Tortorelli 1947). It is a prolific seeder, and at favorable sites it grows rapidly into extensive even-aged stands (Fig. 9.2e–f). Thus postfire regeneration from seed is typically successful as long as seed sources are within about 50 to 100 m (Veblen and Lorenz 1987; Kitzberger and Veblen 1999).

Toward the drier end of its range, *N. dombeyi* jointly colonizes postfire sites with *Austrocedrus chilensis*. However, over most of the moisture gradient where the two species co-occur, *Austrocedrus* is often markedly less abundant in early (<20 years old) postfire stands (Veblen and Lorenz 1987). Age structure analyses of co-dominated postfire stands indicate that the populations of both species are of similar age (Fig. 9.2c–d). As mesic *Austrocedrus–N. dombeyi* postfire stands develop, large gaps (>1000 m^2) are created by the death of large *N. dombeyi* individuals. In gaps of this size, small numbers of both *N. dombeyi* and *Austrocedrus* may establish, and eventually all-aged tree populations develop in older stands (Veblen 1989). At edaphically less favorable sites in valley bottoms, postfire stands may be initially dominated by the short-lived, resprouting *N. antarctica* which is eventually replaced by the long-lived *N. dombeyi* (Veblen and Lorenz 1987). Frequent burning, often followed by livestock browsing, may convert some former *N. dombeyi* sites to long-lasting shrublands (Tortorelli 1947).

Nothofagus Pumilio *Subalpine Forests*

Subalpine *Nothofagus pumilio* forests occur in cooler, more mesic habitats than many of the neighboring vegetation types. This may account for its relatively low contribution to the total area burned despite the great extent of this cover type in the landscape (Fig. 9.3). *N. pumilio* is thin barked, easily killed by fire, and generally does not resprout after fire (Table 9.1). If postfire site conditions are favorable (i.e., not too xeric) and seed sources are available, it can regenerate abundantly following stand-replacing fires (Fig. 9.2g). However, after some fires

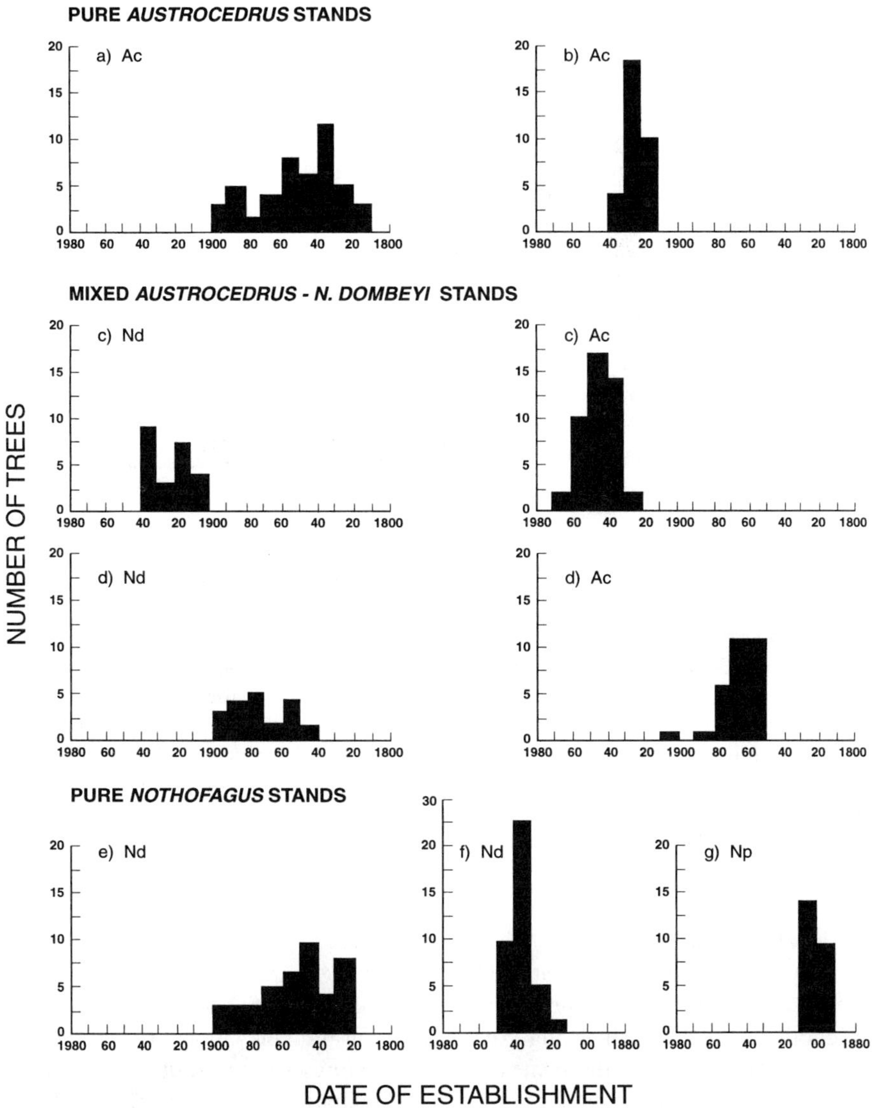

Figure 9.2. Tree age frequency diagrams for postfire stands of pure *Austrocedrus chilensis* (a and b), mixed *Austrocedrus-Nothofagus dombeyi* (c and d), pure *N. dombeyi* (e and f), and pure *N. pumilio* (g). (Data from Veblen and Lorenz 1987; Kitzberger 1994; Veblen et al., unpublished.)

it fails to regenerate (Veblen et al. 1996). In the case of intense fires affecting large surface areas, the absence of surviving seed trees is clearly an important factor in the lack of tree regeneration. For example, following the intense and extensive burning of *N. pumilio* forests in 1999 in southern Nahuel Huapi

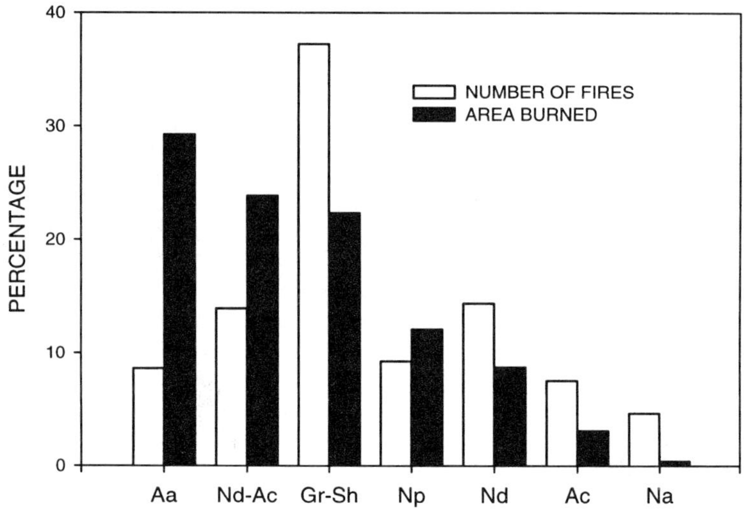

Figure 9.3. Percentages of numbers of fires and area burned by vegetation types in national parks Lanín, Nahuel Huapi, Lago Puelo, and Los Alerces from 1939 through 1997. Vegetation types are Aa, *Araucaria araucana* forest; Nd-Ac, mixed *Nothofagus dombeyi* and *Austrocedrus chilensis* forests; Gr-Sh, grasslands and shrublands (including bamboo thickets); Np, *Nothofagus pumilio* subalpine forests; Nd, *Nothofagus dombeyi*-dominated mesic and rain forests; Ac, *Austrocedrus chilensis* woodlands and forests; and Na, *Nothofagus antarctica*-dominated tall shrublands and low forest. Not shown is a category of miscellaneous minor forest types that accounted for 4.6% of the number of fires and 0.4% of the area burned. Although precise data on the extent of each cover type are not available, the forest cover type of least extent is Aa. Overlapping cover types include Gr-Sh with Na, Nd-Ac with Nd, and Nd-Ac with Ac.

National Park, no seed trees survived the fires over sectors of hundreds of hectares (Veblen et al., unpublished data). However, even in areas of small burns where seed trees are nearby, *N. pumilio* sometimes fails to regenerate. This is most conspicuous on steep, north-facing (xeric) slopes at high elevations. Given the lack of livestock at many of these sites and the proximity of seed sources, the lack of tree regeneration may be due to fire-induced edaphic changes, postfire establishment of a dense cover of herbaceous plants, or possibly unfavorable climatic conditions. The potential for drier climatic conditions to limit postfire regeneration is consistent with seedling survival only at moister micro-sites in small treefall gaps in xeric *N. pumilio* forests (Heinemann, Kitzberger, and Veblen 2000). Soils beneath *N. pumilio* forests that burned in 1996 declined sharply in organic matter, nitrogen and microbial biomass indicating high fire intensity (>300°C), even when small (<40 m^2) patches burned (Alauzis 1999). The reduction in organic matter which affects the availability of nitrogen and moisture in the short and longer term may be a limiting factor for the regeneration of *N. pumilio* (Alauzis 1999).

Nothofagus Obliqua *and* N. Nervosa *Forests*

Both *Nothofagus obliqua* and *N. nervosa* have moderately thick bark and resprout after being cut or burned (Veblen et al. 1996). Most *N. obliqua*-dominated forests are second-growth stands that originate after burning and cutting, and have been subject to heavy, long-term livestock impacts. *N. obliqua* occurs mostly in relatively open stands where multiple fire scars on it indicate its ability to survive surface fires. In the north, *N. obliqua* occurs sporadically at the steppe ecotone and in association with *Araucaria araucana* where fire scars indicate its ability to tolerate surface fires. At more mesic sites (higher elevation and south-facing slopes), *N. nervosa* co-occurs with the evergreen *N. dombeyi* where fires are more likely to be stand replacing.

Austrocedrus Chilensis *Forests and Woodlands*

Austrocedrus is killed by intense fires due to its relatively thin bark and is generally not capable of vegetative reproduction. Stand-replacing fires are typical of dense pure *Austrocedrus* stands and result in even-aged postfire cohorts (Fig. 9.2a–b). Eastward, under more xeric conditions, fuels are discontinuous and low intensity surface fires predominate. Many adult trees survive these surface fires. Toward the steppe, or on xeric slopes, *Austrocedrus* regeneration occurs sporadically in space and time resulting in open woodlands with heterogeneous age distributions (Veblen and Lorenz 1988; Burns 1991; Villalba and Veblen 1997a). In contrast to the mesic, dense stands where regeneration is limited by light conditions and competition, the limiting factors for seedling establishment in the open woodland habitat appear to be the desiccating effects of open sites (Kitzberger, Steinaker, and Veblen 2000). Large herbivores (mainly introduced deer and livestock) also can inhibit the regeneration of *Austrocedrus* (Veblen et al. 1989, 1992; Relva and Veblen 1998). In early postfire stages, *Austrocedrus* seedlings typically establish in association with shrubs such as *Discaria articulata*, *Schinus patagonicus*, *Berberis* spp. and *Lomatia hirsuta*. These resprouting species, many of which have spines or thorns, form dense patches beneath which *Austrocedrus* seedlings occur. This shrub cover offers protection against browsing and provides protection from excessive temperatures and moisture stress in the drier habitats (Kitzberger, Steinaker, and Veblen 2000).

Araucaria Araucana *Forests and Woodlands*

Araucaria araucana occurs in habitats that are prone to fire due to the intensity of the summer drought and frequent ignitions by humans (Tortorelli 1947; Veblen et al. 1995). Although this forest type covers less area than most or all other forest types, it accounts for the largest percentage of area burned in any vegetation type over the 1938 to 1997 record (Fig. 9.3). Fire-promoting characteristics of this forest type include the highly flammable wood and leaf litter of *Araucaria* and rapid accumulation of dead branches from the common crown dieback of the associated *Nothofagus antarctica*. Large *Araucaria* have fire-resistant thick bark and an umbrella crown shape that places foliage out of reach of surface fires. Following fire it sprouts weakly from basal epicormic buds, and the protected terminal buds on branches survive many fires (Tortorelli 1947; Burns 1993).

At the moderately xeric sites occupied by the *Araucaria–N. antarctica* association, postfire sites are initially dominated by the resprouting *N. antarctica* with scattered large *Araucaria* that survive the fire. *Araucaria* seedlings emanating from trees surviving the fire or newly dispersed seed to the site, establish under partial shade and dominance gradually shifts from *N. antarctica* to *Araucaria.* After about 150 years without further fire, *Araucaria* suppresses and excludes the senescent *N. antarctica* subcanopy so that the vegetation develops into an *Araucaria*-dominated forest (Burns 1993). At mesic sites, postfire cohorts of *N. pumilio* or *N. dombeyi* may establish concurrently with *Araucaria*, but typically many large *Araucaria* also survive the fires (Burns 1991).

Nothofagus Antarctica *Woodlands and Other Tall Shrublands*

Given their proximity to the open steppe and their utilization by cattle, *Nothofagus antarctica* woodlands in broad valley bottoms tend to be sites of frequent human-set fires. Heavy epiphyte loads of the highly flammable *Usnea* lichen, as well as abundant partially dead crowns of *N. antarctica*, contribute to the flammability of *N. antarctica*-dominated vegetation. The surprisingly small extent of fire in this type reported in national park data (Fig. 9.3) undoubtedly reflects the inclusion of *N. antarctica*-dominated stands in other vegetation types (mainly grassland-shrubland and *Araucaria araucana* woodlands). Many *N. antarctica*-dominated shrublands are early stages of postfire succession that eventually result in recovery to *N. dombeyi* and/or *Austrocedrus* forests (Seibert 1982; Veblen and Lorenz 1987; Kitzberger and Veblen 1999). Recovery to forest is often retarded by repeated burning, heavy livestock pressure, and severe erosion following intense fires on steep slopes.

Shrublands dominated by *Nothofagus antarctica, Lomatia hirsuta, Schinus patagonicus, Embothrium coccineum, Chusquea culeou*, and/or *Diostea juncea* are highly prone to burning due to fuel and site conditions. In dense shrublands the decurrent multistemmed growth form of the small trees and shrubs provide fuel ladders for crowning of surface fires. Vigorous resprouting of all the tree and shrub species (Table 9.1) allows for rapid fuel recovery that in turn permits short fire return intervals (Veblen, Kitzberger, and Lara 1992). The midslope habitat of shrublands on north-facing slopes is exceptionally dry during the summer months due to higher temperatures and low soil moisture capacity of the thin soils (Rodríguez et al. 1978; Seibert 1982). Climbing daisies, *Mutisia* spp., are abundant in tall shrublands, and the annual dieback of their stems probably enhance fire spread from the ground into shrub crowns.

Steppe

Fire is an important disturbance in the western extent of the Patagonian steppe where plant cover is more extensive than in the interior of the continent. Near the Andean foothills, plant covers of approximately 60% correspond to between 700 and 1100 kg/ha of fine fuels (Bran 1996). Here the steppe consists of a matrix of scattered cushion shrubs *(Mulinum spinosum)* interspersed with tussock grasses (*Stipa speciosa, Festuca pallescens*), forbs (e.g., *Euphorbia collina, Solidago* spp.), and exotic herbaceous species such as *Rumex acetosella*. Most species are flammable due to the consistently warm dry summers of the steppe. Fire is sometimes used for improving forage quality for livestock over small areas; however, steppe fires can also become large and out of control as in the case of a fire that lasted for six days and burned over 8000 ha in Nahuel Huapi National Park in 1996 (Delegación Técnica Regional Patagonia 1996a).

Recovery after fire in the steppe is rapid due the capacity of nearly all the common shrubs and herbs to resprout from basal buds on stems or root crowns, or from rhizomes and other underground organs. At some sites, however, post-fire erosion can impede recovery, as can livestock that are attracted to recently burned steppe by the abundance of tender new shoots.

Human Impacts on Fire Regimes

Pre-1880: Native American Influences

Paleoenvironmental records document fire at least as early as 12,600 BP in the southern Andean region (Heusser 1987, 1994; Markgraf and Anderson 1994). During the prehistoric period in northern Patagonia, fires were ignited both by humans and lightning. The earliest evidence of human occupation in northern Patagonia dates from 8000 to 7000 BP (Crivella and Silveira 1983), and sedimentary records of charcoal have been dated to as early as 3000 years BP (Heusser et al. 1988). Native Americans affected fire regimes and the landscapes of northern Patagonia through intentional burning for hunting purposes, agricultural practices, collecting seed, opening of travel routes through dense forest, and perhaps also for pasture management after European livestock arrived in the seventeenth and eighteenth centuries.

Use of Fire at the Woodland/Steppe Ecotone

Use of fire in northern Patagonia for hunting guanacos (*Lama guanicoe*, an American camelid), rheas (*Pterocnemia pennata*, a 1.5-m-tall ratite bird), and *huemules* (*Hippocamelus bisulcus*, a deer) in the ecotone of open *Austrocedrus* woodlands and steppe is well documented by archeological evidence and by the observations of early explorers and missionaries (Veblen and Lorenz 1988). The association of human remains with guanaco bones dates from ca. 6340 years BP and suggests that the earliest human inhabitants were hunters (Crivella and Silveira 1983). Explorers of northern Patagonia provide eyewitness accounts of the use of fire by native hunters to encircle and drive guanacos and rheas (Cox 1863; Fonck 1896; Musters 1871).

Possession of European livestock by the native inhabitants of northern Patagonia as early as the seventeenth century (Furlong 1964) may have been a motive to burn to create or improve pasture. Cox (1863) noted that the abundant *toldos* (shelters) at Caleufú were periodically moved when the horses and sheep exhausted the local grass supply. Burning of "pasture" is reported for northern Patagonia (Musters 1871), and the same ethnic group (Mapuches) is described as "annually burning the grass" in nearby livestock raising areas in southern Chile (Smith 1855). According to Mapuche oral traditions, fires were intentionally set to improve pasture (Salguero 1998). Signal fires were widely used by the indigenous population of Patagonia, and occasionally they may have escaped to become wildfires (Moreno 1897).

Fire frequencies began to increase at most *Austrocedrus* woodland sites about 1840 and peaked in the late nineteenth century (Fig. 9.4). The midnineteenth century increase in fires is coincident with increased use of the *Austrocedrus* habitat by Native American hunters as a result of immigration across the Andes, stimulated by the European colonization of southern Chile (Cox 1863). A probable increase in burning of *Austrocedrus* woodlands by Native American hunters in the mid-1800s is suggested by contrasting fire-scar records from two adjacent sites separated by the large Limay River (Fig. 9.5). The West Limay site was settled by native hunters in the nineteenth century and earlier (Cox 1863; Musters 1871; Crivella and Silveira 1983) and shows a substantial increase in fire occurrence beginning in the mid-1800s (Fig. 9.5). In contrast, the location of the East Limay site east of the river would have impeded human access and/or fire spread from the western settled area. The East Limay site does not show an increase in burning in the mid-1800s. Prior to 1800 both sites supported similar rates of burning, which suggests that the increased rate of burning in the mid-1800s at West Limay is at least partially due to humans. However, when major fire years (i.e., years in which fire scars occur on more than a single tree) are considered, there is a regionally extensive increase in burning beginning in the mid-1800s that coincides with greater interannual climatic variation (see Kitzberger and Veblen, Chapter 10, this volume). Thus both humans and climatic variation appear to be responsible for increased burning after the mid-1800s.

Fires in the Mesic Forest Zone

Burning believed to have been anthropogenic was also reported for the mesic and rain forest districts. Large burns (*quemazones*) were observed in 1787 by Padre Francisco Menéndez (Fonck 1896) in the Nahuel Huapi region and, in particular, along the *camino de Vuriloche*, the famed Andean crossing southwest of Lake Nahuel Huapi (Fig. 9.1). All mid-nineteenth-century explorers (Emilio Valverde, Oscar de Fischer, Juan Steffen, Francisco Fonck, and Fernando Hess) observed extensive burns in the mesic forests of the Andes, from Lake Nahuel Huapi southward to Rio Puelo (42°S) and Rio Palena (43°40′S; Fonck 1896).

Some of the burns reported along Andean travel routes by eighteenth- and nineteenth-century explorers, may have been intentionally set to keep clear the

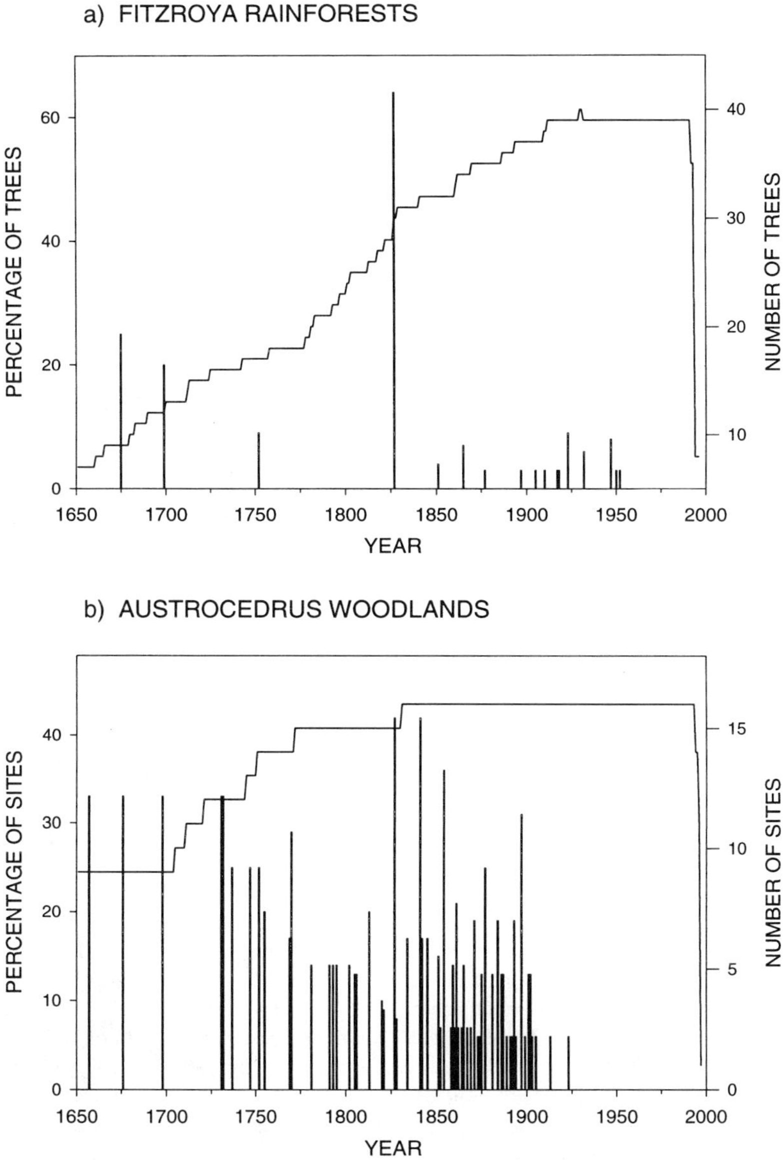

Figure 9.4. Records of regional trends in fire occurrence based on (a) percentage of *Fitzroya cupressoides* rain forest trees recording fire in the same years (40 trees from 3 sites) and (b) percentage of *Austrocedrus* woodland sites at which at least 10% of the recorder trees recorded fire in the same year (16 sites with a total of 331 fire-scarred trees). Sample depth lines give (a) the number of fire-scarred trees alive and (b) the number of sites with fire-scarred trees alive. (Data from Veblen et al. 1999.)

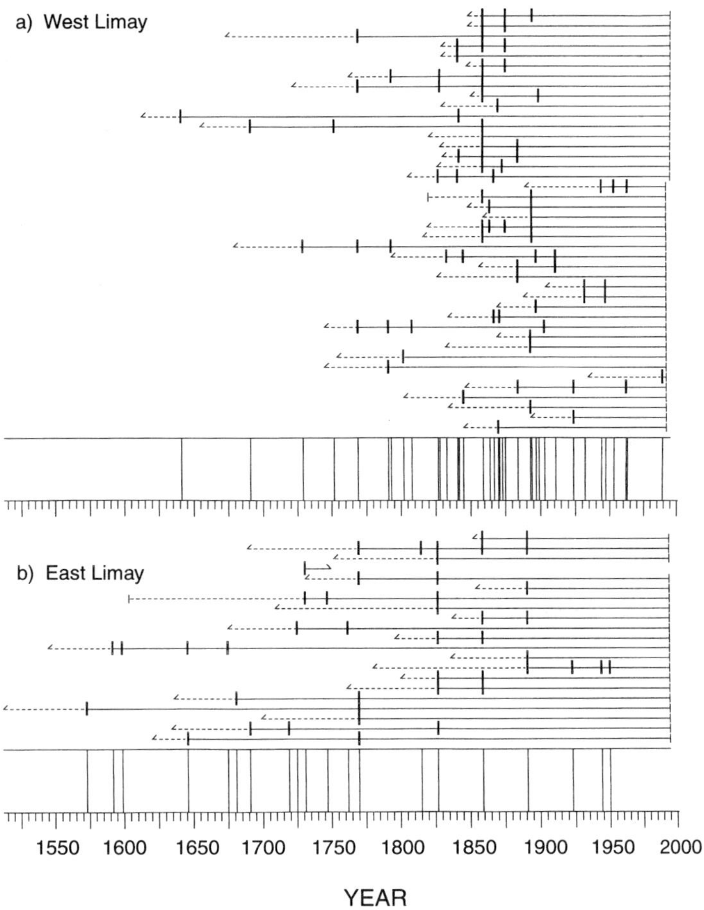

Figure 9.5. Records of fire scars on individual trees (*top*) and composite fire chronologies (*below*) for West Limay (a) and East Limay (b) sample sites in *Austrocedrus* woodland. The two 2-km^2 area sites are adjacent but separated by the large Limay River. Each horizontal line represents an individual tree on which dates of fire scars are indicated by short vertical lines. Pith dates and bark dates are indicated by vertical lines at the beginning or end, respectively, and dates of innermost and outermost rings are indicated by half-arrows, respectively. Dashed lines indicate years prior to the occurrence of the first scar on that tree. Vertical lines drawn to the x-axis indicate occurrence of fire scars on at least one tree in the sample area and are the composite fire chronologies. (Data from Kitzberger and Veblen 1997.)

trading routes across the Andes (Alvarez 1984; Bengoa 1985). Without fire or frequent cutting, the rain forest routes across the Andes become nearly impenetrable to humans because of the dense understories of *Chusquea* bamboos. The native inhabitants of the Lake Nahuel Huapi region also used fire to prepare sites in the humid forest district for crop cultivation (Furlong 1964; Cox 1863). Early twentieth- and nineteenth-century observers in the *Araucaria* forests of northern Patagonia report intentionally set fires for clearing the understory to facilitate collection of the large *Araucaria* seeds (Rothkugel 1916; Tortorelli 1947).

Lightning- Versus Human-Set Fires

Although the native inhabitants undoubtedly set fires throughout their long occupation in northern Patagonia, it is not known what percentage of fires recorded either in sedimentary records or in tree-ring records were ignited by humans or by lightning. Comparison of tree-ring records of fire history with documentary records of lightning-ignited fires suggests that for northern Patagonia, Native Americans increased the rate of burning over the natural rate. In all four northern Patagonian national parks (ca. 1.4×10^6 ha), 46 lightning-ignited fires were recorded in only 22 years from 1938 to 1996 or 1.57 fire years per 100,000 ha (Administración de Parques Nacionales, unpublished data). In contrast, for the 59 years preceding Euro-American settlement (1822–1880), the mean number of fire years is 6.4 for 8 sample areas of about 400 ha each in *Austrocedrus* woodlands, or 1600 fire years per 100,000 ha. During the preceding 59-year period (1764–1822), when fire frequency was lower, 688 fire years occurred per 100,000 ha. Even allowing for omission of some lightning-ignited fires from the park observations, these large differences in numbers of fires suggest that the high rate of burning prior to 1880 could not be accounted for by lightning alone.

European Settlement in the 1880s to 1920s

The 1890s to early 1900s was a period of extensive forest burning by European settlers to create cattle pasture (Moreno 1897; Steffen 1909; Willis 1914; Rothkugel 1916). For the Provinces of Neuquén, Rio Negro, and Chubut, Rothkugel (1916) mapped 692,000 ha (37% of the total forest area) as having burned during the European settlement period prior to 1915. This burning also resulted in the establishment of extensive even-aged, pure *Nothofagus* forests with cohort ages indicating a marked peak in burning ca. 1900 (Fig. 9.6). Similarly a sedimentary record from a bog at 41°16′S contains a 50-cm-thick layer of charcoal dating from this period (Markgraf 1983). In *Fitzroya* rain forests fire records from accessible sites near important trans-Andean travel routes show increased burning in the 1890s and early 1900s (Fig. 9.4a) in contrast to no increase at a remote, inaccessible *Fitzroya* site (Veblen et al. 1999).

Modern Fire Exclusion Period: Post-1920s

Following the 1880s the fire-scar record indicates a sharp decline in fire in *Austrocedrus* woodlands near the steppe (Fig. 9.4). Three factors contribute to the low incidence of fire in the open *Austrocedrus* woodlands during the twentieth century relative to the nineteenth century: (1) decrease in intentionally set fires, (2) fuel reductions due to increased grazing by livestock, and (3) active fire

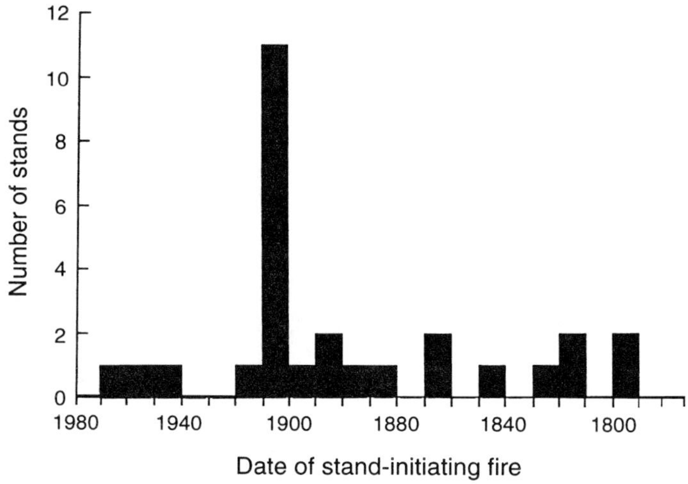

Figure 9.6. Dates of stand-initiating fires in 10-year classes for 31 *Nothofagus dombeyi* and *Austrocedrus chilensis* stands located between Lake Lacar in the north and Lake Nahuel Huapi in the south. (Data from Veblen et al. 1992a.)

suppression. With the demise of the Native American hunting populations, the number of intentionally set fires in the woodland/steppe ecotone declined in the 1890s to early 1900s. This was also a time of increased livestock utilization of this habitat (Willis 1914; Eriksen 1971). Despite the efforts of local authorities to suppress fires as early as 1913, most fires appear to have been extinguished by rain due to the absence of fire-fighting infrastructure (Rothkugel 1916). Relatively rapid access by horse and motor vehicle to sites in the woodland/steppe habitat after about 1920 facilitated fire suppression efforts in that habitat. In contrast, even today, access to much of the wet forest zone is limited, and it is unlikely that active fire suppression has had much impact on fire in these forests. It appears that the decline in fire frequency was primarily the result of a decrease in human-set fires in the *Austrocedrus* woodlands and possibly decreased fuels due to grazing rather than active fire suppression.

National park records since 1938 indicate that in some years, fires continue to be common and extensive despite the adoption of a fire exclusion policy by park authorities in the 1920s (Figs. 9.7 and 9.8). The relative scarcity of fire scars during the post-1938 period (Fig. 9.4) is a reminder that trends in fire-scar data are only relative indicators of past fire. In particular, the extensive zone of monotypic *N. dombeyi* forests lack species that are good recorders of fire scars; thus the absence of fire scars does not necessarily imply absence of fire. The national park records do not include the pre-fire exclusion period which limits evaluation of the effect of fire suppression. However, since 1938 there is no trend toward declining fire frequency (Fig. 9.7), and years of widespread fire continued to occur during the 1980s and 1990s (Fig. 9.8).

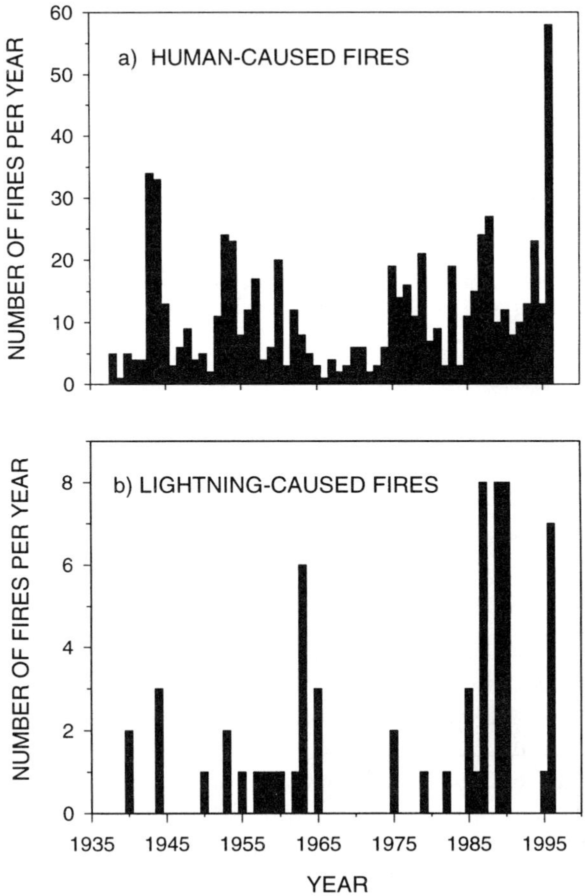

Figure 9.7. Annual documentary fire records for Lanín, Nahuel Huapi, Lago Puelo, and Los Alerces national parks for 1938 to 1996: (a) number of human-caused fires and (b) number of lightning-caused fires. (Data from the Administración de Parques Nacionales.)

Landscape Changes Associated with Human-Caused Changes in Fire Regimes

Consequences of Twentieth-Century Fire Exclusion

The most dramatic temporal change in fire occurrence in northern Patagonia is the abrupt decline in fire scars since the 1890s, in particular, in the *Austrocedrus* woodland habitat where abundant fire-recording trees occur (Fig. 9.4b). During the twentieth-century period of reduced fire frequency, there has been a regionally extensive tendency for the percentage of tree-dominated cover types to increase at the expense of grasslands and shrublands (Veblen and Lorenz 1988; Kitzberger and Veblen 1999). Historical photographs show that during the past

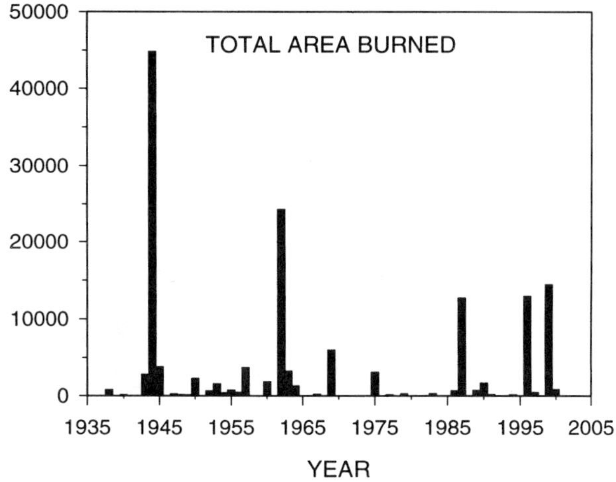

Figure 9.8. Total area burned (hectares) for Lanín, Nahuel Huapi, Lago Puelo, and Los Alerces national parks for 1938 to 2001. (Data from the Administración de Parques Nacionales.)

100 years or so, especially *Austrocedrus*, but also other trees and tall shrubs such as *Lomatia hirsuta, Maytenus boaria*, and *Schinus patagonicus,* have formed dense stands at sites that were formerly covered in grasses and low shrubs (Veblen and Lorenz 1988). At sites of increased tree density near the steppe, absence of burnt spars, logs, and cut stumps indicate that burning or logging had not destroyed a former forest cover in the nineteenth century. Instead, these open vegetation types were maintained by fires occurring frequently enough to prevent development of dense stands. Comparison of historical and modern photographs, as well as age structure data, indicate that abundant establishment of *Austrocedrus* began at the ecotone in the late 1800s (Veblen and Lorenz 1988). Reduction in fire occurrence allowed much greater survival of juveniles of this fire-sensitive species. The rate and timing of *Austrocedrus* establishment have been influenced by interannual and decadal-scale climatic variation (Villalba and Veblen 1997a; Kitzberger, Steinaker, and Veblen 2000) as well as by livestock impacts (Veblen et al. 1992; Relva and Veblen 1998).

The increase in tree density in *Austrocedrus* woodlands and invasion of trees into the steppe have created more contiguous woody fuels (Veblen, Kitzberger, and Lara 1992; Kitzberger and Veblen 1999) so that sites previously supporting only surface fires are now susceptible to stand-replacing fires. In the submesic area, remnant forest patches, resulting from widespread early twentieth-century burning, expanded and in many cases coalesced into continuous forest (Kitzberger and Veblen 1999). Thus, although forest fragmentation is considered a common trend under increasing human influences on wild landscapes elsewhere (e.g., Harris 1984), in protected areas of the forest-steppe ecotone the reverse of forest fragmentation has been the norm over the past approximately 70 years.

Consequences of the Late-Nineteenth-Century Episode of Mesic Forest Burning

Extensive burning of mesic forests by Euro-American settlers in the 1890s to 1920s resulted in vast areas of even-aged, regenerating *Nothofagus-Austrocedrus* forests (Veblen and Lorenz 1987; Veblen, Kitzberger, and Lara 1992; Kitzberger and Veblen 1999). Thus most of the mesic forest zone is in a middle-aged (ca. 80–110 years) postfire stage of stand development during which self-thinning produces abundant intermediate-sized fuels that may favor fire spread. We have observed some fires spreading through nearly 100-year-old postfire stands that did not burn adjacent tall *N. dombeyi* forests. At a regional scale the synchronization of stand development in the *N. dombeyi* forest zone resulting from the late 1800s burning may have increased the potential for fire spread at a regional scale.

During the Euro-American settlement period many forest sites were burned more than once (Rothkugel 1916). The initial burning of wet *N. dombeyi* forests probably increased the subsequent flammability of the site by eliminating the tall shade-producing overstory. Subsequent burns would have reduced seed sources for arboreal species and promoted dominance by bamboo and shrubs that sprout vigorously after being burned (Table 9.1). Multiple burning of the same sites resulted in extensive conversion of forests to shrub-dominated communities, which at some sites recovered to forest but at others have remained in shrublands. Once tall forest is converted to a bamboo thicket or shrubland, it is much more flammable and likely to sustain higher fire frequencies than adjacent tall forests (Veblen et al. 1992a). In general, following forest burning, shrublands tend to persist where moisture conditions are marginal (e.g., on steep sites of easily eroded soils), where repeated fires occurred, or where livestock impacts have been severe (Veblen et al. 1992b; Relva and Veblen 1998).

Configuration of remnant forest patches plays an important role in subsequent changes in landscape pattern through its influence on dispersal of the obligate seed-reproducing *N. dombeyi* and *Austrocedrus* (Kitzberger and Veblen 1999). Postfire forest regeneration, at least over a period of less than a single tree generation, is highly concentrated in a distance of about 25 m from remnant forest patches. In general, in Northern Patagonia the extensive even-aged postfire cohorts of *N. dombeyi* and *N. dombeyi–Austrocedrus* (Veblen and Lorenz 1987; Veblen Kitzberger, and Lara 1992) must have developed within the relatively short range of remnant seed-bearing trees (i.e., within 40 to 80 m radii of the tallest trees).

It has been suggested that some of the strikingly sharp boundaries between subalpine *N. pumilio* forests and adjacent shrublands are maintained by differences in flammability of these communities (Veblen and Lorenz 1988). *N. pumilio* forests have often been noted to burn less frequently than adjacent shrublands (Rothkugel 1916; Tortorelli 1947). The closed-canopies of the subalpine forests produce cool mesic microenvironments with short understory plants, and the morphology of the dominant tree does not provide fuel ladders for easy crowning of fires. In contrast, micro-sites in the midslope shrublands are characterized by high temperatures and dry soils, and the multistemmed habit of the shrubs and small *N. antarctica* facilitates fire-crowning. Furthermore the *Chusquea* bamboos provide continuous fine fuels from near the ground surface to the top of the shrub canopies. Where subalpine *N. pumilio* forests were burned during the early 1900s, in some (but not all) cases they may have been converted to shrublands that are largely self-replacing because of their greater flammability.

The regional trend for grassland and shrubland vegetation types to be replaced by forest has been documented by comparing the 1913 vegetation map of Willis (1914) with modern vegetation maps of northern Patagonia (Kitzberger and Veblen 1999). These regional-scale vegetation transitions reflect both tree regeneration at sites of forests burned in 1880 to 1920 and succession from grassland or shrubland to forest cover during the post-1920 fire exclusion period. Comparison of aerial photographs of mosaics of *N. dombeyi–Austrocedrus* forests, shrublands and grasslands taken in 1940 and 1970 show that shrubland and grassland areas have become more disjunct and in many cases have been completely replaced by tree cover (Kitzberger and Veblen 1999). Many areas that were grasslands in 1940 were replaced by shrublands in 1970, and only areas relatively isolated from tree seed sources remained stable. During this period of reduced fire frequency there was a shift in dominance from species with short life spans and re-sprouting capacity (e.g., shrubs) toward longer-lived species and obligate seed-dispersers (e.g., *Austrocedrus* and *N. dombeyi*). The relatively restricted seed dispersal ability of *N. dombeyi* and *Austrocedrus* may also explain the presence of extensive shrublands of fire-resprouting species (e.g., *N. antarctica*, *Maytenus boaria*, and *Lomatia hirsuta*), where today small numbers of *N. dombeyi* and/or *Austrocedrus* are slowly invading. Similarly dendroecological studies in more mesic stands have demonstrated a shift during long fire-free periods from the short-lived, postfire resprouting *N. antarctica* toward the long-lived, nonsprouting *N. dombeyi* (Veblen and Lorenz 1987).

Synergisms of Natural Variability and Anthropogenic Influences

Synchronous Flowering of *Chusquea* Bamboos

Chusquea bamboos are keystone species in the dynamics of the forests of southern Chile and northern Patagonia due to their importance as fuels and their inhibitory influence on tree regeneration. At long intervals, estimated at 17 to over 70 years, some *Chusquea* species flower synchronously over a two- to three-year period and die massively over areas of many hundreds of square kilometers (Veblen 1982; Pearson, Pearson, and Gomez 1994; Gonzalez and Donoso 1999). Given the slow decay rate of the bamboo leaf litter and culms, such a massive die-off results in an enormous amount of dry understory fuels in these forests for at least four or five years after flowering. When these flowering events coincide with a dry year, the mesic forests of northern Patagonia and southern Chile become much more flammable. A massive flowering of *Chusquea culeou* affecting much of Lanín National Park and adjacent areas in Chile was initiated in November 2000, and it is the first regional-scale *Chusquea* flowering event in northern Patagonia since 1940.

The peak of European burning of mesic forests in the 1890s to 1910s was faciliated by unusually warm-dry weather and probably also by one or more massive flowerings of *Chusquea culeou*. Hosseus (1915) noted that in 1914 some *Chusquea culeou* were in flower around Lake Nahuel Haupi, and he believed that this was a harbinger of a massive flowering. Although we have been unable to unequivocally document this flowering, many long-term residents of northern Patagonia have reported to us that their grand parents or great grand parents described a massive flowering early in this century. Informants interviewed by Pearson et al. (1994) reported a massive flowering between 1900 and 1904. Hosseus (1915) reported that local residents claimed that there had been a massive flowering in 1890 too. It is probable that at least one massive flowering of *Chusquea culeou* occurred in northern Patagonia between 1890 and 1915, which coincides with the period of extensive burning of the mesic *Nothofagus* forests by European settlers.

The late 1800s and early 1900s was also a time of severe droughts. Based on the growth rings of *Austrocedrus* sampled throughout northern Patagonia, 1899 to 1917 is the longest period of below-average tree growth (i.e., moisture deficit) since 1700 (Villalba and Veblen 1997b). Temperature reconstructions from *Nothofagus pumilio* and *Fitzroya* indicate that the first two decades of the twentieth century were among the warmest of the past 230 years (Villalba 1990; Villalba et al. 1997), and precipitation reconstructions from *Austrocedrus* identify 1895 to 1919 as the driest 25-year period since 1599 (Villalba et al. 1998). Thus an abundance of dry fuels resulting from a probable massive flowering of *Chusquea culeou* during a period of prolonged drought coincided with the arrival of European colonists intent on converting forest to pasture by extensive burning.

The 1944 Fire Year

The year 1944 was a year of extraordinary forest burning in northern Patagonia (Fig. 9.8). Since the beginning of the national park records in 1938, the 44,855 ha burned in the four national parks in 1944 is by far the greatest area burned in any single year (Administración de Parques Nacionales, unpublished data). Almost all the area burned in the national parks in that year was due to human-set fires (Administración de Parques Nacionales, unpublished data). Most of the area affected was in the south in Los Alerces National Park where a settler intentionally set a fire that burned 36,200 ha of forest. Tortorelli (1947) reported on the extensive burning that occurred outside of the national parks in 1944 in the province of Chubut and burned an estimated 275,000 ha. Most of the fires were set in Chile as *rozas* (for agricultural clearing) and spread into Argentina, but many of the fires were also set in Argentina (Tortorelli 1947).

Such extensive human-set fires in the wet forest district were promoted by the occurrence of drought and the massive flowering of *Chusquea culeou*. Eleven "credible" long-term residents of northern Patagonia reported a masive flowering of *Chusquea culeou* in about 1940 in around a 230 km south-to-north stretch of the Andes from Lago Futalaufquén (43°S) to north of Lake Nahuel Huapi (41°30′S). Massive flowering of *Chusquea culeou* is also reported for 1940 to 1942 in adjacent parts of Chile (Gonzalez Cangas 1998). Thus extraordinary quantities of drying bamboo fuel characterized the wet forests of northern Patagonia during the early 1940s. The springs and summers of 1942 and 1943 were extremely dry throughout northern Patagonia, and especially near latitude 43°S. For example, Martonne's (1926) aridity index for November through February in 1943–44 indicates that this fire season was the second driest of the century and that no two-year period since 1905 had drier spring–summers than those of 1942 to 1944 (based on the Esquel climate station). As reconstructed from *Austrocedrus* tree rings throughout northern Patagonia, the lowest spring (November–December) precipitation from 1599 to 1988 occurred in 1943 (Villalba et al. 1998). Fire was also widespread in the summer of 1943 in the northern part of northern Patagonia, where Tortorelli (1947) described the rapid spread of fire through 3500 ha of *Araucaria* forest in January 1943 at the northern limit of Lanín National Park. He noted that the rapid spread of the fire was favored by *caña coligue seca* ("dead bamboo"), suggesting that the flowering of *Chusquea culeou* documented for ca. 42°S may have extended as far north as 39°S.

The 1996 and 1999 Fire Years

In 1996 approximately 8000 ha of steppe and shrubland communities burned in the national parks of northern Patagonia. This was the highest amount of burning recorded in these community types since the begining of record keeping in 1938 (Administración de Parques Nacionales, unpublished data). The previous spring (September–December 1995) precipitation was 1.7 standard deviation (SD) below the historical mean (1905–1996, Bariloche weather station), and early summer temperature (December) was 1 SD above the historical mean (1914–1996, Bariloche weather station). In 20 days four major fires burned over 12,000 ha near the resort town of Bariloche. Two major fires occurred in steppe/ecotone areas; one was ignited by lightning and burned nearly 8000 ha mainly in steppe vegetation (Delegación Técnica Regional Patagonia 1996a). This lightning-ignited fire began on a ranch where five years earlier livestock had been removed, resulting in a marked increase in fine fuels (Salguero 1998). Given a trend toward reduced livestock pressure (especially reduced sheep grazing) in northern Patagonia during the 1990s, there may be a general increase in fire hazard due to greater fuel accumulation in the steppe.

The second steppe fire, and two other large events that burned extensive areas of shrubland and xeric woodlands, were set by humans, possibly intentionally. Increased arson in the 1990s in the urban–forest interface may reflect socioeconomic tensions from the juxtaposition of an economically marginalized population surrounding the posh resort city of Bariloche. Accidental fires may also be on the increase due to greater recreational and residential use of the area. Abandoned, unthinned Monterey pine (*Pinus radiata*) plantations contributed to the spread of one of the 1996 fires from its point of origin on the outskirts of the city into Challhuaco Valley in the adjacent national park (Delegación Técnica Regional Patagonia 1996b).

During the 1998–99 fire season more than 14,000 ha burned in Nahuel Huapi National Park alone (Administración de Parques Nacionales, unpublished data). This major fire year followed the driest calendar year (1998) recorded in the 1905 to 1999 Bariloche weather record as well as the warmest spring (October–December) on record. Martonne's (1926) index of aridity for the October 1998 through February 1999 period indicated moisture availability of only 10% of the long-term average. This extraordinary drought was associated with much more widespread burning in the zone of mesic forests than in the 1996 fire season, during which fires occurred mainly in xeric woodland, shrubland, and steppe ecosystems. The susceptibility of mesic forests to widespread burning only during years of exceptional drought is also documented by the tree-ring record of fires (Kitzberger and Veblen, Chapter 10, this volume). The 1998–99 year of drought and widespread burning also coinicided with a La Niña event, which typically promotes dry springs in northern Patagonia (Kitzberger and Veblen, Chapter 10, this volume). Although such intense drought is essential for fire spread in the mesic forest zone, the major forest fires of 1999 were set by humans rather than lightning (Delegación Técnica Regional Patagonia 1999). Preliminary data on fire history in this zone indicate that much of the same area that burned in 1999 also burned in 1908 during a period of severe drought combined with abundant intentional burning by early colonists (Veblen et al., unpublished data).

Nothofagus Dieback

In addition to changes in fuels related to increased or decreased anthropogenic burning, natural stand dieback may contribute to increased fire hazard. *Nothofagus pumilio* and *N. antarctica* stands are characterized by an abundance of trees with partially dead crowns. Although the etiologies of these dieback phenomena are uncertain, contributory factors are believed to include (Veblen et al. 1996) (1) sites of marginal moisture availability, (2) cohort senescence following disturbance-induced regeneration, (3) partial recovery following defoliation by insect outbreaks, and (4) heavy loads of the semiparasitic *Misodendrum* mistletoe. Several or all of these factors may interact synergistically with interannual climatic variability to promote in *Nothofagus* crown dieback. For example, warm winters and dry springs are often associated with outbreaks of defoliating insects from which *Nothofagus* often only partially recover (Veblen et al., unpublished data). Intense drought probably also contributes directly to the dieback in *N. pumilio* and *N. antarctica*.

Drought appears to be a major cause of mortality and dieback in *Nothofagus dombeyi* in the mesic forest zone. For example, following the severe 1998 drought, there has been widespread mortality of *N. dombeyi* in northern Patagonia, especially in southern Nahuel Huapi National Park. Entire stands of large (>1 m diameter) *N. dombeyi*, principally at sites located near the moisture limits of this species, were dead by the summer of 1999–2000. In many cases, however, the tree survived while a few large branches or stem bifurcates died, thus creating a general appearance of stand-level partial dieback. We speculate that earlier droughts contributed to the dieback *N. dombeyi* forests that was widespread prior to the 1998 drought.

Although photographs indicate that dieback in *Nothofagus pumilio* and *N. antarctica* was already widespread by 1900 (Willis 1914; Rothkugel 1916), European impacts on fire regimes may have increased the extent of dieback in two ways. The extensive burning associated with European settlement has created enormous areas of similarly aged cohorts that may dieback synchronously. Furthermore, in the case of *N. antarctica*, fire exclusion has substantially increased the percentage of its population that is in a senescent state. *N. antarctica* becomes markedly senescent at ages of 80 to 100 years and rarely survives beyond ages of around 150 years. However, burning rejuvenates it by promoting vigorous basal sprouting, and younger trees show less incidence of dieback (Veblen and Lorenz 1988). Reduction of fire in *N. antarctica* woodlands may also favor the buildup of large epiphytic loads of the flammable *Usnea* lichen. At the same time, infection by the *Misodendrum* mistletoe probably increases in the absence of fire. Thus flammability and potential fire intensity have probably increased due to the reduction in fire frequency in *N. antarctica* woodlands and shrublands.

Introduced Animals and Plants

Introduced Large Herbivores

Introduced livestock and cervids have greatly affected the vegetation of the northern Patagonian landscape (Martín, Mermoz, and Gallopin 1985; Veblen et al. 1989, 1992; De Pietri 1992b; Relva and Veblen 1998). They have impeded postfire recovery at many sites, and they may have had a significant impact on fuel quality and quantity. Livestock numbers in the region peaked during the 1930s (Ericksen 1971) and locally probably impeded the afforestation of some grassland and shrubland areas (Tortorelli 1947). Although the major tree species are relatively resistant to browsing once they reach sapling stages, exceptionally heavy cattle pressure during early postfire recovery can locally impede tree regeneration and instead result in herbaceous turfs (with abundant exotic species) or shrublands of spiny shrubs and dwarfed trees (Veblen et al. 1992; De Pietri 1992b; Relva and Veblen 1998).

Large livestock populations since around 1890 are believed to have reduced plant cover in the steppe and probably also in open *Austrocedrus* woodlands. For example, overgrazing in some areas of steppe is believed to have reduced plant cover from initial values of 60% to less than 40%, which in turn has probably reduced the spread of fires (D. Bran, personal communication, 1998). In some plant communities, however, livestock browsing may have increased flammability. Heavy pressure from introduced herbivores has shifted dominance toward less palatable species in shrublands (Veblen et al. 1992; Relva and Veblen 1998), and the morphological (rapid resprouting) and chemical features (secondary compounds) associated with defense against herbivory often increase flammability (Bond and Wilgen 1996).

In some northern Patagonian shrublands dominated by the relatively nonflammable but highly palatable *Maytenus boaria*, heavy livestock pressure has shifted the composition toward less palatable but more flammable species such as *Discaria articulata*, *Diostea juncea*, and *Lomatia hirsuta*. Conversely, monitoring of livestock exclosures indicates that there is a shift back toward dominance by the palatable *Maytenus boaria* that has a high moisture content, and other less flammable shrubs such as *Berberis buxifolia* and *Ribes magellanicum* in the absence of cattle (Raffaele and Veblen, 2001). Outside the exclosures the highly flammable *Discaria articulata* remained the dominant shrub in the community. Other shrublands that supported historically high levels of grazing are dominated by the unpalatable and flammable *Diostea juncea* and *Lomatia hirsuta*, both of which are characterized by high foliar lignin content which results in slow decomposition and abundant litter accumulation.

In forests with understories dominated by *Chusquea culeou*, livestock greatly reduce the size and cover of the bamboos so that fuel loads and heights are markedly less. For example, heavy impact of livestock in *Nothofagus dombeyi* and *Austrocedrus* forests creates nearly bare understories where the lack of understory fuels is striking (Veblen et al. 1992; Relva and Veblen 1998). Although much research remains to be done on fuel patterns and their modification by herbivores in northern Patagonia, the overall impact of livestock appears to have been a generalized decrease in fine-fuel quantity in grasslands and forest understories and a possible shift toward more flammable species compositions in some shrublands.

Invasive Plant Species

There are more than 300 exotic vascular plant species that have naturalized in northern Patagonia (Rapoport and Brión 1991). Exotic species are particularly common in habitats severely disturbed by livestock and logging, which have significantly altered natural fuel patterns and/or the capacity of the native vegetation to respond to fire (Veblen et al. 1992b; Gobbi, Puntieri, and Calvelo 1995; Relva and Veblen 1998). *Rumex acetosella* is common in recently burned areas and propagates both vegetatively and from a persistent seed bank (Gobbi, Puntieri, and Calvelo 1995). The European broom (*Sarothamnus scoparius*) is common along roadsides and is highly flammable. Similarly Douglas fir (*Pseudotsuga menziesii*) has naturalized from timber and ornamental plantings and is a common invader along trails and abandoned logging roads in the mesic *Nothofagus dombeyi* forest. Thus Douglas fir is encroaching into high-light sites that otherwise would be occupied by the shade-intolerant *N. dombeyi*, and it is providing more flammable fuels as well as fuel ladders into the tree canopy.

Probably the most conspicuous invading shrub in northern Patagonia is the European rose (*Rosa rubiginosa*) which is especially common in the steppe ecotone but also occurs in anthropogenically disturbed mesic forest habitats. Although it is not particularly flammable, it may be important as a keystone species that alters rates of postfire recovery. *R. rubiginosa* appears to act as a nurse plant for native woody species that are less browse-resistant but are capable of eventually replacing the invader (De Pietri 1992a).

Plantations of Exotic Tree Species

In the 1930s to 1950s, small areas of the national parks were planted to exotic conifers such as *Sequoiadendron giganteum*, *Sequoia sempervirens*, *Picea* spp., and *Pinus* spp. (Dimitri 1972). In recent decades, planting of exotic trees has been limited to the national reserve eastern parts of the parks, where large areas of *Pinus ponderosa* have been planted since about 1980. By 1996 there were about 4000 and 3500 ha of exotic conifer plantation (90% *Pinus ponderosa* and 10% *Pseudotsuga menziesii*) in Nahuel Huapi and Lanín national parks, respectively, and a much larger area has been planted to *Pinus ponderosa* on properties just outside the national parks. The highly flammable pines have been planted in areas that were formerly open woodland or steppe where lack of fuel continuity was an important limitation to fire spread. Today, however, large areas of these exotic conifers have created the potential for extensive crown fires in habitats formerly characterized only by surface fires. Poorly managed plantations that are unthinned and lack fire breaks have further increased the potential for rapid spread of crown fires (e.g., the 1996 Challhuaco fire).

Conclusion and Management Considerations

Human activities and climatic variation are fundamental influences on fire regimes and landscape patterns in northern Patagonia. Although interannual climatic variation has a controlling influence in creating fuel conditions for the spread of fires in northern Patagonia, human activities also have had significant impacts on fire regimes and landscape patterns in this region. Prior to the late 1800s, fires set by Native Americans were important throughout the woodland/steppe hunting grounds and were important locally along trans-Andean travel routes in the mesic forest district. The impacts of increased burning in the mesic forest zone by European settlers in the 1890s to 1910s remains conspicuous in the extensive even-aged *Nothofagus* stands in the modern landscape. The modern fire exclusion period has been a time of transition from seral shrublands to forest and expansion of *Austrocedrus* trees into grasslands. Interannual and decadal-scale climatic variation has been an important preconditioning agent for the spread of fires (Kitzberger, Veblen, and Villalba 1997; Veblen et al. 1999) and for postfire vegetation responses (Villalba and Veblen 1997a; Kitzberger and Veblen 1999).

Major human-caused changes in fire regimes are also important to the spread of fire in northern Patagonia landscapes. Potential fire spread in submesic areas has increased as trees regenerate following episodes of major anthropogenic burning in the early 1900s and early 1940s, and as formerly disjunct patches of forest coalesce. Similarly, near the steppe ecotone, formerly open woodlands of *Austrocedrus* have been replaced by relatively dense stands during nearly 80 years of reduced fire frequency. Thus, even if fewer human-set fires maintain a relatively low fire frequency, the increased connectivity of fire-susceptible vegetation types probably has created a greater potential for high rates of fire spread.

One of the most obvious implications of the documented increase in fuel continuity in the woodland/steppe ecotone is the need for public education of the increased potential for stand-replacing fires. Many areas in this habitat are experiencing rapid residential growth that is exposing humans and property to high fire hazards. A major research need for effective planning of human activities in relation to fire hazard is the development and implementation of a fuels classification and mapping program in northern Patagonia. Experimentation with different mitigation strategies also is needed. For example, experimental prescribed burning should be examined as a technique for reducing fuels and fire hazard in areas of steppe and xeric woodland that have experienced decreases in grazing pressure.

Acknowledgments. This review is based on research funded by the National Science Foundation of the United States, the Fundación Nacional de Ciencia y Tecnología of Argentina, the National Geographic Society, the Universidad Nacional del Comahue, and the Council for Research and Creative Work of the University of Colorado. For critically commenting on the manuscript, we thank L. Daniels. For sharing insights about fire ecology and fire management and for facilitating our research, we thank Mónica Mermoz, Juan Salguero, and Carlos Martín of Argentine National Parks.

References

Alvarez, G. 1984. *Donde Estuvo el Paraíso del Tronador a Copahue*. Siringa Libros, Neuquén, Argentina.

Alauzis, M.V. 1999. Cambios en la fertilidad química, físico-químico y biológica del suelo en parches incendiados de un bosque de *Nothofagus pumilio*. Thesis, Universidad Nacional del Comahue, Bariloche, Argentina.

Attiwell, P.M. 1994. The disturbance of forest ecosystems: The ecological basis for conservative management. *For. Ecol. Manag.* 63:247–309.

Barros, V., Cordón, V., Moyano, C., Méndez, R., Forquera, J., and Pizzio, O. 1983. Cartas de precipitación de la zona oeste de las provincias de Rio Negro y Neuquén. Report to the Facultad de Ciencias Agrarias, Universidad Nacional del Comahue, Cinco Saltos, Neuquén, Argentina.

Bengoa, J. 1985. *Historia del Pueblo Mapuche (Siglo XIX y XX)*. Santiago, Chile: Ediciones Sur.

Bond, W.J., and van Wilgen, B.W. 1996. *Fire and Plants*. London: Chapman and Hall.

Bran, D. 1996. El fuego en las estepas de la Patagonia Norte. *Patagonia Silvestre* 3:7–8.

Burns, B.R. 1993. Fire-induced dynamics of *Araucaria araucana—Nothofagus antarctica* forest in the southern Andes. *J. Biogeogr.* 20:669–685.

Burns, B.R. 1991. Regeneration dynamics of *Araucaria araucana*. Ph.D. thesis. University of Colorado, Boulder.

Cox, G. 1863. Viajes a las regiones septentrionales de Patagonia 1862–1863. *An. Univ. Chile* 23:3–239, 437–509.

Crivella, M.E.A., and Silveira, M.J. 1983. Radiocarbon chronology of a tephra layer in Rio Traful Valley, Province of Neuquén, Argentina. *Quat. S. Am. Antarc. Penin.* 1:135–150.

Delegación Técnica Regional Patagonia. 1996a. Incendios Rincón Grande y Villa Lanquín, Reserva Nacional Nahuel Huapi. Unpublished Rep. Delegación Técnica Regional Patagonia, Parques Nacionales, Bariloche, Argentina.

Delegación Técnica Regional Patagonia. 1996b. Incendio Forestal Valle del Challhuaco, Reserva Nacional Nahuel Huapi. Informe de consecuencias ecológicas. Unpublished Rep. Delegación Técnica Regional Patagonia, Parques Nacionales, Bariloche, Argentina.

Delegación Técnica Regional Patagonia. 1999. Informe Sobre Las Consecuencias Ecológicas de los Incendios Forestales. Unpublished Rep. Delegación Técnica Regional Patagonia, Parques Nacionales, Bariloche, Argentina.

De Pietri, D.E. 1992a. Alien shrubs in a national park: Can they help in the recovery of natural degraded forest? *Biol. Conserv.* 62:127–130.

De Pietri, D.E. 1992b. The search for ecological indicators: Is it possible to biomonitor forest system degradation caused by cattle ranching activities in Argentina? *Vegetatio* 101:109–121.

Dezzotti, A. 1996. A*ustrocedrus chilensis* and *Nothofagus dombeyi* stand development during secondary succession, in northwestern Patagonia, Argentina. *For. Ecol. Manag.* 89:125–137.

Dimitri, M.J. 1972. *La Región de los Bosques Andino-Patagónicos*. Colección Científica. 10. Buenos Aires: Instituto Nacional de Tecnología Agropecuario.

Eriksen, W. 1971. Betriebsformen und Probleme der Viehwirtschaft am Rande der Argentinischen Südkordillere. *Zeit. Ausländ. Landwirts.* 10:24–27.

Eriksen, W. 1975. Disruptions in ecosystems of the steppe and forest regions of Patagonia by climate and man. *Appl. Sci. Dev.* 6:127–142.

Eskuche, U. 1968. Fisionomía y sociología de los bosques de *Nothofagus dombeyi* en la región de Nahuel Huapi. *Vegetatio* 16:192–204.

Fonck, F. 1896. *Libro de Los Diarios de Fray Francisco Menéndez*. Valparaiso, Chile: Niemeyer.

Furlong, G. 1964. *Nícolas Mascardi, S.J. y Su Carta Relación (1670)*. Buenos Aires: Ediciones Theoria.

Ghermandi, L. 1992. Caracterización del banco de semillas de una estepa del Noroeste de Patagonia. *Ecologia Austral* 2:39–46.

Glenn-Lewin, D.C., Peet, R.K., and Veblen, T.T., eds. 1992. *Plant Succession: Theory and Prediction*. London: Chapman and Hall.

Gobbi, M. 1994. Regeneración de la vegetación en incendios recientes de bosques de "Cipres de la Cordillera" (*Austrocedrus chilensis*) en el area del Parque Nacional Nahuel Huapi. *Medio Ambiente* 12:9–15.

Gobbi, M., and Sancholuz, L. 1992. Regeneración post-incendio del ciprés de la cordillera (*Austrocedrus chilensis*) en los primeros años. *Bosque* 13:25–32.

Gobbi, M., Puntieri, J., and Calvelo, S. 1995. Post-fire recovery and invasion by alien plant species in a South American woodland-steppe ecotone. In *Plant Invasions: General Aspects and Special Problems*, eds. P. Pyšek, K. Prach, M. Rejmánekm, and Wade, M. pp. 105–115. Amsterdam, The Netherlands: SPB Publishing.

Gonzalez Cangas, Y. 1998. Memoria historica y saber cotidiano: Validación del conocimiento en el florecimiento de la *Chusquea quila* en el sur de Chile (X Región). M.S. thesis. Universidad de la Frontera, Temuco, Chile.

Gonzalez, M.E., and Donoso, C. 1999. Producción de semillas y hojarasca en la bambúcea *Chusquea quila* (Kunth) (Poaceae: Bambusoideae), posterior a su floración sincrónica en la zona centro-sur de Chile. *Rev. Chil. Hist. Nat.* 72:169–180.

Harris, L.D. 1984. *The Fragmented Forest: Island Biogeography and the Presevation of Biotic Diversity*. Chicago: University of Chicago Press.

Heinemann, K., Kitzberger, T., and Veblen, T.T. 2000. Influences of gap microheterogeneity on the regeneration of *Nothofagus pumilio* in a xeric old-growth forest of northwestern Patagonia, Argentina. *Can. J. For. Res.* 30:25–31.

Heusser, C.J. 1987. Fire history of Fuego-Patagonia. *Quat. S. Am. Antarc. Penin.* 5:93–109.

Heusser, C.J. 1994. Paleoindians and fire during the late Quaternary in southern South America. *Rev. Chil. Hist. Nat.* 67:435–442.

Heusser, C.J., Rabassa, J., Brandani, A., and Stuckenrath, R. 1988. Late-Holocene vegetation of the Andean *Araucaria* region, Province of Neuquén, Argentina. *Mount. Res. Dev.* 8:53–63.

Hosseus, C.K. 1915. Las cañas de bambú en las cordilleras del Sud. *Bol. Minis. Agric.* (Buenos Aires) 19:195–208.

Kalela, E.K. 1941. Über die Holzarten und die durch die klimatischen Verhältnisse verursachten Holzartenwechsel in den Wäldern Ostpatagoniens. *Ann. Acad. Sientar. Fennicae* (ser. A) 2:5–151.

Kitzberger, T. 1994. Fire regime variation along a northern Patagonian forest-steppe gradient: stand and landscape response. Ph.D. thesis. University of Colorado, Boulder.

Kitzberger, T., and Veblen, T.T. 1997. Influences of humans and ENSO on fire history of *Austrocedrus chilensis* woodlands in northern Patagonia, Argentina. *Ecoscience* 4: 508–520.

Kitzberger, T., and Veblen, T.T. 1999. Fire-induced changes in northern Patagonian landscapes. *Landscape Ecol.* 14:1–15.

Kitzberger, T., Steinaker, D.F., and Veblen, T.T. 2000. Establishment of *Austrocedrus chilensis* in Patagonian forest-steppe ecotones: Facilitation and climatic variability. *Ecology* 81:1914–1924.

Kitzberger, T., Veblen, T.T., and Villalba, R. 1997. Climatic influences on fire regimes along a rain forest-to-xeric woodland gradient in northern Patagonia, Argentina. *J. Biogeogr.* 24:35–47.

Markgraf, V. 1983. Late and postglacial vegetational and paleoclimatic changes in subantarctic, temperate, and arid environments in Argentina. *Palynology* 7:43–70.

Markgraf, V., and Anderson, L. 1994. Fire history of Patagonia: Climate *versus* human cause. *Rev. Instit. Geográf. Sao Paulo* 15:33–47.

Martín, C.D., Mermoz, M., and Gallopín, G. 1985. Impacto de la ganadería en la cuenca del Rio Manso Superior. Report, Administración de Parques Nacionales, Buenos Aires.

Martonne, E. de. 1926. Une nouvelle fonction climatologique: L'indice d'aridité. *Météorologie* 2:449–458.

McQueen, D.R. 1976. The ecology of *Nothofagus* and associated vegetation in South America. *Tuatara* 22:38–68.

Molina, R., and M. Correa. 1996. *Territorios y Comunidades Pehuenches del Alto Bio Bio.* Santiago, Chile: Corporación Nacional de Desarrollo Indígena.

Moreno, F.P. 1897. Reconocimiento de la región andina de la República Argentina. Apuntes preliminares sobre una excursión a los Territorios de Neuquén, Rio Negro, Chubut y Santa Cruz. *Rev. Museo Plata* 8:1–180.

Musters, G.C. 1871. *At Home with Patagonians: A Year's Wandering over Untrodden Ground from the Straits of Magellan to the Rio Nergo.* London: Murray.

Pearson, A.K., Pearson, O.P., and Gomez, I.A. 1994. Biology of the bamboo *Chusquea culeou* (Poacaeae: Bambusoideae) in southern Argentina. *Vegetatio* 111:93–126.

Raffaele, E., and Veblen, T.T. 1998. Facilitation by nurse shrubs on resprouting behavior in a post-fire shrubland in northern Patagonia, Argentina. *J. Veg. Sci.* 9:693–698.

Raffaele, E., and Veblen, T.T. 2001. Effects of cattle grazing on early postfire regeneration of matorral in northwest Patagonia, Argentina. *Nat. Areas J.* 21:243–249.

Rapoport, E.H., and Brión, C. 1991. Malezas exóticas y plantas escapadas de cultivo en el noroeste patagónico: segunda aproximación. *Cuad. Alternat.* (Bariloche) 1:1–19.

Relva, M.A., and Veblen, T.T. 1998. Impacts of introduced large herbivores on *Austrocedrus chilensis* forests in northern Patagonia, Argentina. *For. Ecol. Manag.* 108:27–40.

Rodríguez, D., Sourrouille, A., Gallopín, G.C., and Montaña, C. 1978. Estudio ecológico integrado de la cuenca del Rio Manso Superior (Rio Negro, Argentina). II. Tipos de vegetación. *An. Parq. Nacion.* (Argentina) 14:231–248.

Rogers, P. 1996. Disturbance ecology and forest management: A review of the literature. USD Agriculture Forest Service, Gen. Tech. Rep. INT-GTR-336.

Rothkugel, M. 1916. *Los Bosques Patagónicos*. Buenos Aires: Ministerio de Agricultura.

Salguero, J. 1998. Subprograma: Ecología del Fuego. Unpublished rep. Delegación Técnica Regional Patagonia, Parques Nacionales, Bariloche, Argentina.

Seibert, P. 1982. Carta de vegetación de la región de El Bolsón, Rio Negro y su aplicación a la planificación del uso de la tierra. *Doc. Phytosociol.* 2:1–120.

Singer, R. 1971. Forest mycology and forest communities in South America. II. Mycorrhiza sociology and fungus succession in the *Nothofagus dombeyi-Austrocedrus chilensis* woods of Patagonia. USDA Miscellaneous Pub. 1189.

Smith, E.R. 1855. *The Araucanians*. New York: Harper.

Steffen, H. 1909. *Viajes de Exploracion: Estudio en la Patagonia Occidental*. Santiago, Chile: Imprenta Cervantes.

Tortorelli, L.A. 1947. *Los Incendios de Bosques en la Argentina*. Buenos Aires: Ministerio de Agricultura.

Tortorelli, L.A. 1956. *Maderas y Bosques Argentinos*. Buenos Aires, Argentina: Editorial Acme.

Veblen, T.T. 1982. Growth patterns of *Chusquea* bamboos in the understory of Chilean *Nothofagus* forests and their influences in forest dynamics. *Bull. Torrey Botan. Club* 109:474–487.

Veblen, T.T. 1989. *Nothofagus* regeneration in treefall gaps in northern Patagonia. *Can. J. For. Res.* 19:365–371.

Veblen, T.T., Kitzberger, T., and Lara, A. 1992. Disturbance and forest dynamics along a transect from Andean rain forest to Patagonian shrubland. *J. Veg. Sci.* 3:507–520.

Veblen, T.T., and Lorenz, D.C. 1987. Post-fire stand development of *Austrocedrus-Nothofagus* forests in Patagonia. *Vegetatio* 73:113–126.

Veblen, T.T., and Lorenz, D.C. 1988. Recent vegetation changes along the forest/steppe ecotone in northern Patagonia. *Ann. Assoc. Am. Geogr.* 78:93–111.

Veblen, T.T., Burns, B.R., Kitzberger, T., Lara, A., and Villalba, R. 1995. The ecology of the conifers of southern South America. In *Ecology of the Southern Conifers*, ed. N.J. Enright and R.S. Hill, pp. 120–155. Melbourne: Melbourne University Press.

Veblen, T.T., Donoso, C., Kitzberger, T., and Rebertus, A.J. 1996. Ecology of southern Chilean and Argentinean *Nothofagus* forests. In *The Ecology and Biogeography of* Nothofagus *Forests*, eds. T.T. Veblen, R.S. Hill, and J. Read, pp. 293–353. New Haven: Yale University Press.

Veblen, T.T., Kitzberger, T., Villalba, R., and Donnegan, J. 1999. Fire history in northern Patagonia: The roles of humans and climatic variation. *Ecol. Monogr.* 69:47–67.

Veblen, T.T., Mermoz, M., Martín, C., and Ramilo, E. 1989. Effects of exotic deer on forest regeneration and composition in northern Patagonia. *J. Appl. Ecol.* 26:711–724.

Veblen, T.T., Mermoz, M., Martin, C., and Kitzberger, T. 1992. Ecological impacts of introduced animals in Nahuel Huapi National Park, Argentina. *Conserv. Biol.* 6:71–83.

Villalba, R. 1990. Climatic fluctuations in northern Patagonia during the last 1000 years as inferred from tree-ring records. *Quat. Res.* 34:346–360.

Villalba, R., and Veblen, T.T. 1997a. Regional patterns of tree population age structures in northern Patagonia: Climatic and disturbance influences. *J. Ecol.* 85:113–124.

Villalba, R., and Veblen, T.T. 1997b. Spatial and temporal variation in tree growth along the forest-steppe ecotone in northern Patagonia. *Can. J. For. Res.* 27:580–597.

Villalba, R., Boninsengna, J.A., Veblen, T.T., Schmelter, A., and Rubulis, S. 1997. Recent trends in tree-ring records from high elevation sites in the Andes of northern Patagonia. *Clim. Change* 36:225–254.

Villalba, R., Cook, E.R., Jacoby, G.C., D'Arrigo, R.D., Veblen, T.T., and Jones, P.D. 1998. Tree-ring based reconstructions of precipitation in Patagonia since A.D. 1600. *Holocene* 8:677–692.

Willis, B. 1914. *El Norte de la Patagonia*. Buenos Aires: Dirección de Parques Nacionales.

Wu, J., and Loucks, O.L. 1995. From balance of nature to hierarchical patch dynamics: A paradigm shift in ecology. *Quart. J. Biol.* 70:439–466.

10. Influences of Climate on Fire in Northern Patagonia, Argentina

Thomas Kitzberger and Thomas T. Veblen

One of the major challenges in ecology is to identify and quantify the ecological mechanisms that control ecosystem responses to climatic variation. Such studies are required to understand how present landscape patterns have been influenced by past climatic variation and to predict how landscapes may change in response to future climatic variation. Climate-induced vegetation changes result from both direct effects of climatic variation on individual species' performances (Körner 1996; Lloyd and Gramulich 1997; Pederson 1998) and indirect effects mediated by climatically altered disturbance regimes (Gardner et al. 1996; Larsen and MacDonald 1998). Climate-model simulations of vegetation under a 2 × CO_2 scenario suggest that increased disturbance by drought, fire, and wind storms will significantly accelerate rates of forest change compared to the rates that would result from climatic change alone (Overpeck, Rind, and Goldberg 1990; Alaback and McClellan 1993; Franklin et al. 1991; Price and Rind 1994).

Climatically altered fire regimes, in particular, are expected to be important proximate causes of source of climatically driven vegetation change because most of the factors that control fire regimes are directly or indirectly controlled by climate (Chandler et al. 1983). Understanding and separating influences of long-term versus high-frequency climatic variability is critical in predicting the effects altered climate on vegetation change (Baker 1990; Baker et al. 1991; Bergeron and Archambault 1993; Johnson and Larsen 1991; Malanson and Westman 1989; Sirois and Payette 1991; Gardner et al. 1996).

By varying the spatial scale of interest, it is possible to distinguish responses of fire to environmental variations occurring on hemispheric, continental, regional, landscape, or local scales. At the broadest scale, mean positions of atmospheric circulation features, such as subtropical jets and semipermanent subtropical anticyclones, influence temperature, precipitation, and lightning patterns that control the timing and nature of the fire season in a particular region. Anomalies of large-scale climatic features driven by global phenomena such as the El Niño–Southern Oscillation (ENSO) can produce climatic anomalies that synchronize fire regimes over regional to global scales (Swetnam and Betancourt 1990; 1992; 1998; Johnson and Wowchuck 1993; Kitzberger and Veblen 1997; Veblen et al. 1999; Kitzberger, Swetnam, and Veblen 2001). At finer spatial scales, fire regimes may be more strongly influenced by local land-use patterns (fire suppression, logging) and less controlled by regional synoptic climatic patterns (Swetnam 1993; Kitzberger and Veblen 1997; Veblen et al. 1999).

In this chapter we provide an overview of the current knowledge of climatic influences on fire regimes in northern Patagonia along the gradient from temperate rain forest to steppe (see Veblen et al., Chapter 9, this volume, for a description of the vegetation). We emphasize seasonal, annual, and multi-annual variability in regional climatic patterns and atmospheric circulation features. The steep west-to-east rainfall gradient from the humid Andes to the xeric steppe offers a unique opportunity for analysis of how different vegetation types respond to the same pattern of regional climatic variation. Recent development of networks of tree-ring records of climatic variation (Villalba 1995; Villalba and Veblen 1997a) and of fire history (Kitzberger 1994; Kitzberger and Veblen 1997; Veblen et al. 1999) from ca. 39° to 43°S latitude allows analysis of within region spatial variability of climate and fire history and linkages to large-scale atmospheric circulation features.

Regional Climate and Synoptic Influences

The climate of the mid-latitudes of southern South America is most proximately controlled by the mid-latitude westerlies with their cyclonic storms, the southeast Pacific subtropical high-pressure cell, and the topographic barrier of the Andes (Miller 1976), but also shows significant relationships to higher-latitude circulation patterns and southeastward movement of maritime and continental subtropical air masses (Taljaard 1972; Villalba et al. 1998). The Andean Cordillera reaches elevations of more than 2000 m and is an effective barrier to moisture-laden storms that flow westerly from the Pacific into the continent at ca. 35° and higher latitudes. Most of the precipitation is discharged in the coastal mountains of Chile and on western slopes of the Andes. In the rain shadow of the Andes, precipitation declines dramatically from west to east. For example, at ca. 41°S mean annual precipitation declines along nearly a 100-km west–east transect from about 4000–6000 mm in the Chilean Andes to about 200–300 mm in the Patagonian plains (Barros et al. 1983). The Andes are also important in funneling humid subtropical air masses southward from Brazil and sometimes bringing convective storms to northern Patagonia (Taljaard 1972).

The southeast Pacific anticyclone is most intensively developed between 27° and 38°S off the coast of Chile, and seasonally shifts poleward about 4° to 7° during the summer (Taljaard 1972). In northern Patagonia autumns and winters are wet when westerly storm tracks are at their most equatorward position. Relatively dry springs and summers result from the poleward shift of the anticyclone which effectively blocks the westerly flow of moisture into the continent (Schwerdtfeger 1976).

Interannual variability of rainfall over southwestern South America is closely controlled by variations in the latitudinal position and intensity of the southeast Pacific anticyclone. A stronger and more poleward located cell produces negative precipitation anomalies between ca. 35° and 45°S (Pittock 1980; Villalba 1990a). In turn, the strength and latitudinal position of the subtropical anticylcone is closely related to anomalies in the Pacific tropical convection associated with the ENSO. During the positive (La Niña) phase of the SO, the southeast Pacific high tends to be intensified and displaced poleward during the austral winter (Aceituno 1988). Thus during La Niña events negative rainfall anomalies occur over south-central Chile during the winter and spring (May–November) (Rutllant and Fuenzalida 1991; Aceituno 1988). In northern Patagonia during the La Niña phase, winter–spring precipitation is below average and temperature is above average (Aceituno 1988; Fig. 10.1). During the El Niño phase, summer precipitation is below average and temperature is above average. Rainfall in northern Patagonia also is influenced by high-latitude circulation features. Blocking high-pressure events at ca. 60°S over the Antarctic Peninsula sector of the Southern Ocean drive westerly storms northward into South America, resulting in positive precipitation anomalies in northern Patagonia (Villalba et al. 1998).

In northern Patagonia positive temperature anomalies also result from incursions of subtropical air masses from northern Argentina and Brazil. When the

Figure 10.1. Correlations of spring (October; *upper*) and summer (January; *lower*) precipitation (*left*) and temperature (*right*) with the Southern Oscillation Index (SOI; 1882–1996). SOI is the standardized sea level pressure difference between Tahiti and Darwin, Australia (Ropelewski and Jones 1987). Isolines indicate points of equal correlation based on a network of 12 weather stations located between ca. 36° and 46°S latitude and 68° and 76°W longitude in south-central Chile and northern Patagonia. Correlations are significant ($P < 0.05$) when >0.20 or <–0.20 (*hatched isolines*). Dark areas indicate negative correlations or high/low values of the variable related to El Niño/La Niña phase, respectively. Conversely, light areas indicate high/low values of the variable related to La Niña/El Niño phase, respectively. Weather stations and lengths of their records are (precipitation/temperature) Concepción (1876–1968/1951–1990). Temuco (1951–1990). Valdivia (1853–1973/1941–1990), Pt. Montt (1862–1993/1919–1990), Is. Guafo (1908–1996/1910–1986), Pt. Aysén (1931–1990/1953–1990), Esquel (1896–1993/1901–1990), Sarmiento (1904–1961), Bariloche (1905–1990/1914-1990), Collún Co (1912–1989), Neuquén (1900–1993/1957–1993), and Chosmalal (1904–1961/1931–1960).

SOI – SPRING RAINFALL

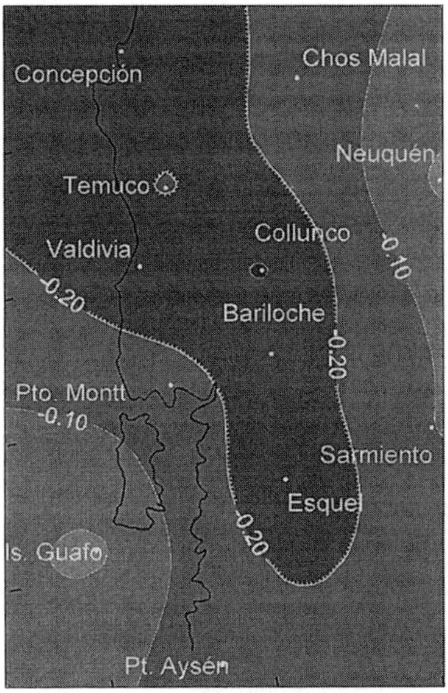

SOI – SPRING TEMPERATURE

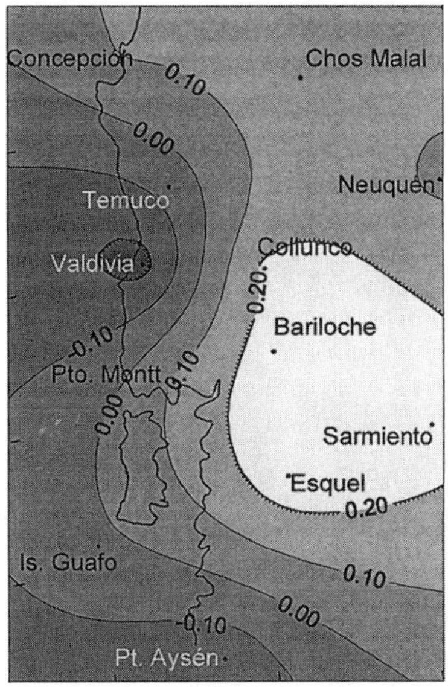

SOI – SUMMER RAINFALL

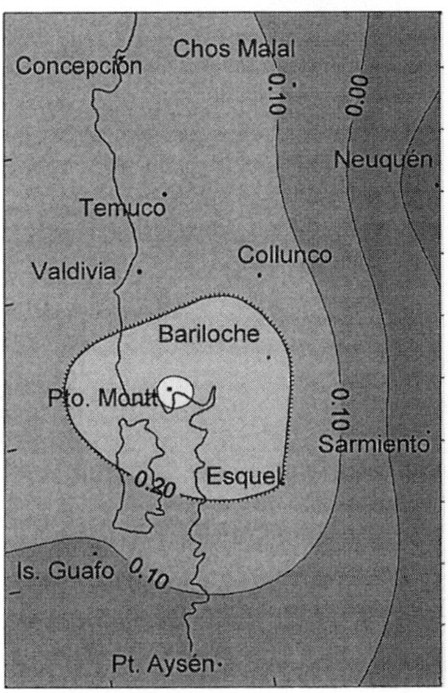

SOI – SUMMER TEMPERATURE

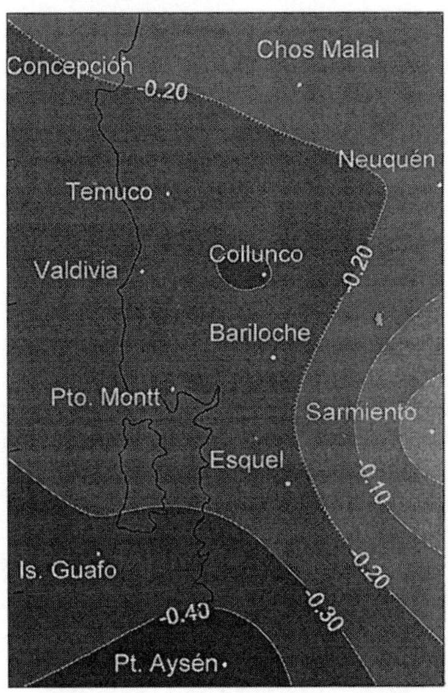

southeastern Pacific anticyclone is intensified and more southerly located, these subtropical air masses are more likely to reach mid-latitudes (Taljaard 1972). A strong Atlantic influence on temperature variability in northern Patagonia is also evident from high correlations found between tree-ring reconstructed summer temperature and temperature records from stations located eastward across Patagonia to the Atlantic coast at 38° and 50°S (Villalba 1990b). Correlations to the west fall sharply along the continental divide suggesting that the Atlantic influence, either related to the South Atlantic Polar front or to a southern incursion of subtropical air masses, is restricted by the Andes to the eastern sector of southern South America.

Fire Occurrence, Extent, and Behavior

Fire Seasonality

A documentary record of fire beginning in 1938 is available from the four large National Parks covering much of the Andes and foothills from ca. 37° to 43°S (see map in Veblen et al., Chapter 9, this volume). Based on this documentary record, the fire season in northern Patagonia coincides with the period of greatest water deficit, which extends from October through April. Larger fires are concentrated in the summer months of January through March, and only one-fourth of fires occur in spring and early summer (October–December). Seasonality of fire based on *Austrocedrus chilensis* fire-scar samples in which the intra-ring position of the fire-scar tip could be determined indicates that for the period 1573 to 1944, 41% of fires (n = 23) occurred during the dormant season and the season of earlywood formation (early spring); 59% (n = 33) occurred during middle earlywood, late earlywood, and latewood formation (summer).

Fire Extent and Climatic Control Along the Precipitation Gradient

In the mesic *Nothofagus*-dominated forests, years of high fire activity occurred at an average rate of two fire years per decade from 1940 to 1996, and years of low fire activity occurred at rate of 5.6 years per decade over this 57-year period (Administración de Parques Nacionales, unpublished data). In the dry woodlands and grasslands further east, the most frequent type of fire year was one of intermediate fire activity (10–1000 ha burned), which occurred at a rate of 4.7 years per decade; in this vegetation zone, years of high fire activity occurred at a rate of only 0.9 years per decade. In comparison, the strongly bimodal distribution of fire activity in mesic *Nothofagus* forests may reflect climatic thresholds that dramatically increase the flammability of these forests under relatively infrequent weather conditions. In fact, the relationship of area burned in the wet forest zone increases exponentially with increasing spring–early summer water deficit (Fig. 10.2a and b). Long and pronounced drought periods including months that are normally moist (e.g., August–October) appear to be important in desiccating the coarse fuels characteristic of wet forests (Kitzberger, Veblen, and Villalba 1997). In the mesic forest zone, soil moisture storage is high in the deep porous

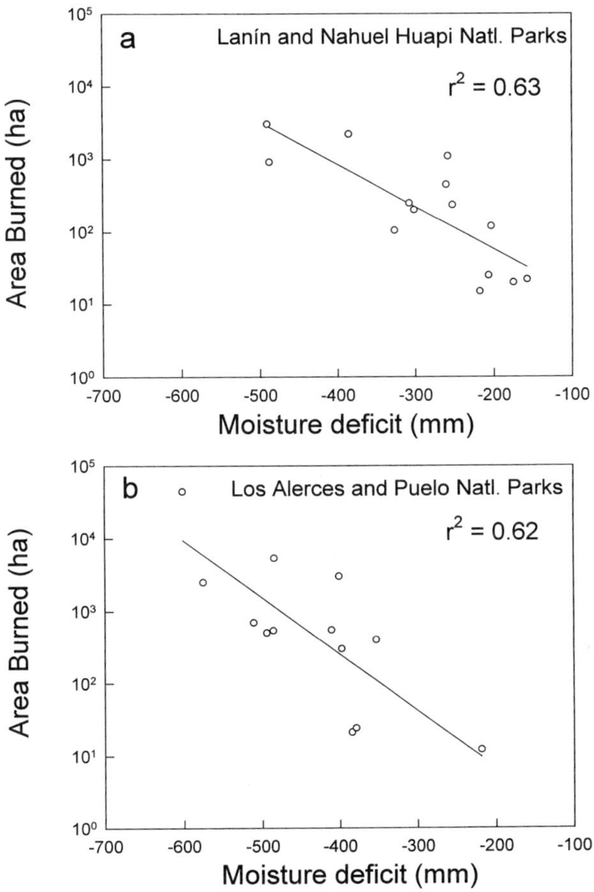

Figure 10.2. Annual area (log transformed) burned of *Nothofagus*-dominated mesic forests in relation to spring and early summer integrated water deficit for (a) Lanin and Nahuel Huapi national parks (ca. 39–41°30′S) and (b) Los Alerces and Lago Puelo national parks (ca. 42–42°30′S). Climate data are from the Bariloche Airport weather station and Esquel, and are for the months of September through December and October through January, respectively. Water deficits were computed as precipitation minus potential evapotranspiration in mm (Thornthwaite 1948). Years are plotted only if >10 ha of forest were burned. Adjusted exponential regressions were significant. $p < 0.01$).

soils derived from volcanic ash, and desiccation of the vegetation requires prolonged drought. In contrast, in the dry woodlands and grasslands nearly all summers are dry enough for adequate fuel desiccation. Years of widespread fire in this vegetation type tend to lag anomalously wet springs by one year, suggesting that above-average production of fine fuels is important to fire occurrence in the drier habitats (Kitzberger, Veblen, and Villalba 1997; Veblen et al. 1999).

The fire-climate relationships suggested by the relatively short documentary record of fire history are confirmed by tree-ring records of fire and climatic variation over the past several hundred years (Kitzberger, Veblen, and Villalba 1997; Veblen et al. 1999). For example, fire history data from 10 sites ranging from rain forest to xeric woodland (Kitzberger, Veblen, and Villalba 1997) indicate that for all vegetation types considered together, fire years ($n = 74$) and the year prior to fire occurrence are characterized by below-average spring–summer moisture availability over the period 1820 to 1974 (Fig. 10.3a). In contrast, years in which no fires were detected in scar samples ($n = 82$) had above-average moisture availability during the fire year and the previous year ($p < 0.05$; Fig. 10.3b). These strong climatic relationships held true only when the analysis included major fire years as indicated by a regional fire index (RFI). RFI of 3 and 4 includes years in which fire scars occurred over areas >1000 ha and in large disjunct areas, respectively. RFI of 1 and 2 are years in which fire scars were limited to a single area or scarred only one or a few trees (Kitzberger, Veblen, and Villalba 1997). For years of major fire (RFI = 3 or 4; $n = 53$) moisture availability is well below average, but for years of minor fire occurrence (RFI = 1 or 2; $n = 21$), it is not significantly different from the long-term average (Fig. 10.3c–d).

Analogous to the results based on fire reports, tree-ring dated fire years in mesic *Nothofagus* forest ($n = 27$) were associated with greater moisture deficits than were fire years in the dry vegetation types (−1.22 SD and −0.80 SD, respectively; Fig. 10.3e–f). Over the period of the tree-ring index of moisture availability (1722–1974), the moisture availability index fell only below −1.22 SD in only 55 of 252 years, suggesting that there had been 2.2 potential opportunities per decade for fire in the wet forests. This contrasts with 76 years over 252 years, or 3.0 years per decade, during which the moisture index fell below −0.83, potentially creating fire opportunities in the dry vegetation types (Kitzberger, Veblen, and Villalba 1997). Analogous to the results from the documentary fire record, however, in dry habitats fire years tend to lag years of significantly above-average moisture by two years.

Regional Fire Synchrony

A more extensive network of 21 fire history sites located between 39° and 43°S latitude (Veblen et al. 1999) permits a more regionally extensive analysis of climatic variability and regional patterns of fire synchrony over the period 1600 to 1988. Synchronous occurrence of fires in the same years over extensive areas indicates a strong influence of interannual climatic variation on fire occurrence. For example, in 1827, tree-ring fire histories indicate that 11 of 21 sites burned synchronously spanning a N–S distance of nearly 300 km. Similarly, in 1897, 10 of 21 sites burned simultaneously over a N–S distance of nearly 380 km.

Both spring (November–December) and spring–summer (October–March) rainfall as reconstructed from *Austrocedrus* tree-rings (Villaba and Veblen 1997) decline sharply for years of increasing regional fire synchrony (Fig. 10.4a–b). Summer temperature, as reconstructed from *Fitzroya* tree-rings over the period,

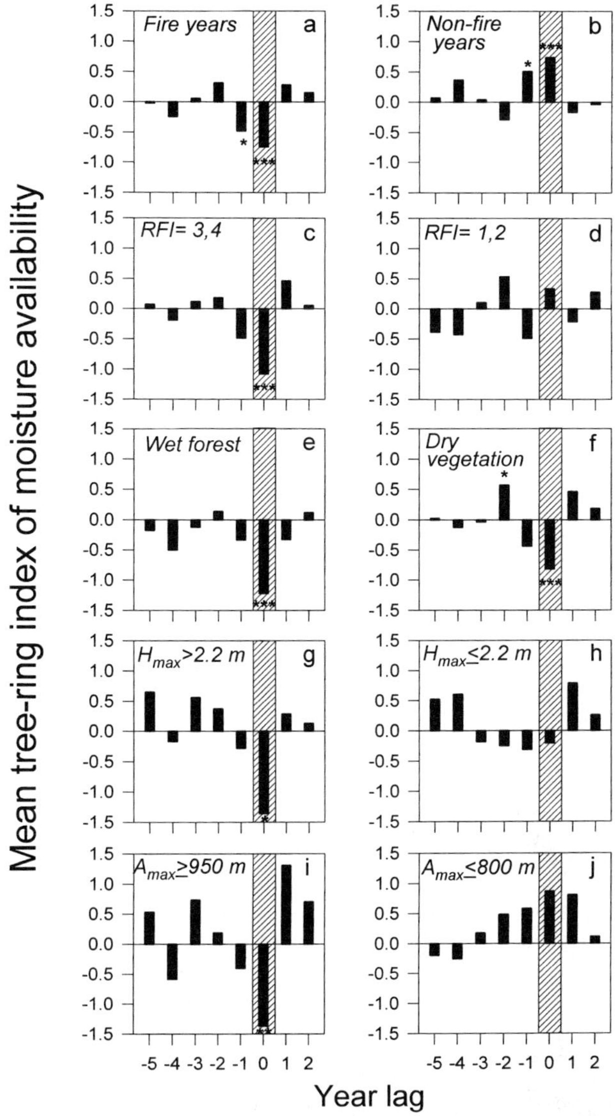

Figure 10.3. Mean tree-ring index moisture availability for all vegetation types during fire years (a); non-fire years (b); years of extensive fires, when the regional fire index (RFI) is ≤3 (c); and years of localized fire, when the regional fire index (RFI) is ≥2 (d); fire years in wet *Nothofagus* forests (e); fire years in dry vegetation types (f); years in which the upper edge of the tallest scar (H_{max}) was >2.2 m above the ground (g); years in which the upper edge of the tallest scar (H_{max}) was ≤2.2 m above the ground (h); years in which the elevation of the highest trees scarred (A_{max}) was located ≤950 m in elevation (i); and years the elevation of the highest trees scarred (A_{max}) was located ≥800 m in elevation (j). The eight-year window includes values for five years prior to and two years after the fire season. Bootstrap 95%, 99%, and 99.9% confidence intervals derived from Monte Carlo simulations indicate the significance of departures from the long-term mean (1820–1974) (*$p < 0.05$, **$p < 0.01$, ***$p < 0.001$. Sample sizes are 74 in (a), 81 nn (b), 27 in (c), 60 in (d), 53 in (e), 21 in (f) 12 in (g), 12 in (h) 20 in (i), and 10 in (j). (Data are from Kitzberger, Veblen, and Villaba 1997.)

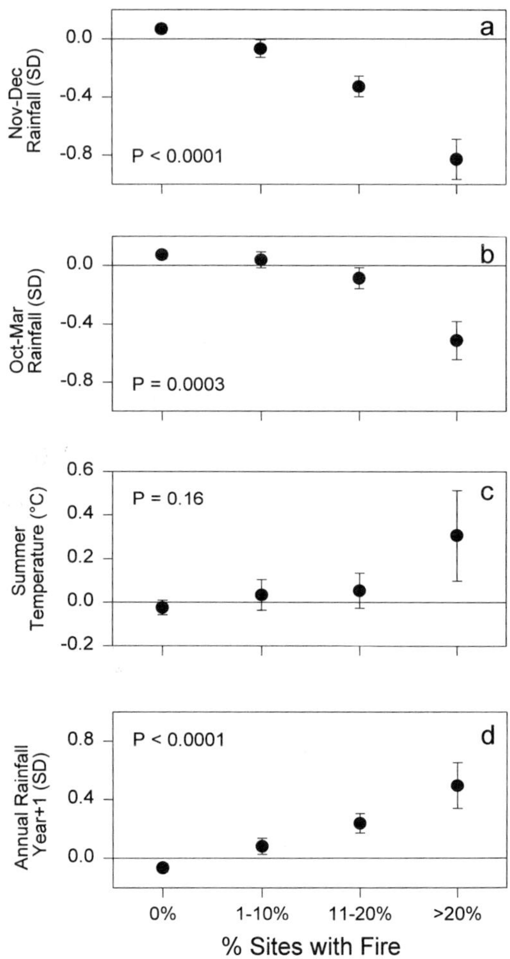

Figure 10.4. Relationships between fire synchrony expressed as the percentage of sites recording fire in a particular year based on a network of 21 fire history sites located between 39° and 43°S in northern Patagonia (Veblen et al. 1999) and mean (±SE) values of tree-ring reconstructions of late spring (November–December) rainfall (a), spring-summer (October–March) rainfall (b), summer temperature (c), and annual rainfall for the year after the occurrence of fire (d). Precipitation reconstructions are based on a network of 25 *Austrocedrus chilensis* tree-ring chronologies located of northern Patagonia (Villalba and Veblen 1997) and summer temperature is reconstructed from a *Fitzroya cuppressoides* chronology located at ca. 41°10′S (Villalba 1990b). Probability levels indicate the significance of the effect of classifying into synchroneity classes defined as 0%, 1–10%, 11–20%, and >20% of the sites recording fire in the same year (based on one-way ANOVA).

appears to be an important influence only for years of the most widespread fire (i.e., years with >20% of the sites recording fire; Fig. 10.4c). Warm and dry summers are probably especially critical to fuel desiccation in the otherwise moist western forests. In these forests with their large leaf areas and biomass, prolonged warm temperatures in the absence of precipitation induce high transpiration rates of live fuels and eventually desiccate the coarse dead fuels. As discussed below, warm summers in northern Patagonia are also associated with enhanced lightning activity.

Somewhat surprisingly, the instrumental climate record shows that over the 1938 to 1996 period the winters following years of major forest burning are anomalously high in precipitation (Veblen et al. 1999). Tree-ring reconstructed annual precipitation, which is mainly influenced by variability in winter–spring precipitation, over the period 1600 to 1988 also increases following years when large percentages of the 21 fire history sites recorded fire scars (Fig. 10.4d). This consistent pattern is explained by the influences of the ENSO cycle on climate and fire in northern Patagonia as discussed below.

In contrast to the clear influence of interannual climatic variability on fire regimes in northern Patagonia, over longer time periods the relationship of fire synchrony to mean climatic conditions is weaker and less consistent (Veblen et al. 1999). For example, over the period 1599 to 1989, the five driest single years (derived from tree-ring reconstructions) coincided with positive departures (91–445%) from the long-term mean number of sites recording fire. Analogously, the 5 single years of wettest springs and spring–summers in the record were years of little or no fire occurrence. At the pentad scale, climatic control on fire synchrony was weaker; only 3 of the 5 driest pentads coincided with positive departures (62–118%) from the long-term mean number of sites recording fire, and during the 5 wettest pentads very few sites recorded fire. Association of fire extent with mean climatic conditions at 25- and 50-year scales is weak or inconsistent (Veblen et al. 1999). For instance, 1843 to 1892 is one of the three wettest 50-year periods in the record, but shows an 88% positive departure from the long-term mean number of sites recording fire. The lack of consistent patterns of fire occurrence and mean climatic conditions at 25- and 50-year time periods is at least partially explained by changes in land use (see Veblen et al., Chapter 9, this volume). However, as explained below, changes in interannual climatic variability, in contrast to multidecadal mean conditions, at a 50-year time scale also appears to influence fire regimes in northern Patagonia.

Fire Behavior

Analysis of fire-scar heights on trees and the elevations of fire-scarred trees from four nearby fire history sites of *Austrocedrus*-dominated woodlands and shrublands permit some tentative inferences about changes in fire behavior in relation to interannual climatic variability. As flame height is proportional to fire intensity (Chandler et al. 1983), higher scars on trees generally indicate more intense fires that presumably resulted from drier or more abundant fuels. Even allowing for uncontrolled influences such as changes in wind speed that also affect scar heights, years with taller mean maximum scar heights (H_{max}) are believed to be years of more intense fires permitted by fuel conditions and/or quantities. Mean tree-ring reconstructed moisture availability is significantly below average for years when mean maximum scar heights were >2.2 m above the ground ($n = 12$); in contrast, years in which mean maximum scar heights were ≤220 cm tall ($n = 12$) showed no significant climatic anomalies (Fig. 10.3g–h). This suggests that greater desiccation of coarser fuels during drier years promote more intense fires.

Similarly, annual variation in the mean maximum elevation (A_{max}) at which individual fire-scarred trees record fires shows a strong climatic influence. Years during which trees recorded fires at elevations above 950 m ($n = 20$) are years of drought, whereas years in which fires remained below 800 m in elevation ($n = 10$) did not differ significantly from the long-term mean moisture index (Fig. 10.3i–j). Historical and modern observations in northern Patagonia indicate a tendency for many fires to burn upslope from *Austrocedrus*-dominated vegetation but to often extinguish themselves when they reach the more mesic subalpine forests that occur above 1000 m (Rothkugel 1916; Tortorelli 1947; Veblen and Lorenz 1988; Veblen, Kitzberger, and Lara 1992). Generally, fuel structure is coarser and fuels have higher moisture contents at higher elevations due to reduced water demand (see Veblen et al., Chapter 9, this volume). Thus, similarly to mesic western rain forest, burning of subalpine forests appears to be dependent on more severe drought.

Lightning

Although most modern fires are set by humans, lightning in Patagonia is an important source of ignition. From 1938 to 1996 in the four national parks of northern Patagonia (1,400,000 ha), lightning accounted for 64 ignitions or 8.9% of the 722 ignitions for which cause was reported (Bruno and Martin 1982; Administración de Parques Nacionales, unpublished data). More significantly however, lightning-ignited fires accounted for 16.5% of the total area burned (119,469 ha), which suggests that lightning coincides with weather that creates fuel conditions conducive to extensive spread of fire. Over the 1938 to 1996 period, 64% of lightning ignitions occurred during the summer months of January and February, and approximately 31% occurred in the late spring and late summer months of December and March (Bruno and Martin 1982; Administración de Parques Nacionales, unpublished data).

Although lightning-ignited fires are not frequent, single thunderstorm events can ignite fires over relatively large areas. For example, on February 24, 1987, a single storm event ignited several fires over 150 km of north–south distance from Lake Tromen (ca. 39°30′S) to Volcano Puyehue (ca. 40°42′S). Three days later, the same weather pattern resulted in a 2000 ha lightning-ignited fire at Brazo Tristeza (Lake Nahuel Huapi, ca. 41°04′S), 50 km further south. Similarly, on December 26, 1995, a thunderstorm ignited at least four fires over a 100 km distance from Lake Ruca Choroi (ca. 39°14′S) to Lake Lacar (ca. 40°09′S), and on January 13–14, 1989, lightning-caused fires extended more than 130 km from Lake Quillén (ca. 39°22′S) to Lake Filo Hua Hum (ca. 40°29′S; Bruno and Martin 1982; Administración de Parques Nacionales, unpublished data). Thus it is possible that these storm events contributed together with regional drought conditions to produce synchronous fires over extensive regions such as that which occurred in 1827.

Lightning ignitions are strongly associated with hot relatively dry summers (Kitzberger, Veblen, and Villalba 1997). Over the period 1940 to 1988, during years of lightning-ignited fires ($n = 16$), December to February temperatures were above average and December to February precipitation was below average. During years of average to below-average summer temperatures the probability of a lightning-ignited fire is almost nil, but increases dramatically as summer temperatures increase (Fig. 10.5).

At a decadal scale there are also tentative trends in the frequency of lightning ignitions and summer temperatures in northern Patagonia. For example, mean summer temperatures were higher after 1978 when compared to the previous 1938 to 1977 period ($p < 0.01$; Fig. 10.6). This long-lasting temperature anomaly has been accompanied by a threefold increase in the rate of lightning ignitions (from 0.6 ignitions/year to 1.95 ignitions/year; $p < 0.02$). As discussed below, the post-1978 warmer and drier conditions in northern Patagonia are associated with changes in large-scale circulation features.

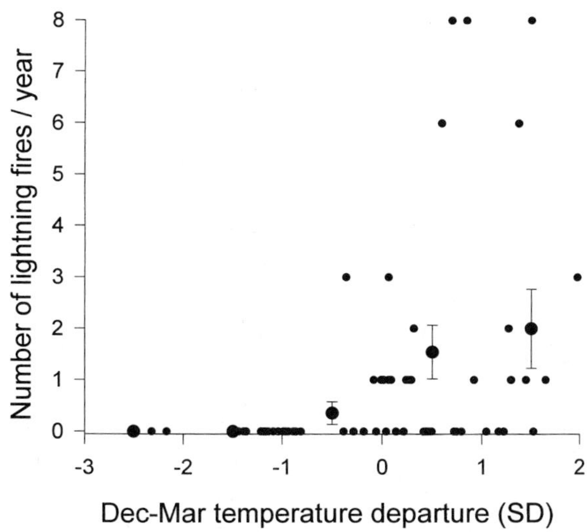

Figure 10.5. Number of lightning ignitions (*small dots*) reported in Lanín, Nahuel Huapi, Lago Puelo, and Los Alerces national parks between 1938 and 1996 (Bruno and Martin 1982; Administración de Parques Nacionales, unpublished data) in relation to summer (December–March) mean temperature (based on Bariloche Airport weather station). Means (*large dots*) (±SE) were calculated for intervals of one SD of summer temperature.

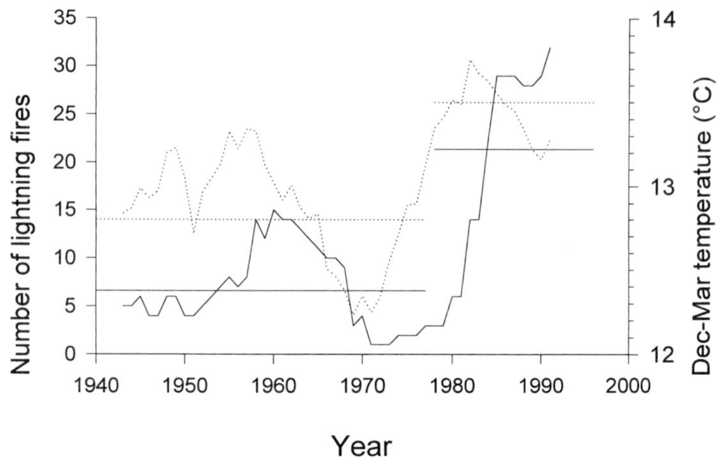

Figure 10.6. Eleven-year moving sum of lightning ignitions (*solid line*) reported in Lanín, Nahuel Huapi, Lago Puelo, and Los Alerces national parks between 1938 and 1996 (Bruno and Martin 1982; Administración de Parques Nacionales, unpublished data) and 11-year moving mean summer temperature (*dotted line*) (based on Bariloche Airport weather station). Horizontal lines are means for the 1938–1977 and 1978–1996 periods.

Large-Scale Circulation Anomalies

Influences of the Southeastern Pacific Subtropical Anticyclone

Years of relatively high fire activity in wet *Nothofagus* forests are years when the southeastern Pacific anticyclone is strong and displaced towards the south during winter and spring (Fig. 10.7; Kitzberger, Veblen, and Villalba 1997). One year prior to the summers of high fire activity, the Pacific anticyclone is also displaced southward during spring (Fig. 10.7). Thus a stronger and more southerly located anticyclone is important in blocking westerly cyclonic storms and creating dry conditions conduce to widespread burning in the mesic *Nothofagus* forests (Kitzberger, Veblen, and Villalba 1997).

Precipitation anomalies in northern Patagonia are also associated with high latitude blocking events at 60°S in the Antarctic Peninsula–South America sector of the Southern Ocean. These blocking events drive westerly storms northward into Patagonia and are associated with positive precipitation departures based on tree-ring reconstructions of pressure and precipitation for the period 1746 to 1984 (Villalba et al. 1998). Thus years of synchronous fire in northern Patagonia are associated with below-average summer atmospheric pressure in the Antarctic Peninsula sector due to the association of less precipitation with an absence of blocking highs (Veblen et al. 1999). The strength of the relationship between precipitation in northern Patagonia and summer atmospheric pressure at ca. 60°S, however, has been greater during the twentieth century than during the preceding 150 years (Villalba et al. 1998). The strength of the teleconnections between middle and high latitudes in the South American sector varies with changes in the degree of zonal versus meridional airflow. Increased precipitation variability and stronger correlations of precipitation with high latitude pressure since 1900 may reflect stronger meridional circulation across South America and stronger interaction between mid- and high-latitude circulation features (Villalba et al. 1998).

Years in which lightning-ignited fires occur in northern Patagonia are associated with changes in circulation features that are manifested in the southern Atlantic Ocean. Such years ($n = 7$; considering only lightning fires that occurred in February) are associated with above-average sea-level atmospheric pressure at

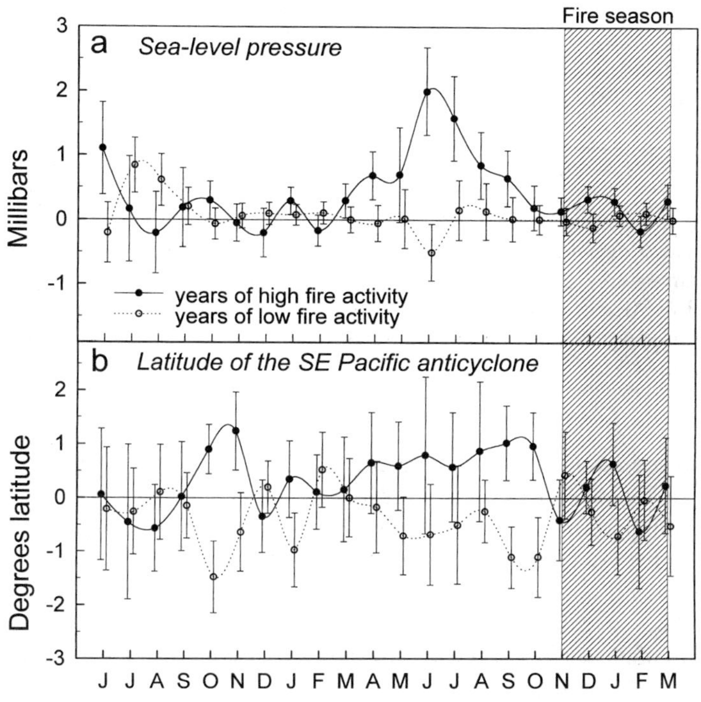

Figure 10.7. Intensity and location of the southeast Pacific subtropical anticyclone during years of high *versus* low fire activity in wet *Nothofagus* forests in Nahuel Huapi and Lanín National Parks, northern Patagonia. (a) Mean (±SE) deviations from the long-term mean sea-level atmospheric pressure (millibars) at Punta Galera Chile (1911–1960) over 23 months prior to the fire season. (b) Mean (±SE) deviations from the long-term mean latitudinal position of the southeast Pacific anticyclone (1943–1962) along coastal Chile over the 23 months prior to the fire season. High fire activity years had >100 ha burned (1940–1988) or a regional fire index of 4. Low fire activity years had <10 ha burned or a regional fire index of 0. Sample sizes in (a) and (b), respectively, are 15 and 9 for high fire years and 20 and 9 for low fire years. (Pressure data are from Pittock 1980.)

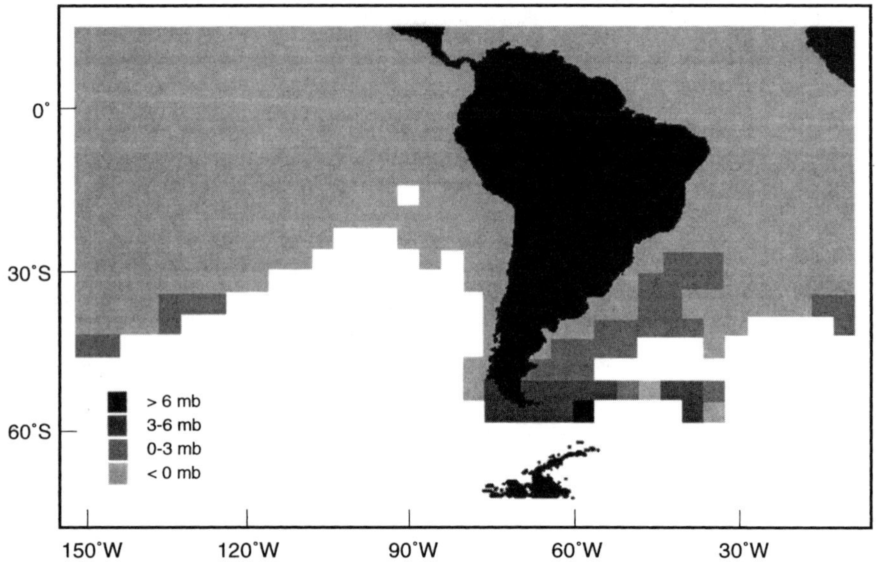

Figure 10.8. February mean sea-level pressure anomalies over the ocean near southern South America over seven years (1950, 1953, 1957, 1982, 1987, 1989, 1990) in which lightning fires were reported during the month of February in Lanín, Nahuel Huapi, Lago Puelo or Los Alerces national parks. Sea-level pressure data are taken from the global 4 × 4° grid of ocean sea level pressure anomalies (Lamont Doherty Earth Observatory IRI RSA COADS).

high latitude east of southern South America during February (Fig. 10.8). During summers of anomalously high pressure at 45° to 55°S in the Atlantic Ocean, the Atlantic portion of the subtropical high-pressure belt is located further south, which allows Atlantic subtropical air to flow southwestward into northern Patagonia (van Loon, Kidson, and Mullan 1993). These moist, warm air masses spawn thunderstorms and lightning through convective uplift over the heated Patagonian plains or by advective or orographic uplift when they reach cooler air in the Andean foothills.

ENSO Influences

High-Frequency Climatic Variability

Both the documentary and tree-ring records of fire in northern Patagonia reflect strong influences of ENSO activity (Kitzberger and Veblen 1997; Veblen et al. 1999). The area of forest burned annually in northern Patagonia is strongly associated with variations in the standardized Southern Oscillation Index (SOI; Ropelewksi and Jones 1987) and sea surface temperatures (SST) in the eastern Ecuatorial Pacific (Niño regions 1 + 2). Over the period 1882 to 1989, years of extensive fire in northern Patagonia tend to be associated with late stages of the

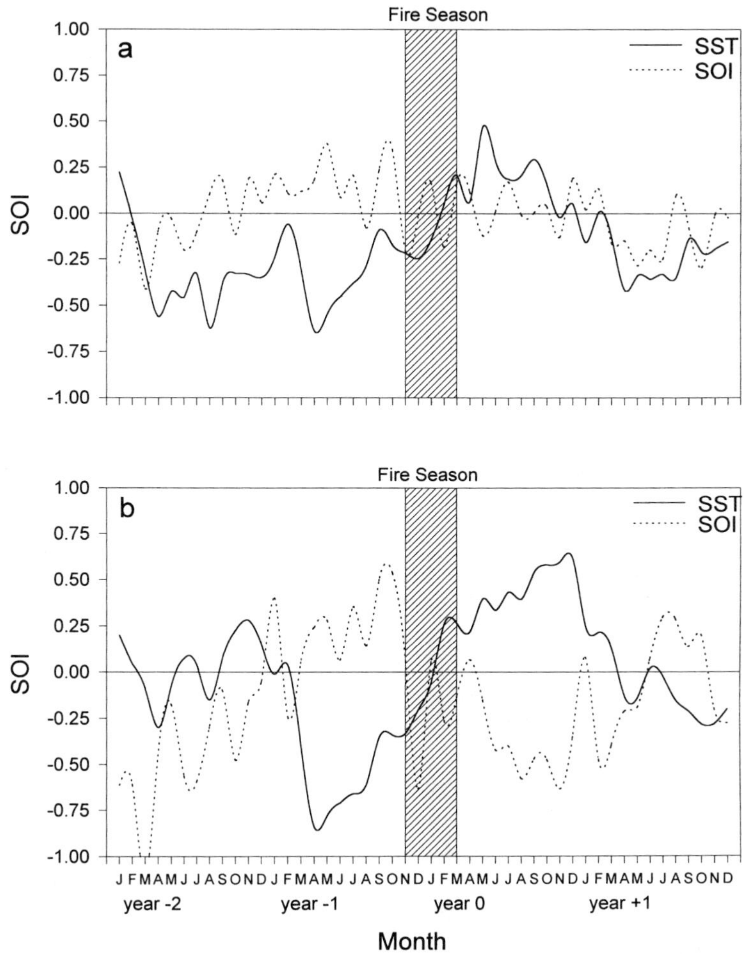

Figure 10.9. Mean sea surface temperature (SST) in the eastern equatorial Pacific (Niño Regions 1 + 2; *solid line*) and mean Southern Oscillation Index (SOI) calculated for a moving window of 48 months centered (a) on years of extensive fire ≤2000 ha burned (1950–1996) ($n = 14$) and (b) on years when two or more lightning fires occurred February in Lanín, Nahuel Huapi, Lago Puelo or Los Alerces national parks ($n = 9$; *dotted line*). The shaded area is the fire season of the fire-event year (i.e., year 0). (Fire reports are based on Bruno and Martin 1982 and Administración de Parques Nacionales, unpublished data.)

positive phase of the Southern Oscillation or cold La Niña stage indicated by positive SOI anomalies and negative SST anomalies about two years before the fire season (Fig. 10.9a). Years of lightning-ignited fires show similar timing relationships with SST and SOI. However, the cold La Niña preceding period tends to be shorter, and the strong El Niño pattern (negative SOI anomalies and positive SST anomalies) after the fire season strongly suggests that lightning ignitions may also be due to late-developing El Niño events that are related to warmer summers (Fig. 10.9b).

Although the most common pattern is for years of major fire activity to occur during the late stages of La Niña events, major fire years also can coincide with El Niño events (Veblen et al. 1999). Among the 10 years of greatest fire occurrence between 1740 and 1995, as determined from percentages of 21 fire-history sample sites recording fire, six years coincide with moderate to very strong El Niño events (Quinn 1992; Ortlieb and Macharé 1993). The timing of ENSO within the annual cycle is critical in determining its influence on fire occurrence. El Niño events most frequently begin to develop in March to May (Kiladis and Diaz 1989) which for northern Patagonia tends to increase winter–spring (June–November) precipitation during the same calendar year. However, El Niño events that begin to develop after the winter rainy season (e.g., after October) do not result in increased winter–spring precipitation until after the summer dry season. Thus the warm temperatures associated with late-developing El Niño events more effectively desiccate fuels and can promote widespread fire. For example, the El Niño events of 1965 and 1972 began to develop in April, and by the winter rainy season (July–August) were well developed; consequently little or no burning occurred during the summer fire seasons (Fig. 10.10). In contrast, the 1969 and 1986 El Niño events did not develop early enough in the preceding calendar year to enhance winter precipitation prior to the fire season of each respective year. By January and February of 1969 and 1986 each event was well developed and increased burning is associated with the warm El Niño summers (Fig. 10.10). Thus 1969 and 1986 are classified as years of El Niño events (Díaz and Kiladis 1992; Quinn 1992) and are also years of high fire activity. The preceding years, 1968 and 1985, are not classified as La Niña events (Díaz and Kiladis 1992), but each pair of years could be considered a transition from "Niña-like conditions" to El Niño conditions under which warm summers tend to follow winter–springs of normal or below-average precipitation.

There are two ENSO-related patterns associated with years of extreme burning: warmer summers associated with El Niño events that develop after the preceding winter rainy season, and reduced winter–spring precipitation during La Niña events preceding the summer fire season. The latter is the most common pattern because most El Niño events start early enough in the calendar year to increase winter–spring precipitation prior to the summer fire season. Despite the association of 6 of the 10 most extreme fire years from 1520 to 1929 with El Niño events (that probably developed late in the calendar year), most major fire years between 1520 and 1929 (n = 88) are the year prior to the beginning of El Niño events in Quinn's (1992) record (Veblen et al. 1999). This is consistent with the tendency of ENSO to switch from one extreme to the other in consecutive years (Díaz and Kiladis 1992), so many of these major fire years would have followed dry La Niña winter–springs. Some of the others would have been associated with early stages of late-developing El Niño events. This pattern is also consistent with the occurrence of above-average precipitation during the year following major fire years based both on the instrumental record as well as tree-ring records of pre-

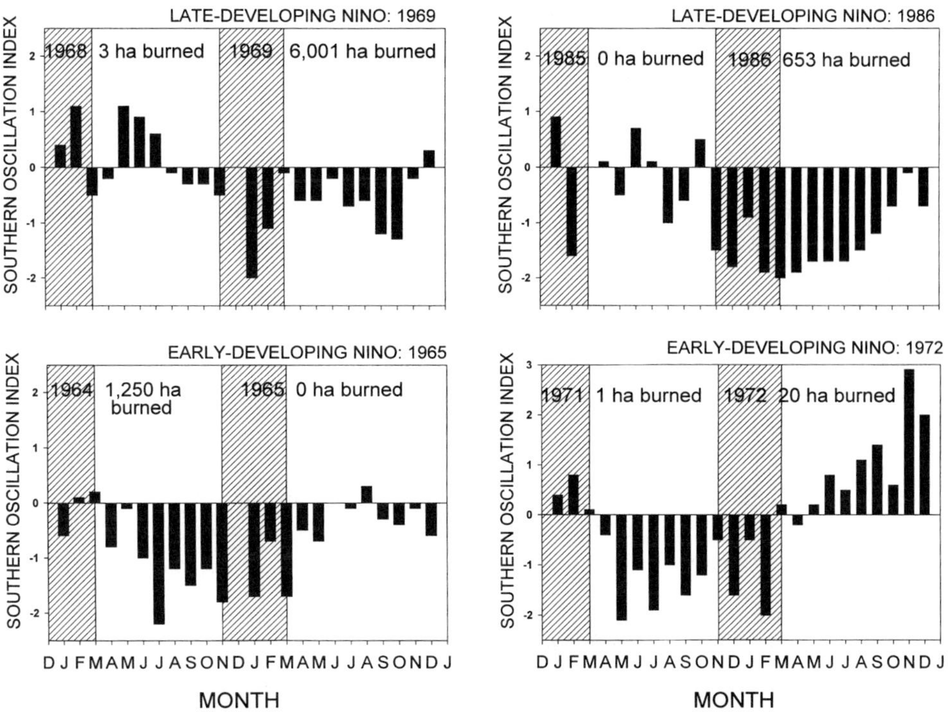

Figure 10.10. Monthly Southern Oscillation Indexes for early-developing (1965 and 1972) and late-developing (1969 and 1986) El Niño events and areas (in hectares) burned in Lanin, Nahuel Huapi, Lago Puelo and Los Alerces national parks (Bruno and Martin 1982; Administración de Parques Nacionales, unpublished data). Only the late-developing El Niño events are associated with high rates of burning.

cipitation (Veblen et al. 1999; Fig. 10.4). Given the greater abundance of El Niño events that start in March through May (Kiladis and Díaz 1989), but the association of widespread fire with El Niño years is less frequent than for La Niña years.

Fire records from 1938 to 1996 indicate that during middle to late La Niña stages (i.e., during the middle to final months of extended periods of positive departures of SOI) there was a higher relative percentage of spring as opposed to summer fires (ca. 80% vs. 50%, respectively). In contrast, a much larger percentage of summer as opposed to spring fires occurred during early to middle El Niño stages (i.e., early to middle months of extended periods of negative SOI anomalies; ca. 40% vs. 8%, respectively). Season of fire occurrence and stage of ENSO (mid to late La Niña vs. early to mid El Niño were significantly associated ($p < 0.05$ Fisher exact test).

Lightning activity in northern Patagonia can also be linked to ENSO events. Although lightning activity is strongly associated with pressure anomalies in the southern Atlantic Ocean (Fig. 10.8), the incursion of tropical air masses into the mid-latitudes of South American may be favored by a strong and southerly located southeast Pacific anticyclone that shifts the zone of westerly flow farther south. Thus the interaction of Atlantic and Pacific circulation features appears to influence lightning ignitions in northern Patagonia. Of the seven years of major lightning-ignited fire between 1938 and 1996, five coincided with late La Niña to early El Niño transitions. This association probably reflects a combination of enhanced lightning activity associated with warmer El Niño summers and/or greater fuel desiccation related to drier La Niña springs or early El Niño summer drought.

Precipitation anomalies in northern Patagonia are linked both to ENSO events and high-latitude circulation features, which themselves may be coupled. Variation in the southeast Pacific anticyclone and high-latitude atmospheric circulation is also linked to ENSO events (Díaz and Kiladis 1992). Major blocking highs southwest of South America at ca. 55°S, 90°W tend to coincide with warm SO events (Rutllant and Fuenzalida 1991). Furthermore there are similar periodicities (3 to 5 years) of interannual variations in the circumpolar flow at ca. 55°S, interannual variations in sea-level atmospheric pressure, and sea-ice extent (White and Peterson 1996; Villalba et al. 1998).

Low-Frequency Changes in ENSO

Given the strong association of years of widespread fire in northern Patagonia with interannual climatic variability, it is likely that ENSO-induced changes in interannual climatic variability at multidecadal scales would also influence fire regimes. Periods of greater frequency and/or amplitude of ENSO events are likely to be periods of greater fire occurrence due to the more frequent alternation between fuel-enhancing wet periods and fuel-desiccating dry periods.

Tree-ring derived fire histories from 1650 to 1990 based on hundreds of widely distributed fire-scarred trees from 39° to 43°S in northern Patagonia indicate important decadal-scale fluctuations in fire frequency that closely mirror variations in ENSO activity (Kitzberger and Veblen 1997; Veblen et al. 1999). Frequency of years of widespread fire (i.e., years in which >30% of the trees recorded fire) is relatively high in the mid-1700s, reaches a nadir about 1800, and increases to a peak in the late 1800s (Fig. 10.11a). Years of less widespread fire (>15% of the trees recorded fire), which would be expected to be somewhat less controlled by climate and perhaps are more responsive to changes in human-set ignitions, show less variation in frequency. In particular, the greater reduction in the frequency of widespread fires (>30% scarred) from ca. 1780 to 1830s relative to the decline in years of moderate fire occurrence (>15% scarred) may be a response to decadal-scale change in ENSO activity. The variation in frequency of widespread fires closely tracks variation in several independently derived reconstructions of ENSO activity (Fig. 10.11b, c, and d). These include tree-ring calibrated reconstructions of Southern Oscillation indexes from regional tree-ring networks (Villalba 1994), records of El Niño/La Niña events from Spanish archival documents (Quinn and Neal 1992), and $\delta^{18}O$ time series from tropical coral (Dunbar

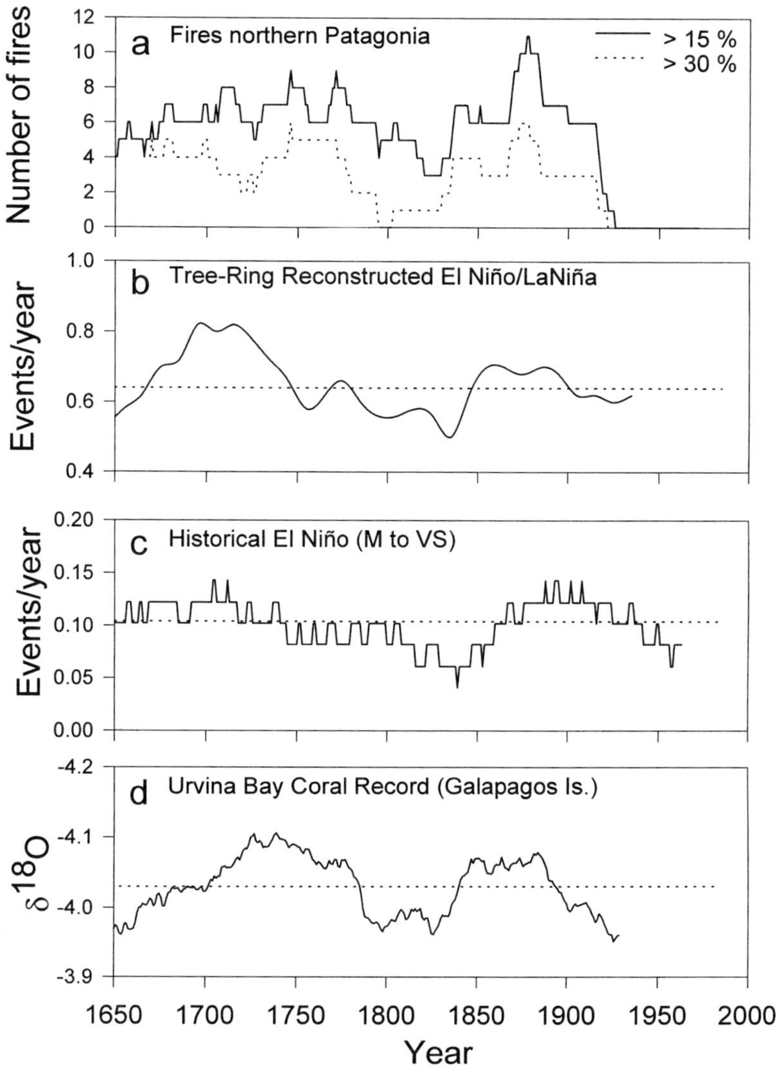

Figure 10.11. Number of regional-scale fire years over a moving 49-year window in northern Patagonia (a), and multi-proxy reconstructions of low-frequency changes in ENSO activity between 1650 and 1990 based on (b) La Niña and El Niño events reconstructed from tree-ring chronologies in Patagonia and central Chile (Villaba 1994) and (c) moderate to very strong El Niño events reconstructed from archival documents (Quinn and Neal 1992), (d) record of ENSO-related central Pacific upwelling based on the $\delta^{18}O$ (%0) coral record from Urvina Bay, Galapagos Islands (Dunbar et al. 1994). In (a) fire years are years in which more than 15% (*solid line*) or more than 30% (*dotted line*) of all trees in five sites recorded fire (data from Kitzberger and Veblen 1997). Plots in (b) and (c) are mean number of events per year based on moving 49-year sums, and in (d) is the 49-yr running mean of $\delta^{18}O$ (%0) coral. In all cases the horizontal solid line represents long-term mean values.

et al. 1994). Reduced amplitude of the ENSO during 1780 to 1830 is indicated by all these records (Fig. 10.11). This pattern, in combination with the previously documented association of fire and ENSO-induced climatic variation (Fig. 10.8; Kitzberger and Veblen 1997; Veblen et al. 1999), suggests that fire regimes in northern Patagonia reflect long-term changes in the amplitude and/or frequency of ENSO events.

Conclusion

In northern Patagonia interannual variations in fire regimes closely track regional climatic variability, which is linked to large-scale atmospheric circulation anomalies. Although climatic variability overrides human influences on fire regimes at an interannual scale, human activity can be of equal or greater importance in determining fire frequency at multidecadal scales (Veblen et al., Chapter 9, this volume). However, by focusing on years of widespread fire, which are mainly controlled by climate, it is feasible to relate changes in fire regimes and climate at decadal to centennial scales.

In northern Patagonia years of widespread burning in mesic forests coincide with drier and warmer than average spring–summers, but in the grassland zone summer drought is severe enough in normal years to permit burning. Years of extensive grassland burning, however, do tend to follow wetter than normal springs one year prior to the fire season, which may increase the availability of fine fuels through enhanced growth of grasses. Years in which the southeast Pacific subtropical anticyclone is more intense and located further south are years of greater drought and fire. Climatic conditions conducive to widespread fire in both rain forests and xeric woodlands are also closely related to ENSO events. Despite the significant influence of tropical Pacific atmospheric phenomena, ENSO activity is not the sole determinant of fire weather in northern Patagonia. Years of widespread fire are also associated with an absence of atmospheric blocking events at ca. 50 to 60°S that would otherwise steer cyclonic storms northward into northern Patagonia.

The strength of the relationship between ENSO events and climate is known to have varied at hemispherical and global scales over decadal and centennial time scales (Díaz and Pulwarty 1994). In northern Patagonia, although spring and summer temperature and precipitation variations are significantly correlated with the SOI over the full instrumental record (ca. 1915–1997), correlations are nearly absent during the 1930s and 1940s (Villalba and Veblen 1998; Daniels and Veblen 2000). The relationship between climate and ENSO-forcing in northern Patagonia is highly variable according to the timing and strength of events (Villalba 1994). Thus, despite the statistically significant associations demonstrated here, variation in fire regimes in northern Patagonia can only be partially explained by ENSO forcing.

Analyses of fire–ENSO relationships between widely separated ENSO-sensitive regions such as the southwestern United States and northern Patagonia show similar interannual and decadal changes. In both regions there was a decline in widespread burning from 1780 to 1830 that coincides with reduced amplitude and/or strength of the teleconnections of ENSO (Swetnam and Betancourt 1998; Kitzberger, Swetnam, and Veblen 2001). Multicentury time series of regional fire activity in the two regions are also spectrally coherent within the dominant ENSO frequency band (i.e., 2–7 years; Kitzberger, Swetnam, and Veblen 2001). These synchronous changes suggest that regional forest fire regimes in these regions may be phase-locked with the Southern Oscillation and may be responding synchronously to long-term changes in the modal frequencies or amplitudes of the Southern Oscillation (Kitzberger, Swetnam, and Veblen 2001).

The relationships of fire and ENSO summarized for northern Patagonia are of potential value in forecasting fire hazards and planning mitigation activities a year or more in advance. Furthermore, in the context of longer-term modeling of the ecological effects of global waring, these results indicate the importance of considering year-to-year variability rather than just long-term mean climatic conditions. At much longer time scales, increased fire has also been linked to periods of greater climatic variability. Comparison of sedimentary charcoal records with fossil pollen records from different environments in southern South America indicate increased fire occurrence for periods of greater climatic variability during the late-Glacial and late-Holocene periods (Heusser 1987; Markgraf and Anderson 1994). The greater late-Glacial variability has been attributed to fluctuations in the extent of Antarctic sea ice, which, in turn, influence the latitudinal position of the westerly storm tracks. The variability of the late Holocene appears to be related to the onset of ENSO as an important influence on mid-latitude climates along the west coast of South America (McGlone, Kershaw, and Markgraf 1992; Markgraf and Anderson 1994).

Similar to the association of drought and fire demonstrated here, other studies in northern Patagonia (Villalba and Veblen 1997b; Villalba and Veblen 1998) show that the establishment of seedlings and mortality of adult trees of *Austrocedrus* are strongly associated with variations in ENSO and in the strength and position of the southeastern Pacific anticyclone. For example, the predominance of the negative mode of the Southern Oscillation (i.e., El Niño conditions) since the late 1970s is reflected by warmer summers and a lack of *Austrocedrus* seedling survival in dry habitats (Villalba and Veblen 1997b). Analogously, the stepped increase in the frequency of lightning-ignited fires since the mid-1970s (Fig. 6) also coincides with the increase in El Niño events. However, tree-ring proxy records indicate that over the past 250 years or so there have been important variations at decadal- to centennial-time scales in major circulation features, such as ENSO activity and blocking events at high latitudes, and also in the relationships of climate in northern Patagonia to these circulation features. For understanding possible impacts of global climate change on regional fire regimes and forest dynamics, it is important to consider past variations in large-scale atmospheric circulation features and fluctuations in the strengths of their influences on regional climates.

Acknowledgments. This review is based on research funded by the National Science Foundation of the United States, the National Geographic Society, and the Council for Research and Creative Work of the University of Colorado. For providing unpublished data, we thank R. Villalba, and for assistance with the figures, we thank D.C. Lorenz.

References

Aceituno, P. 1988. On the functioning of the Southern Oscillation in the South American sector. Part 1. Surface Climate. *Mon. Wea. Rev.* 116:505–524.

Alaback, P., and McClellan, M. 1993. Effects of global warming on managed coastal ecosystems of western North America. In *Earth System Response to Global Change: Contrasts between North and South America*, eds. H.A. Mooney, E. Fuentes and B.I. Kronberg, pp. 299–327. New York: Academic Press.

Baker, W.L. 1990. Climatic and hydrologic effects on the regeneration of *Populus angustifolia* James along the Animas River, Colorado. *J. Biogeogr.* 17:59–73.

Baker, W.L., Egbert, S.L., and Frazier, G.F. 1991. A spatial model for studying the effects of climatic change on the structure of landscape subject to large disturbances. *Ecol. Model.* 56:109–125.

Barros, V., Cordón, V., Moyano, C., Méndez, R., Forquera, J., and Pizzio, O. 1983. Cartas de precipitación de la zona oeste de las provincias de Rio Negro y Neuquén. Report to the Facultad de Ciencias Agrarias, Universidad Nacional del Comahue, Cinco Saltos, Neuquén.

Bergeron, Y., and Archambault, S. 1993. Decreasing frequency of forest fires in the southern boreal zone of Quebec and its relation to global warming since the end of the "Little Ice Age." *Holocene* 3:255–259.

Bruno, J., and Martin, G. 1982. Los incendios forestales en los Parques Nacionales. Unpublished report, Administración de Parques Nacionales, Buenos Aires.

Chandler, C., Cheney, P., Thomas, P., Trabaud, L., and Williams, D. 1983. *Fire in Forestry. Volume I: Forest Fire Behavior and Effects.* New York: Wiley.

Daniels, L.D., and Veblen, T.T. 2000. ENSO effects on temperature and precipitation of the Patagonian-Andean region: Implications for biogeography. *Phys. Geogr.* 21: 223–243.

Díaz, H.F., and Kiladis, G.N. 1992. Atmospheric teleconnections associated with extreme phases of the Southern Oscillation. In *El Niño: Historical and Paleoclimatic Aspects of the Southern Oscillation*, eds. H.F. Díaz and V. Markgraf, pp. 7–28. Cambridge: Cambridge University Press.

Díaz, H.F., and Pulwarty, R.S. 1994. A comparison of the Southern Oscillation and El Niño signal in the tropics. In *El Niño: Historical and Paleoclimatic Aspects of the Southern Oscillation*, eds. H.F. Díaz and V. Markgraf, pp. 175–192. Cambridge: Cambridge University Press.

Dunbar, R., Wellington, G.M., Colgan, M.W., and Glynn, P.W. 1994. Eastern Pacific sea surface temperature since 1600 A.D.: The $d^{18}O$ record of climate variability in Galapagos corals. *Paleoceanography* 9:291–316.

Franklin, J.F., Swanson, F.J., Harmon, M.E., Perry, D.A., Spies, T.A., Dale, V.H., McKee, A., Ferrell, W.K., Means, J.E., Gregory, S.V., Lattin, J.D., Schowalter, T.D., and Larsen, D. 1991. Effects of global climatic change on forests in northwestern North America. *Northwest. Environ. J.* 7:233–254.

Gardner, R.H., Hargrove, W.W., Turner, M.G., and Romme, W.H. 1996. Climate change, disturbances and landscape dynamics. In *Global Change and Terrestrial Ecosystems*, eds. B. Walker and W. Steffen, pp. 149–172. Cambridge: Cambridge University Press.

Heusser, C.J. 1987. Fire history of Fuego-Patagonia. *Quaternary of South America and Antarctic Peninsula* 5:93–109.

Johnson, E.A., and Larsen, C.P.S. 1991. Climatically induced change in fire frequency in the southern Canadian Rockies. *Ecology* 72:194–201.

Johnson, E.A., and Wowchuk, D.R. 1993. Wildfires in the southern Canadian Rocky Mountains and their relationship to mid-tropospheric anomalies. *Can. J. For. Res.* 23: 1213–1222.

Kiladis, G.N., and Díza, H.F. 1989. Global climatic anomalies associated with extremes in the Southern Oscillation. *J. Climate* 2:1069–1090.

Kitzberger, T. 1994. Fire regime variation along a northern Patagonian forest-steppe ecotone: Stand and landscape response. PhD. dissertation. University of Colorado, Boulder.

Kitzberger, T., and Veblen, T.T. 1997. Influences of humans and ENSO on fire history of *Austrocedrus chilensis* woodlands in northern Patagonia, Argentina. *Ecoscience* 4: 508–520.

Kitzberger, T., Veblen, T.T., and Villalba, R. 1997. Climatic influences on fire regimes along a rainforest-to-xeric woodland gradient in northern Patagonia, Argentina. *J. Biogeogr.* 23:35–47.

Kitzberger, T., Swtenam, T.W., and Veblen, T.T. 2001. Inter-hemispheric synchrony of forest fires and the El Niño–Southern Oscillation. *Global Ecol. Biogeogr.* 10:315–326.

Körner, C. 1996. The response of complex multispecies systems to elevated CO_2. In *Global Change and Terrestrial Ecosystems*, eds. B. Walker and W. Steffen, pp. 20–42. Cambridge: Cambridge University Press.

Larsen, C.P.S., and MacDonald, G.M. 1998. An 840-year record of fire and vegetation in a boreal white spruce forest. *Ecology* 79:106–118.

Lloyd, A.H., and Graumlich, L.J. 1997. Holocene dynamics of treeline forests in the Sierra Nevada. *Ecology* 78:1199–1210.

Malanson, G.P., and Westman, W.E. 1989. Modeling the interactions of fire regime, air pollution, and CO_2-induced climate change on Californian coastal sage scrub. *Clim. Change* 18:363–376.

Markgraf, V., and Anderson, L. 1994. Fire history of Patagonia: Climate *versus* human cause. *Rev. Instit. Geogr. Sao Paulo* 15:35–47.

McGlone, M.S., Kershaw, A.P., and Markgraf, V. 1992. El Niño/Southern Oscillation climatic variability in Australasian and South American paleoenvironmental records. In *El Niño. Historical and Paleoclimatic Aspects of the Southern Oscillation*, eds. H.F. Díaz and V. Markgraf, pp. 435–462. Cambridge: Cambridge University Press.

Miller, A. 1976. The climate of Chile. In *World Survey of Climatology*, ed. W. Schwerdtfeger, pp. 113–145. Amsterdam: Elsevier.

Ortlieb, L., and Macharé, J. 1993. Former El Niño events: records from western South America. *Global Planet. Change* 7:181–202.

Overpeck, J.T., Rind D., and Goldberg, R. 1990. Climate-induced changes in forest disturbance and vegetation. *Nature* 343:51–53.

Pederson, B.S. 1998. The role of stress in the mortality of Midwestern oaks as indicated by growth prior to death. *Ecology* 79:79–93.

Pittock, A.B. 1980. Patterns of climatic variation in Argentina and Chile. I. Precipitation, 1931–60. *Mon. Wea. Rev.* 108:1347–1361.

Price, A.J., and Rind, D. 1994. The impact of $2 \times CO_2$ climate on lightning-caused fires. *J. Clim.* 7:1484–1494.

Quinn, W.H., and Neal, V.T. 1992. The historical record of El Niño events. In *Climate since A.D. 1500*, eds. R.S. Bradley and P.D. Jones, pp. 623–646. London: Routledge.

Quinn, W.H. 1992. A study of the Southern Oscillation-related climatic activity for A.D. 622–1900 incorporating Nile river flood data. In *El Niño: Historical and Paleoclimatic*

Aspects of the Southern Oscillation, eds. H.F. Dĩaz and V. Markgraf, pp. 119–149. Cambridge: Cambridge University Press.

Ropelewksi, C.F., and Jones, P.D. 1987. An extension of the Tahiti–Darwin Southern Oscillation Index. *Mon. Wea. Rev.* 115:2161–2165.

Rothkugel, M. 1916. *Los Bosques Patagónicos.* Ministerio de Agricultura, Buenos Aires.

Rutllant, J., and Fuenzalida, H. 1991. Synoptic aspects of the central Chile rainfall variability associated with the Southern Oscillation. *Int. J. Climatol.* 11:63–76.

Schwerdtfeger, W. 1976. Introduction. In *World Survey of Climatology*, ed. W. Schwerdtfeger, pp. 1–12. Amsterdam: Elsevier.

Sirois, L., and Payette, S. 1991. Reduced postfire regeneration along a boreal forest–tundra transect in northern Quebec. *Ecology* 72:619–629.

Swetnam, T.W. 1993. Fire history and climate change in giant sequoia groves. *Science* 262:885–889.

Swetnam, T.W., and Betancourt, J.L. 1990. Fire–Southern Oscillation relations in the southwestern United States. *Science* 249:1017–1020.

Swetnam, T.W., and Betancourt, J.L. 1992. Temporal patterns of El Niño/Southern Oscillation—Wildfire teleconnections in the southwestern United States. In *El Niño: Historical and Paleoclimatic Aspects of the Southern Oscillation*, eds. H.F. Díaz and V. Markgraf, pp. 259–270. Cambridge: Cambridge University Press.

Swetnam, T.W., and Betancourt, J.L. 1998. Mesoscale disturbance and ecological response to decadal climatic variability in the American Southwest. *J. Clim.* 11:3128–3147.

Taljaard, J.J. 1972. Synoptic meteorology of the Southern Hemisphere. *Meteorol. Monogr.* 13:139–213.

Thornthwaite, C.W. 1948. An approach toward a rational classification of climate. *Geogr. Rev.* 38:55–94.

Tortorelli, L.A. 1947. *Los Incendios de bosques en la Argentina.* Ministerio de Agricultura, Buenos Aires.

van Loon, H., Kidson, J.W., and Mullan, A.B. 1993. Decadal variation of the annual cycle in the Australian data set. *J. Clim.* 6:1227–1231.

Veblen, T.T., and Lorenz, D.C. 1988. Recent vegetation changes along the forest/steppe ecotone in northern Patagonia. *Ann. Assoc. Am. Geogr.* 78:93–111.

Veblen, T.T., Kitzberger, T., and Lara, A. 1992. Disturbance and forest dynamics along a transect from Andean rain forest to Patagonian shrubland. *J. Veg. Sci.* 3:507–520.

Veblen, T.T., Kitzberger, T., Villalba, R., and Donnegan, J. 1999. Fire history in northern Patagonia: The roles of humans and climatic variation. *Ecol. Monogr.* 69:7–67.

Villalba, R. 1990a. Latitude of the surface high-pressure belt over western South America during the last 500 years as inferred from tree-ring analysis. *Quat. S. Am. Antarc. Penin.* 7:273–303.

Villalba, R. 1990b. Climatic fluctuations in northern Patagonia during the last 1000 years as inferred from tree-ring records. *Quat. Res.* 34:346–360.

Villalba, R. 1994. Tree-ring and glacial evidence for the Medieval Warm Epoch and the Little Ice Age in southern South America. *Clim. Change* 26:183–197.

Villalba, R. 1995. Climatic influences on forest dynamics along the forest-steppe ecotone in northern Patagonia. Ph.D. dissertation. Department of Geography, University of Colorado, Boulder.

Villalba, R., and Veblen, T.T. 1997a. Spatial and temporal variation in tree growth along the forest-steppe ecotone in northern Patagonia. *Can. J. For. Res.* 27:580–597.

Villalba, R., and Veblen, T.T. 1997b. Regional patterns of tree population age structures in northern Patagonia: Climatic and disturbance influences. *J. Ecol.* 85:113–124.

Villalba, R., and Veblen, T.T. 1998. Influences of large-scale climatic variability on episodic mortality at the forest-steppe ecotone in northern Patagonia. *Ecology* 79: 2624–2640.

Villalba, R., Cook, E.R., Jacoby, G.C., D'Arrigo, R., Veblen, T.T., and Jones, P.D. 1998. Tree-ring based reconstructions of northern Patagonia precipitation since A.D. 1600. *Holocene* 8:677–692.

White, W.B., and Peterson, R.G. 1996. An Antarctic circumpolar wave in surface pressure, wind, temperature and sea-ice extent. *Nature* 380:699–702.

11. Fire Regimes and Forest Dynamics in the Lake Region of South-Central Chile

Antonio Lara, Alexia Wolodarsky-Franke, Juan Carlos Aravena, Marco Cortés, Shawn Fraver, and Fernando Silla

Fire is one of the major disturbances shaping the vegetation and landscape patterns in the Lake Region of south-central Chile (39°30′–43°30′S). Most of these fires occurred after the European settlement in the area, which started ca. 1750, but it was not until the 1850s that extensive settlement took place which led to massive burning and clearing of forests for agriculture and pasture land (Elizalde 1970; Wilhelm 1968). Recent research from tree rings in the Cordillera Pelada, (ca. 40° S) has documented fires in the last 600 years, that may be attributed to both lightning and the native human population (Lara et al. 1999a). Research from pollen records and Quaternary stratigraphy indicates the extensive occurrence of fire in southern South America, since about 13,000 BP (Heusser 1994). A long history of fire occurrence has also been found in central Chile (see Aravena et al., Chapter 12, this volume) and in Patagonia, Argentina (Veblen et al., Chapter 9, Kitzberger and Veblen, Chapter 10, Huber and Markgraf, Chapter 13, this volume).

Forest dynamics of various vegetation types in the region and their relation to different kinds of disturbances—especially volcanism, landslides, logging, and fire—have been described by several studies (Veblen and Ashton 1978, 1982; Veblen et al. 1981; Veblen 1983, 1985; Veblen et al. 1996). Nevertheless, the detailed study of fire regimes, and their relation to forest dynamics is only incipient in the Chilean Lake Region (Lara et al. 1999a). In contrast, the ecological role of fire has received substantial research attention in the forests of northern Patagonia, Argentina (Veblen et al. 1995, Veblen et al., Chapter 9, this volume).

In this chapter we describe the environmental and vegetation patterns in the Lake Region; we analyze fire regimes from recent fire records and the influence of fire on the dynamics of *Fitzroya cupressoides* forests. We also analyze the influence of fires on forest conservation in the region, and we make some recommendations for future research.

The Lake Region of Chile

The Lake Region of Chile extends from ca. 39°30′ to 43°30′S and corresponds to the Xth Administrative Region of the country. Three main physiographic features characterize the region: the coastal range, the Central Depression, and the Andean Range (Fig. 11.1). The coastal range is a relatively low, narrow mountain range with rounded tops and gentle slopes, reaching up to 1048 m elevation at Cerro Mirador (40°10′S), and decreasing toward the south. The Central Depression represents an extensively glaciated low and relatively flat area, with elevations under 150 m (Fig. 11.1). The Andean Range is characterized by frequent steep slopes and several peaks and volcanoes above 2200 msl, with its eastern slopes located in Argentina.

Geology, Soils, and Climate

Most of the Central Depression and Andean Range of the Lake Region was covered by ice during the Quaternary glaciations, until ca. 15,000 to 13,000 BP when glaciers retreated (Mercer 1976; Porter 1981; Denton 1993; Clapperton 1994). These glaciations originated most of the lakes in the area, and left an extensively glaciated landscape.

The geology and soils of the region vary with the previously mentioned physiographic features. The coastal range is a chain of metamorphic bedrock of Paleozoic to Precambrian age. Soils vary from moderately deep and well drained at mid-elevations to thin, acidic, sandy, and poorly drained soils with varying degrees of formation of a gley horizon toward the flat tops (Lusk 1996). In the Central Depression, soils are developed from thick layers of Quaternary fluvioglacial and volcanic sediments. Soils developed from old tephra on morraines are typically deep (80–120 cm), loamy or clay in texture, and well drained. Soils, developed on outwash plains of fluvioglacial pavement and other flat areas, called *ñadis* are thin (20–30 cm deep), poorly drained or seasonally flooded (INIA 1985). The Andean Range in the Lake Region is a geologically complex system dominated by granitic and sedimentary rocks, with the local presence of metamorphic rocks (Levi, Aguilar, and Fuenzalida 1966; Servicio Nacional de Geología y Minería 1982; Kühne 1985). Plio-pleistocene and Holocene volcanic sediments are widespread, and glacial and fluvioglacial sediments are common (Levi, Aguilar, and Fuenzalida 1966; Mercer 1976; Servicio Nacional de Geología y

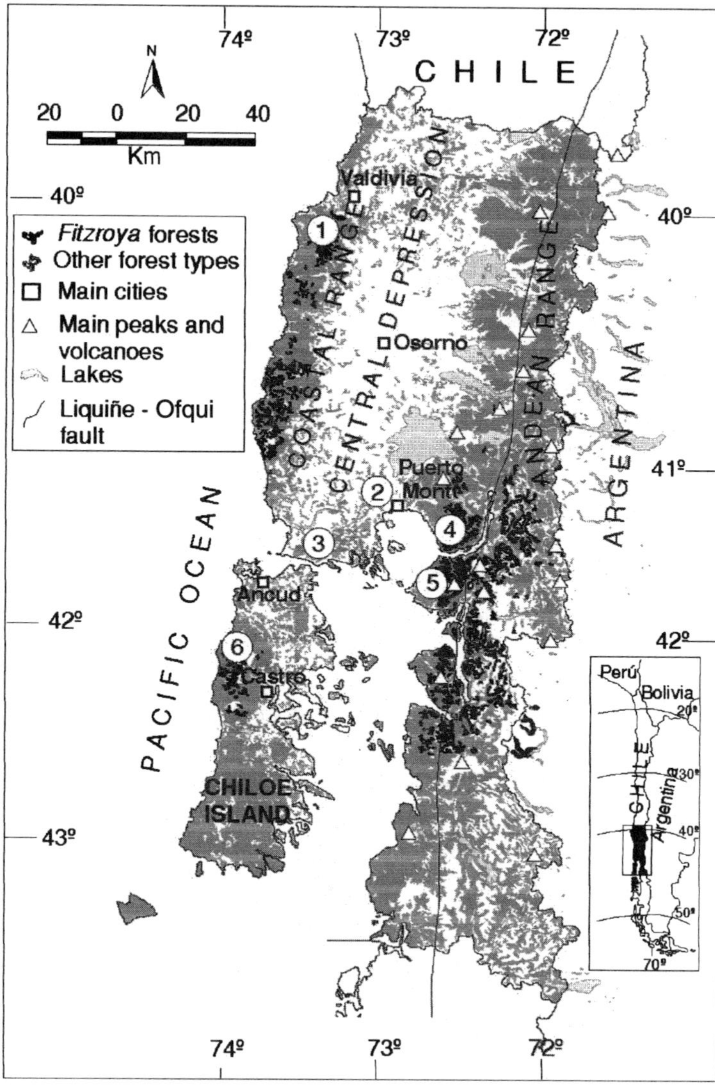

Figure 11.1. Location map of the study and sampling areas in the Lake Region in south-central Chile. Sampling sites: (1) Cordillera Pelada, (2) Pto. Montt, (3) Astilleros, (4) Contao, (5) Alerce Andino, and (6) Abtao. (Cover of *Fitzroya* and other forest types developed from CONAF et al. 1999.)

Minería 1982). The Lake Region is an active tectonic zone, with the Liquiñe-Ofqui fault running from north to south along the region and further south (Fig. 11.1; Hauser 1984). Soils below 1500 m elevation, where most forests occur, are called *trumaos*. These are volcanic soils, generally deep (80–150 cm), loamy, and well drained (INIA 1985).

Climate of the Lake Region in Chile is characterized by high annual precipitation, with a somewhat lower rainfall in summer. It is classified as oceanic wet temperate with mild Mediterranean influence (Fuenzalida 1950; Di Castri and Hajek 1976). There is a general increase in precipitation and decrease in seasonality toward the south. Since moisture is brought by the westerly winds, there is an important west-to-east gradient, with a strong rainshadow effect in the Central Depression and in the eastern slopes of the Andes in Argentina. Annual rainfall ranges from 1800 mm in the Central Depression to more than 4000 mm at the tops of the coastal and the Andean Ranges. Average July and January temperatures at the Central Depression are 8° and 16°C, respectively (Almeyda and Saez 1958).

Vegetation and Disturbance Regimes

Vegetation varies dramatically within the Chilean Lake Region, according to the north-to-south and west-to-east physiographic zones and environmental gradients, and the degree of human disturbance. The total area covered by native forests is 3.6 million hectares, representing 54% of the Lake Region. Forest plantations (mainly *Pinus radiata* and to a less extent *Eucalyptus* spp.) cover 117,000 ha (CONAF et al. 1999).

Most of the native forests are concentrated in the Andean Range (59% of the total), where the influence of human disturbance by fire and clearing for agriculture and pasture has been more restricted. Conversely, only 10% of the native forests are currently located in the Central Depression because of extensive clearing that began in the 1850s. The coastal range has an intermediate situation with a 31% of the native forests cover (Lara 1991).

In the Lake Region, forests are dominated by the Valdivian rain forest (tipo *forestal siempreverde* according to the current classification of forest cover types, Donoso 1981), representing 54% of the native forests in the region (CONAF et al. 1999). This forest type occurs at low and mid-elevations across the region, and is characterized by mixed forests with a high vascular plant diversity. The flora of these forests includes 155 woody species, 44 tree species, and 28 genera, 28% of which are endemic to Chile and the adjacent area of Argentina (Kalin et al. 1996). The main tree species are *Nothofagus dombeyi*, *N. nitida*, *Eucryphia cordifolia*, *Laureliopsis philippiana*, *Weinmannia trichosperma*, as well as several species in the myrtaceae family such as *Amomyrtus luma*, *A. meli*, *Myrceugenia planipes*, and *Tepualia stipularis* (Donoso 1981; Donoso 1993).

Other important forest types in the region are the *Nothofagus dombeyi–N.alpina–Laureliopsis philippiana* forests, occurring mainly as old-growth at mid-elevations (500–1000 m), as well as *Nothofagus obliqua–N. alpina–N. dombeyi* forests, mainly as second-growth forests at low and mid-elevations (50–1000 m); each forest type covers 8.3% and 7.8% of the forests in the Lake Region, respectively (CONAF et al. 1999). In the Andes, *Nothofagus pumilio* subalpine forests dominate from ca. 1000 m, and form the treeline at ca. 1400 to 1600 m in elevation, representing 15.9% of the forests in the region (CONAF

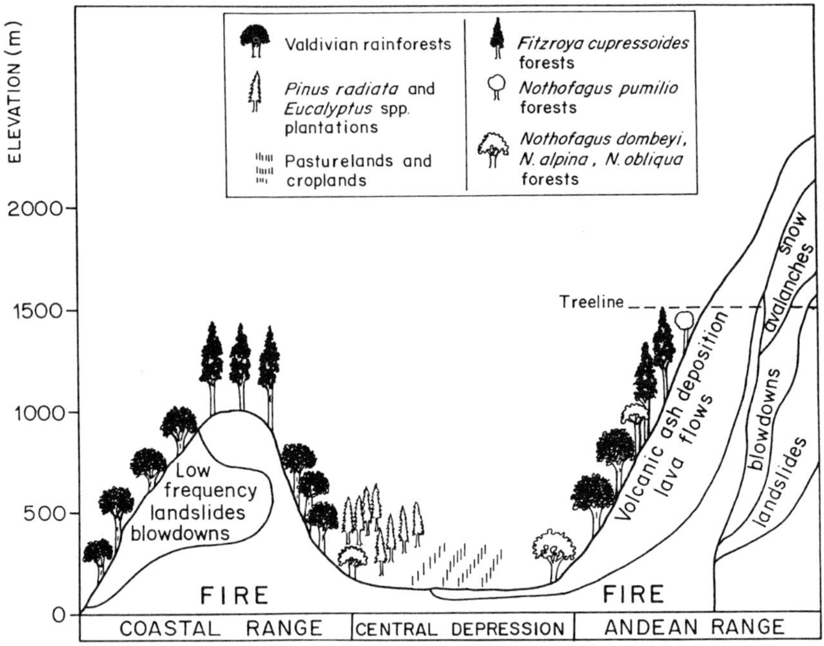

Figure 11.2. Present vegetation and disturbance regimes across the Lake Region in Chile. The relative widths of the different kinds of disturbance indicate their relative importance at a given position in the west-to-east and elevation gradients.

et al. 1999). *N. betuloides* forests also grow in the subalpine zone, and this species is also mixed with *N. pumilio*. Among the conifer forests the most extensive are the *Fitzroya cupressoides* forests that grow from 600 to 1000 m elevations in the coastal range, and from 400 to 1200 m elevation in the Andes. Some small remnant populations are also found in the Central Depression (Lara et al. 1999b; Silla 1997; Fraver et al. 1999).

Disturbance regimes vary significantly with vegetation along the environmental gradients created by the coastal range, the Central Depression, and the Andean Range (Fig. 11.2). Fire is a widespread disturbance across the region. It is the main disturbance in the Central Depression and toward the summits of the coastal range, and at lower elevations in the Andes (Fig. 11.2). In the Andes there are several other kinds of disturbances, such as volcanism, landslides, wind throw, and snow avalanches, of varying relative importance according to elevation (Fig. 11.2).

Fire Regimes from Recent Fire Records

The Chilean Forest Service (CONAF) is in charge of fire suppression, and has kept reliable fire records in the Lake Region since 1979 (CONAF 2000). These records are organized according to vegetation cover type where they occur: native forests (including old-growth and second-growth forests of various heights and crown cover classes, under the categories *arbolado* and *matorrales*), forest plantations, and grasslands. For the purpose of this analysis, we considered the fires that have affected native forests in the Lake Region (Xth Administrative Region). Following Schulman's (1956) convention for tree rings in the Southern Hemisphere, the fire seasons (October–March of the following year) were named according to the calendar year in which the fire season began (e.g., 1979 for the fire season that starts in 1979 and ends in 1980).

Available records about fire origin cover the 1985 to 1999 period (Table 11.1). All fires in the Lake Region are attributed to human action (CONAF 2000). Although lightning and volcanism have been documented as sources of ignition in this region (Veblen et al. 1996; Lara et al. 1999a), these natural fires are less frequent and are not recognized as a separate ignition cause by the available records. The main causes of fires are classified as intentional (i.e., started with the purpose of forest clearing) and forest activities (i.e., started from logging operations, burning of slash for establishing plantations) accounting for 30% and 24% of the number of fires in the period 1985 to 1999, respectively (Table 11.1; CONAF 2000).

Forest fires show great annual variability, related to summer precipitation (Fig. 11.3). The annual area of native forests burned in the Lake Region ranges between 69 and 38,387 ha, with a mean area of 4969 ha (SD = 8930 ha). The correlation coefficient (r^2) of the logarithm of December through February precipitation with the logarithm of the burned area in the 1979 to 1999 period is 0.42, which is statistically not significant. The forest area burned annually in the period 1979 through 1999 shows a flat curve with most years below the mean, and five outlier years which match dry summers (December–February precipitation <45 mm: 1982, 1986, 1987, 1995 and 1997, Fig. 11.3). Three of these dry summers, with a large burned area, are related with very strong ENSO warm events (1982–83, 1986–87, and 1997), whereas the 1991–92 ENSO warm event did not cause a decrease in the precipitation in southern Chile (Thudhope et al. 2001). However, dry summers occurred in 1988, 1990, and 1998 do not show an increase in the

Table 11.1. Number of fires in the Lake Region classified by their origin

Fire origin	Number of fires (%)[a]
Intentional	30
Forest activities	24
Unknown causes	19
Agriculture activities	12
Transportation	10
Recreation	5

Source: CONAF 2000.
[a] Includes a total of 8624 fires recorded from 1985 to 1999 and all vegetation cover types (native forests, forest plantations, grasslands, and pastures).

area of forests burned (Fig. 11.3), indicating that fire occurrence is influenced by the variability of both climate and ignition by humans, and probably by other factors that have not been identified yet. The total number of fires affecting native forests in the period 1979 to 1999 is 8624 with an average of 410.7 fires per year (CONAF 2000). The mean area per fire is 9.9 ha (SD = 14.5 ha), with extreme values of 0.6 and 63.4 ha per fire for 1999 and 1997, respectively (Fig. 11.4; CONAF 2000). Years with a large area burned also have large mean area per fire (Figs. 11.3 and 11.4), with a positive correlation between both variables ($r^2 = 0.96$, $p < 0.001$).

Spatially, there is also a wide range of variation in the rate of forest destruction by fire in the 1979 to 1999 period among the counties in the Lake Region (Fig. 11.5). This rate for a given county was defined as the ratio between the area of native forests burned in the 1979 to 1999 period and the area of native forests at the beginning of the period multiplied by 100. The latter was estimated as the area of native forests in 1999 plus the area of native forests burned in the period in a given county. Data were taken from CONAF et al. (1999) and CONAF

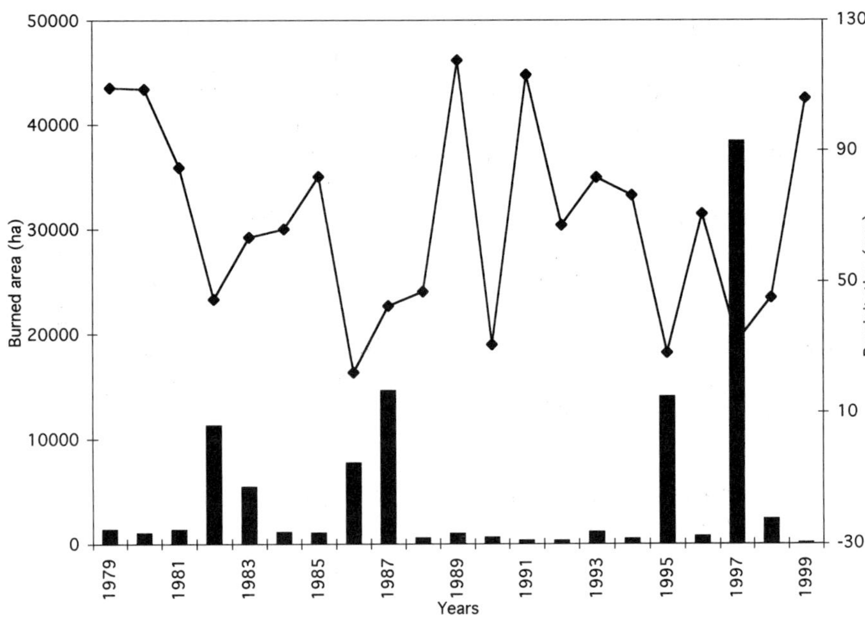

Figure 11.3. Area of native forests burned per fire season (hectares) in the Lake Region (bars; CONAF 2000). December through February summer precipitation in Valdivia for the 1979 to 1999 period (lines; Huber, unpublished data). Fire seasons include October through March, and both fire and precipitation in a particular summer are named according to the calendar year in which the fire season began. Native forests include dense and open old-growth forests as well as second-growth forests of various crown cover and height classes.

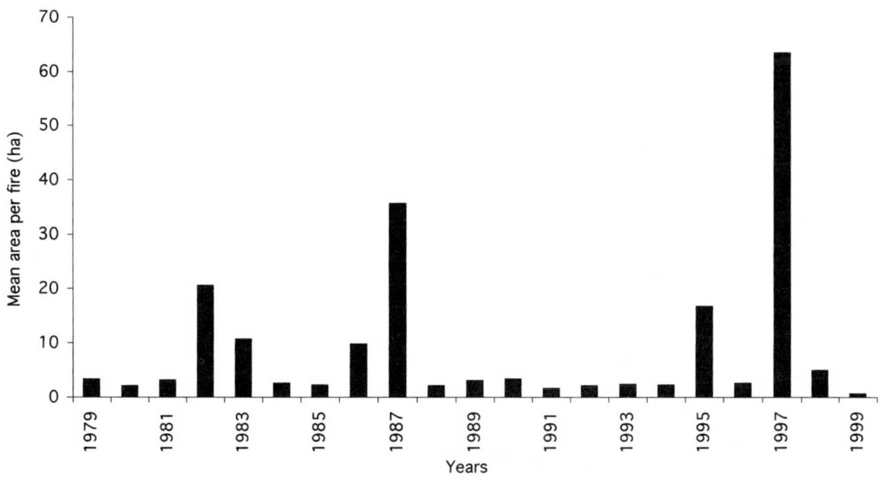

Figure 11.4. Mean area burned of native forests per fire (hectares) during the 1979 to 1999 period in the Lake Region. (Area burned according to CONAF 2000.)

(2000). The county of Fresia has the highest rate of forest destruction by fire (>15%), followed by San José de la Mariquina (North of Valdivia) and Castro (in the center of Chiloé Island) both in the category 11.01–15.0% (Fig. 11.5). Counties near the cities of Valdivia, Osorno, and Puerto Montt have intermediate rates of fire occurrence (5.01–11%, Fig. 11.5). Counties located in the Andean Range and the southern portion of Chiloé Island show low rates of forest destruction by fire (Fig. 11.5).

Fire Regimes and Dynamics of *Fitzroya cupressoides* Forests

Fitzroya cupressoides is the largest and longest-lived conifer that grows in Chile and Argentina; it reaches up to 5 m in diameter and 50 m in height, living up to 3600 years (Veblen, Delmastro, and Schlatter 1976; Lara and Villalba 1993). In Chile it grows as discontinuous populations from 39°50′ to 43°30′S in humid areas on nutrient-poor soils. The dynamics of *Fitzroya* forests are related to disturbances, mainly volcanism, landslides, and fire (Veblen and Ashton 1982; Lara 1991; Donoso et al. 1993; Lara et al. 1999a). Extensive logging and destruction from human-set fires have reduced the natural range of the species (Veblen, Delmastro, and Schlatter 1976; Donoso 1993).

The main regeneration mechanism in *Fitzroya* is by root sprouting on sites affected by low-intensity fires in the coastal range, and by layering of low branches in the Central Depression (Cortés 1990; Silla 1997; Silla et al. 2001). In the Chilean Andes regeneration is both by root sprouting and seeds (Lara 1991).

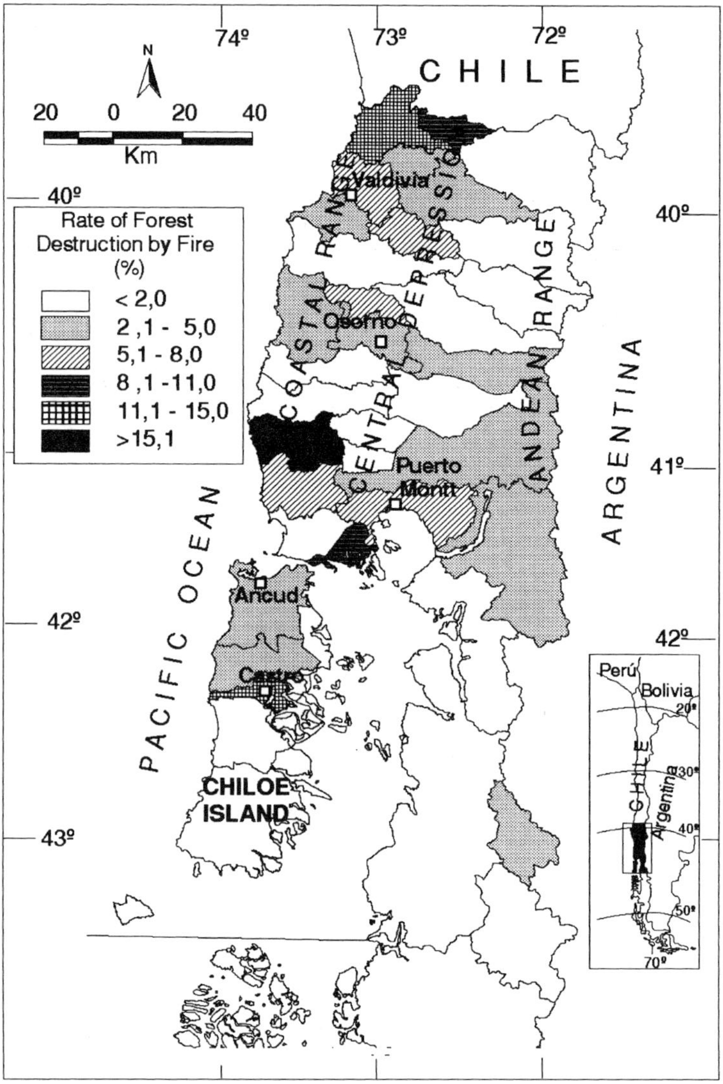

Figure 11.5. Map indicating the rate of forest destruction by fire in the Lake Region, during the 1979 to 1999 period, calculated as (burned area / (burned area + area of native forest in 1997)) × 100. (Area burned from CONAF 2000; area of native forests from CONAF et al. 1999.)

Coastal Range

On the coastal range we studied *Fitzroya* stands in two different areas: Cordillera Pelada (40°10′S) and Abtao (42°30′S) (Fig. 11.1). In the upper gentle slopes and flat mountain plateaus (800–900 m of elevation in Cordillera Pelada), we selected four stands that differ in the time since the last stand-devastating fire: from stand A (most recently disturbed) through D (free of disturbances for a long period). Detailed methods and site descriptions are given in Lara et al. 1999a). In Abtao we analyzed stand E, located on the flat tops of Cordillera de Piuchué within Chiloé National Park at 650m of elevation. Further descriptions for this stand are given in Armesto et al. (1996).

The stands and the area around Piedra del Indio (elevation 900m) and other neighboring areas in Cordillera Pelada were searched for stumps with fire scars, following methods described by Dietrich and Swetnam (1984). We used standard

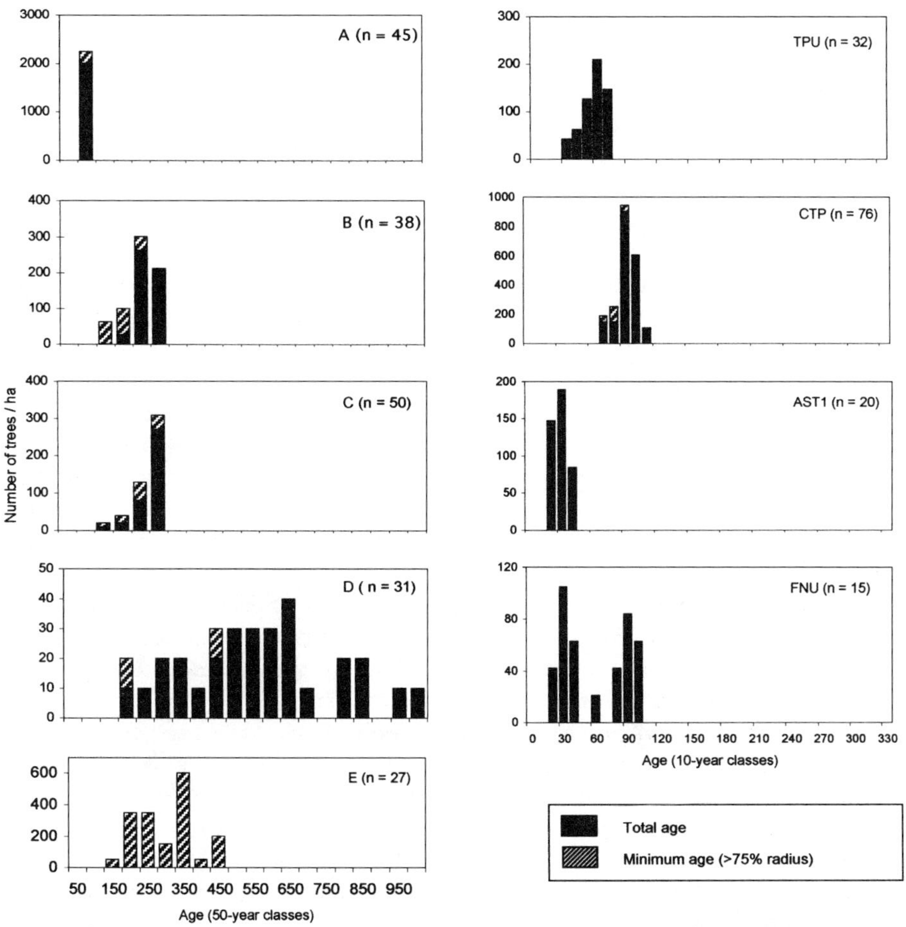

Figure 11.6. Age structures at coring height. Stands A, B, C, and D are located in Cordillera Pelada; stand E is located in Abtao. Stands TPU, CTP, AST1, and FNU are located in the Central Depression. (Data from Lara et al. 1999a; Aravena, unpublished data; and Silla 1997.)

dendrochronological methods to produce tree-ring-width chronologies for each area in order to date the fire scars (Stokes and Smiley 1968; Fritts 1976). We also determined the age structure at coring height (ca. 30 cm) for each stand, assuming that relatively even-aged stands were established following a stand-devastating fire (Veblen 1985). Duncan's (1989) method was used to estimate the missing rings to the center, as most tree cores did not reach the pith. This method has certain limitations, since it assumes that rings form concentric circles around the pith and that ring width is constant in the missing part of the sample (Villalba and Veblen 1997; Kitzberger, Veblen, and Villalba 2000). From our tests on *Fitzroya* cross sections we accepted a maximum estimate of 25 rings to center (Lara et al. 1999a). Ages are estimated at coring height (ca. 30 cm), and a period of 10 to 20 years since germination is estimated for a *Fitzroya* seedling to reach coring height in the sites studied in Cordillera Pelada (Lara et al. 1999a).

Age-class distributions for *Fitzroya* in each stand are shown in Figure 11.6. Stands A, B, and C are even-aged single cohort stands. In Stand A, *Fitzroya* regenerated after a stand-devastating fire (not dated), as indicated by the presence of dead and charred *Fitzroya* trees 35 to 70 cm in diameter at breast height (dbh) compared to the 5 to 21 dbh of the living trees in this stand. Stand B was a young postfire stand, also presenting charred snags, 84% of living trees with fire scars, fallen logs, and large dead trees (Lara et al. 1999a). The presence of large dead trees in stand C, together with its age-class distribution, indicate that it rapidly became established after a stand-devastating disturbance, perhaps fire. The lack of fire scars on living trees reflects the absence of recent fire.

Stand D was an old-growth mixed-species stand where *Fitzroya* shares its dominance with *Nothofagus betuloides*, *Pilgerodendron uvifera*, and *Drimys winteri*. The age-class distributions for *Fitzroya* ranges from 150–199 to 950–999 age classes, indicating slow or sporadic regeneration in this open-canopy stand due to a poorly drained site (Lara et al. 1999a; Fig. 11.6). Although the origin of this stand is not clear, the presence of *Fitzroya* snags with diameters much larger than the living trees may indicate a postfire origin.

In Abtao, stand E shows a broadly even-aged character (Fig. 11.6). All the trees in this stand are dead, the outer sapwood rotten or partially burned or charred. Therefore determination of the date when the trees were killed, probably by an intense stand-devastating fire, was not attempted. The age structure of this stand shows a single cohort, indicating that probably it was originated following a stand-devasting fire, emphasizing the repeated occurrence of this kind of disturbance. In nearby stands in Abtao, seedlings of *Fitzroya* and *Pilgerodendron uvifera* have been described as being abundant in areas with low canopy cover where dead standing trees are predominant (Armesto et al. 1996).

Based on the fire dates determined from fire-scarred *Fitzroya* stumps, we produced a fire chronology for stand B and Piedra del Indio (Fig. 11.7). Our results indicate that *Fitzroya* can survive low-intensity fires, forming up to four fire scars on the same tree. In Piedra del Indio, the oldest dated fire occurred in 1397, and other fires were dated in 1539, 1643, and 1750. In stand B, the oldest fire was

Figure 11.7. *Fitzroya cupressoides* dead standing trees killed by fire in Cordillera Pelada, near stand B and Piedra del Indio. The background shows even-aged *Fitzroya* stands developed after fire. (Photograph: Carlos Le Quesne 2001.)

dated in 1739 followed by fires in 1876 and 1943, dated from 1, 9, to 3 trees, respectively (Lara et al. 1999a; Fig. 11.8). The postfire origin of stand B becomes clear from the pith dates at ground level determined for eight trees that formed the main cohort. These dates indicated that these trees became established between 1753 and 1756 (14–17 years after the 1739 fire, Fig. 11.8; Lara et al. 1999a).

The relationship between drought and fire occurrence found from fire and instrumental precipitation records since 1979, already discussed, can also be found in the fire and precipitation records reconstructed from tree rings. Interestingly the years 1876 and 1943, when fires were dated in stand B, are among the driest for the last centuries from climatic reconstructions from *Austrocedrus chilensis* tree-ring chronologies for Argentinean northern Patagonia (Villalba et al. 1998), and from *Nothofagus pumilio* tree-ring chronologies for the central Andes in Chile (Lara et al. 2001).

Central Depression

In the Central Depression we studied *Fitzroya* stands in the area near Puerto Montt (41°15′S), in flat sites at 100 to 150 m of elevation, growing over *ñadi* poorly drained soils (Silla 1997; Silla et al. 2001; Fig. 11.1). We applied methods similar to those described for the coastal range. Here we describe four stands—TPU, CTP, AST1, and FNU—representing a range of time since last stand-devastating disturbance. These small stands are remnants of the extensive *Fitzroya*

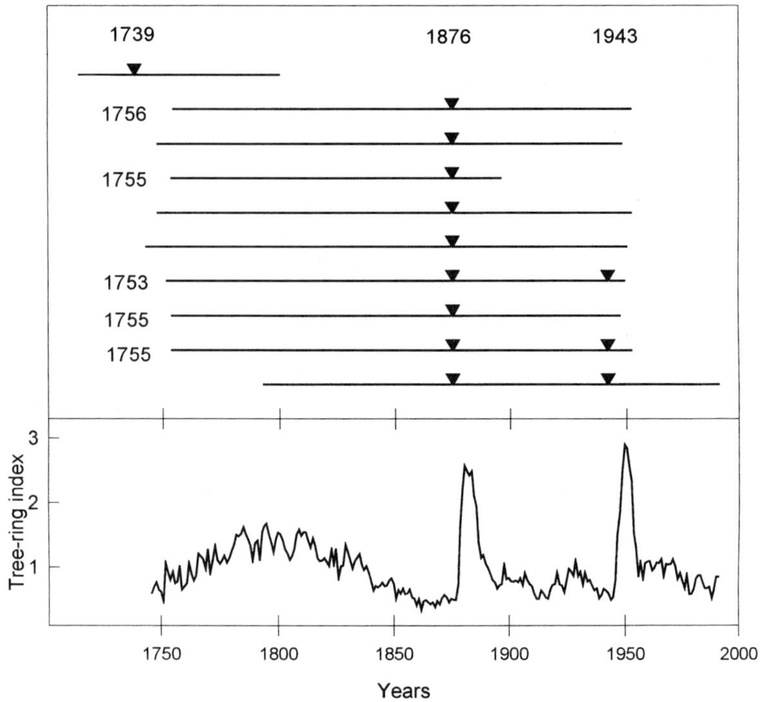

Figure 11.8. Fire chronology (*top*) and tree-ring-width chronology (*bottom*) for stand B in Cordillera Pelada. The horizontal lines represent the lifespan of individual trees, indicating the pith date. Black triangles are fires from scars, indicating their date on top. The tree-ring chronology used 10 to 15 trees and a horizontal standardization. Tree-ring indices provide a dimensionless indicator of radial growth.

forests that covered this area before the European settlement that started in the 1850s (Wilhelm 1968; Donoso 1993; Fraver et al. 1999). Despite the devastating fires to which these forests were exposed, *Fitzroya* was capable of colonizing certain sites, creating dense even-aged stands (Silla 1997; Silla et al. 2001).

Stands TPU, CTP, and AST1 show bell-shaped age structures, which indicate even-aged cohorts and a rapid establishment of *Fitzroya* after a stand-devastating disturbance. Presence of charred snags and stumps indicate that this disturbance probably was fire (Fig. 11.6). Stand FNU shows two cohorts with ages at coring height ranging from 80 to 109 years for the oldest one and 20 through 49 years for the youngest one, according to the age classes that are present (Fig. 11.6). The oldest cohort was originated after a stand-devastating disturbance, probably fire. The youngest cohort seems to have been established after a low intensity fire, which many older trees survived. This latter interpretation is supported by the presence of a growth release of many of the older surviving trees starting in 1943 and of abundant charred older living trees (Silla 1997).

Andean Range

The most important disturbances in the *Fitzroya* forests in the Andes are tephra deposition, landslides, lava flows, and logging (Schmidt and Burgos 1977; Rodríguez 1989; Lara 1991; Fig. 11.2). Fire is a minor type of disturbance in this area (Lara 1991). Available data indicate that in two study areas in the Andes (Contao and Alerce Andino National Park, Fig. 11.1), over a total area of 26,900 ha, human-set fires represent 0.45% of the total disturbed area (13,260 ha) in the 1943 to 1990 period (Lara 1991).

Fitzroya regeneration in areas affected by clear-cutting, selective logging, or human-set fires in the Andes is absent or extremely scarce (Veblen, Delmastro, and Schlatter 1976; Schmidt and Burgos 1977; Rodríguez 1989; Lara 1991; Donoso et al. 1993). Nevertheless, there are no specific studies addressing the influence of fire in the dynamics of *Fitzroya* forests in the Chilean Andes.

Fires and Forest Conservation

Human-set fires for clearing of forests for the development of pasture and agriculture land has been a major disturbance and the main cause of reduction of forest cover in the Lake Region since the extensive European settlement starting in the 1850s (Wilheim 1968; Elizalde 1970; Donoso 1983). This settlement of the Lake Region resulted in one of the most massive and rapid deforestation processes recorded in Latin America, which prevailed until the early 1980s (Veblen 1983).

The reconstruction of forest cover prior to the European settlement from historical documentary data, and potential sites using Geographic Information System (GIS) estimates that prior to the European settlement, native forests covered 5.6 million hectares in the Lake Region (Lara et al. 1999b). This means that the present native forest cover of 3.6 million hectares in this region represents 62% of the presettlement condition. The forest cover types that were more dramatically affected are *Pilgerodendron uvifera*, *Nothofagus* spp., and *Fitzroya*, with remaining fractions of 22%, 39%, and 46%, respectively, compared to the presettlement condition (Lara et al. 1999b). At the same time the area of grasslands, shrublands, and agriculture land increased from covering less than 1% of the region to 29% after the European settlement (Lara et al. 1999b).

By the turn of the nineteenth century extensive areas formerly covered by native forests in the Lake Region had been burned by human-set fires and converted to pasture and agriculture land, especially in the Central Depression (Elizalde 1970; Donoso 1983). Although reliable data are not available, the rate of forest destruction by human-set fires probably decreased through the twentieth century. Nevertheless, as previously discussed, fire records demonstrate that human-set fires have continued as an important disturbance and cause of forest destruction in the Lake Region in the last two decades until present. Other important causes of native forest destruction and degradation in this recent period have been the conversion to *Pinus radiata* and *Eucalyptus* spp. plantations and logging through high-grading (Lara, Donoso, and Aravena 1996).

Fire records for the last two decades indicate that there is a high spatial heterogeneity in the rate of forest destruction by fire through the Lake Region (Fig. 11.5). Rate of forest destruction by fire in the 1979 to 1999 period varies from <2% of the area of native forests existing in 1979 in 16 counties to 11.1–15% in two other counties (Fig. 11.5). If this latter rate is maintained in the future, it would take between 140 and 191 years to burn an area equivalent to the total area of native forests existing in 1979 in these counties (i.e., rotation period; *sensu* White and Pickett 1985). These are rough estimates limited by the short period of observations, the high annual variability of the area burned, and the uncertainty of how the area burned each year will vary in the future. Nevertheless, these estimates indicate that human-set fires are an important threat to native forests in some counties, and that action should be taken for the conservation of the remaining native forests in these counties.

The impact of fire on forest conservation is higher in the counties located in the coastal range and the Central Depression compared to that in the Andes (Fig. 11.5). The areas that show low fire incidence coincide with those located within national parks, which are mainly concentrated in the Andes. The insufficient amount of protected areas in the Central Depression and the coastal range (with less than 1.4% and 4.4% of forest area being protected, respectively; Lara 1991) adds an extra pressure over these forests. This contrasts with the Andean Range, where 17% of the forests are within national parks and reserves.

Although fire records are not separated by forest type, these records have been kept for *Fitzroya* forests since 1987. These records indicate that human-set fires are an important threat to the conservation of *Fitzroya* forests. Due to its high-quality wood which is resistant to decay, *Fitzroya* forests have been extensively logged since the European settlement until now (Elizalde 1970; Veblen, Delmastro, and Schlatter 1976; Lara 2000). Exports of *Fitzroya* timber have contributed to increase the pressure over these forests. The awareness of the international community about these issues determined that in 1973 *Fitzroya* became listed in Appendix I under the Convention on International Trade of Endangered Species (CITES), and its international trade was forbidden until today. *Fitzroya* has also been included as a threatened species in the U.S. Endangered Species Act since 1979, which prohibits its importation into the United States (Anonymous 1979). The Chilean law has forbidden cutting of living *Fitzroya* trees through the Supreme Decree 490 since 1976. Nevertheless, this law permits cutting and trading of timber coming from trees that have been killed by fire, cut or were dead by natural causes before 1976 (Anonymous 1976). This weakness in the law has permitted, and to a certain extent promoted, intentional fires of *Fitzroya* forests with the purpose of getting dead trees for which a logging permit may be obtained, circumventing the law since the trees were killed after 1976. The law is also circumvented through the cutting of living trees and trading the wood as if it came from previously dead trees. Law enforcement by CONAF (Chilean Forest Service) in the remote areas where *Fitzroya* forests occur has been problematic and insufficient. In addition legal actions taken by the Local Courts of the Counties have been weak. Only in a small percentage of the cases denounced by CONAF have these Courts determined fines and sanctions against the violators of the law protecting *Fitzroya* (Lara 2000). The international protection of this species has also been difficult. Different interpretations of CITES regarding the restrictions on *Fitzroya* exports as well as limitations in the enforcement of the treaty have determined that the timber of this species continues to be exported mainly to Japan and Argentina and also to Australia, Spain, the Netherlands, and the United States (Lara 2000).

An extreme example of the threat to *Fitzroya* forests by human-set fires to kill the trees and then log them, circumventing the law is the case of the Fresia county. This county has the highest rate of forest destruction by fire in the Lake Region (28.8% between 1979 and 1999, which provides an estimation of the rotation period of 73 years). In the dry summer of 1997 in the Lake Region produced by an ENSO warm event, 9477 ha of *Fitzroya* forests were burned in the Fresia county, representing 68% of the total forested area burned that year in the county. In 1987, in another dry summer associated with an ENSO warm event, 11,000 ha of native forests were burned in Fresia, 12% of which corresponded to *Fitzroya* forests.

The large surfaces of burned forests in the coastal range and Central Depression, together with the scarce protection of forests in these areas, indicate the need to carry out programs toward the protection and conservation of these forests. Threatened forest types such as *Fitzroya* forests in the Central Depression and the coastal range should be considered a priority. Changes in the law protecting *Fitzroya* and the improvement of law enforcement are needed as well as the creation of new protected areas, especially in the coastal range and the Central Depression (Lara 2000). Studies on the genetic variability using DNA markers (RAPDs) have demonstrated important genetic variations of *Fitzroya* through its geographic range in Chile and Argentina (Alnutt et al. 1999). Ongoing ecological restoration efforts in the Central Depression using nursery-produced seedlings of local provenances to promote the genetic conservation of few remnant fragmented *Fitzroya* populations need to be strengthened and expanded to other areas (Gardner et al. 1999; Silla et al. 2001). The development of alternatives for the owners (e.g., use of other species for wood or nontimber forest products, eco-tourism) should also contribute to the conservation of these forests (Lara 2000).

Conclusion and Future Research

Vegetation patterns at a regional scale are a response to the combined effect of the environmental gradients (physiography, soils, moisture, temperature, etc.) and the disturbance regimes (Fig. 11.2). Since the disturbance regimes vary gradually along these environmental gradients, we propose that the physical environment controls vegetation both directly and indirectly through the disturbance gradient.

Fire has been an important disturbance throughout the Lake Region. Fire dates reported in this chapter, starting in AD 1397, indicate that some fires occurred prior to the European settlement of the Chilean Lake Region. In this period (prior to ca. 1850) fires might have been started by the native people, who traveled through the coastal range, or by lightning, which occasionally occurs during spring and summer storms. Several authors have presented evidence, from historical documents, of the influence of Native American hunters on fire ignition, in northern Patagonia, Argentina, prior to ca. 1890–1900 (Veblen and Lorenz 1988; Veblen and Markgraf 1988; Markgraf and Anderson 1994).

Most of the fires that have affected natural forests in the Lake Region in the 1979 to 1999 period are intentionally set by people or due to forest use and other activities, and are concentrated during the summer months. The annual variability of the burned area coincides with five dry summers, but there are three dry summers in which the burned area did not increased. This indicates that climate and human ignition variability, both factors and probably others that need to be identified, influence the fire regime of the Lake Region. Similarly detailed studies in northern Patagonia, Argentina, have shown important influences of climate and human activities on fire regimes along the gradient from the xeric woodlands near the Patagonian steppe to *Nothofagus* and *Fitzroya* rain forests in the Andes (Kitzberger and Veblen 1997; Kitzberger, Veblen, and Villalba 1997; Veblen et al. 1999; Veblen et al., Chapter 9, this volume, Kitzberger and Veblen, Chapter 10, this volume).

Repeated fires have played a major role in the dynamics of *Fitzroya* forests in the coastal range and Central Depression during the last 600 years. In contrast, fires do not seem to have played a crucial role in the forest dynamics of this forest type in the Chilean Andes. Thus the *Fitzroya* forests in the coastal range differ from those of the Chilean Andes in both their disturbance regimes and their ability to regenerate following fire. These differences in *Fitzroya* response to fire may be explained by reduced competition from other species under the lower soil nutrient availability of the coastal range compared to the Andes, but this hypothesis needs to be further investigated.

Our results indicate that there is a significant potential for the development of a network of fire chronologies from tree rings for the last 1000 years or more. Other potential studies on fire regimes could focus on the distribution of these records along latitudinal, longitudinal, and elevation gradients in order to produce a long-term, regional view and a better understanding of the relationships among the climate variability, fire regimes, and vegetation responses to both. These studies could be compared to similar ones already done in drier areas in central Chile and Argentinean Patagonia, and in North America. Specific aspects that need to be addressed are (1) parameters of the fire regimes such as intensity, mean return interval, and ignition sources; (2) the variation of these parameters along environmental gradients; and (3) the combined effect of different types of disturbance (i.e., fire, logging, grazing) on the regeneration and dynamics of different forest types, such as *Fitzroya cupressoides*, *Pilgerodendron uvifera*, *Araucaria araucana*, and *Nothofagus* spp. forests. Future research on fire regimes should be taken as a basis for adequate planning and for improving decision making related to forest management and conservation.

Acknowledgments. Financial support for this work was provided by FONDECYT (Project 1-93-0049), the National Geographic Society (Project 4987-93), Fundación Andes (Project C12600/9), a Darwin Initiative for the Protection of Species grant administered through the Royal Botanical Garden Edinburgh, European Commission DGXII (Contract ERBIC18CT970146), the CRN03 project of the Inter-American Institute for Global Change Research (IAI), and various grants from WWF. We are grateful to CONAF for providing permits for sample collection, for the use of the forest cover GIS database, and for their support during fieldwork. We thank J. Bosnich, L. Escandar, and S. Mendoza for providing the fire records database and E. Neira for preparing the figures.

References

Allnutt, T.R., Newton, A.C., Lara, A., Premoli, A., Armesto, J.J., Vergara, R., and Gardner, M. 1999. Genetic variation in *Fitzroya cupressoides* (alerce), a threatened South American conifer. *Mol. Ecol.* 8:975–987.

Almeyda, A.E., and Sáez, S.F. 1958. Recopilación de datos climáticos de Chile y mapas sinópticos respectivos. Ministerio de Agricultura, Santiago.

Anonymous. 1976. Decreto Superemo 490 que declara Monumento Natural al Alerce. Santiago: Ministerio de Agricultura.

Anonymous. 1979. Determination that *Fitzroya cupressoides* is a threatened species. *Fed. Reg.* 44:64730–64733.

Armesto, J.J., Aravena, J.C., Villagrán, C., Pérez, C., Parker, G.G., and Villagrán, C. 1996. Bosques templados de la Cordillera de la Costa. In *Ecología de los Bosques Nativos de Chile*, eds. J.J. Armesto, C. Villagrán, and M.K. Arroyo, pp. 199–212. Santiago: Editorial Universitaria.

Clapperton, C.M. 1994. The quaternary glaciation of Chile: a review. *Rev. Chil. Hist. Nat.* 67:369–383.

CONAF, CONAMA, Universidad Austral de Chile, P. Universidad Católica de Chile and Universidad Católica de Temuco. 1999. *Catastro y Evaluación de los Recursos Vegetacionales Nativos de Chile. Informe Final.* Santiago: Corporación Nacional Forestal.

CONAF 2000. *Información estadística histórica de ocurrencia y daño de los incendios forestales: Período 1979–1999. Décima Región de Los Lagos.* Puerto Montt: Corporación Nacional Forestal.

Cortés, M.A. 1990. Estructura y dinámica de los bosques de alerce (*Fitzroya cupressoides*) en la Cordillera de la Costa de la Provincia de Valdvia. M.S. thesis. Facultad de Ciencias Forestales, Universidad Austral de Chile, Valdivia.

Denton, G.H. 1993. Chronology of late Pleistocene glaciation near Lago Llanquihue between Puerto Varas and Puerto Octay. In *El Cuaternario de la Región de los Lagos del Sur de Chile*, ed. C. Villagrán, pp. 53–63. Taller Internacional "El Cuaternario de Chile," Santiago, November 1–9, 1993. Guía de Excursión.

Di Castri, F., and Hajek, E. 1976. *Bioclimatología de Chile.* Santiago: Vicerrectoría Académica de la Universidad Católica de Chile.

Dieterich, J.H., and Swetnam, T.W. 1984. Dendrochronology of a fire-scarred ponderosa pine. *For. Sci.* 30:238–247.

Donoso, C. 1981. Tipos forestales de los bosques nativos chilenos. Proyecto CONAF/FAO/PNUD. Documento de Trabajo 38. Santiago.

Donoso, C. 1983. Modificaciones del paisaje forestal chileno a lo largo de la historia. In *Proceedings of the Symposium Desarrollo y perspectivas de las disciplinas forestales de la Universidad Austral de Chile*, pp. 365–438. Valdivia.

Donoso, C. 1993. *Bosques templados de Chile y Argentina: Variación, estructura y dinámica.* Santiago: Editorial Universitaria.

Donoso, C., Sandoval, V., Grez, R., and Rodríguez, J. 1993. Dynamics of *Fitzroya cupressoides* forests in southern Chile. *J. Veg. Sci.* 4:303–312.

Duncan, R.P. 1989. An evaluation of errors in tree age estimates based on increment cores in kahikatea (*Dacrycarpus dacrydioides*). *New Zealand Nat. Sci.* 16:31–37.

Elizalde, R. 1970. *La sobrevivencia de Chile*. Santiago: Ministerio de Agricultura, Servicio Agrícola y Ganadero.

Fraver, S., González, M.E., Silla, F., Lara, A., and Gardner, 1999. Composition and structure of remnant *Fitzroya cupressoides* forests of southern Chile's Central Depression. *J. Torrey Bot. Soc.* 126:49–57.

Fritts, H. 1976. *Tree Rings and Climate*. London: Academic Press.

Fuenzalida, H. 1950. Biogeografía. In *Geografía Económica de Chile*, ed. CORFO, pp. 371–428. Santiago: Editorial Universitaria.

Gardner, M.F., Thomas, P., Lara, A., and Escobar, B. 1999. *Fitzroya cupressoides* (Cupressaceae). *Curti's Bot. Mag.* 16:229–240.

Hauser, A. 1984. *Consideraciones Geológicas y, Geotécnicas en Relación con la Construcción del Camino Longitudinal Austral, X y XI Regiones*. Santiago: Servicio Nacional de Geología y Minería.

Heusser, C.J. 1994. Paleoindians and fire during late Quaternary in southern South *America. Rev. Chil. Hist. Nat.* 67:435–443.

INIA. 1985. *Referencia suelos volcánicos de Chile*. Santiago: Ministerio de Agricultura.

Kalin, M.T., Cavieres, L.L., Peñaloza, A., Riveros, M., and Faggi, A.M. 1996. Relaciones fitogeográficas y patrones regionales de riqueza de especies en la flora del bosque lluvioso templado de Sudamérica. In *Ecología de los Bosques Nativos de Chile*, eds. J.J. Armesto, C. Villagrán, and M.K. Arroyo, pp. 71–99. Santiago: Editorial Universitaria.

Kitzberger, T., and Veblen, T.T. 1997. Influences of humans and ENSO on fire history of *Austrocedrus chilensis* woodlands in northern Patagonia, Argentina. *Écoscience* 4: 508–520.

Kitzberger, T., Veblen, T.T., and Villalba, R. 1997. Climatic influences on fire regimes along a rain forest-to-xeric woodland gradient in northern Patagonia, Argentina. *J. Biogeogr.* 24:35–47.

Kitzberger, T., Veblen T.T., and Villalba, R. 2000. Metodos deudrocromológicos Suy sus aplicaciomes eu estudios de dinamica de bosques templados de d'américa. In Dendrocronologia ou America Latina, ed. F.A. Roig, EDIUNC, Mendoza, Argentina.

Kühne, A. 1985. *Estudio pedológico y geomorfológico de Contao a Río Negro en la X Región de Los Lagos. Boletín Técnico 20.* Santiago: Corporación Nacional Forestal.

Lara, A. 1991. *The dynamics and disturbance regimes of Fitzroya cupressoides forests in the south central Andes of Chile*. Ph.D. dissertaiton. University of Colorado, Boulder.

Lara, A. 2000. Importancia Científica, protección legal y uso destructivo de los bosques de alerce (*Fitzroya cupressoides*): Una contradición que debe resolverse. *Bosque Nativo* 27:3–13.

Lara, A., and Villalba, R. 1993. A 3620-year temperature record from *Fitzroya cupressoides* tree rings in southern South America. *Science* 260:1104–1106.

Lara, A., Donoso, C., and Aravena, J.C. 1996. La conservación del bosque nativo de Chile: problemas y desafíos. In *Ecología de los Bosques Nativos de Chile*, eds. J.J. Armesto, C. Villagrán, and M. K. Arroyo, pp. 335–362. Santiago: Editorial Universitaria.

Lara, A., Fraver, S., Aravena, J.C., and Wolodarsky-Franke, A. 1999a. Fire and dynamics of *Fitzroya cupressoides* forests of Chile's Cordillera Pelada. *Ècoscience* 6:100–109.

Lara, A., Solari, M.E., Rutherford, P., Thiers, O., and Trecamán, R. 1999b. *Cobertura de la vegetación original de la Ecoregión de los Bosques Valdivianos de Chile hacia 1550.* Informe Técnico. Valdivia: Universidad Austral de Chile-World Wildlife Fund.

Lara, A., Aravena, J.C., Villalba, R., Wolodarsky-Franke, A., Luckman, B., and Wilson, R. 2001. Dendroclimatology of high-elevation *Nothofagus pumilio* forests at their northern distribution limit in the Central Andes of Chile. *Can. J. For. Res.* 31:925–936.

Levi, B., Aguilar, A., and Fuenzalida, R. 1966. *Reconocimiento geológico de las Provincias de Llanquihue y Chiloé. Boletín N°. 19.* Santiago: Instituto de Investigaciones Geológicas.

Lusk, C.H., 1996. Gradient analysis and disturbance history of temperate rain forests of the coast range summit plateau, Valdivia, Chile. *Rev. Chil. Hist. Nat.* 69:401–411.

Markgraf, V., and Anderson, L. 1994. Fire history of Patagonia: Climate *versus* human cause. *Rev. Instit. Geogr. Sao Paulo* 15:35–47.

Mercer, J.H. 1976. Glacial history of southernmost South America. *Quat. Res.* 6:125–166.

Porter, S.C. 1981. Pleistocene glaciation in the southern Lake Region of Chile. *Quat. Res.* 16:263–292.

Rodríguez, J.P. 1989. *Estrategias regenerativas de Alerce (Fitzroya cupressoides (Mol.) Johnston) en el sector de Contao, Cordillera de los Andes, Provinicia de Palena.* M.S. thesis. Facultad de Ciencias Forestales, Universidad Austral de Chile Valdivia.

Schmidt, H., and Burgos, P. 1977. Estructura y desarrollo natural del bosque de Alerce. In *Informe Forestal de las Areas de Futaleufú y Contao en la X Región.* Facultad de Ciencias Forestales, pp. 57–64. Santiago: Universidad de Chile.

Schulman, E. 1956. *Dendroclimatic change in semiarid America.* Tucson: University of Arizona Press.

Servicio Nacional de Geología y Minería. 1982. *Mapa Geológico de Chile Escala 1 : 1.000.000.* Santiago: SERNAGEOMIN.

Silla, F. 1997. *Dinámica regenerativa del alerce (Fitzroya cupressoides) de la Depresión Intermedia.* M.S. thesis. Facultad de Ciencias. Universidad Austral de Chile, Valdivia.

Silla, F., Shawn, F., Lara, A., Allnut, T., and Newton, A. 2001. Regeneration and stand dynamics of *Fitzroya cupressoides* (*Cupressaceae*) forests of Southern Chile's Central Depression. *For. Ecol. Manag.*, in press.

Stokes, M.A., and Smiley, T.L. 1968. *An Introduction to Tree-Ring Dating.* Chicago: University of Chicago Press.

Tudhope, A.W., Chilcott, C.P., McCullock, M.T., Cook, E., Chapell, J., Ellam, R.M., Lea, D.W., Lough, J.M., and Shimmield, G.B. 2001. Variability in the El Niño–Southern Oscillation through a Glacial-Interglacial cycle. *Science* 291:1511–1517.

Veblen, T.T. 1983. Degradation of native forest resources in southern Chile. In *History of sustained-yield forestry: A symposium*, pp. 344–352. Durham, NC: Forest History Society.

Veblen, T.T. 1985. Stand dynamics in Chilean Nothofagus forests. In *The Ecology of Natural Disturbance and Patch Dynamics*, eds. S.T.A. Pickett and P.S. White, pp. 35–51. New York: Academic Press.

Veblen, T.T., and Ashton, D.H. 1978. Catastrophic influence on the vegetation of the Valdivian Andes. *Vegetatio* 36:149–167.

Veblen, T.T., and Ashton, D.H. 1982. The regeneration status of *Fitzroya cupressoides* in the Coastal Range, Chile. *Biolog. Conserv.* 23:141–161.

Veblen, T.T., and Lorenz, D.C. 1988. Recent vegetation changes along the forest–steppe ecotone in northern Patagonia. *Ann. Assoc. Am. Geogr.* 78:93–111.

Veblen, T.T., and Markgraf, V. 1988. Steppe expansion in Patagonia? *Quat. Res.* 30: 331–338.

Veblen, T.T., Delmastro, R.J., and Schlatter, J.E. 1976. The conservation of *Fitzroya cupressoides* and its environment in southern Chile. *Environ. Conserv.* 3:291–301.

Veblen, T.T, Donoso, C., Schlegel, F.M., and Escobar, B. 1981. Forest dynamics in south-central Chile. *J. Biogeogr.* 8:211–247.

Veblen, T.T., Burns, B.R., Kitzberger, T., Lara, A., and Villalba, R. 1995. The ecology of conifers of southern South America. In *Ecology of the Southern Conifers*, eds. N.J. Enright and R.S. Hill, pp. 120–155. Melbourne: Melbourne University Press.

Veblen, T.T., Kitzberger, T., Burns, B.R., and Robertus, A.J. 1996. Perturbaciones y dinámica de regeneración en bosques andinos del sur de Chile y Argentina. In *Ecología de los Bosques Nativos de Chile,* eds. J.J. Armesto, C. Villagrán, and M.K. Arroyo, pp. 169–198. Santiago: Editorial Universitaria.

Veblen, T.T., Kitzberger, T., Villalba, R., and Donnegan, J. 1999. Fire history in northern Patagonia: The roles of humans and climatic variation. *Ecol. Monogr.* 69:47–67.

Villalba, R., and Veblen, T.T. 1997. Improving estimates of total tree ages based on increment core samples. *Écoscience* 4:534–542.

Villalba, R., Cook, E.R., Jacoby, G.C., D'Arrigo, R.D., Veblen, T.T., and Jones, P.D. 1998. Tree-ring based reconstructions of norhern Patagonian precipitations since AD 1600. *Holocene* 8:659–675.

White, P.S., and Pickett, S.T.A. 1985. Natural disturbance and patch dynamics: an introduction. In *The ecology of natural disturbance and patch dynamics*, eds. S.T.A. Pickett and P.S. White, pp. 3–13. New York: Academic Press.

Wilhelm, E.J. Jr. 1968. Fire ecology of the Valdivian rain forest. *Proceedings 8th Tall Timbers Fire Ecology Conference*, Tallahasee, FL, Tall Timbers Research, Inc. pp. 55–70.

12. Fire History in Central Chile: Tree-Ring Evidence and Modern Records

Juan Carlos Aravena, Carlos LeQuesne, Héctor Jiménez, Antonio Lara, and Juan J. Armesto

Wildfires in Chile are believed to have originated primarily from fires set by humans to clear natural vegetation to permit agriculture (also see Montenegro et al., Chapter 14, this volume). Besides the intentionally set fires, rapid population growth during the last several decades of the twentieth century has further contributed to an exponential increase in fires set accidentally by motor vehicles along roadsides and by careless campers, especially in central Chile where the human population is concentrated (CONAF 2000). Of course, some fires are ignited naturally by lightning and volcanic eruptions, but the frequency of such ignitions is relatively low in comparison with human-set fires. It is likely that such a low-frequency fire regime may have prevailed before the arrival of indigenous populations to this region, as suggested by sedimentary records of charcoal and fossil pollen covering the past 40,000 years (Heusser 1994).

Effects of fires on the sclerophyllous vegetation of central Chile have been thoroughly studied from a physiological and ecological point of view (Montenegro et al., Chapter 14, this volume). However, long-term records of fire history, derived either from sedimentary charcoal or from tree rings are extremely scarce for Chile, in general, including central Chile which is the focus of this chapter (Heusser 1994). Tree rings permit the dating of past fires to an annual resolution and thus can provide useful information on the frequency and recurrence intervals of past fires. Additional information on fire extent, intensity, and direction of spread may be derived from data on spatial patterns of fire-scarred trees. In central Chile (32° to 38°S) several woody species produce annual tree rings that are appropriate for the study of fire history. In particular, montane and subalpine conifer forests of *Austrocedrus chilensis*, discontinuously distributed throughout central Chile, are dominated by long-lived trees highly resistant to low-intensity fires (Veblen et al. 1995). These trees can provide valuable information about fire regimes associated with human occupation of this region during the past 100 to 1000 years, as well as the association of fire with particular weather conditions. This chapter is the first attempt to reconstruct fire histories in forests of *Austrocedrus chilensis* in central Chile. It is based on data from our ongoing dendrochronological studies and the analysis of modern historical records of fire.

Modern Records

Historical Data

Complete statistics of the occurrence and intensity of wildfires in Chile are available only for the last three decades (CONAF 2000). For earlier periods the historical record is incomplete and fragmentary. Although historical records for the pre-1970 period do not provide a complete record, they do allow the detection of some qualitative patterns of fire occurrence in relation to human activities, and permit tentative interpretations of the impacts of changes in fire regimes on the vegetation.

Archaeological data document the beginning of human occupancy in Chile as late Glacial, as occurring 14,000 BP. The first inhabitants were hunters of megafauna who reached the southernmost tip of the continent at the start of the Holocene period (Mostny 1994; Dillehay 1988; Nuñez et al. 1994). Pollen records show remarkable increases of charcoal traces that coincide with this early Paleo-Indian occupation (Heusser 1994). This evidence suggests that the setting of fires was common practice among early inhabitants; presumably it served the purpose of opening hunting grounds. Heusser (1994) postulates an anthropogenic cause for the charcoal evidence in the period from 44,000 to 15,000 BP, and assumes intermittent human presence during interglacials periods. This hypothesis, however, is not supported by any archaeological evidence. After the extinction of the megafauna, the human populations survived as hunter-gatherers during the Holocene until ca. 2300 BP. The effect of these activities on natural vegetation is considered to be negligible given the low human population density of this period.

In the following cultural period (from 2300 BP), called *agroalfarero*, the higher number of settlement sites implied an increase in human population density. Encina (1940–52) estimated that the number of inhabitants living during that period between Aconcagua and Seno Reloncaví (32° to 42°S) reached one million, and concomitantly, there was an increase in the use of fire to eliminate native vegetation. Fire was used to prepare croplands and grazing areas for camelids in the Aconcagua and Maipo Valleys under the Inca dominion (until the beginning of sixteenth century).

During the Spanish Conquest and Colonial Period (1540–1810), the use of fire was intensified in the central valley between Aconcagua and Bío Bío (32° to 38°S). This increase in burning indicates more intensive agriculture, cattle ranching, and mining practices (Mooney et al. 1972). In particular, in the late 1700s wheat was introduced, and its cultivation became widespread to meet an export market created by the mining operations of neighboring colonies (Amunategui 1940). In addition there was a selective exploitation of commercially valuable tree species such as *Quillaja saponaria*, *Austrocedrus chilensis*, *Cryptocarya alba*, and the endemic Palm *Jubaea chilensis*. Despite this early exploitation of timber resources in central Chile, the period from the mid-1800s to the late twentieth century is clearly when the most intensive and devastating exploitation of forests has occurred. Since the second half of the 1800s, the deciduous coastal forest of this region dominated by many species of the genus *Nothofagus*, was cleared to open lands for wheat crops to be exported to Perú, to northern Chile whose growth was due to silver and copper mining, and to California during the nineteenth-century gold rush (Amunategui 1940; Pérez-Rosales 1980).

Recent Fire Records

The statistics for the last 25 years (CONAF 2000) on the number of forest fires recorded for central Chile (32° to 38°S) show an increasing trend in wildfires (Fig. 12.1a). Fire seasons (October–March of the following year) were named according to the calendar year in which the fire season began, according to Schulman's (1956) convention for tree rings in the Southern Hemisphere. In contrast to the number of fires, the total area burned by wildfires each year shows large fluctuations from year to year (Fig. 12.1b). Thus, although the number of fires has increased over the past few decades, the land area burned by fires has fluctuated widely during the same period (Fig. 12.1c). Fluctuations in the area burned may be related to dry years which occur in central Chile in association with the cool phase of the Southern Oscillation (i.e., La Niña events; Aceituno 1988; Rutlland and Fuenzalida 1991).

The origins of all these fires have been attributed to human activity (Fig. 12.2). The two main causes were careless activity from roadside motor vehicles (transportation = 27%) and intentional ignitions (26%). Forestry, agriculture, and recreation activities together accounted for almost one-third of the wildfires. One interesting point is the different trends of these fire-causing factors during the last three decades of the twentieth century (Fig. 12.2). While the fires set by activities related to forestry and agriculture show a clear descending trend toward the present ($r^2 = 0.71$, slope $= -1.07$), both intentionally set fires and fires due to transportation increased remarkably. Intentionally set fires increased sevenfold ($r^2 = 0.81$, slope $= 1.56$), whereas transportation fires more than doubled in an 18 year period ($r^2 = 0.30$, slope $= 0.51$). This reinforces the idea that the increase in the human population is the major cause of the increasing fire frequency.

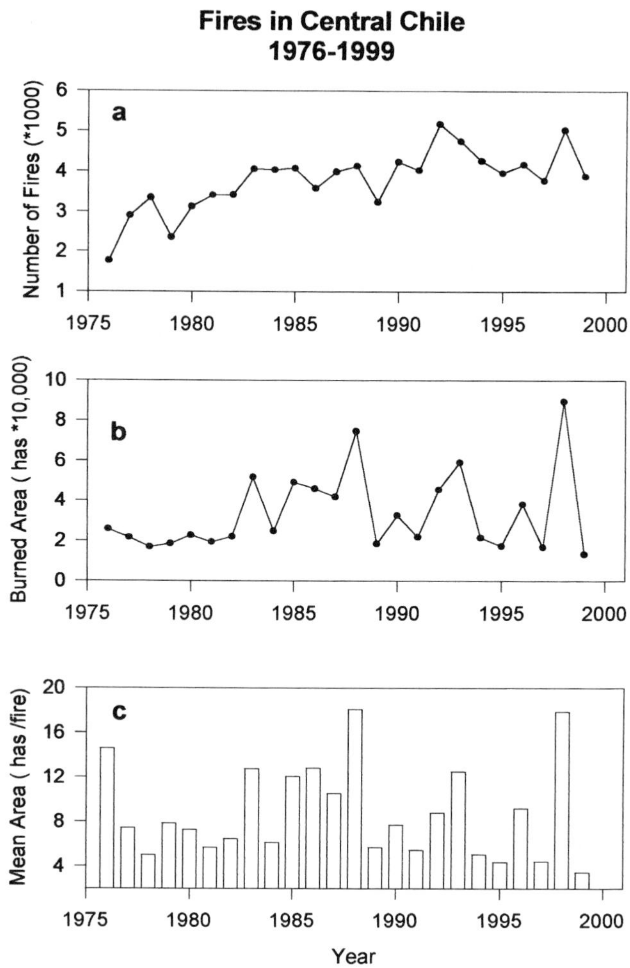

Figure 12.1. (a) Number of fires, (b) area affected by fires, and (c) mean area per fire between 1976 and 1999 in central Chile (CONAF 2000).

Tree-Ring Evidence

Austrocedrus chilensis Forests

Austrocedrus chilensis ("ciprés de la cordillera") is a dioecious tree of pyramidal habit that grows on steep slopes in highly eroded and rocky substrates. This species, which is morphologically similar to the genus *Libocedrus* extant in New Zealand, has an austral-antarctic origin and is endemic to the temperate montane forests of Chile and Argentina (32°39′–44°S; Fig. 12.3). The northern *Austrocedrus* populations are isolated patches of long-lived individuals occurring in the arid margin of the Mediterranean-climate region, and are subjected to strong

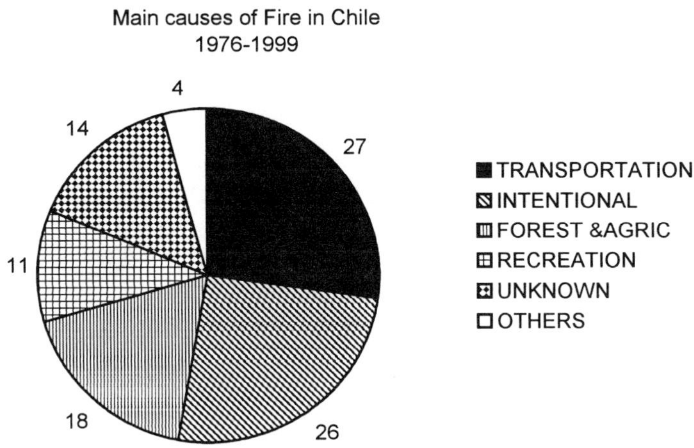

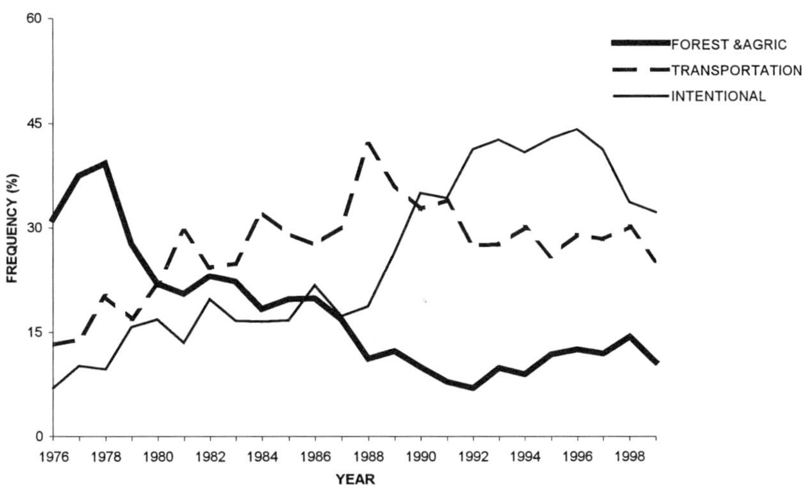

Figure 12.2. Main causes and trends of fires between 1976 and 1999 (CONAF 2000).

climatic stress. Northern populations are widely scattered in the Andes of central Chile between 900 and 2200 m. The southernmost populations are found more continuously in the Andes of southern Chile and Argentina, between 600 and 1000 m elevation. Occasionally, it is possible to find small, isolated stands of *Austrocedrus* in the Coastal Range (Fig. 12.3) in Nahuelbuta (37°10–37°50′S) and Río Bueno (40°30′S; Veblen and Schlegel 1982).

On the eastern side of the Andes, fire seems to be the most prevalent disturbance agent in *Austrocedrus* forests, giving origin to more or less even-aged stands depending on the position of the site in a dry-wet gradient from east to west in northern Patagonia (Veblen, Kitzberger, and Lara 1992; Kitzberger, Veblen, and Villalba 1997). For this region complete chronologies have been

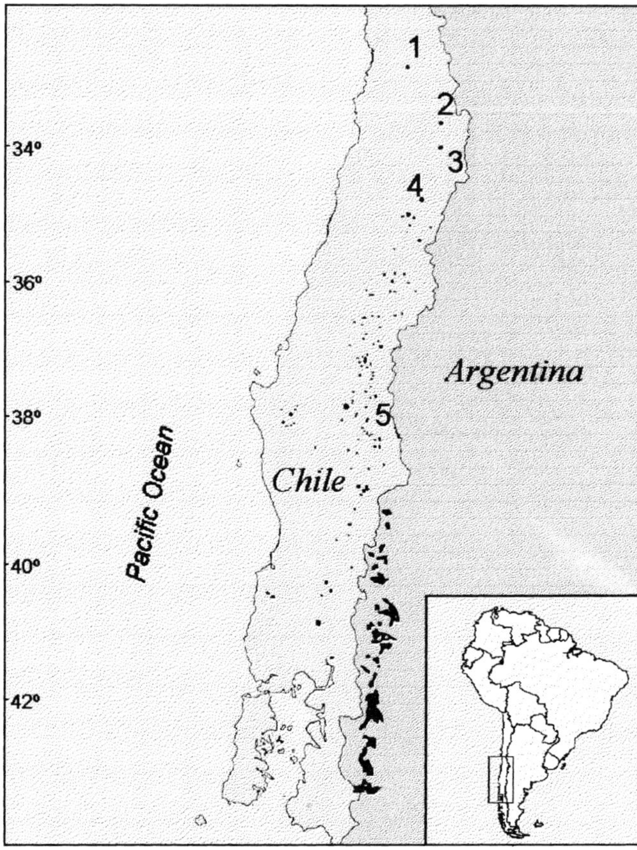

Figure 12.3. Geographical distribution of *Austrocedrus chilensis* forests and dendroecological study sites in central Chile: (1) El Asiento, (2) San Gabriel, (3) Río Clarillo, (4) Río Cipreses, and (5) Alto Bío Bío.

developed that document fire occurrence over the last 450 years (Kitzberger, Veblen, and Villalba 1997; Kitzberger and Veblen 1997; Veblen et al. 1999). These authors have studied the influence of climate on fire regimes of wet *Nothofagus*-dominated forests and *Austrocedrus chilensis* dry woodlands. For the wet forests there is a strong relationship between total annual area burned and drought during spring and summer of the same year, whereas for xeric *Austrocedrus* woodlands fire recurrence appears related not only to droughts during the fire season but also to the precipitation conditions during the preceding 1 or 2 growing seasons (see Kitzberger and Veblen, Chapter 10, this volume).

Although detailed studies of the role of fire in the Austrocedrus forests of central Chile have not yet been conducted, preliminary evidence and field observations indicate an important role for fire in these forests (Donoso 1982; LeQuesne 1988). Veblen et al. (1995) suggested that the diameter growth curves of *Austrocedrus* trees presented by LeQuesne (1988) might be the result of postfire sapling establishment and initial suppression of growth by the shrub layer. We are currently studying *Austrocedrus* sites in the northern portion of its distribution range (Fig. 12.3). From north to the south, our study sites are El Asiento (32°40′S, 70°49′W), where lies the northernmost population of the species, with scattered trees located on a polar slope between 1700 and 2200 m; San Gabriel, in the Maipo River valley, between 1100 and 1400 m; Rio Clarillo Forest Reserve (33°55′S, 72°25′W), 20 km south-east of Santiago, at 2200 m of altitude; Rio Cipreses Forest Reserve (34°49′S, 70°51′W), with populations occurring in a glacial valley near Los Cipreses Glacier front, between 1500 and 1900 m.a.s.l. The southernmost *Austrocedrus* site studied by us is Alto Bio Bio (38°00′S, 71°42′W), located at 900-m elevation. Precipitation in these localities increases from north to south from about 800 mm annually to more than 3000 mm in Alto Bío Bío. The *Austrocedrus* populations that we studied in all sites are near treeline and hence are exposed to low winter temperatures and snowfall.

Evidence of Fire in Central Chile

In all the above-mentioned sites in central Chile the same methodology was used: sampling was conducted in plots of 80 × 40 m placed perpendicular to the direction of the slope, and all the trees reaching at least 1.3 m in height were mapped, cored (using increment borers), and their diameter at breast height (dbh) recorded. Cross sections were obtained from dead *Austrocedrus* present within the plot or in its neighborhood in order to obtain an expanded record of fires and other disturbances that had occurred in the area.

In all study sites we found abundant evidence of fire, such as presence of charcoal in standing or fallen dead trees, and/or fires scars in dead and living trees. In Río Cipreses and Alto Bío Bío we were able to date fire scars observed in cross sections of stumps using standard dendrochronological methods (Fritts 1976; Schweingruber 1988; Stokes and Smiley 1968). These methods included crossdating (matching ring-width patterns) ring-width patterns in fire-scar samples in dead specimens with tree-ring-width chronologies developed for each study site.

In the Alto Bío Bío study site it was possible to date and estimate the area affected by fire disturbance. Here the positions of trees that were fire scarred in combination with the positions of trees showing changes in growth rates due to the fire were used to estimate the direction and extent of the fire (Fig. 12.4; Jiménez 1995). A large section of this stand was affected by more than one fire front. Unfortunately, only one of these fronts could be effectively dated, and consequently it was not possible to estimate the frequency of fire. Sharp reductions of tree growth rates are often due to damages to tree crowns or the phloem when trees survive fires (Schweingruber 1988). Surviving trees may also increase their growth rates after recovery from the initial fire damage if the fire kills neighboring trees and reduces competition (Kitzberger, Veblen, and Villalba 2000). Such patterns of tree growth were sought for trees within the plot as well as in the neighboring area. Radial growth increments of four trees located near the boundary of the fire front (labeled 1 to 4), as reconstructed from fire scars in Alto Bío Bío, showed a pronounced growth decrease in the year 1893, which supports

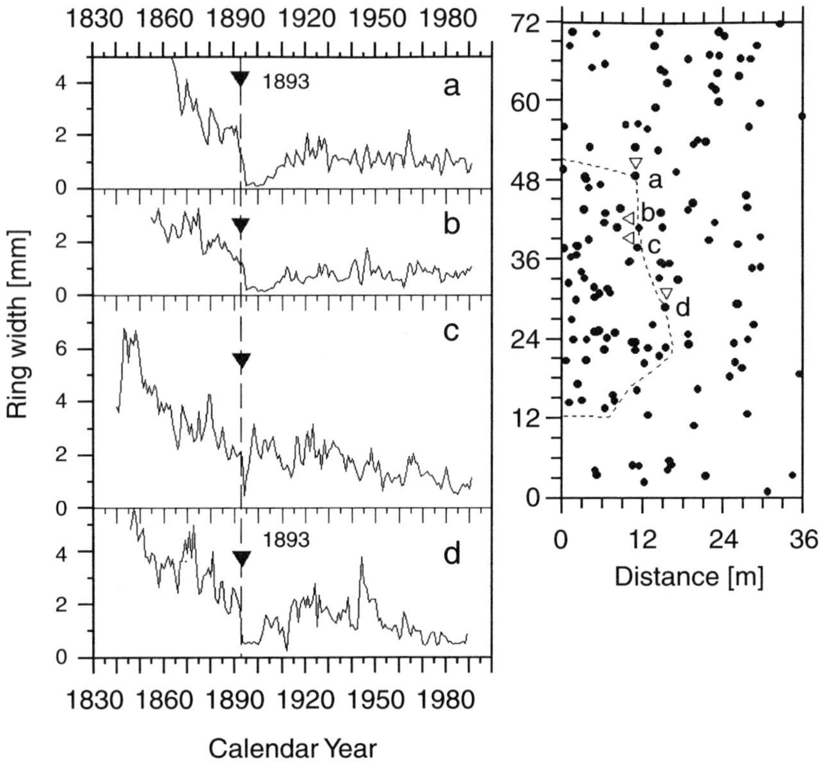

Figure 12.4. Map of the Alto Bío Bío study plot delimiting a gap (*dashed line*) created by a fire in 1893, and showing the locations of individuals of *Austrocedrus chilensis* (*black circles*). The four graphs show radial growth patterns of the four *Austrocedrus chilensis* individuals (indicated by arrows). The *y*-axes of the graphs give the ring width in millimeters (From Jiménez 1995).

our interpretation of fire disturbance of the stand in 1893 (Fig. 12.4). In addition lower growth rates between 1895 and 1900 in all surviving trees in the stand is consistent with a long-lasting effect of fire on tree growth rates (Aravena et al., unpublished data).

At the Río Cipreses site, tree-ring series along the margin of a stand of *Austrocedrus chilensis* affected by fire shows how past fire can be identified from tree-growth patterns (Fig. 12.5; Le Quesne 1999). Based on a cross section of a dead tree located inside the study plot, we dated two fire events, 1716 and 1845 (Fig. 12.5h). Tree-ring series show a synchronous growth release in the 1820s, especially evident in panels a, b, e, and h of Figure 12.5. An index of growth release (Kitzberger, Veblen, and Villalba 1995), calculated for the tree-ring chronology of Río Cipreses, detected an abrupt growth increase for the year 1825 (Fig. 12.6). This release occurs after a growth decrease in the year 1823 especially evident in the trees with fire scars (Fig. 12.5). These rapid changes in radial growth patterns may be associated with a fire event that produced an initial

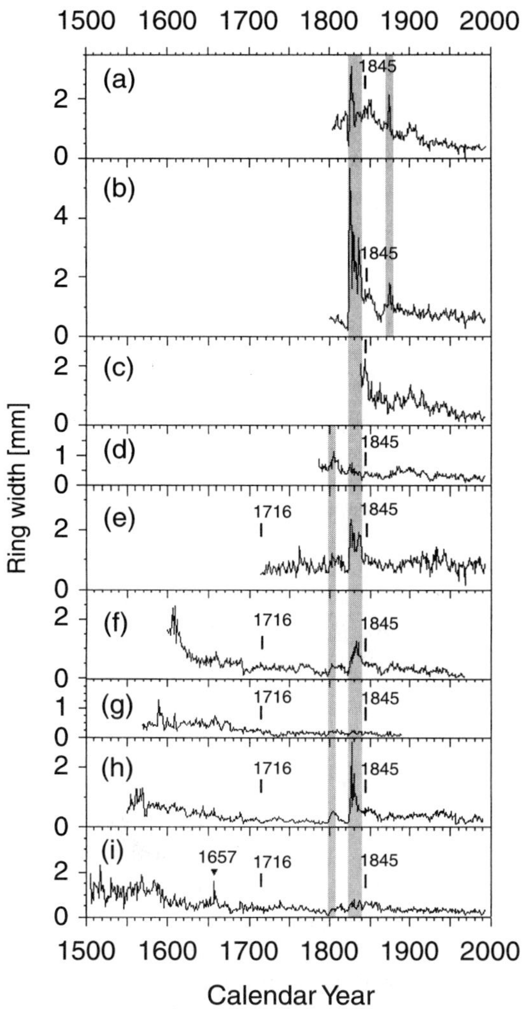

Figure 12.5. Tree-ring series from living trees with fire scars (a, b, f, g, i), living trees at the limit of the fire front (c, d, e), and a dead tree with fire scars (h). From this dead tree we dated two fires: 1716 and 1845 (LeQuesne 1999). Shaded bars indicate periods of synchronous growth release.

growth suppression followed by a growth increase due to a higher level of resources availability (Le Quesne 1999).

Examination of the age structure of *Austrocedrus* in many stands provides another way to detect the effect of recurrent fire events on forest dynamics (Fig. 12.7). In Alto Bío Bío the age structure suggests that a regeneration pulse was initiated nearly 100 years ago. This pulse probably followed fire disturbance, and the recruitment period lasted for the next 60 years (Fig. 12.7a). These younger individuals are primarily located within the area affected by the 1893 fire (Fig.

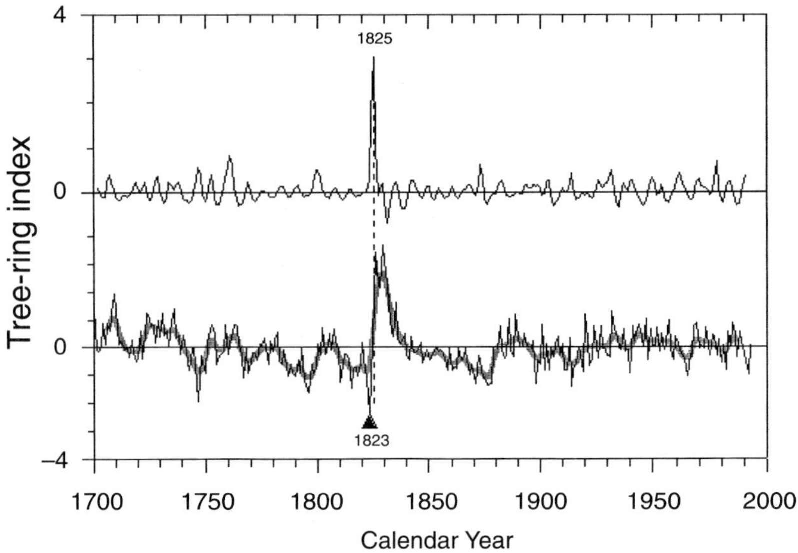

Figure 12.6. A tree-growth release index (Kitzberger et al. 1995) computed on the Rio Cipreses tree-ring chronology (*above*), and a difference chronology (*below*) developed by subtracting the tree-ring indexes of fire-scarred trees from the stand chronology for the site (LeQuesne 1999).

12.4), suggesting that fire disturbance is the cause of the regeneration pulse in this *Austrocedrus* stand. Both in Rio Cipreses and Rio Clarillo the age structures also show pulses that indicate longer recruitment periods (Fig. 12.7b and c). In particular, for Río Cipreses (Fig. 12.7b) the age structure indicates an abundant cohort younger than 300 years old that is distinct from a small group of trees older than 360 years. This supports the idea that a fire disturbance was the cause for this regeneration pulse, since most of the cross sections obtained from dead trees in this study plot indicated that these trees died around 300 years ago (Fig. 12.8). Thus the stand represents a population established after a high-intensity fire that killed most of trees except for a few survivors. On the other hand, the pulses of regeneration in Río Clarillo (Fig. 12.7c) are not so clearly associated with fire events. Here, because of the steep slopes, other disturbance agents in addition to fire may be operating, such as landslides, which produce a more complicated regeneration pattern (Aravena et al. 1994).

Conclusion

Potential of *Austrocedrus* Forests for Studies of Fire Regimes

Our preliminary results on the reconstruction of past fires from fire scars, radial growth patterns, and age structures of populations of *Austrocedrus chilensis* in central Chile demonstrate the feasibility of using tree-ring methods for better understanding of the fire regimes in this geographic area. *Austrocedrus* has a high

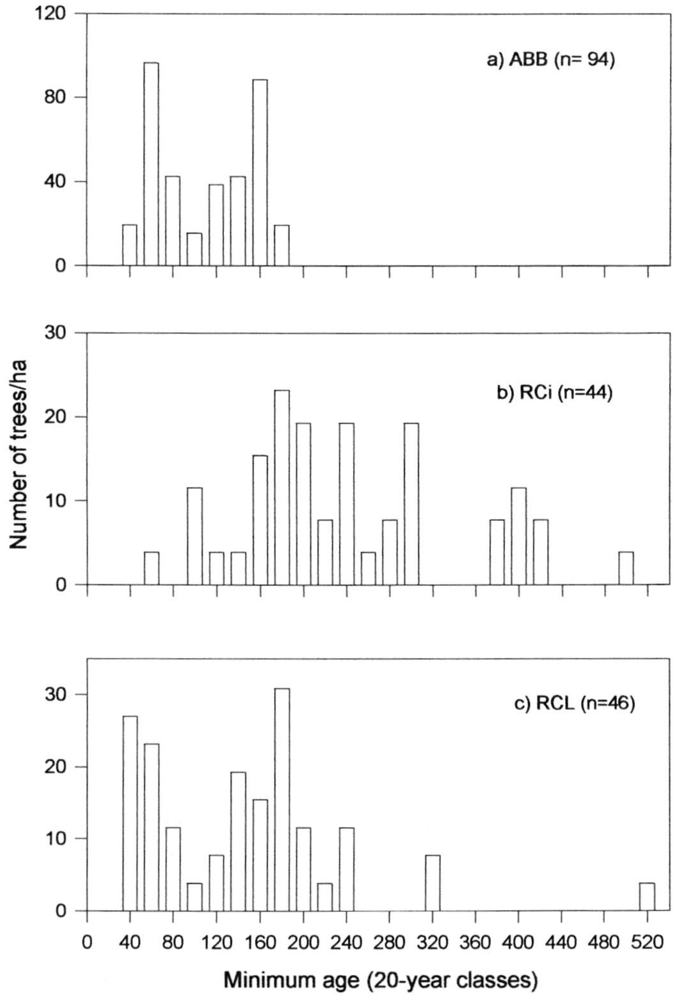

Figure 12.7. Age structures of *Austrocedrus chilensis* stands in (a) Alto Bio Bio (Jiménez 1995), (b) Rio Cipreses (Rci), and (c) Rio Clarillo (RCL). The number of trees is given by *n*.

potential for use in fire history studies because the species is long-lived (up to 1000 years old) and many trees survive fire. Furthermore tree-ring widths are sensitive to climatic variations, which permits analysis of climatic influences on fire regimes. The distribution of *Austrocedrus* forests in central Chile, especially toward their northern limit, is particularly attractive for studies of climatic influences on fire regimes because of the sensitivity of this region to moisture variability related to ENSO (El Niño–Southern Oscillation) episodes (Aceituno 1988). Fire regimes in Austrocedrus forests in northern Patagonia, Argentina, are significantly linked to ENSO activity at annual and multi-decadal time scales (Kitzberger and Veblen 1997; Veblen et al. 1999). However, the linkage of cli-

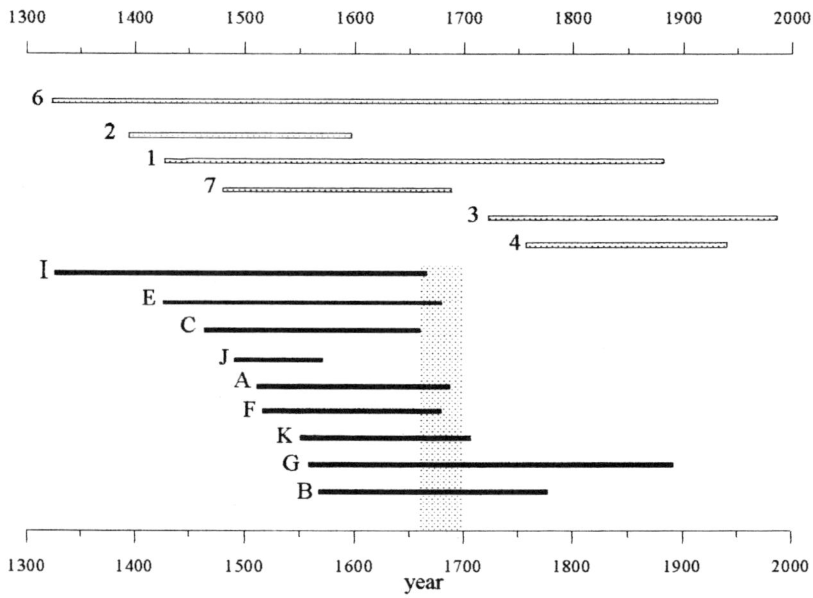

Figure 12.8. Dating of dead stems (n = 15) of *Austrocedrus chilensis* from Río Clarillo (1 to 7) and Río Cipreses (A to K). The horizontal bars approximate the life of the tree. The shaded area shows overlapping range for the outermost rings in several trees in Rio Cipreses, which implies an episode of synchronous tree mortality.

matic variation in central Chile to ENSO activity is stronger and more predictable than in the case of the climate-ENSO linkages for northern Patagonia. Thus the potential for relating past fire to ENSO activity in central Chile is high.

Future Research Needs

The quantitative study of forest dynamics and disturbance history in the *Austrocedrus* forests of central Chile is in an incipient stage. In our studies we need now to include more analyses of stand regeneration, age structure, tree sex ratios, and spatial distribution of trees. This information is essential for better understanding the effects of disturbance regimes and climatic fluctuations in the history of these stands. We have regarded factors such as climatic trends, reproduction costs, and intraspecific interactions in our study of stand dynamics and fire regimes. We hope to expand our scope to include evidence of fire occurring over the entire landscape and to obtain more precise dating of outlying fires in the sampled stands.

Another line of research would be comparative studies of fire history patterns in the *Austrocedrus* of central Chile with those of nearby regions in southern South America and then with distant regions (e.g., western North America) where the local climate is teleconnected to ENSO. Already for northern Patagonia there are abundant fire history data and related climatic analyses for *Austroce-*

drus forests (see Kitzberger and Veblen, Chapter 10, this volume) as well as documents on human activity affecting the patterns of stand and landscape dynamics (Veblen et al., Chapter 9, this volume). Larger databases for *Austrocedrus* forests in central Chile might include comparisons of the effects of fire in these two regions and significant differences in human history and ENSO-related climatic patterns. Analogously, such data on fire history in the *Austrocedrus* forests of central Chile would permit comparisons with other forest types in southern South America, for example, *Fitzroya cupressoides* whose fire history is currently under study (see Veblen et al., Chapter 9, this volume, Lara et al., Chapter 11, this volume). Future research on fire scars in nearby forests of *Araucaria araucana*, which also forms fire scars and where fire has been a major agent of disturbance (Veblen et al. 1995), could also be useful in providing a regional history of forest fires in the southern cone of South America.

Although we have stressed tree-ring studies of fire history, sedimentary records of fire in central Chile could also yield promising results (see Whitlock and Anderson, Chapter 1, this volume; Huber and Markgraf, Chapter 13, this volume). Together, dendrochronologic data and sedimentary charcoal analyses can extend our knowledge of fire history in the Holocene period. Finally there are the archaeological studies of the evidence of prehistoric human settlement and resource use to consider. Such studies would add much to our understanding the causes of changes in fire regimes in southern South America.

Acknowledgments. We are grateful to T. T. Veblen for the opportunity to present a previous version of this work at the IAI Fire Meeting, Silver Falls, Portland, Oregon, in 1996, from which this chapter was developed. Funding was provided by projects UE CI1*CT93-0336, Sarec/Conicyt "Climate change during Holocene in Chile," and CRN03 project of the Inter American Institute for Global Change Research (IAI). We would like to thank CONAF (Corporación Nacional Forestal) for permission to collect samples and logistic support. Support to JJA was provided by the A. W. Mellon Foundation, and to C.L.Q. through a Doctoral Grant from CONAF and Universidad de Oviedo, Spain.

References

Aceituno, P. 1988. On the functioning of the Southern Oscillation in the South American sector. Part I: Surface climate. *Mon. Wea. Rev.* 116:505–524.

Amunategui, D. 1940. *Estudios históricos*. Santiago de Chile: Ediciones de la Universidad de Chile.

Aravena, J.C., LeQuesne, C., Jiménez, H., Hinojosa, L.F., and C. Peña. 1994. Estudio de un rodal de *Austrocedrus chilensis* (D.Don) Pic.Ser. et Bizz. en la Reserva Nacional Río Clarillo. (Antecedentes preliminares). CONAF, Santiago, Chile.

CONAF, 2000. Manejo del fuego. Resultados Temporada 1999–2000. Unidad de Gestión Manejo del fuego. Santiago, Chile.

Dillehay, T.D. 1988. Early rain-forest archaeology in Southwestern South America: Research context, design an data at Monte Verde. In *Wet Site Archaeology*, ed. B. Purdy. pp. 177–206. NJ: Telford Press.

Donoso, C. 1982. Reseña ecológica de los bosques mediterráneos de Chile. *Bosque* 4: 117–146.

Encina, F.A. 1940–1952. *Historia de Chile*, 20 vols. Santiago, Chile: Editorial Nascimento.

Fritts, H.C. 1976. *Tree Rings and Climate*. San Diego, CA: Academic Press.

Heusser, C.J. 1994. Paleoindians and fire during the late Quaternary in southern South America. *Rev. Chil. Hist. Nat.* 67:435–442.

Jiménez, H. 1995. Reconstrucción dendroecológica de la historia de un rodal de *Austrocedrus chilensis* (D.Don.) Pic. Ser. et Bizz. en la cuenca superior del Río Bio Bio. M.S. thesis. Facultad de Ciencias, Universidad de Chile.

Kitzberger, T., and Veblen, T.T. 1997. Influences of humans and ENSO on fire history of *Austrocedrus chilensis* woodlands in northern Patagonia, Argentina. *Ecoscience* 4: 508–520.

Kitzberger, T., Veblen, T.T., and Villalba, R. 1995. Tectonic influences on tree growth in northern Patagonia, Argentina: the roles of substrate stability and climatic variation. *Can. J. For. Res.* 25:1684–1696.

Kitzberger, T., Veblen, T.T., and Villalba, R. 1997. Climatic influences on fire regimes along a rain forest-to-xeric woodland gradient in northern Patagonia, Argentina. *J. Biogeogr.* 24:35–47.

Kitzberger, T., Veblen, T.T., and Villalba, R. 2000. Métodos dendroecológicos y sus aplicaciones en estudios de dinámica de bosques templados de Sudamérica. In *Dendrocronología en América Latina*, ed. F. Roig, pp. 17–78. Mendoza, Argentina: Editorial de la Universidad del Cuyo.

LeQuesne, C. 1988. Caracterización de los bosques de ciprés de la cordillera (*Austrocedrus* chilensis (D.Don.) flor. et Bout.), en Radal Siete Tazas, Séptima región, Chile. M.S. thesis. Universidad Austral de Chile, Valdivia.

LeQuesne, C. 1999. Dendrocronología de Austrocedrus chilensis (D.Don.) Pic. Ser. et Bizz. (Cupressaceae) en el límite norte de su distribución, Chile. Ph.D. dissertation. Universidad de Oviedo.

Mooney, H.A., Dunn, E.L., Shropshire, L., and Song, Jr. 1972. Land use history of California and Chile as related to the structure of the sclerophyll scrub vegetations. *Madroño* 21:305–319.

Mostny, G. 1994. Prehistoria de Chile. Santiago: Editorial Universitaria.

Núñez, L., Varela, J., Casamiquela, R., and Villagrán, C. 1994. Reconstrucción multidisciplinaria de la ocupación prehistórica de Quereo, centro de Chile. *Lat. Am. Antiquity* 5(2):99–118.

Pérez-Rosales, V. 1980. Recuerdos del pasado. Editorial Andrés Bello. Santiago, Chile.

Rutlland, J., and Fuenzalida, H. 1991. Synoptic aspects of the central Chile rainfall variability associated with the Southern oscillation. *Int. J. Climatol.* 11:63–76.

Schulman, E. 1956. *Dendroclimatic Change in Semiarid America*. Tucson: University of Arizona Press.

Schweingruber, F.H. 1988. *Tree-Rings: Basics and Applications of Dendrochronology*. Dordrecht: Riedel.

Stokes, M.A., and Smiley, T.L. 1968. *Introduction to Tree-Ring Dating*. Chicago: University of Chicago Press.

Veblen, T.T., and Lorenz, D.C. 1987. Post-fire stand development of *Austrocedrus-Nothofagus* forests in Patagonia. *Vegetatio* 73:113–126.

Veblen, T.T., and Schlegel, F.M. 1982. Reseña ecológica de los bosques del sur de Chile. *Bosque* 2:73–115.

Veblen, T.T., Kitzberger, T., and Lara, A. 1992. Disturbance and vegetation dynamics along a transect from rain forest to Patagonian shrublands. *J. Veg. Sci.* 3:507–520.

Veblen, T.T., Burns, B.R., Kitzberger, T., Lara, A., and Villalba, R. 1995. The ecology of the conifers of southern South America. In *Ecology of the Southern Conifers*, eds. N.J. Enright and R.S. Hill, pp. 120–155. Melbourne: Melbourne University Press.

Veblen, T.T., Kitzberger, T., Villalba, R., and Donnegan, J. 1999. Fire history in northern Patagonia: The roles of humans and climatic variation. *Ecol. Monog.* 69:47–67.

13. Holocene Fire Frequency and Climate Change at Rio Rubens Bog, Southern Patagonia

Ulli M. Huber and Vera Markgraf

Over the last decade, growing concern over the ecological effects of global warming has fueled interest in the mechanisms of climate-induced vegetation change. Vegetation model results indicate that global warming may favor increased frequency and severity of forest disturbance, thus increasing the rate at which vegetation responds to climate change and the magnitude of this response (Overpeck, Rind, and Goldberg 1990; Gardner et al. 1996). Fire is one of the most important climatically linked disturbance agents in temperate forest systems (Fig. 13.1). Climate variability on different time scales can influence fire regimes through its effects on both ignition sources (lightning frequency) and fuel characteristics (fuel type, fuel structure, fuel accumulation, fuel desiccation) (Renkin and Despain 1992). Understanding the links among climate variability, fire regimes, and vegetation in different environments requires a long-term perspective. Sedimentary records of macroscopic charcoal can provide important information concerning submillennial and millennial scale changes in local fire frequency during periods of major reorganizations in climate and vegetation (e.g., Clark 1990; Clark and Royall 1996; Long et al. 1998; Millspaugh, Whitlock, and Bartlein 2000).

Patagonia and Tierra del Fuego are well suited for studies of past climate and the paleoenvironmental role of fire. Fire has been an important disturbance agent in many bioclimatic regimes of this region (Veblen, Kitzberger, and Lara 1992; Veblen et al. 1996). Tree-ring research in northern Patagonia (Kitzberger, Veblen, and Villalba 1997; Veblen et al. 1999) has confirmed a strong relationship

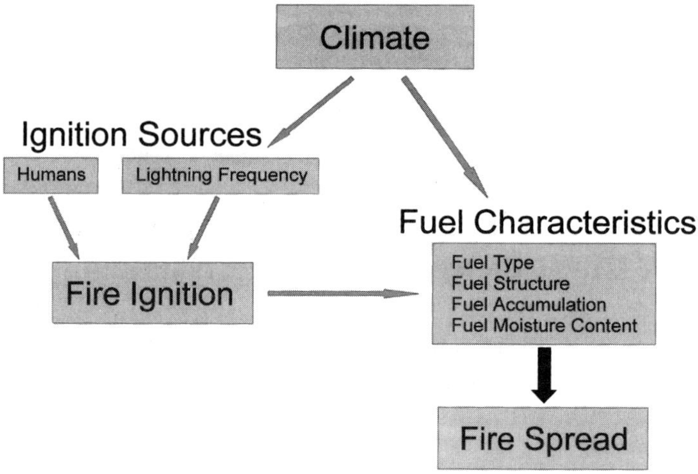

Figure 13.1. Effects of climate on fire ignition and fire spread. Climate variability can influence *ignition* by changing lightning frequency. Changes in climate influence fuel type, accumulation, structure, and moisture content, and hence the spread of fires. Even in the presence of ignition sources, fuel characteristics have to be conducive for fires to spread.

between interannual climate variability and the occurrence of widespread fires during the last ca. 550 years. On decadal time scales, human activity seems to strongly impact fire frequency, although longer-term changes in atmospheric circulation may still have a significant impact (Veblen et al. 1999). Sedimentary charcoal records from Patagonia and Tierra del Fuego indicate that fires have been important during different periods throughout the late Quaternary (Heusser 1987, 1994, 1995b, 1998; Markgraf and Anderson 1994). On these longer time scales, the interactions among climate variability, fire regimes, and vegetation are not well understood. In addition, the causes for fires in southern Patagonia and Tierra del Fuego in the late Quaternary have been, and continue to be, an issue of debate. Both human impact (Heusser 1987, 1994, 1998) and climate (Markgraf and Anderson 1994) have been invoked to explain the observed temporal and spatial patterns of fire occurrence. Heusser (1987, 1994, 1998) attributed the occurrence of fires to the impact of Paleo-Indian hunters, European settlers, and volcanic activity, and suggested that lightning as an alternative ignition source has been virtually nonexistent in southern high latitudes. In contrast, Markgraf and Anderson (1994) proposed that although lightning is uncommon at present, it might have been more important under different climate conditions in the past. Furthermore, whereas humans may have been critical as initiators of fires in many instances, the determining factor in frequency, spatial extent, and intensity of fires has been climate.

Although multi-proxy paleoenvironmental reconstructions do not directly address the issue of ignition sources, they provide important information for understanding how climate variability influenced vegetation and fire occurrences over a range of temporal scales in the presence of ignition sources (human fire use and/or lightning). In southern Patagonia and Tierra del Fuego, the links between century and millennial scale climate variability and fire frequency have not been addressed adequately. First, the available sedimentary charcoal records lack high temporal resolution. Second, microscopic charcoal has not been separated into different size fractions and therefore represents a range of source areas. These factors preclude assessments of local versus regional fire activity and quantitative estimates of fire frequency. Furthermore these low-resolution records cannot address century-scale changes in fire occurrence.

The aim of this multi-proxy study is to examine the relationships between past variability in moisture regimes, fire frequency, and vegetation in southern Patagonia on different time scales. Here we present a summary of our results from the Holocene section of the Rio Rubens peat core (52°08′15″S, 71°52′53″W) (Huber 2001). In this study, analyses of peat macrofossils and macroscopic sedimentary charcoal particles >125 μm in diameter in contiguous samples allow an assessment of relationships between changes in local effective moisture and fire frequency on orbital (multimillennial) and shorter (century-to-millennial) time scales.

Site Location

Rio Rubens Bog (52°08′15″S, 71°52′53″W, elevation ca. 220 m) is located east of the Andean cordillera in southern Patagonia, Chile (Fig. 13.2). The size of the mire is approximately 25 ha. *Empetrum rubrum* and the moss *Polytrichum strictum* dominate the bog surface. Mires in this area are typically located in elongate depressions that appear to be part of a system of former glacial meltwater channels.

The regional climate of southern Patagonia is characterized by a steep west-to-east precipitation gradient, which is related to the orographic effects of the Andes and is strongly reflected in the vegetation. With decreasing precipitation, evergreen rain forests are replaced by mixed evergreen-deciduous forests, deciduous forests, open woodlands, and finally steppe (Fig. 13.2). Prior to European settlement, Rio Rubens Bog was situated in the deciduous *Nothofagus* forest formation (Huber 2001) in close proximity to the steppe-forest ecotone. In the deciduous forest region, mean annual precipitation ranges from ca. 650 to 450 mm/yr (Tuhkanen et al. 1989–1890). Precipitation is fairly evenly distributed throughout the year with a slight maximum in fall and minimum in spring. *Nothofagus pumilio* dominates the deciduous forests, which have been strongly affected by recent burning and logging (Cruz and Lara 1987). The immediate vicinity of Rio Rubens Bog is heavily impacted by human disturbance, and open *Nothofagus antarctica* woodlands, *Chiliotrichium* shrub, and grasslands dominate. The mean annual temperature at the site is ca. 5.2°C (interpolated temperature from the climate station in Torres del Paine National Park, ca. 120 km northeast of the site, applying an environmental lapse rate of 0.55°C/100 m). Interpolated mean temperatures for the coldest and warmest month in the Rio

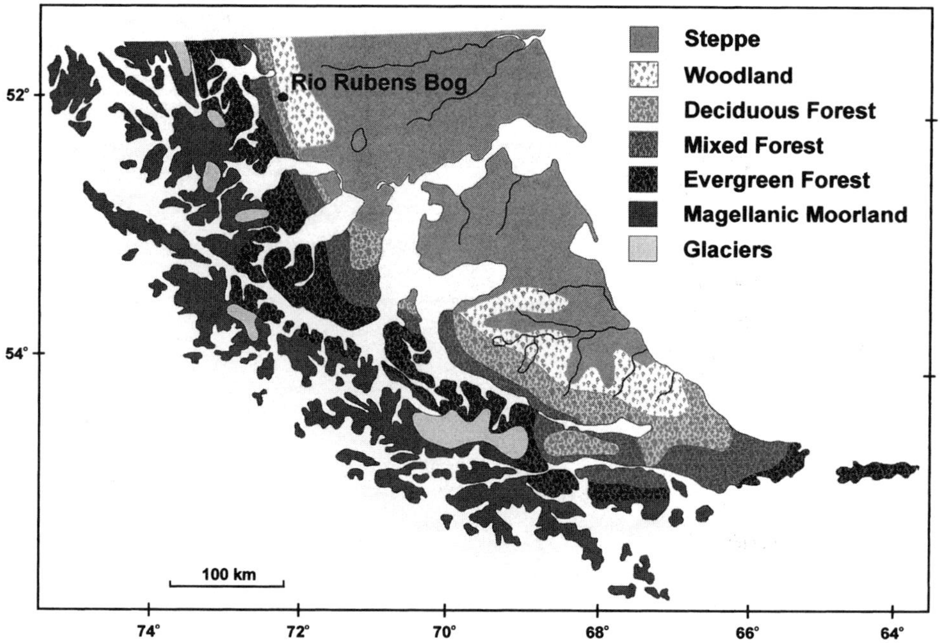

Figure 13.2. Location of Rio Rubens Bog in relation to major vegetation zones in southern Patagonia and Tierra del Fuego. Rio Rubens Bog is situated in the steppe-forest ecotone. (Based on field observations from Tuhkanen 1992.)

Rubens region are ca. –2°C and 11°C, respectively (Tuhkanen 1992). During short periods, especially in winter, temperatures can be influenced by northward intrusions of cold Antarctic air masses, which are accompanied by southerly winds (Zamora and Santana 1979; Tuhkanen et al. 1989–1990). However, throughout most of the year, strong westerly winds associated with moderate temperatures prevail.

Environmental conditions at Rio Rubens Bog make this site ideal for investigating the links between past changes in effective moisture, fire regimes, and vegetation. Local mire hydrology in this moisture-limited region should have reacted quickly to variability in effective moisture. In addition, the location of the steppe-forest ecotone in Patagonia is highly sensitive to changes in effective moisture and fire regimes (Veblen and Markgraf 1988; Villalba and Veblen 1997a, 1997b).

Interpretation of Charcoal Data from Peat Cores

Sedimentary charcoal records can provide a long-term perspective of fire frequency changes and their relationship to climate. However, prior to using these records as a proxy of local fire frequency, two fundamental issues must be addressed: (1) at what spatial scales are fires recorded in peat sediments, and (2) how reliable are peat sediments as recorders of local fires?

At What Spatial Scales Are Fires Recorded in Peat Sediments?

A major assumption in sedimentary charcoal analysis is that charcoal from local source areas can be distinguished from regional charcoal input. Charcoal transport modeling results, experimental burn data, and comparisons of sedimentary charcoal profiles with known fire events suggest that both total charcoal accumulation rates and particle size distributions may be used to distinguish between local and regional charcoal input (e.g., Clark 1988, 1990; Whitlock and Millspaugh 1996; Clark et al. 1998; Long et al. 1998; Ohlson and Tryterud 2000; Gardner and Whitlock 2001; Whitlock and Anderson, Chapter 1, this volume). Models of charcoal transport (Clark 1988) indicate that the majority of charcoal particles >50 μm in diameter are deposited in close proximity to a burn. Also, charcoal accumulation rates have been shown to decrease sharply at the edge of high-intensity experimental burns in west-central Siberia and Scandinavia, and large particles are more abundant closer to the burn (Clark et al. 1998; Ohlson and Tryterud 2000). A comparison of dendrochronologically dated fire-scar records with sedimentary charcoal data from lakes in northwestern Minnesota suggests that charcoal particles >80 μm in diameter originate primarily from fires that occur within the catchment of a lake basin (Clark 1990). A study examining charcoal deposition associated with fires in Yellowstone National Park indicates that charcoal particles >125 μm in diameter are deposited within a 10 km radius of a fire (Whitlock and Millspaugh 1996), and fire events are expressed as distinct peaks in sedimentary charcoal in small lakes within the burnt watersheds (Millspaugh and Whitlock 1995).

Charcoal accumulation rates and particle size distributions likely vary for different types of fires (Clark et al. 1998), and sharp thresholds between local and regional sources do not exist (Clark and Patterson 1997). However, macroscopic charcoal analysis of peat sediments can, in part, circumvent the problem of source area. Peat sediments often contain charred peat macrofossils, which indicates that fires spread onto the peatland surface. The actual location of the burn can therefore be determined, which provides the spatial precision otherwise lacking in lacustrine sedimentary charcoal records (Tolonen 1983; Clark and Richard 1996).

How Reliable Are Peat Sediments as Recorders of Local Fires?

In order to recognize individual fires in sedimentary charcoal records, high temporal resolution is required. The sampling has to be continuous, and sampling increments have to be shorter than fire return intervals. If sampling intervals are too large, a single charcoal peak may represent more than one fire event. Whether or not individual fires are recorded in the sediment may also be related to charcoal deposition processes. In lakes, charcoal is deposited to deep-water sediments by atmospheric fallout, saltation, surface runoff, stream input, and sediment focusing within the lake basin itself (Clark and Patterson 1997). These processes may concentrate charcoal in the lake (Clark and Patterson 1997), and fires burning within a catchment are likely to be recorded as distinct charcoal peaks.

In peatlands, charcoal is predominantly derived from atmospheric fallout and/or in-situ production as fires spread onto the wetland surface (Tolonen 1986). Smoldering peat fires may destroy some evidence of previous fire events (Clark and Richard 1996). Also, the absence of processes that concentrate charcoal (i.e., lacustrine sediment focusing, input from streams and runoff) probably diminishes the expression of fires that did not burn across the wetland surface. Together, these factors suggest that peat charcoal records may underestimate local fire occurrence and that fires recorded in peat sediments may represent a subset of fires in the catchment.

Interpretation of Macrofossil Data from Peat Cores

Mires are sensitive to changes in hydrology, which in turn can lead to changes in peatland vegetation (e.g., Barber 1981; Moore 1986; Barber et al. 1994; Glaser et al. 1996). The hydrology of mires is controlled by the complex interplay between regional climate, local geomorphology, and site history (Almquist-Jacobson and Foster 1995). Major climatic controls on mire hydrology are changes in temperature and precipitation, which in turn impact both the evapotranspiration regime and the surface and groundwater flow (e.g., Moore 1986; Gignac, Halsey, and Vitt 2000). Modifications of these moisture fluxes influence the effective moisture that is available to mire vegetation. In addition, nonclimatic processes may lead to changes in mire hydrology (Moore 1986). Peat accumulation and erosion, and changes in the vegetation of the catchment and the mire surface itself, can influence both influx and efflux terms of the water balance. Ombrotrophic bogs obtain water predominantly through precipitation. In contrast, minerotrophic fens receive water primarily through groundwater input, and secondarily through precipitation and surface runoff. The response of fens to climatic change may be less direct because time lags generally exist between groundwater recharge and discharge, whose length depends on the size and physiography of the catchment.

Peat macrofossil stratigraphy can be used to detect past changes in mire hydrology (e.g., Barber et al. 1994; Hughes et al. 2000). The primary assumption in the use of peat macrofossils as paleoclimate indicators is that climate plays an overriding role in mire hydrology. In southern Patagonia and Tierra del Fuego, the present-day geographic distribution of different bog and fen types is closely related to climate (Roivanen 1954; Auer 1963; Moore 1979, 1983; Tuhkanen et al. 1989–1990; Tuhkanen 1992). Mire types change along a west-to-east gradient in effective moisture. Ombrotrophic *Sphagnum* bogs are dominant in the deciduous forest zone (Fig. 13.2) but extend into the evergreen forest region (Roivanen 1954; Moore 1983). Mean annual precipitation in the deciduous forest zone ranges from ca. 450 to 650 mm/yr (Tuhkanen 1992). In areas of decreasing precipitation toward the eastern limit of the deciduous forest zone,

Marsippospermum bogs become characteristic (Moore 1983). At the steppe-forest ecotone and in the steppe region, ombrotrophic mires are replaced by minerotrophic fens that are restricted to valleys and depressions with groundwater influence. Mean annual precipitation in this zone ranges from ca. 250 to 500 mm/yr (Tuhkanen et al. 1989–1990). Sedges and grasses are common on the drier fens and mesic grasslands of the steppe region and the steppe-forest ecotone (Roivanen 1954; Moore 1983). Fen mosses, such as *Drepanocladus* spp., become more dominant with increasing wetness toward the deciduous forest zone (Roivanen 1954).

These observed spatial differences in mire vegetation can also be recognized in down-core changes in mire stratigraphy. Considering the strong climatic influence on mire vegetation at present, it is likely that climate has also been a significant factor for peatland differentiation in the past. However, climatic response thresholds may vary between mires in different geologic and geomorphic settings and in different climate regimes. The climatic impact on mire hydrology should be particularly pronounced in moisture-limited areas, where mires may have very low response thresholds for changes in effective moisture. Throughout the Holocene, Rio Rubens Bog has been located in the drier region of the deciduous forest formation or at the steppe-forest ecotone (Huber 2001), where even minor changes in temperature and precipitation probably had large effects on mire hydrology. Hence, the response of peatland vegetation to climate change should have been particularly pronounced and rapid.

Methods

Chronology

A 716-cm-long sediment core of 5-cm diameter was retrieved from the center of Rio Rubens Bog with a Livingstone piston corer (Wright, Mann, and Glaser 1983). The chronology for the last ca. 100 cal yr is based on ^{210}Pb age determinations (Huber 2001). A total of 12 AMS radiocarbon (^{14}C) ages and the Hudson tephra layer (Stern 1992) provide the chronological control for the last 13,000 years (ca. 460 cm) of the Rio Rubens core (Fig. 13.3), which are the focus of this chapter. Wherever possible, mosses (*Drepanocladus* spp., *Sphagnum magellanicum, Polytrichum strictum*) and aboveground parts of macrofossils (wood, leaves) were picked for dating to avoid contamination with younger carbon through roots. Radiocarbon years were converted to calendar years using the calibration program INTCAL98 (Stuiver et al. 1998). The age model for the entire Rio Rubens record is based on 9 ^{210}Pb and 19 ^{14}C ages from both the Holocene and the late-glacial sections of the core (Huber 2001). A weighted sixth-order polynomial curve fit was applied to the core section between ca. 16,900 and 3500 cal yr BP. The age model between ca. 3500 cal yr BP and AD 1995 (year of core retrieval) is based on a third-order polynomial equation. Details of the age model are described in Huber (2001).

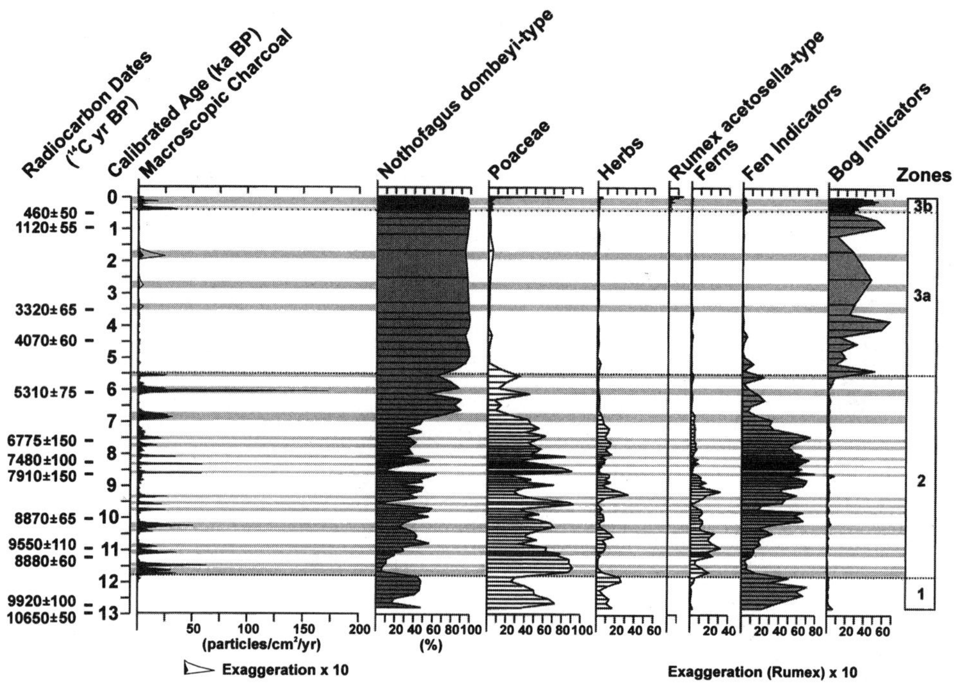

Figure 13.3. Summary diagram of pollen percentages and macroscopic charcoal accumulation rates (CHAR) from Rio Rubens Bog for the last ca. 13,000 cal yr (Huber 2001). Gray bars indicate the width of charcoal peaks. Herb taxa (grasses excluded) primarily consist of Asteraceae tubuliflorae, Caryophyllaceae, and *Acaena*. Ferns mainly comprise Polypodiaceae. Fen taxa consist of Cyperaceae and *Eleocharis*-type. Bog indicators include Ericaceae (primarily *Empetrum*) and *Nanodea* pollen, and spores of *Sphagnum* and *Tilletia sphagni*.

Macroscopic Charcoal Analysis

The Rio Rubens peat core was sampled continuously at 0.5- to 1-cm increments for charcoal analysis. Sample preparation followed methods described in Millspaugh and Whitlock (1995) and Whitlock and Anderson (Chapter 1, this volume), adjusted for peat sediments (Huber 2001). Subsamples of 1 cm^3 were dispersed in a 5% solution of hot KOH for about 30 minutes, and then gently washed through a set of nested sieves with 125 and 250 μm screens. The sieved residues were dispersed in water and placed in a gridded petri dish. Charcoal pieces in the size classes 125–250 μm and >250 μm were counted separately under a stereomicroscope at 40× and 10× magnification, respectively. Charcoal concentrations were divided by the deposition time to obtain charcoal accumulation rates (number of charcoal particles/cm^2/cal yr). Charcoal accumulation rates were corrected for tephra dilution in some sections of the core (Huber 2001).

Temporal resolution in the Rio Rubens record is high with continuous sampling increments of ca. 10 to 160 cal yr per sample and a mean resolution of ca. 25 cal yr (Huber 2001). In most sections of the core, sampling resolution ranges between ca. 10 and 35 cal yr. Low temporal resolution of >100 cal yr per sample only occurs between 48 and 59 cm depth (ca. 3000 to 1600 cal yr BP), during a time period when charcoal peaks are very rare. Charcoal peaks are in most cases distinct and narrow with very low background values between peaks. Thus, the sampling appears to be at sufficiently high resolution to discern individual fires in most sections of the core. A calibration of the Rio Rubens charcoal record with historical fire data is not possible. Forestry fire records only date back to 1986 and are spatially not very specific, and dendrochronologic records of fire do not exist for the region.

Charcoal size fractions >125 μm were analyzed to emphasize the local scale of the recorded fires (Millspaugh and Whitlock 1995; Long et al. 1998; Whitlock and Anderson, Chapter 1, this volume). Many charcoal peaks contain charred peat macrofossils, which indicates that fires spread onto the wetland surface (Huber 2001). Consequently, distinct charcoal peaks in the size fractions >125 μm are interpreted as local fire events. Estimates of fire occurrence in the Rio Rubens core are considered minimum estimates of fires in the catchment, because (1) the sampling resolution may not be high enough in all sections of the core, (2) smoldering peat fires may destroy some evidence of previous fires (Clark and Richard 1996), and (3) peat sediments may primarily record peat fires and therefore a subset of all catchment fires (Tolonen 1983; Huber 2001).

Peat Macrofossil Analysis

Sieve residues (>250 μm), retrieved for charcoal analysis, were also analyzed for macrofossil composition (Huber 2001). Relative abundances of the primary organic peat constituents in the >250 μm size fraction were estimated on a 1 to 5 scale (0, 25, 50, 75, and 100 volume %) by scanning the entire petri dish under a dissecting scope at 10× magnification. Volumetric estimates were assigned to the nearest relative abundance increment. Major peat components include roots of vascular plants, and fen and bog mosses (*Drepanocladus* spp., *Sphagnum* spp., and *Polytrichum strictum*). This semiquantitative approach only records major changes in peat stratigraphy but enables continuous analysis of macrofossils at the same temporal resolution as macroscopic charcoal data. It would be prohibitively time-consuming to achieve decadal-scale resolution over a ca. 13,000 cal yr record with a more detailed approach (e.g., Janssens 1983; Barber et al. 1994; Kuhry 1997).

The Rio Rubens peat stratigraphy is divided into minerotrophic fen peat and ombrotrophic bog peat, based on pollen assemblages of mire plants (Fig. 13.3) and peat macrofossil data (Fig. 13.4). Further, peat macrofossils are grouped into dry and wet fen and bog indicators as a proxy for local effective moisture changes. In the Rio Rubens record, the moss *Drepanocladus* spp. is considered a wet-fen indicator, whereas root-rich fen peat represents drier conditions. Pollen data from

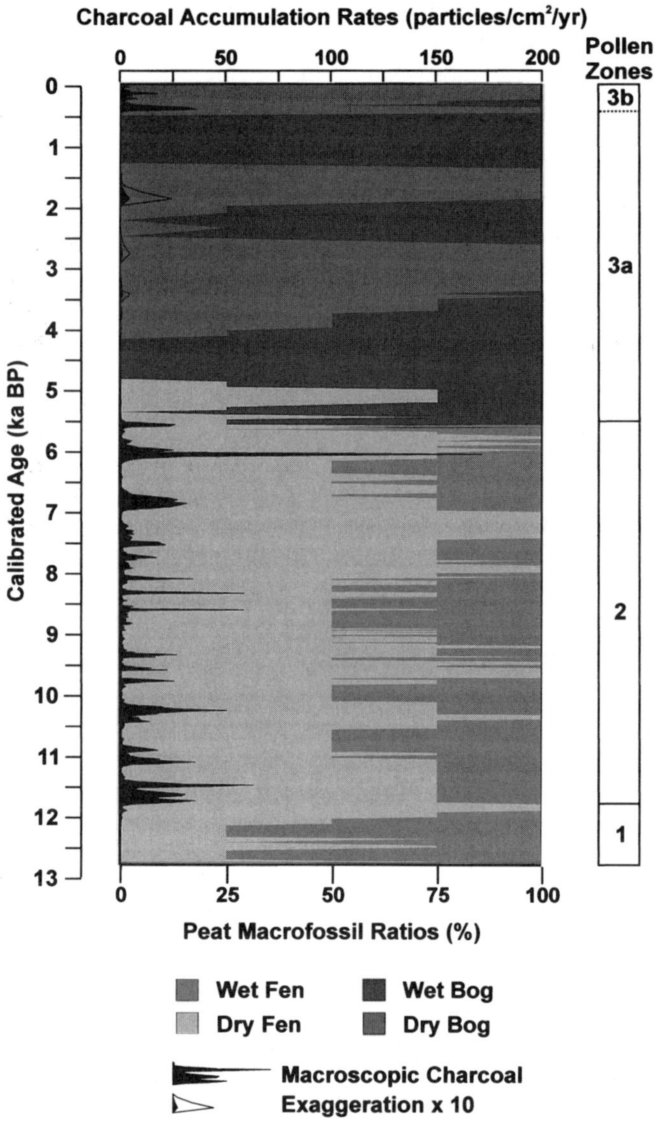

Figure 13.4. 13,000 cal yr record of bog hydrology and macroscopic (>125 μm) charcoal from Rio Rubens Bog (Huber 2001). The macrofossil diagram shows the relative percentages of wet versus dry bog and fen indicators. An exaggeration factor of 10× was applied to charcoal peaks between ca. 5500 and 400 cal yr BP.

the fen section of the core (Huber 2001) suggest that rootlets are most likely of the genus *Cyperaceae* but may also originate from *Poaceae* and other vascular plants. Sedges and grasses dominate the present-day mesic grasslands in the valleys and depressions of the steppe region (Moore 1983). With increasing moisture, fen mosses (*Drepanocladus* spp.) become more prevalent (Roivanen 1954).

In the Rio Rubens record, *Sphagnum* (primarily *S. magellanicum*) is the primary wet-bog indicator. Dry-bog indicators include roots of vascular plants, the moss *Polytrichum strictum*, and strongly decomposed organic matter. Bog pollen assemblages (Huber 2001) indicate that unlike the situation in fen peat, rootlets are not predominantly of Cyperaceae and Poaceae. Instead, root-rich ombrotrophic peat may form in *Marsippospermum* bogs (Moore 1983) that are characteristic of the drier areas of the deciduous *Nothofagus* forest zone and the woodlands of the steppe-forest ecotone (Fig. 13.2) (Tuhkanen 1992). *Marsippospermum grandiflorum* (Juncaceae) produces abundant roots but is virtually absent in the pollen record. The dominance of the moss *Polytrichum strictum* is characteristic of the very dry bog surfaces at the eastern limit of bog growth at the steppe-forest ecotone (Roivanen 1954; Oberdorfer 1960; Tuhkanen et al. 1989–90). Sections of the ombrotrophic peat in the Rio Rubens core (ca. 500–400 and 1300–1800 cal yr BP) are strongly decomposed and contain very few identifiable macrofossil remains in the size range >250 μm. In these intervals relative percentages of peat constituents could not be reliably estimated. Strong decomposition of peat is likely related to increased microbial activity during periods of strong drying of the bog surface (Kuhry 1997). Highly decomposed peat sections that lack macrofossils in the >250-μm-size class are therefore assigned a value of 100% dry-bog indicators.

To test whether local changes in mire hydrology have been predominantly dictated by regional climate variability rather than autogenic processes, the macrofossil data from Rio Rubens Bog are compared to pollen data from the site (Huber 2001) and other paleoclimate data from the region.

Results

Based on peat macrofossil assemblages and macroscopic charcoal data the Rio Rubens Bog profile is divided into three major zones (Fig. 13.4): zone 1 prior to ca. 11,700 cal yr BP, zone 2 from ca. 11,700 to 5500 cal yr BP, and zone 3 from ca. 5500 cal yr BP to present. Details of the peat macrofossil and macroscopic charcoal data from Rio Rubens Bog are described in Huber (2001).

Zone 1: Prior to ca. 11,700 cal yr BP

From prior to 12,700 through ca. 11,700 cal yr BP, charcoal peaks are absent, and charcoal accumulation rates are ≤3 particles/cm^2/cal yr (Fig. 13.4). Wet-fen indicators dominate the sediment, representing up to 75% of the macrofossils. However, century scale variability in peat macrofossil stratigraphy is high, and periods with high percentages of wet-fen macrofossils are repeatedly interrupted by periods dominated by dry-fen indicators.

Zone 2: ca. 11,700 to 5500 cal yr BP

Macroscopic charcoal accumulation rates increase abruptly at ca. 11,700 cal yr BP, and large charcoal peaks (ca. 15–175 particles/cm^2/cal yr) become frequent. Charcoal peaks are most frequent between ca. 11,700 and 7500 cal yr BP and become less frequent thereafter. Century-to-millennial scale variability of dry-fen and wet-fen macrofossils is high. Charcoal peaks concentrate in core sections with ≥75% dry-fen indicators and rarely occur in intervals with high amounts (50%) of wet-fen indicators.

Zone 3: ca. 5500 cal yr BP to present

After ca. 5500 cal yr BP and prior to ca. 400 cal yr BP (zone 3a), charcoal peaks become infrequent and are much smaller (ca. 1–5 particles/cm^2/cal yr) than prior to ca. 5500 cal yr BP. One prominent charcoal peak occurs at ca. 1600 cal yr BP. After ca. 400 cal yr BP, charcoal peaks are again more frequent, and peak sizes increase to between 5 and 40 particles/cm^2/cal yr. At ca. 5500 cal yr BP, macrofossils switch abruptly from fen to bog indicators, and bog macrofossils are subsequently dominant throughout the entire zone. Changes in peat macrofossil stratigraphy occur on century-to-millennial time scales but are markedly less frequent than prior to 5500 cal yr BP. An exception is the transition period from fen to bog peat between ca. 5500 and 5000 cal yr BP when variability is high. Intervals with high amounts (100%) of dry-bog indicators occur between ca. 3500 and 2500 cal yr BP, 1900 and 1300 cal yr BP, and after 500 cal yr BP. Distinct charcoal peaks are limited to those intervals.

Discussion

Peat macrofossil and macroscopic charcoal data from Rio Rubens Bog show distinctly different early and late Holocene climate and fire histories that are separated by a rapid transition at ca. 5500 cal yr BP (Fig. 13.4, zones 2 and 3). Short-term variability in effective moisture and fire frequency is superimposed on these multi-millennial trends and occurs on century to millennial time scales.

Early and Mid Holocene (11,700 to 5500 cal yr BP)

Prior to ca. 11,700 cal yr BP, very low macroscopic charcoal accumulation rates indicate the absence of fires at Rio Rubens Bog (Fig. 13.4, zone 1). Near the transition to the Holocene (zone 2), charcoal abundance increases sharply, and frequent large charcoal peaks suggest that fires became an important disturbance factor. Charred peat moss fragments in these layers demonstrate that the fen surface itself burned repeatedly, and large charcoal peaks are primarily associated with peat fires (Huber 2001).

Pollen data from Rio Rubens Bog (Huber 2001) show that, with the onset of frequent fires, vegetation switched repeatedly between grass-dominated steppe, and open herb- and fern-rich *Nothofagus* woodlands (Fig. 13.3, zone 2). Furthermore, high grass and low tree pollen percentages are frequently associated with peaks in charcoal accumulation rates, implying that distinct reductions in tree cover were associated with many of the local fires. However, not all local fire events show a clear vegetation response. The temporal resolution of the pollen data may not be high enough to register a response for every fire. Also, distinct reductions in woodland cover were likely associated with stand-devastating crown fires, whereas low-intensity surface fires may have left little trace in the pollen record.

A number of records from the modern mixed evergreen-deciduous and deciduous forest zones in Tierra del Fuego and southern Patagonia show increased charcoal levels between ca. 11,700 and 5500 cal yr BP (Heusser 1987, 1990, 1994, 1995a, 1995b, 1998; Markgraf and Anderson 1994). The concurrent increase in fire activity over a large area of Fuego–Patagonia at the late-glacial to Holocene transition suggests a large-scale climatic forcing. Warmer-than-present conditions in the early Holocene (Markgraf and Kenny 1997; Grimm et al. 2001) could have decreased effective moisture in the region. In the Rio Rubens macrofossil record, high percentages of wet-fen indicators suggest that effective moisture was high before ca. 12,000 cal yr BP (Fig. 13.4, zone 1). Subsequently, an abrupt increase in dry-fen indicators implies a pronounced drying of the fen surface. This local, and possibly regional, decrease in effective moisture started about three centuries prior to the first occurrence of local fires at Rio Rubens Bog (Fig. 13.4). A change to drier, more fire-conducive climate conditions approximately coincidental with the start of the Holocene may be responsible for "synchronizing" the onset of high fire activity in Fuego–Patagonia. Increased aridity throughout the early Holocene likely maintained a low fuel moisture content in the xeric woodlands around Rio Rubens Bog, and frequent drying of the fen surface would have enabled the spread of fires onto the mire surface. Consequently, in the presence of ignition sources (humans and/or lightning), the probability of fire occurrence was high during this interval.

Paleo-Indian hunters may have been important initiators of fires (e.g., Heusser 1999). Archaeological evidence suggests the presence of humans in southern Patagonia since at least ca. 13,000 cal yr BP (Dillehay et al. 1992; Borrero and McEwan 1997). Early explorers describe the use of fires by northern Patagonian Indians for hunting of guanacos and rheas in the steppe and steppe-forest ecotone (Cox 1863; Musters 1871; Fonck 1900), although earlier use of fire by prehistoric people is not known. In addition to human ignition sources, warmer-than-present conditions during the early Holocene could have favored convective storms and increased lightning strikes in a region where lightning-caused fires are rare at present (Markgraf and Anderson 1994).

In the early Holocene, when fires were widespread in southern Patagonia and Tierra del Fuego, pollen records indicate that open *Nothofagus* woodlands were much more extensive than at present (Heusser 1987, 1990, 1994, 1995a, 1995b, 1998; Schäbitz 1991; Markgraf 1993; Huber 2001). In Tierra del Fuego the strongly moisture-limited steppe-forest ecotone was located west of its present-day location throughout the early Holocene, and only shifted eastward when moisture levels increased and fire frequency declined in the late Holocene (Heusser 1993, 1994). Drier-than-modern climate during the early Holocene may be related to differences in the width and seasonal migration patterns of the southern westerly belt relative to today. Under modern conditions, the zone of maximum westerly precipitation migrates seasonally and extends from ca. 55° to 45°S in summer to ca. 55° to 35°S in winter (Lawford 1993). During the early Holocene, regional moisture patterns, based on pollen and lake-level data, indicate that the southern westerlies may have been focused between latitudes 45° to 50°S year-round (Markgraf et al. 1992). A focusing of the westerly stormtracks in a narrower latitudinal band may have been related to changes in the seasonal cycle of insolation associated with variations in the earth's orbital parameters (Markgraf et al. 1992; Whitlock et al. 2001). In the early Holocene, the amplitude of the seasonal cycle was smaller than at present at southern high latitudes because perihelion occurred during the southern hemisphere winter and the earth's axial tilt was greater (Berger 1978). Reduced seasonality, in turn, could have caused a reduction in the seasonal migration of the westerly stormtracks (Whitlock et al. 2001). More narrowly focused westerlies would have kept moisture levels low at southern high latitudes (Whitlock et al. 2001), therefore providing ideal conditions for the persistence of fire-prone xeric woodland environments.

The Rio Rubens peat macrofossil data indicate that local effective moisture throughout the early Holocene was highly variable on century-to-millennial time scales (Fig. 13.4). This short-term variability in moisture is superimposed on climatic conditions that were drier than today between 11,700 and 5500 cal yr BP (Fig. 13.5). Local fires cluster in century-to-millennial scale intervals with relatively low effective moisture, as evidenced by 75% to 100% dry-fen indicators. These long intervals of dry conditions and high fire frequency were repeatedly interrupted by century-scale periods of increased effective moisture and reduced fire activity. Century-to-millennial scale dry periods that favored low fuel moisture in the woodlands around Rio Rubens Bog would have enhanced the rapid spread of fires. Under these circumstances, the occurrence of fires was determined by fuel availability and ignition sources. In contrast, during century-scale intervals of overall increased wetness, sufficient fuel desiccation would have been achieved less frequently, reducing the likelihood of fires. Several centuries of increased effective moisture and decreased fire frequency appear to have favored the expansion of woodland cover at the steppe-forest ecotone (Fig. 13.3), and these prolonged wet periods may have been essential for the buildup of coarse woody fuel over grassy fine fuel in these moisture-limited environments.

Century-to-millennial scale variability in effective moisture in southern Patagonia during the early Holocene could have been caused by small temperature fluctuations and/or changes in precipitation. Under modern climate conditions, an increase in the meridionality of the southern westerlies leads to decreased precipitation in southern Patagonia and increased precipitation in northern Patagonia (Pittock 1980; Rutllant and Fuenzalida 1991; Villalba et al.

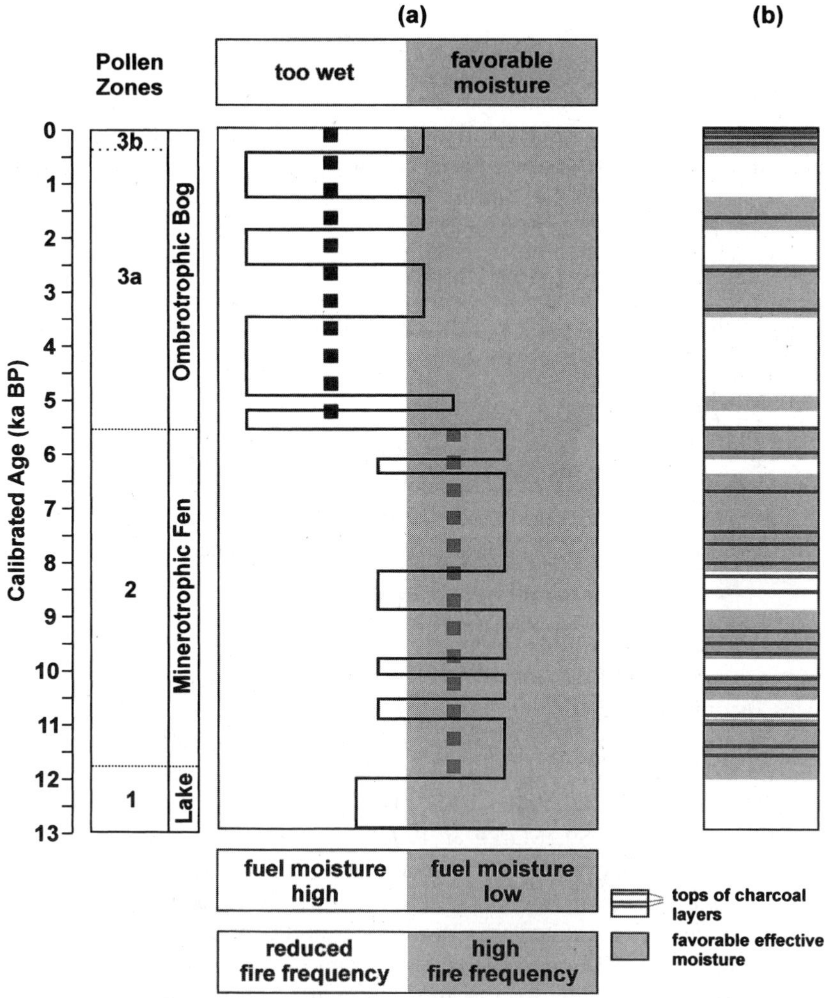

Figure 13.5. Conceptual model showing the relationship between local effective moisture changes, fuel conditions, and fire frequency at Rio Rubens Bog for the last ca. 13,000 years. Interpretation of moisture regimes and fire occurrence are based on macrofossil and charcoal data in Fig. 13.4. (a) Effective moisture and fuel conditions at Rio Rubens Bog. Multi-millennial trends in effective moisture are indicated by bold stippled lines (*gray*: dry early Holocene, *black*: mesic late Holocene). Superimposed multicentury-to-millennial scale variability in effective moisture is shown by black solid line. Shaded gray block indicates the effective moisture range that is favorable for the spread of fires. White block shows the moisture range unfavorable for fire occurrence. "Favorable" moisture conditions correspond to 75–100% dry indicators in the (dry) fen section of the core and to 100% dry indicators in the (wet) bog section (Fig. 13.4). Fuel conditions conducive to fire spread were more frequently reached in the drier climate of the early Holocene and less frequently in the wetter climate of the late Holocene. (b) Minimum estimate of local fire events at Rio Rubens Bog in relation to multicentury-to-millennial scale intervals with, on average, favorable moisture conditions. Assuming that the top of a charcoal layer represents the time of fire in peat sediments (Huber 2001), fire events for the most part cluster in dry intervals lasting several centuries to a millennium. One exception to this pattern are two fires that occur during a generally wet interval between ca. 8900 and 8200 cal yr BP. These fires are, however, associated with dry episodes lasting approximately one century or less.

1997, 1998). More meridional westerly circulation, in turn, is associated with a weak and more northerly located southeast Pacific anticyclone and/or the increased occurrence of blocking highs at southern high latitudes (Villalba et al. 1998; Veblen et al. 1999). In contrast, strongly zonal westerlies, associated with the absence of high-latitude blocking highs and a more intense southeast Pacific anticyclone, cause higher precipitation in Fuego–Patagonia, while decreasing precipitation in northern Patagonia (Villalba et al. 1998; Veblen et al. 1999). Such changes in the zonality of the westerlies provide a possible explanation for short-term variations in effective moisture and fire frequency at Rio Rubens Bog during the early Holocene.

After ca. 6700 cal yr BP (Fig. 13.3), sharply increasing southern beech pollen percentages indicate a shift from open woodlands to closed *Nothofagus* forests (Huber 2001). Decreasing fire frequency (Fig. 13.5) and increasing forest density (Fig. 13.3) mark the transition from warmer and drier conditions in the early Holocene to cooler and wetter conditions in the late Holocene.

Late Holocene (ca. 5500 cal yr BP to present)

Between ca. 5500 and 400 cal yr BP, charcoal peaks are infrequent in the Rio Rubens record (Fig. 13.4, zone 3a), suggesting a distinct decrease in local fire activity. Many sedimentary charcoal records from southern Patagonia and Tierra del Fuego exhibit a similar pattern of lower charcoal concentrations after ca. 5500 cal yr BP (Heusser 1987, 1989, 1993, 1994, 1995a, 1995b, 1998; Rabassa, Heusser, and Rutter 1989; Markgraf 1993). However, the presence of widely spaced charcoal peaks in the Rio Rubens record demonstrates that local fires did occur in this time period, although with greatly reduced frequency. The abrupt decrease in fire frequency at Rio Rubens and over a large area of Fuego–Patagonia after ca. 5500 cal yr BP is likely caused by a pronounced and prolonged regional increase in effective moisture. Further, charcoal peaks during the late Holocene are substantially smaller than during the early and mid Holocene (Fig. 13.4). A distinct decrease in charcoal peak sizes may be primarily related to shallower burning of the bog surface due to a higher water table under more mesic climate conditions (Huber 2001).

Contemporaneous with the decrease in frequency and size of macroscopic charcoal peaks, peat macrofossil (Fig. 13.4) and pollen (Fig. 13.3) assemblages indicate that the Rio Rubens wetland changed from a minerotrophic fen to an ombrotrophic bog. Several peat records from the present-day deciduous and mixed forest zones in southern Patagonia and Tierra del Fuego show that dates for *Sphagnum* peat inception cluster around 5500 cal yr BP (e.g., Heusser 1989, 1995b, 1998; Rabassa, Heusser, and Rutter 1989), approximately synchronous with the switch from fen to bog conditions at the Rio Rubens site. A shift from minerotrophic to ombrotrophic peatlands over a large area is likely related to broad-scale climatic forcing rather than autogenic peatland processes. Increased glacial activity in the southern Patagonian Icefields, located >120 km northwest of Rio Rubens Bog, indicate decreased temperatures and/or increased precipitation after ca. 5400 cal yr BP (e.g., Mercer 1970, 1976, 1982; Aniya 1995; Clapperton and Sugden 1988; Wenzens 1999; Porter 2000). This change in precipitation and/or temperatures may have increased effective moisture enough to pass a critical threshold for ombrotrophic bog establishment (Heusser 1998). At present, the northeastern limit of *Sphagnum* bog distribution in Fuego–Patagonia approximately follows the ecotone between *Nothofagus pumilio* forests and steppe (Roivanen 1954; Tuhkanen et al. 1989–90), where mean annual precipitation ranges between ca. 450 and 650 mm.

The increase in effective moisture after ca. 5500 cal yr BP may have been associated with changes in the average position and the seasonal migration patterns of the southern westerlies. In the late Holocene, seasonality at southern high latitudes increased because perihelion occurred during the Southern Hemisphere summer and the earth's axis was less tilted (Berger 1978). Increased seasonality, in turn, could have led to a more pronounced seasonal migration of the westerly stormtracks (Markgraf et al. 1992; Whitlock et al. 2001), causing higher precipitation and decreased fire frequency in Fuego–Patagonia.

The Rio Rubens pollen data (Huber 2001) suggest that, after ca. 5500 cal yr BP, tree cover in the vicinity of the bog increased rapidly (Fig. 13.3). Pollen assemblages are characteristic of closed *Nothofagus* forests with limited understory (Markgraf, D'Antoni, and Ager 1981; Heusser 1989, 1995b). At about the same time, forest cover expanded over a large region in Tierra del Fuego and southern Patagonia. Closed forests were established in the present-day evergreen, mixed evergreen-deciduous and deciduous forest zones (Markgraf 1983, 1993; Heusser 1995a, 1995b, 1998). This expansion of *Nothofagus* forests was approximately synchronous with the oligotrophication of peatlands and the decrease in fire activity. In addition, the steppe-forest ecotone in southern Patagonia and Tierra del Fuego migrated eastward after ca. 5500 cal yr BP (e.g., Heusser 1993, 1994; Huber 2001). Increased forest cover at the expense of steppe and woodland vegetation is likely due to a combination of higher effective moisture and reduced fire frequency. Whereas closed *Nothofagus* forests would have provided ample coarse fuel, fuel desiccation was likely the limiting factor for the spread of fires under the wetter climate conditions of the late Holocene. Also, cooler climate would have greatly reduced the likelihood of convective storms and thus lightning-ignited fires, although human ignition sources were probably present throughout the Holocene.

The Rio Rubens pollen data (Huber 2001) do not register a strong vegetation response to fires between ca. 5500 cal yr BP and the onset of European settlement in the early 1900s (Fig. 13.3). Increased effective moisture likely favored rapid tree regeneration after fire events. Also higher resolution pollen data may be necessary to record the effects of infrequent fire events on late Holocene vegetation. Furthermore, wind-dispersed *Nothofagus* pollen is strongly overrepresented in pollen records (e.g., Markgraf, D'Antoni, and Ager 1981). Thus, fires in dense *Nothofagus* forests would have to be very large-scale in order to be registered in the pollen record.

Rio Rubens peat macrofossil data indicate that century-to-millennial scale variations in effective moisture were also superimposed on the generally wetter climate conditions of the late Holocene (Figs. 13.4 and 13.5). Drying of the bog surface (100% dry-bog indicators) occurred between approximately 3500 to 2500 cal yr BP, 1900 to 1300 cal yr BP, and after 500 cal yr BP. Distinct charcoal peaks only occur during these relatively drier intervals (Fig. 13.4). The continuous presence of dense forest cover after 5500 cal yr BP and prior to the time of European settlement (Fig. 13.3) suggests that late-Holocene moisture decreases were moderate in magnitude and had limited impacts on the vegetation surrounding the bog. In the closed *Nothofagus* forests, fuel accumulation was not limiting, and fuel desiccation alone likely determined the frequency of fires at Rio Rubens Bog. Episodes of moderate drying appear to have been necessary in order to raise the probability of fires in the mesic *Nothofagus* forests and allow fires to spread onto the bog surface. These prolonged periods of lower effective moisture would have led to more frequent desiccation of the peatland surface and of coarse woody fuels in the surrounding forests.

Late-Holocene periods of increased effective moisture at Rio Rubens Bog are approximately coeval with intervals of Neoglacial ice advances in the southern Patagonian Icefields (Mercer 1970, 1976, 1982; Clapperton and Sugden 1988; Aniya 1995; Wenzens 1999). Increased moisture and lower peat fire frequency may have been caused by increased precipitation and/or cooling. In contrast, intervals with relatively higher temperatures and/or decreased precipitation may have caused negative glacier mass balance, while also increasing the likelihood of peat fires at Rio Rubens Bog. As in the early Holocene, variable moisture at Rio Rubens Bog on century-to-millennial time scales may be related to changes in the zonality of the southern westerlies. Westerly flow may have been more zonal during wet intervals with low fire activity. In contrast, intervals with decreased moisture and increased fire frequency may have been associated with more meridonal westerly flow.

After ca. AD 1600 (zone 3b), fire frequency increased abruptly (Fig. 13.4). Contemporaneously, European weeds occurred for the first time at the site (Fig. 13.3), suggesting that increased fire activity was associated with early European contact. The opening of the *Nothofagus* forests in the early 1900s (Fig. 13.3) was likely associated with European settlement in the region (Huber and Markgraf in press; Huber 2001). European settlement was accompanied by widespread burning, logging, and the introduction of livestock (Butland 1957; Martinic 1997). These combined effects of human activity were likely responsible for the rapid and drastic reduction in forest cover that culminated in the replacement of the previously dense forests with a mosaic of grass steppe and small remnants of *Nothofagus* woodlands (Huber and Markgraf in press; Huber 2001).

Conclusion

Peat macrofossil and macroscopic charcoal data from Rio Rubens Bog, southern Patagonia, suggest a strong relationship between past variability in effective moisture and fire frequency at the eastern limit of the deciduous forest zone. This relationship seems to hold on different temporal scales. Fires could have been ignited by lightning and/or humans, but it appears that climate had to be favorable for fires to spread in deciduous *Nothofagus* woodlands and forests and to burn the peatland surface.

On multi-millennial timescales, increased aridity appears to have favored fire occurrence at Rio Rubens Bog. Regionally low effective moisture levels between ca. 11,700 to 5500 cal yr BP were associated with frequent fires (Figs. 13.4 and 13.5, zone 2). In the early and mid Holocene, increased aridity combined with high fire frequency likely maintained open woodland and steppe vegetation (Fig. 13.3, zone 2). Under drier-than-present climate conditions, fuel moisture content probably remained low for extended periods, and in the presence of ignition sources, fires would have spread rapidly. In contrast, regionally wetter climate conditions from ca. 5500 cal yr BP to present (Figs. 13.4 and 13.5, zone 3) were, prior to European impact, associated with infrequent fires. The combined effects of increased effective moisture and reduced fire frequency were probably essential for the development of closed *Nothofagus* forests near the site. Under generally wetter climate conditions, fuels would have frequently been too moist for fires to spread in the dense *Nothofagus* forests and to affect the bog surface. Hence fires would have been infrequent even in the presence of ignition sources

At Rio Rubens Bog, century-to-millennial scale variability in effective moisture is superimposed on the long-term climatic trends (Fig. 13.5). This short-term moisture variability had important effects on fire frequency near the steppe-forest ecotone. During the early Holocene, century-to-millennial scale dry intervals, inferred from peat macrofossils, were associated with frequent fires, and fire frequency declined when climate became too wet. Increased aridity likely maintained the fuel moisture content of the xeric *Nothofagus* woodlands close to the critical threshold for fire spread, and frequent drying of the fen surface would have enabled the spread of fires onto the mire surface. In contrast, during century-scale periods with relatively high effective moisture levels, coarse woody fuel would have less frequently reached low desiccation levels, leading to a lower probability of widespread fires (Fig. 13.5). These wetter periods were likely important for the expansion of woodlands at the steppe-forest ecotone, whereas frequent fires in combination with dry climate conditions led to the expansion of the steppe. Under the markedly wetter climatic conditions of the late Holocene after ca. 5500 cal yr BP, infrequent fire events occurred during century-to-millennial scale intervals of moderately wet conditions and fires were absent during the wettest periods (Fig. 13.5). In the closed *Nothofagus* forests of the late Holocene, fuel desiccation likely was the limiting factor for fire occurrence.

The combination of peat macrofossil and macroscopic charcoal records allows independent reconstructions of local moisture conditions and fire frequency and therefore provides a powerful tool for evaluating the relationship between climate variability and fire frequency on a range of timescales.

Acknowledgments. M. Reasoner, C. Whitlock, and T. Veblen provided suggestions which greatly improved this chapter. J. Turnbull, at the INSTAAR radiocarbon laboratory, and J. Southon at Lawrence Livermore National Lab, are gratefully acknowledged for their support with radiocarbon dating. The University of Arizona's AMS facility provided two radiocarbon dates for the Rio Rubens core. We thank D. Engstrom at the St. Croix Watershed Research Station, Science Museum of Minnesota, for support with ^{210}Pb-dating and M. Reasoner and P. Bradbury for help with fieldwork. Research for this chapter was supported by National Science Foundation grants NSF-ATM 9321857 and NSF-EAR 9709145 to V. Markgraf, and two Geological Society of America student research grants and a University of Colorado Dean's Small Grant to U. Huber.

References

Almquist-Jacobson, H., and Foster, D.R. 1995. Toward an integrated model for raised-bog development: Theory and field evidence. *Ecology* 76:2503–2516.

Aniya, M. 1995. Holocene glacial chronology in Patagonia: Tyndall and Upsala glaciers. *Arct. Alp. Res.* 27:311–322.

Auer, V. 1963. Die geographischen Gebiete der Moore Feuerlands. *Mitteil. Fränk. Geograph. Gesells.* 10:31–38.

Barber, K.E. 1981. *Peat Stratigraphy and Climatic Change: A Palaeoecological Test of the Theory of Cyclic Peat Bog Regeneration.* Rotterdam: Balkema.

Barber, K.E., Chambers, F.M., Maddy, D., Stoneman, R., and Brew, J.S. 1994. A sensitive high-resolution record of late Holocene climatic change from a raised bog in northern England. *Holocene* 4:198–205.

Berger, A.L. 1978. Long-term variations of caloric insolation resulting from the earth's orbital elements. *Quat. Res.* 9:139–167.

Borrero, L.A., and McEwan, C. 1997. The peopling of Patagonia: The first human occupation. In *Patagonia. Natural History, Prehistory and Ethnography at the Uttermost End of the Earth*, eds. C. McEwan, L.A. Borrero, and A. Prieto, pp. 32–45. Princeton: Princeton University Press.

Butland, G.J. 1957. The human geography of southern Chile. *Instit. Br. Geogr. Pub.* 24: 1–132.

Clapperton, C.M., and Sugden, D.E. 1988. Holocene glacier fluctuations in South America and Antarctica. *Quat. Sci. Rev.* 7:185–198.

Clark, J.S. 1988. Particle motion and the theory of charcoal analysis, source area, transport, deposition and sampling. *Quat. Res.* 30:81–91.

Clark, J.S. 1990. Fire and climate change during the last 750 yr in northwestern Minnesota. *Ecol. Monogr.* 60:135–159.

Clark, J.S., and Patterson, W.A. 1997. Background and local charcoal in sediments: Scales of fire evidence in the paleorecord. In *Sediment Records of Biomass Burning and Global Change*, eds. J.S. Clark, H. Cachier, J.G. Goldammer, and B. Stocks, pp. 23–48. Berlin: Springer-Verlag.

Clark, J.S., and Richard, P.J.H. 1996. The role of paleofire in boreal and other cool-coniferous forests. In *Fire in Ecosystems of Boreal Eurasia*, eds. J.G. Goldammer, and V.V. Furyaev, pp. 65–89. Dordrecht: Kluwer Academic.

Clark, J.S., and Royall, P.D. 1996. Local and regional sediment charcoal evidence for fire regimes in presettlement northeastern North America. *J. Ecol.* 84:365–382.

Clark, J.S., Lynch, J., Stocks, B.J., and Goldammer, J.G. 1998. Relationships between charcoal particles in air and sediments in west-central Siberia. *Holocene* 8:19–29.

Cox, G. 1863. Viajes a las regiones septentrionales de Patagonia 1862–1863. *An. Univ. Chile* 23:3–239, 437–509.

Cruz, G.M., and Lara, A.A. 1987. *Regiones naturales del area de uso agropecuario de la XII region, Magallanes y de la Antartica chilena.* Santiago, Chile: Instituto de Investigaciones Agropecuarias (INIA).

Dillehay, T.D., Calderon, G.A., Politis, G., and da Conceicao de Moraes Coutinho Beltrao, M. 1992. Earliest hunters and gatherers of South America. *J. World Prehist.* 6:145–204.

Fonck, F. 1900. *Viajes de Fray Francisco Menéndez a Nahuelhuapi*. Valparaiso, Chile: C.F. Niemeyer.

Gardner, J.J., and Whitlock, C. 2001. Charcoal accumulation following a recent fire in the Cascade Range, northwestern USA, and its relevance for fire-history studies. *Holocene* 11:541–549.

Gardner, R.H., Hargrove, W.W., Turner, M.G., and Romme, W.H. 1996. Climate change, disturbances and landscape dynamics. In *Global Change and Terrestrial Ecosystems*, eds. B. Walker and W. Steffen, pp. 149–172. Cambridge: Cambridge University Press.

Gignac, L.D., Halsey, L.A., and Vitt, D.H. 2000. A bioclimatic model for the distribution of *Sphagnum*-dominated peatlands in North America under present climatic conditions. *J. Biogeogr.* 27:1139–1151.

Glaser, P.H., Bennett, P.C., Siegel, D.I., and Romanowicz, E.A. 1996. Palaeo-reversals in groundwater flow and peatland development at Lost River, Minnesota, USA. *Holocene* 6:413–421.

Grimm, E.C., Lozano-Garcia, S., Behling, H., and Markgraf, V. 2001. Holocene vegetation and climate variability in the Americas. In *Interhemispheric Climate Linkages*, ed. V. Markgraf, pp. 325–370. San Diego, CA: Academic Press.

Heusser, C.J. 1987. Fire history of Fuego–Patagonia. *Quat. S. Am. Antarct. Penin.* 5: 93–109.

Heusser, C.J. 1989. Late Quaternary vegetation and climate of southern Tierra del Fuego. *Quat. Res.* 31:396–406.

Heusser, C.J. 1990. Late-Glacial and Holocene vegetation and climate of subantarctic South America. *Rev. Palaeobot. Palynol.* 65:9–15.

Heusser, C.J. 1993. Late Quaternary forest-steppe contact zone, Isla Grande de Tierra del Fuego, subantarctic South America. *Quat. Sci. Rev.* 12:169–177.

Heusser, C.J. 1994. Paleoindians and fire during the late Quaternary in southern South America. *Rev. Chilena Hist. Nat.* 67:435–443.

Heusser, C.J. 1995a. Palaeoecology of a *Donatia–Astelia* cushion bog, Magellanic moorland–subantarctic evergreen forest transition, southern Tierra del Fuego, Argentina. *Rev. Palaeobot. Palynol.* 89:429–440.

Heusser, C.J. 1995b. Three late Quaternary pollen diagrams from southern Patagonia and their palaeoecological implications. *Palaeogeogr. Palaeoclim. Palaeoecol.* 118:1–24.

Heusser, C.J. 1998. Deglacial paleoclimate of the American sector of the Southern Ocean: Late Glacial–Holocene records from the latitude of Canal Beagle (55°S), Argentine Tierra del Fuego. *Palaeogeogr. Palaeoclim. Palaeoecol.* 141:277–301.

Heusser, C.J. 1999. Human forcing of vegetation change since the last Ice Age in southern Chile and Argentina. *Bamb. Geograph. Schrif.* 19:211–231.

Huber, U.M. 2001. *Linkages among climate, vegetation and fire in Fuego–Patagonia during the late-Glacial and Holocene*. Ph.D. dissertation. University of Colorado, Boulder.

Huber, U.M., and Markgraf, V. In press. European impact on fire regimes and vegetation dynamics at the steppe-forest ecotone of southern Patagonia.

Hughes, P.D.M., Mauquoy, D., Barber, K.E., and Langdon, P.G. 2000. Mire-development pathways and palaeoclimatic records from a full Holocene peat archive at Walton Moss, Cumbria, England. *Holocene* 10:465–479.

Janssens, J.A. 1983. A quantitative method for stratigraphical analysis of bryophytes in Holocene peat. *J. Ecol.* 71:189–196.

Kitzberger, T., Veblen, T.T., and Villalba, R. 1997. Climatic influences on fire regimes along a rainforest-to-xeric woodland gradient in northern Patagonia, Argentina. *J. Biogeogr.* 24:35–47.

Kuhry, P. 1997. The palaeoecology of a treed bog in western boreal Canada: A study based on microfossils, macrofossils and physicochemical properties. *Rev. Palaeobot. Palynol.* 96:183–224.

Lawford, R.G. 1993. Regional hydrologic response to global change in western North America. In *Earth System Responses to Global Change: Contrasts between North and South America*, eds. H.A. Mooney, E.R. Fuentes, and B.I. Kronberg, pp. 73–99. San Diego, CA: Academic Press.

Long, C.J., Whitlock, C., Bartlein, P.J., and Millspaugh, S.H. 1998. A 9000-year fire history from the Oregon Coast Range, based on a high-resolution charcoal study. *Can. J. For. Res.* 28:774–787.

Markgraf, V. 1983. Late and postglacial vegetational and paleoclimatic changes in subantarctic, temperate, and arid environments in Argentina. *Palynology* 7:43–70.

Markgraf, V. 1993. Paleoenvironments and paleoclimates in Tierra del Fuego and southernmost Patagonia, South America. *Palaeogeogr. Palaeoclim. Palaeoecol.* 102:53–68.

Markgraf, V., and Anderson, L. 1994. Fire history of Patagonia: Climate versus human cause. *Rev. Instit. Geograf. São Paulo* 15:35–47.

Markgraf, V., and Kenny, R. 1997. Character of rapid vegetation and climate change during the late-Glacial in southernmost South America. In *Past and Future Rapid Environmental Changes: The Spatial and Evolutionary Responses of Terrestrial Biota*, eds. B. Huntley, W. Cramer, A.V. Morgan, H.C. Prentice, and J.R.M. Allen, pp. 81–90. Berlin: Springer-Verlag.

Markgraf, V., D'Antoni, H.L., and Ager, T.A. 1981. Modern pollen dispersal in Argentina. *Palynology* 5:43–63.

Markgraf, V., Dodson, J.R., Kershaw, A.P., McGlone, M.S., and Nicholls, N. 1992. Evolution of late Pleistocene and Holocene climates in the circum–South Pacific land areas. *Clim. Dyn.* 6:193–211.

Martinic, M.B. 1997. The meeting of two cultures. Indians and colonists in the Magellan region. In *Patagonia. Natural History, Prehistory and Ethnography at the Uttermost End of the Earth*, eds. C. McEwan, L.A. Borrero, and A. Prieto, pp. 110–126. Princeton: Princeton University Press.

Mercer, J.H. 1970. Variations of some Patagonian glaciers since the late-Glacial. *Am. J. Sci.* 269:1–25.

Mercer, J.H. 1976. Glacial history of southernmost South America. *Quat. Res.* 6:125–166.

Mercer, J.H. 1982. Holocene glacier variations in southern South America. *Striae* 18: 35–40.

Millspaugh, S.H., and Whitlock, C. 1995. A 750-yr fire history based on lake sediment records in central Yellowstone National Park. *Holocene* 5:283–292.

Millspaugh, S.H., Whitlock, C., and Bartlein, P.J. 2000. Variations in fire frequency and climate over the past 17,000 yr in central Yellowstone National Park. *Geology* 28:211–214.

Moore, D.M. 1979. Southern oceanic wet-heathlands (including Magellanic Moorland). In *Heathlands and Related Shrublands: Descriptive Studies*, ed. R.L. Specht, pp. 489–497. Amsterdam: Elsevier.

Moore, D.M. 1983. *Flora of Tierra del Fuego*. Oswestry: Nelson.

Moore, P.D. 1986. Hydrological changes in mires. In *Handbook of Holocene Palaeoecology and Palaeohydrology*, ed. B.E. Berglund, pp. 91–107. New York: Wiley.

Musters, G.C. 1871. *At Home with the Patagonians: A Year's Wanderings over Untrodden Ground from the Straits of Magellan to the Rio Negro*. London: Murray.

Oberdorfer, E. 1960. *Pflanzensoziologische Studien in Chile. Ein Vergleich mit Europa*. Weinheim: Cramer.

Ohlson, M., and Tryterud, E. 2000. Interpretation of the charcoal record in forest soils: Forest fires and their production and deposition of macroscopic charcoal. *Holocene* 10:519–525.

Overpeck, J.T., Rind, D., and Goldberg, R. 1990. Climate-induced changes in forest disturbance and vegetation. *Nature* 343:51–53.

Pittock, A.B. 1980. Patterns of climatic variation in Argentina and Chile—I. Precipitation, 1931–60. *Mon. Wea. Rev.* 108:1347–1361.

Porter, S.C. 2000. Onset of neoglaciation in the Southern Hemisphere. *J. Quat. Sci.* 15: 395–408.

Rabassa, J., Heusser, C.J., and Rutter, N. 1989. Late-Glacial and Holocene of Argentine Tierra del Fuego. *Quat. S. Am. Antarct. Penin.* 7:327–351.

Renkin, R.A., and Despain, D.G. 1992. Fuel moisture, forest type and lightning-caused fire in Yellowstone National Park. *Can. J. For. Res.* 22:37–45.

Roivanen, H. 1954. Studien über die Moore Feuerlands. *Ann. Bot. Soc. Zool. Bot. Fenn. "Vanamo"* 28:1–205.

Rutllant, J., and Fuenzalida, H. 1991. Synoptic aspects of the central Chile rainfall variability associated with the Southern Oscillation. *Int. J. Climatol.* 11:63–76.

Schäbitz, F. 1991. Holocene vegetation and climate in southern Santa Cruz, Argentina. *Bamb. Geograph. Schrif.* 11:235–244.

Stern, C.R. 1992. Tefrocronología de Magellanes: Nuevos datos e implicaciones. *An. Instit. Patagonia* 21:129–141.

Stuiver, M., Reimer, P.J., Bard, E., Beck, J.W., Burr, G.S., Hughen, K.A., Kromer, B., McCormac, F.G., van der Pflicht, J., and Spurk, M. 1998. INTCAL98 radiocarbon age calibration, 24,000–0 cal B.P. *Radiocarbon* 40:1041–1083.

Tolonen, K. 1983. The post-glacial fire record. In *The Role of Fire in Northern Circumpolar Ecosystems*, eds. R.W. Wein, and D.A. MacLean, pp. 21–44. New York: Wiley.

Tolonen, K. 1986. Charred particle analysis. In *Handbook of Holocene Palaeoecology and Palaeohydrology*, ed. B.E. Berglund, pp. 485–496. New York: Wiley.

Tuhkanen, S. 1992. The climate of Tierra del Fuego from a vegetation geographical point of view and its ecoclimatic counterparts elsewhere. *Acta Bot. Fenn.* 145:1–62.

Tuhkanen, S., Kuokka, I., Hyvönen, J., Stenroos, S., and Niemelä, J. 1989–1990. Tierra del Fuego as a target for biogeographical research in the past and the present. *An. Instit. Patagonia, Ser. Cie. Nat.* 19:1–107.

Veblen, T.T., and Markgraf, V. 1988. Steppe expansion in Patagonia? *Quat. Res.* 30: 331–338.

Veblen, T.T., Kitzberger, T., and Lara, A. 1992. Disturbance and forest dynamics along a transect from Andean rain forest to Patagonian shrubland. *J. Veg. Sci.* 3:507–520.

Veblen, T.T., Donoso, C., Kitzberger, T., and Rebertus, A.J. 1996. Ecology of southern Chilean and Argentinean *Nothofagus* forests. In *Ecology and Biogeography of Nothofagus Forests*, eds. T.T. Veblen, R.S. Hill, and J. Read, pp. 293–353. New Haven: Yale University Press.

Veblen, T.T., Kitzberger, T., Villalba, R., and Donnegan, J. 1999. Fire history in northern Patagonia: The roles of humans and climatic variation. *Ecol. Monogr.* 69:47–67.

Villagrán, C. 1980. Vegetationsgeschichtliche und pflanzensoziologische Untersuchungen im Vincente Perez Rosales Nationalpark (Chile). *Dissert. Bot.* 54:1–165.

Villalba, R., and Veblen, T.T. 1997a. Regional patterns of tree population age structures in northern Patagonia: Climatic and disturbance influences. *J. Ecol.* 85:113–124.

Villalba, R., and Veblen, T.T. 1997b. Spatial and temporal variation in tree growth along the forest–steppe ecotone in northern Patagonia. *Can. J. For. Res.* 27:580–597.

Villalba, R., Jones, P.D., Salinger, M.J., Palmer, J., Cook, E.R., D'Arrigo, R.D., and Jacoby, G.C. 1997. Sea-level pressure variability around Antarctica since AD 1750 inferred from subantarctic tree-ring records. *Clim. Dyn.* 13:375–390.

Villalba, R., Veblen, T.T., Jones, P.D., Cook, E.R., Jacoby, G.C., and D'Arrigo, R.D. 1998. Tree-ring based reconstructions of northern Patagonia precipitation since AD 1600. *Holocene* 8:659–674.

Wenzens, G. 1999. Fluctuations of outlet and valley glaciers in the southern Andes (Argentina) during the past 13,000 years. *Quat. Res.* 51:238–247.

Whitlock, C., and Millspaugh, S.H. 1996. Testing the assumptions of fire-history studies: An examination of modern charcoal accumulation in Yellowstone National Park, USA. *Holocene* 6:7–15.

Whitlock, C., Bartlein, P.J., Markgraf, V., and Ashworth, A.C. 2001. The midlatitudes of North and South America during the last Glacial maximum and early Holocene: Similar paleoclimatic sequences despite differing large-scale controls. In *Interhemispheric Climate Linkages*, ed. V. Markgraf, pp. 391–416. San Diego, CA: Academic Press.

Wright, H.E. Jr., Mann, D.H., and Glaser, P.H. 1983. Piston cores for peat and lake sediments. *Ecology* 65:657–659.

Zamora, E., and Santana, A. 1979. Caracteristicas climaticas de la costa occidental de la Patagonia entre las latitudes 46°30 y 56°30 S. *An. Instit. Patagonia* 10:109–144.

14. Regeneration Potential of Chilean Matorral After Fire: An Updated View

Gloria Montenegro, Miguel Gómez, Francisca Díaz, and Rosanna Ginocchio

Mediterranean-type ecosystems, such as the Mediterranean Basin, California, central Chile, South Africa, and Southwest Australia, represent important hot spots for plant diversity as they harbor 20% of the world's flora in only 5% of the earth's land surface (Cowling et al. 1996; Davis et al. 1997). These regions also have been major centers of human population growth (Cincotta, Wisnewski, and Engelman 2000), and thus human impacts on natural ecosystems have been many and varied. For instance, the Mediterranean-type climate area of central Chile supports 53% of the total population of continental Chile (INE 1995), 50% of the total plant species, and 45% of endemic plant species described for the continental territory (Arroyo and Cavieres 1997). Therefore the long history of human occupation has led to a highly altered landscape and an important reduction of the land occupied by wild vegetation (Fuentes, Avilés, and Segura, 1990; Fuentes et al. 1995).

Besides the direct impacts of human populations on Mediterranean-type ecosystems at the local level, human activities may also have indirect impacts on Mediterranean ecosystems due to large-scale changes, such as global climate change. Climate change and local human activities may thus result in land degradation and desertification of Mediterranean-type ecosystems. Therefore the high human potential for directly and indirectly altering ecosystems or for introducing new unnatural disturbances to these natural systems are a priority of concern among ecologists (Fuentes et al. 1995; Mooney, Hamburg, and Drake 1986; Montenegro et al. 2001). An important human impact on some Mediterranean-type ecosystems, such as central Chile and central and southern California, has been an increasing rate of human-induced fires in recent decades (Zunino and Riveros 1990; Keeley Fotheringham, and Morais 1999), as shown by the closely parallel increase between fire frequency and population growth in these two regions over the same period (Palmer 1993). Desertification due to global climate change may also have an important effect on the increasing rate of human-induced fires as a longer dry season would lead to longer periods of dry standing biomass, thus increasing the fire risk. Although fire is an important natural disturbance that has long played an important role in the ecology and evolution of Mediterranean floras, with the exception of the matorral in central Chile, its role has been modified as consequence of increased human activities in these ecosystems. Human impacts on natural Mediterranean fire regimes is evident in many ways, although their net effect on fire regimes is still a matter of some debate (Minnich 1989; Keeley et al. 1989).

Natural fires seem much less common in Chile than in other Mediterranean regions, such as California, the southwestern Cape, southwestern Australia and the Mediterranean Basin (Aschmann 1991; Aschmann and Bahre 1977; Keeley and Johnson 1977; Rundel 1981a; Araya and Ávila 1981; Ávila, Montenegro, and Aljaro 1988). Convective thunderstorms and associated lightning are uncommon in central Chile, thereby providing few ignition sources under natural circumstances (Rundel 1981a). However, an increasing rate of human-induced fires has been also detected in the Chilean Mediterranean region since the Spanish Conquest in the sixteenth century (Bahre 1979). Since then, fires are quite common in the natural vegetation, known as *matorral*, particularly during the spring and summer (Araya and Ávila 1981; Ávila, Aljaro, and Silva 1981, 1988).

In the context of global climate change, how are Mediterranean-type ecosystems likely to respond to changes in fire regime? Although similar patterns of climatic change might be assumed to lead to similar changes in fire regimes, the ecological consequences of altered fire regimes are not necessarily the same for all Mediterranean-type ecosystems. To assess the implications of climatic change for ecological processes and patterns in Mediterranean-type ecosystems requires a fine-scale understanding of the current and historical role of fire in these ecosystems. The strategy of this chapter is to compare the roles of fire in the regeneration ecology of California chaparral and Chilean matorral. Although Mediterranean-type ecosystems are generally regarded as being fire-dependent ecosystems, this chapter identifies important differences in the nature of fire adaptations and the history of fire between California chaparral and Chilean matorral. These differences are potentially important to the prediction of future ecological patterns in these regions.

Natural Vegetation in Mediterranean-Type Ecosystems of Central Chile: The Matorral

The matorral is the natural shrubby sclerophyll vegetation growing in the semi-arid Mediterranean region of central Chile, that dominates on the slopes of the coastal range (coastal matorral) and the Andean foothills (mid-elevation mator-

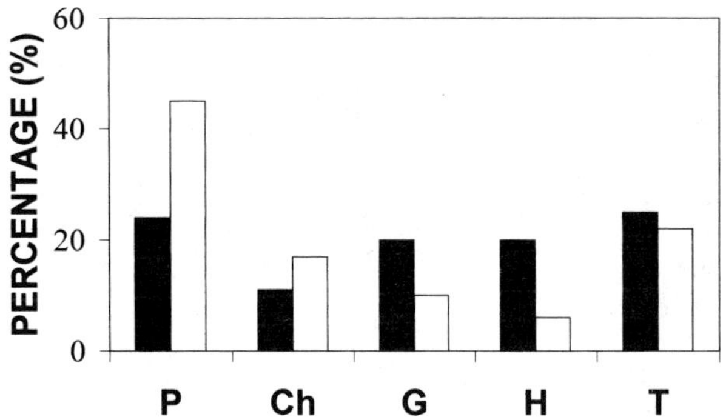

Figure 14.1. Percentage of growth life forms present in coastal (■) and mid-elevation (□) matorral in central Chile. P, phanerophyte; Ch, chamaephyte; G, geophyte; H, hemicryptophyte, T, therophyte.

ral) between 32° and 36° South latitude (Arroyo et al. 1995). This vegetation is adapted to a severe environment, which includes extended drought, unstable land forms, desiccating winds, and low nutrient availability in the soil (Aljaro and Montenegro 1981; Miller 1981; Montenegro et al. 1989). Matorral vegetation is diverse in growth forms (Fig. 14.1), and ancient woody groups (e.g., phanerophytes) of tropical origin coexist today with many short-lived herbaceous species. The opening up of vegetation in the Tertiary and the establishment of a drier climate in central Chile selected for a variety of drought-tolerant (e.g., geophytes) and drought-evading (e.g., therophytes or annuals) plants (Arroyo and Cavieres, 1997).

Matorral shrubs tend to grow less densely than shrubs in the chaparral of California or maquis of the Mediterranean Basin (Thrower and Bradbury 1977), particularly on sunny, equatorial-facing slopes, where open spaces between clumps of shrubs and succulent plants characterize the landscape (Fuentes and Muñoz 1995). Only on the moister, shady, polar-facing slopes do shrub clumps overlap leading to a closed canopy. Variations also occur in species diversity, dominance and cover along an altitudinal transect from the coast up to the 2200 m above sea level in the Andes mountains. Evergreen sclerophyllous shrubs and trees, succulents, and drought-evading herbs predominate along this gradient, from the coast to about 1000-m elevation (Mooney et al. 1970; Mooney 1977; Montenegro, Aljaro, and Kummerow 1979a). Evergreen shrubs predominate on polar-facing slopes, while drought-deciduous shrubs and succulents are mostly found on equatorial-facing slopes (Rundel 1975; Parsons 1976; Mooney 1977; Armesto and Martínez 1978). The coastal matorral and the mid-elevation sclerophyllous scrub in the foothills of the Andes are replaced at about 1850 m by a montane evergreen scrub community (Mooney et al. 1970; Rundel and Weisser 1975; Hoffmann and Hoffmann 1978, 1982; Montenegro, Aljaro, and Arrieta 1979b). There are also some changes in plant growth forms with altitude (Fig. 14.1), from a coastal matorral where all growth forms are well represented to a

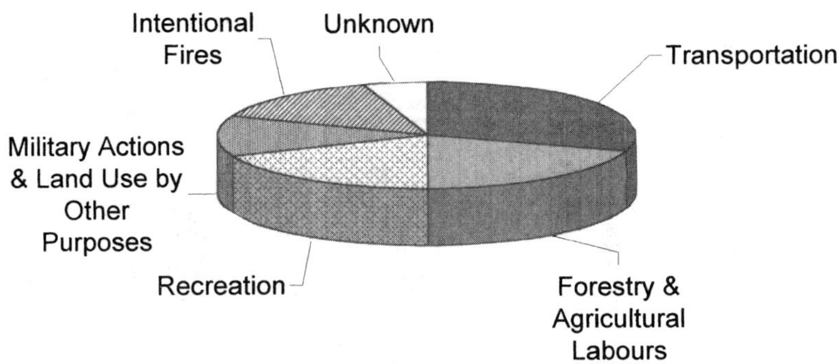

Figure 14.2. Percentage distribution of fires in Chile in the period 1963 to 1998. (Data from CONAF, 1998.)

mid-elevation matorral with increased dominance by phanerophyes and chamaephytes and a decreased dominance in geophytes and hemicriptophytes (Fig. 14.1 and Appendix). However, the transition from coastal matorral to mid-elevation one is more gradual than from coastal sage scrub to chaparral in California (Dallman 1998).

Fires in Central Chile

Almost all literature indicates that wildfires are essentially the result of human causes in central Chile because natural lightning-ignited fires are rare and absent from official records (Fig. 14.2). The high Andean Cordillera protects central Chile from humid subtropical air masses with convectional storms with lightning. This is an important difference with other Mediterranean-type ecosystems around the world such as California, where natural lightning-ignited fires are a common phenomenon (Table 14.1) (Aschmann 1991; Keeley 1977, 1981; Rundel 1981a; Araya and Ávila 1981; Ávila, Montenegro, and Aljaro 1988).

Nevertheless, Fuentes and Espinoza (1986), using published botanical, palynological, and geomorphological evidence, argued that volcanism, a frequent phenomenon in Chile, could have been a nonhuman ignition source in the

Table 14.1. Frequency and extent of burning by natural lightning fires and human causes on state and federal wildlands in California, 1970 to 1979

	Fires, 10^6 ha/yr		Hectares burned, 10^6 ha/yr	
Jurisdiction	Humans	Lightning	Humans	Lightning
State of California, Division of Forestry	541	31	3347	416
U.S. Forest Service	134	129	669	189

Source: From Keeley 1982.

Mediterranean region of central Chile. However, the relative paucity of volcanism compared to lightning implies much less selection of plant traits and therefore lack of plant–fire-dependence as observed in all other Mediterranean-type ecosystems. Although this is an interesting hypothesis, humans historically account for a substantial amount of the ignition sources in central Chile.

A further manifestation of the importance of the human component in fire regimes detected in central Chile is the dramatic increase in fire incidence observed in recent decades. Historical data from the Chilean Forestry Service (Corporación Nacional Forestal, CONAF 1998) indicates that total fire frequency was quite low at the country level in the 1960s but has increased in the last three decades from 500 fires per year in 1963–64 to 5500 fires per year in 1997–98. The area with the most accentuated Mediterranean climate of the country (central Chile) follows the same national trend (Zunino and Riveros 1990), and the increase in fire frequency closely parallels the rapid population growth produced in central Chile over the same period of time (Palmer 1993).

Fire disturbance in central Chile has not had, however, the same effect on natural ecosystems as on agricultural lands. Wild areas have been more affected by human-induced fires in comparison with agricultural ones, such as croplands and commercial forest. This difference may be the result of high human pressure on matorral areas, such as agricultural and urban expansion pressures on natural ecosystems through controlled fires and more diverse human use of wild areas with higher fire risks.

The seasonality of the Mediterranean climate strongly influences the seasonal distribution of fire (Fig. 14.3). High fire frequencies, as high as 90% in January, occur in summer, whereas cold and wet months of winter and autumn reduce fire risk to almost zero. This pattern may not only be explained by changes in climate and natural fuel load accumulation but also by higher human-matorral interactions during the summer period. Higher fire frequencies in summer also may have strong influences on plant regeneration capabilities after fire. Plant reproductive processes in woody matorral species, such as flowering (Fig. 14.3), are concentrated in those months with higher fire risk, which may greatly reduce sexual regeneration possibilities and mechanisms of temporal fire avoidance.

Despite increased incidence of human-induced wildfires in central Chile, there is no evidence that the total area burned has increased over the same period (CONAF 1998). Reasons for the lack of congruence between incidence of fires and area burned are multiple and complex, but they clearly point to the obvious conclusion that average fire size has declined over this period. Paramount among the reasons is the increased fragmentation of habitats that has accompanied accelerated population growth and development. Landscapes have been altered by replacing vast stretches of continuous matorral fuels with patches of vegetation dispersed in a mosaic interspersed with less flammable agricultural and suburban vegetation as well as nonflammable urban developments.

Coupled with the greater presence of humans in these regions is the increasing concern for fire detection and fire suppression. Therefore, despite the increased fire incidence by human activities, there has not been a net impact on

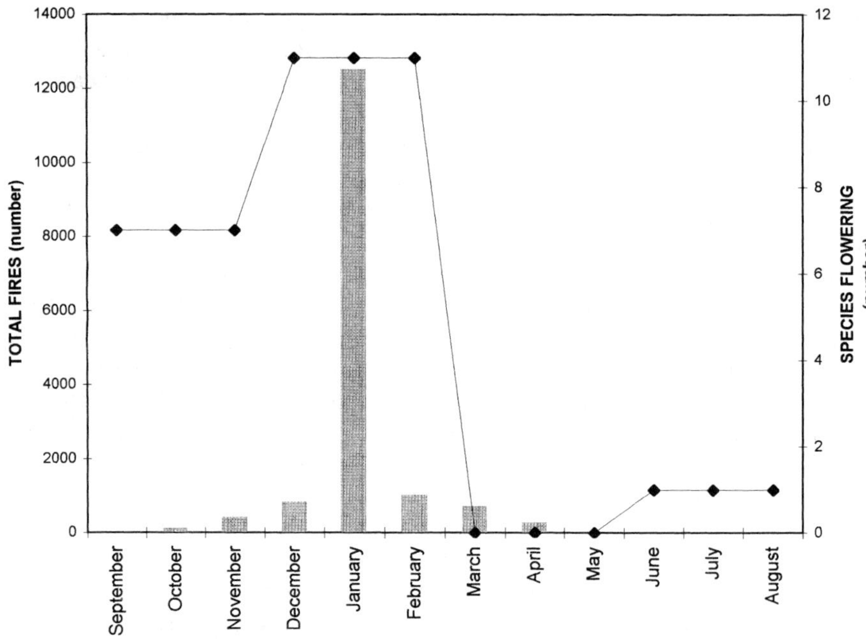

Figure 14.3. Seasonal distribution of fire frequency in Chile in the period 1978 to 1995 (*bars*) and seasonal distribution of flowering of woody species (*lines*) in matorral vegetation of central Chile.

the extent of burned surface. However, this impression needs to be tempered by recognition that while area burned may, broadly speaking, be roughly the same over time, the spatial extent of natural vegetation has declined. As a consequence, for any given parcel of natural vegetation, the relative proportion burned has likely increased in recent decades. As a consequence fuel accumulation has likely declined, reducing flammability of stands and thus further contributing to a reduction in fire spread and ultimate fire size.

Vegetation Response to Fire in Chile and California at the Life-Forms Level

The marked similarity in vegetation structure between Chile and California has been well described (Mooney and Parsons 1973; Parsons and Moldenke 1975; Parsons 1976; Mooney 1977). At the landscape scale, similar semideciduous shrublands dominate at low latitudes or low elevation coastal sites while taller evergreen sclerophyllous shrublands and woodlands dominate at higher latitudes or elevations (Mooney et al. 1970). Similar differences have been also described between plant species growing on polar- and equator-facing slopes (Armesto and Martinez 1978).

Even though there are marked similarities in both plant structure and function between central Chile and California, one may expect that different fire disturbance histories may have led to different plant responses to fire. Plant responses to fire in central Chile can be seen as strategies to cope with a rather novel evolutionary challenge whereas fires have been a long-term natural disturbance for Californian vegetation. Since the sclerophyllous shrublands in these two regions are best known in terms of their response to fire, we have focused on a comparison of chaparral versus matorral. When comparing plant evolutionary responses to natural fires at the growth form level, interesting differences can be found (Table 14.2).

Matorral in central Chile has a high diversity of succulent plants, most notably members of the Cactaceae, from near sea level to more than 2000 msl (Rundel 1975, 1977). On the other hand, the chaparral ecosystem of California has a low diversity of Cactaceae because fire kills cacti and other succulent plants (Nierig and Lowe 1984). The relatively low natural fire frequency in central Chile compared to California may well be the critical element promoting the survival of large cacti in matorral vegetation (Fuentes et al. 1995).

Table 14.2. Regeneration responses to fire disturbance and relative importance* in Californian chaparral and Chilean matorral

Growth forms	Regeneration response after fire	Chaparral	Matorral
Phanerophytes Chamaephytes	(1) Resprouting from epicormic stem buds	++++	++++
	(2) Resprouting from rootcrown	++++ ++	++++ ++++
	(3) Resprouting from lignotuber	++ +	— —
	(4) Fire-stimulated flowering		
	(5) Release of seeds from serotinous cones or fruits	+++	+
	(6) Germination of dormant soil-seed banks stimulated by heat shock	+++	—(?)
	(7) Germination of dormant soil-seed banks stimulated by chemicals from smoke or charred wood		
Geophytes	(1) Resprouting from deeply buried bulbs or corms	+++	+++
	(2) Fire-stimulated flowering	++	++(?)
Hemicryptophytes	(1) Resprouting from roots or rootcrowns	(?)	(?)
Therophytes	(1) Germination of soil-seed banks stimulated by heat shock	++++	—

*+ rare, ++ common, +++ abundant, and ++++ very abundant.

Geophytes are a diverse and abundant component of all mediterranean-climate regions (Dafni, Cohen, and Noy-Meir 1981; Le Maitre and Brown 1992; Rundel 1993, 1996). For instance, along the north-to-south climatic gradient observed in Chile, the highest diversity of geophytes corresponds to the Mediterranean-type region (Hoffman 1989; Hoffmann, Liberona, and Hoffmann 1998) whereas along the east-to-west altitudinal gradient, the highest diversity of geophytes corresponds to coastal areas (Fig. 14.1; Montenegro, unpublished data). Moreover Chile and California are remarkably similar in the proportion of their flora comprised by geophytes, 5.4% in both regions (Rundel 1996). This is a growth form that enables plants to avoid water stress; however, in some respects it also preadapts them to avoid fires. In the absence of fire their normal seasonal cycle involves a dormancy period where the aboveground vegetative material dies back. Typically this resting period coincides with the fire season in these regions and thus geophytes are well buffered against fires as it seems likely that the deeply buried corms and bulbs are not negatively affected from fire. However, surficial buried underground structures, such as rhizomes, may be strongly affected by high intensity fires (Table 14.2). Montenegro (unpublished data) found high geophyte survival in coastal matorral after a fire, were six geophyte species are well represented (*Tecophilaea violaeflora*, *Conanthera trimaculata*, *Dioscorea humifusa*, *Pasithea coerulea*, *Fortunatia biflora*, and *Alstroemeria pulcra*).

The geophyte life form may also respond favorably to fire because of the much greater frequency of flowering apparently stimulated by enhanced nutrients and light (Table 14.2; Stone 1951; Rundel 1981b,c; Le Maitre and Brown 1992). In one extreme example of flowering stimulation by fire, the geophyte *Cyrtanthus ventricosum*, which flowers immediately (and only) after fire, is triggered by smoke (Keeley 1993). Smoke-triggered flowering has never been described for any Chilean geophyte, but enhanced flowering after fire has been mentioned by Hoffmann, Liberona, and Hoffmann (1998) for some geophytes belonging to the *Alstroemeria* genus.

Hemicriptophytes are well represented in the Chilean matorral (Fig. 14.1). This plant growth form is also well adapted to seasonal climates such as the Mediterranean-type climate. It has a herbaceous root crown with dormant buds at ground level that are protected by a dense rosette of death leaves. Lack of hardbark formation in this structure with proper cell suberization may lead to its complete death under high-intensity fires, but they may be able to survive under low-intensity fires. There is no published information about the fire survival capability of hemicriptophytes in central Chile and in California (Table 14.2).

Although herbaceous annuals are diverse in the matorral, they are less diverse than in California (Fuentes et al. 1995). It has been shown that seed germination of Chilean herbaceous annual species is significantly decreased by fires, while "fire-endemic" annuals are common in chaparral (Carter 1973; Keeley and Johnson 1977; Ávila, Aljaro, and Silva 1981). Again, differences in natural fire frequency may have played a determining role in this phenomenon.

Phanerophyte adaptative traits to fire in California are more diverse and complex than the other growth forms perhaps due to their longer generational times. Phanerophyte survival after fire in chaparral is mainly the result of the following mechanisms: (1) vegetative regeneration from buds buried in underground structures (root crown and lignotuber) or epicormic stem buds and (2) sexual regeneration due to fire-stimulated flowering, fire-stimulated seed release from fruits, or heat-shock-enhanced germination (Trabaud 1987). These contrasting strategies, although not mutually exclusive, have been emphasized in the literature of almost all fire-disturbed Mediterranean-type ecosystems (Keeley 1977; Cody and Mooney 1978; Kruger 1983; Keeley 1984; Keeley and Keeley 1984; Keeley et al. 1986), but they have not been described for matorral (Table 14.2).

It has been shown that most matorral species are able to resprout after fire (Cody and Mooney 1978; Araya and Ávila 1981; Montenegro, Ávila, and Schatte 1983; Ginocchio, Holmgren, and Montenegro 1994) as in chaparral (Kruger 1983). However, matorral significantly differs from chaparral because all woody species can resprout after fire (Parsons 1976; Mooney 1977; Araya and Ávila 1981; Montenegro, Ávila, and Schatte 1983), whereas a substantial portion of the Californian chaparral shrub species fail to resprout, even following low-intensity fires (Table 14.2; Wells 1969).

Montenegro and Ginocchio (1995) found that a common ecomorphological character shared by shrubs of matorral and chaparral is the development of underground lignified stems or lignotubers. Lignotubers have been defined as a source of dormant epicormic buds buried in a modified stem (Montenegro, Ávila, and Schatte 1983). Such buds are capable of resprouting after the crown is killed, consumed by fire, or removed by mechanical means, regenerating the aerial part of the plant. Another shared character in both regions is the presence of epicormic buds that can sprout after fire. Although both ecomorphological characteristics have not been the result of fire selective pressure in matorral as in chaparral, they may have been the result of seasonality in climate (Montenegro and Ginocchio 1995) and thus represent a "pre-adaptation" to human-induced fires.

Presence of adventitious buds in root crowns of woody plants is nearly a ubiquitous trait in dicotyledonous plants (Wells 1969). Although wildfires are an important feature of many ecosystems worldwide and have been present since the early evolution of angiosperms, there are other disturbance factors that could select for this trait, such as seasonal climate. Storage of carbohydrates is another important role of adventitious buds, and there is strong circumstantial evidence from comparison of burning and cutting experiments that seasonal depletion of carbohydrates may strongly affect regenerative capacity (Rundel et al. 1987).

Mediterranean-climate ecosystems are unusual in having a large percentage of the landscape dominated by lignotuberous species. These structures are ontogenetic features that are adapted to initiate development early in seedling growth (Wells 1969; Montenegro, Ávila, and Schatte 1983). This is in contrast to lignotuber development in most non–Mediterranean-climate species where these basal swellings are a wound response to having the aboveground stems destroyed (Keeley 1981). A recent interesting anatomical study has shown that buds are developed in lignotubers from the vascular cambium, and they have not seen to

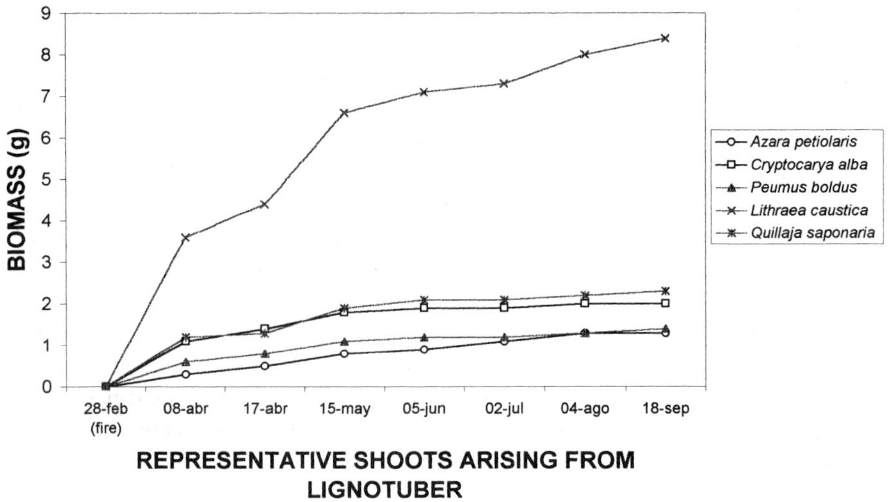

Figure 14.4. Dynamics of resprouts in some burned matorral shrubs after fire.

be developed from cortical parenchyma (Montenegro, personal observations). Therefore bark thickness is of great importance in protecting these buds as well as the cambium during fire. Some bark loss is usual during fires, but restoration may take place if fire interval is long enough.

Our evidence suggests that resprouting capability differs among species, both in the proportion of individuals that exhibit the response and in the amount of foliage they produce (Ginocchio et al. 1994). Resprouting can immediately occur after a fire independently of the time of the year (Montenegro, Díaz, Lewin, and Gómez, unpublished data). Figure 14.4 shows biomass change in time of resprouts generated from lignotubers in several woody species after a fire produced in late summer in central Chile. It is clear that although all species were able to resprout from lignotubers in autumn, there were interspecific differences. *Lithrea caustica* showed a significantly higher biomass recovery than the other four species. However, these species normally start their vegetative growth later in the season when mean temperatures reach higher values in summer (Montenegro et al. 1981, 1989). Rapid shoot production from lignotubers may be triggered by changes in water balance due to higher root to shoot ratios typical at burned shrubs, and high nutrient availability from lignotubers.

Another interesting aspect of resprouting capability from lignotubers relates to the age of the plant, and therefore the age of its underground structure. Some evidence suggests that older plants have larger lignotubers, and therefore increased resprouting capabilities (Montenegro, Díaz, Lewin, and Gómez, unpublished data) may be due to higher starch reserves in parenchymatic cells of larger underground woody structures (Montenegro, Ginocchio, and Segura 1996). Carbohydrate levels have been detected as high as 4.5% to 10.2% dry weight in lignotubers of *Erica australis* in Mediterranean-type ecosystems of Spain

(Cruz and Moreno 1997a,b,c). Resprouting capability after fire is only affected by very low carbohydrate levels in lignotuber that are not observed in the field, at least in this plant species.

Besides observed changes in plant productivity, leaf size and secondary products are other plant characteristics that can also be affected by fires. Montenegro and collaborators (unpublished data) found important changes in leaf size of shoots generated after fire in four evergreen matorral shrubs species when compared with normal leaves present in adult shrubs that have not been affected by fire. Leaves generated after fire are larger than unburned adult shrub leaves (Fig. 14.5) leading to a rapid recovery of plant photosynthetic structure. These young large leaves formed after fire may be heated by higher solar radiation observed in open areas produced by fires and may also be a good food source to herbivores. However, preliminary data indicate chemical protection against both factors through increased phenol concentrations in new formed leaves after fire (Gómez 2000).

A second general phanerophyte survival mechanism involves fire-stimulated flowering, fire-stimulated seed release from fruits, or heat-shock-enhanced germination. This is an uncommon plant responses in matorral (Table 14.2). In California, nonsprouters or obligate-seeders recruit massive seedling populations after fire and are considered highly specialized "fire-dependent" species (Keeley 1986a, 1994). This phenomenon has not been observed in Chile (Muñoz and Fuentes 1989).

Fire-stimulated flowering has not been detected in Chile, although it is a phenomenon described in other Mediterranean-type ecosystems, such as southwestern Australia (Specht 1988). The rapid recovery of sexual reproduction observed in *Trevoa trinervis* motivated its classification as a pyrogenic species

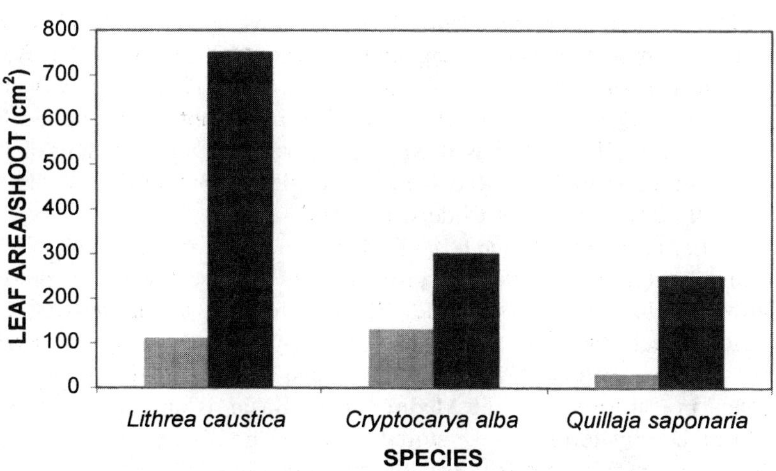

Figure 14.5. Leaf area per shoot formed by some woody matorral shrubs 240 days after a summer fire (■) and in control shoots (■).

(Montenegro and Teillier 1988). However, results of Ginocchio et al. (1994) indicate that flowering is not really stimulated after fire; it only reaches similar levels observed in control plants growing in undisturbed sites. Rapid flowering recovery may be the result of branch types of *T. trinervis*. In comparison with most evergreen species, the canopy of *T. trinervis* is mainly built by short shoots or brachyblasts where reproductive and vegetative plant functions cannot be separated and therefore are not constrained (Ginocchio and Montenegro 1992; Montenegro and Ginocchio 1993). In other words, stem production in *T. trinervis* not only led to leaf production but also to inflorescences in the same growing season. The fact that resprouting shrub of *T. trinervis* can generate and disperse seeds faster than other matorral species is another indicator of the potential of this species to establish new individuals at burned sites. This may explain the relatively large cover of *T. trinervis* seedlings observed at burned sites one year after the fire.

From both laboratory and field experiments it has been shown that unlike the findings in California (Keeley 1986b, 1987; Keeley et al. 1986), neither fires nor ash-enriched soils induce the germination of seeds in the matorral (Muñoz and Fuentes 1989). It appears that high seed mortality occurs with intense fires, with the exception of *Muehlenbeckia hastulata* and *Trevoa trinervis* (Keeley and Johnson 1977; Muñoz and Fuentes 1989). In the former species, however, seedling frequency observed under burned shrubs is the same as under cut shrubs (Muñoz and Fuentes 1989). Therefore it cannot be said that fire specifically stimulated their germination. All fire-stimulated *Ceanothus* species in chaparral and *Phylic* species in South African fynbos as well as *T. trinervis* belong to the family Rhamnaceae (Keeley and Bond 1997). This pattern may be the result of some similarity in seed coat characteristics in plants belonging to Rhamnaceae that allow them to survive and germinate after fire, such as hard and thick coats that can be scarified by fire.

Germination triggered by chemicals from charred wood or smoke is widespread in Californian chaparral, South African fynbos, and Australian kwongan, heath, and other associations (de Lange and Boucher 1990; Brown 1993; Keeley 1994; Dixon et al. 1995; Keeley and Fotheringham 1997). This is clearly the most specialized postfire germination pattern. To date it has not been reported from Chile (Table 14.2). However, based on the relatively depauperate flora of postfire species that recruit from seed banks in Chile, it seems unlikely that this response will be widespread in Chilean matorral.

Species that accumulate seed banks between fires and produce a pulse of seedling recruitment in the first growing season after fire are common in Mediterranean-climate regions (Keeley 1994). However, in this respect, Chile differs greatly from the other four regions. Canopy storage of dormant seeds in serotinous cones or fruits, which is well developed in South Africa and Australia, and present in California and the Mediterranean Basin, is unknown from Chile (Table 14.2). Deeply dormant seeds stored in the soil are common and widespread in the other four regions but only weakly developed in Chile.

We can conclude, in this section, that a large proportion of the growth forms of the matorral flora fails to recruit seedlings immediately after fire, as is clear

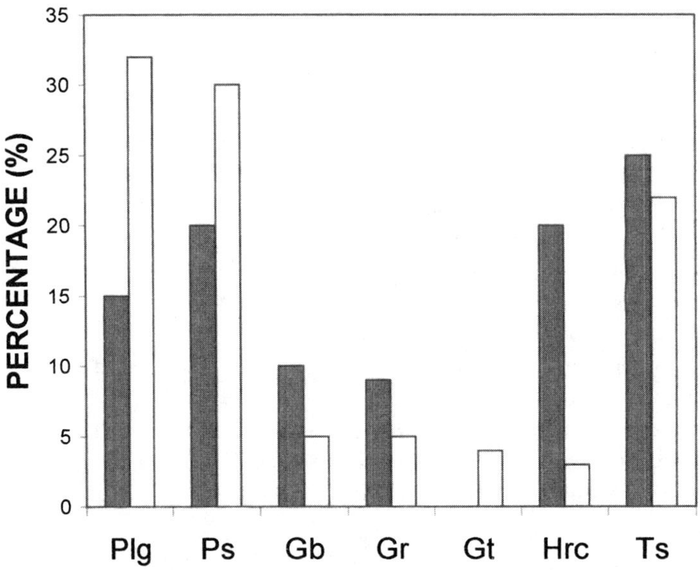

Figure 14.6. Percentage distribution of natural regeneration capabilities by growth form in coastal (■) and mid-elevation (□) matorral of central Chile. Plg, phanerophyte with lignotuber; Ps, phanerophyte that regenerates from seeds; Gb, geophyte that regenerates from bulbs; Gr, geophyte that regenerates from rhizome; Gt, geophyte that regenerates from tuber; Hrc, hemicryptophyte that regenerates from root crown; T, therophyte that regenerates from seeds.

from Figure 14.6 (see also Appendix). For instance, approximately half of phanerophyes present either in coastal or mid-elevation matorral are able to resprout after fire while the other half, which cannot regenerate vegetatively after fire, is eliminated from the burned site. Exceptions are *T. trinervis* and *M. hastulata*. Most coastal and mid-elevation matorral plant species would be locally extirpated from burned sites if it were not for their vegetative regeneration from epicormic buds or underground structures (lignotubers, woody root crowns, bulbs). While resprouting is clearly adaptive in this context, it is a matter of some debate as to whether resprouting was initially selected in response to low fire frequencies produced by volcanism (Fuentes and Espinoza 1986) or whether it was mainly selected in response to seasonality and water stress.

Vegetation Response to Fire in Chile and California at Community and Landscape Levels

Fire not only initiates cycles of vegetation succession in chaparral of California but contributes to the maintenance of ecosystem structure and function, provided that its occurrence does not greatly exceed the natural fire frequency (Pickett and

White 1985). Natural selection has acted to maintain relatively flammable vegetation structure and chemistry to favor high combustibility (Rundel 1981b; Keeley et al. 1989) and to select adequate plant regeneration mechanisms as discussed in previous section. Hanes (1971) has used the term "auto-succession" to describe postfire community-level response in Californian chaparral, where the pre-fire flora is fully represented in the immediate postfire flora. Nevertheless, fire does generate major structural and floristic changes in this community. In particular, dominant shrubland communities are typically replaced by short-lived herbaceous or subligneous vegetation. The woody flora itself changes in structure because shrubs are replaced by either seedlings or herbaceous resprouts from basal lignotubers or root crowns (Fig. 14.7a and b). Therefore chaparral is resilient to natural fires. Vegetation recovery after fire involves endogenous processes of local plant species which restore burned areas to a similar state as that before the fire (Fig. 14.7a and b), with very few pioneer species colonizing such systems (Trabaud 1987). Recuperation of the plant community in California is achieved within the first 10 years after the fire (Keeley 1986a).

In Chile very limited research has been conducted on matorral resilience to fire. However, there are two observations that may suggest that this plant community is not necessarily resilient to human-induced fires. The first is that coastal matorral recovers less rapidly than chaparral. Recovery may require from 25 to 30 years, or the vegetation maybe never recover to the pre-fire community, leading to an alternative plant community dominated by *T. trinervis* (Lazo, unpublished data). The second is that pioneer species colonizing burned areas are different from those found in undisturbed matorral communities and from "natural" pioneer matorral succession processes described by Armesto and Pickett (1985). Consequently the plant community becomes dominated by nonnative annual grasses and forbs (Montenegro, personal observation).

An important characteristic of human-induced fires in central Chile is the high variability in intensity. Not all fires affecting matorral are intense enough to consume all the aboveground biomass on a slope. Fires usually have patchy effects that leave a slope as a mosaic of consumed shrubs mixed with lightly burned ones. High- and low-intensity human-made fires can produce ecologically different effects in the Chilean matorral, determining species distribution patterns (Segura et al. 1998).

High-intensity fires tend to destroy the seed bank in the matorral and thus eliminate the possibility of recolonization by this mechanism (Muñoz and Fuentes 1989). In such cases resprouting from underground structures (lignotubers, woody rootcrowns, bulbs) allows the maintenance of previously colonized space (Fig. 14.8a and b). However, only some matorral species are able to resprout after fire and not all have the same resprouting capability. *Trevoa trinervis* and *M. hastulata* show the lowest resprouting capability when compared with other matorral shrub species (Segura et al. 1998). In addition high rates of human-induced fires in the same area can limit resprouting capability and generate a complete change in matorral structure and functioning. This contributes to

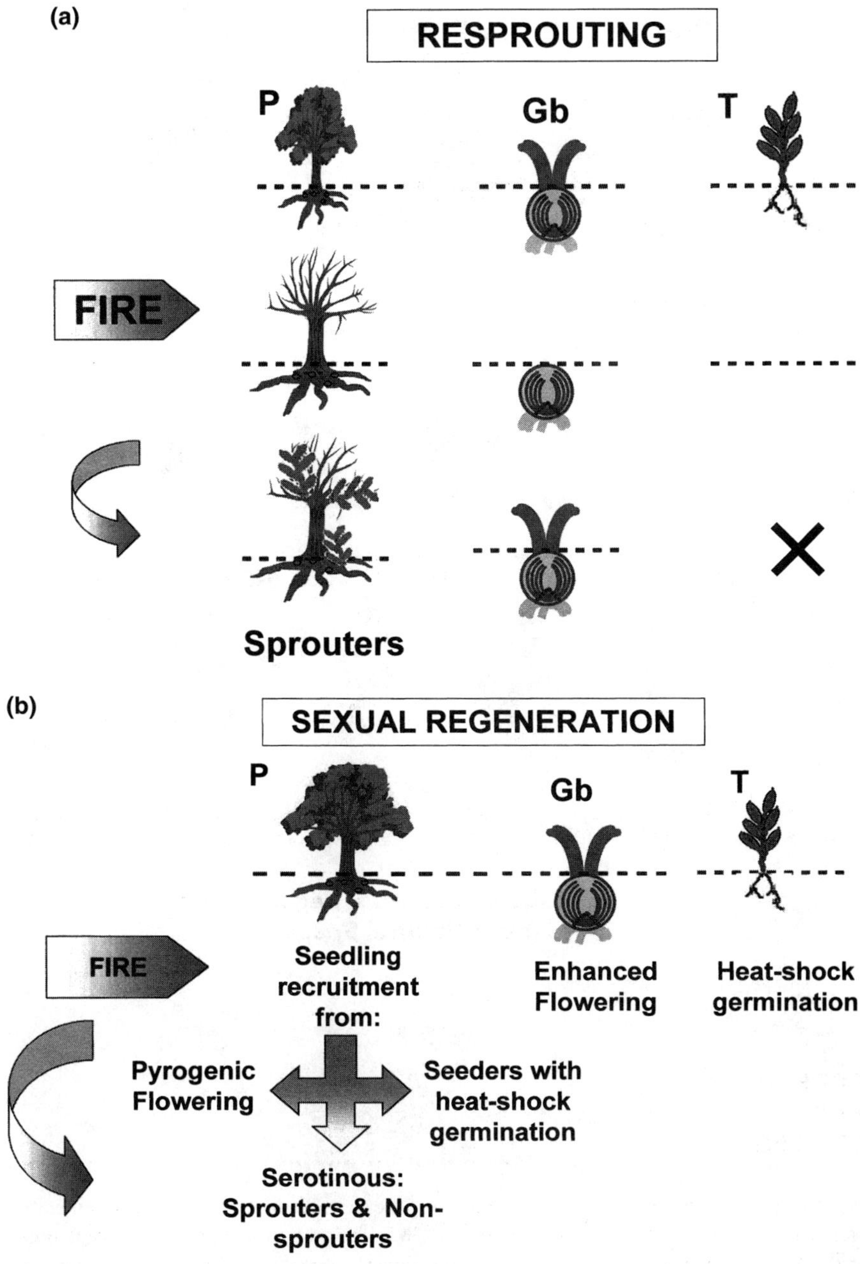

Figure 14.7. Vegetative (a) and sexual (b) regeneration models for chaparral vegetation after fire. P, phanerophyte; Gb, geophyte that regenerates from bulbs; T, therophyte.

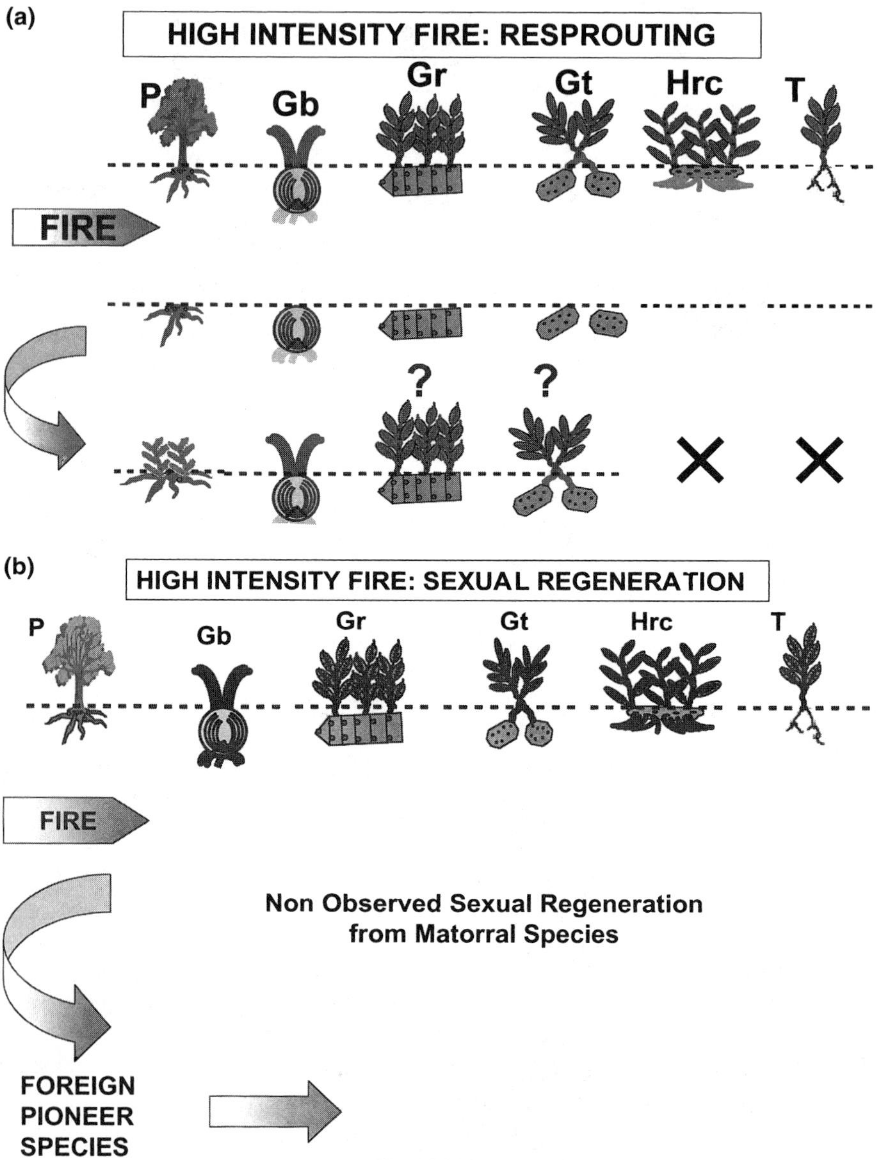

Figure 14.8. Vegetative (a) and sexual (b) regeneration models for matorral vegetation after a high-intensity fire. P, phanerophyte; Gb, geophyte that regenerates from bulbs; Gr, geophyte that regenerates from rhizome; Gt, geophyte that regenerates from tuber; Hrc, hemicryptophyte that regenerates from root crown; T, therophyte that regenerates from seeds.

landscape fragmentation that can lead to plant extintion and therefore to landscape desertification (Armesto and Gutiérrez 1978).

Low-intensity fires leave lightly burned shrubs and some soil seed bank that allow seedling establishment from seeds and resprouting from epicormic stem buds besides resprouting from underground structures (Fig. 14.9a and b). However, seedling establishment under lightly burned shrubs differs among species. *Trevoa trinervis* and *Muehlenbeckia hastulata* show the strongest response among all woody species (Segura et al. 1998). It is clear that not all matorral plant species are equally adapted to negative effects produced by fires. Different fire frequencies and fire intensities also can change from place to place and time to time. This can result in vegetation mosaics and landscape heterogenity (Keeley and Swift 1995) of greater biological diversity at broad spatial scales, but due to colonization by foreign species.

Fires were a natural part of the Californian Mediterranean-climate ecosystems prior to human influences, but in contrast, Chilean matorral appears to have had little evolutionary exposure to fires. As a consequence we see a substantial portion of the Californian flora as "fire dependent." Specific fire-related chemical germination cues are required for many species to complete their life cycles. Such is not the case in Chile because all species appear to have some significant capacity for regeneration in the absence of fire, and thus represent a very different successional model than chaparral (Armesto and Pickett 1985). Models of plant community-level response to human-induced fire disturbances in central Chile may vary from situations where vegetation is totally replaced by temporary successional stage of exotic species, to communities where immediate postdisturbance regeneration is from the original vegetation, depending on fire intensity and frequency.

The present human impact on these regions appears to be one of increasing fire frequency. In California, the impact of increasing fire frequency is a function of whether or not fires occur frequently enough to prevent the nonsprouting shrub element from establishing a seed bank sufficient to regenerate the population. In Chile, this is apparently not an issue. In addition fire frequency may reduce survivorship of resprouting species and over time thin perennial plant populations both in matorral and chaparral.

In contrast to chaparral, matorral suffers from pressures in addition to fire, such as intensive browsing by domestic goats and wood gathering for charcoal production. The main consequence of these diverse and intense human impacts in central Chile is that areas near urban developments often exhibit marked declines in the woody component and an increase in nonnative annual grasses and forbs. For instance, Matthei (1995) has recently shown that central Chile is a strong focus for the concentration of invasive and native weedy species.

Conclusion

At a global scale it is attractive to assume that global warming would have highly similar effects on ecosystem structure and function in the five regions of Mediterranean-type ecosystems. Indeed, the similarity of these ecosystems is

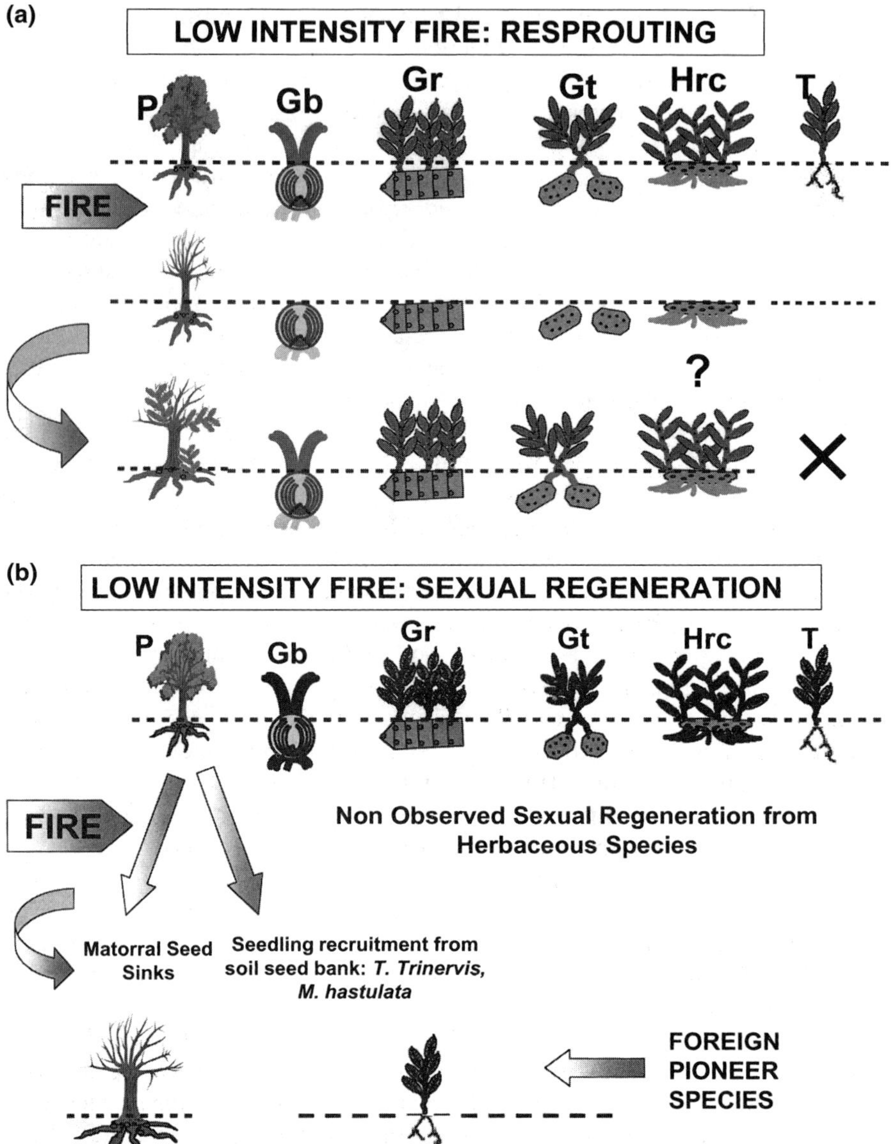

Figure 14.9. Vegetative (a) and sexual (b) regeneration models for matorral vegetation after a low-intensity fire. P, phanerophyte; Gb, geophyte that regenerates from bulbs; Gr, geophyte that regenerates from rhizome; Gt, geophyte that regenerates from tuber; Hrc, hemicryptophyte that regenerates from root crown; T, therophyte that regenerates from seeds.

commonly attributed to the similarity of their climates, and it is not unreasonable to assume that similar patterns of climate change will lead to similar ecological changes in these ecosystems. However, the comparison in this chapter of Chilean matorral and Californian chaparral demonstrates the importance of fine-scale differences in the plant adaptations and history of human impact that are likely to affect future responses to climate change. Differences in the role of fire in the regeneration ecology of these two regions are likely to result insignificantly different outcomes of climatically induced ecological change.

Acknowledgments. The work was supported by grant NIH-NSF-USDA 2UO1 TW 00316-09 to B. N. Timmermann and Fundación Andes Research Fellow to R. Ginocchio. We thank also Corporación Nacional Forestal, CONAF, for the access to the fire data and to grant FIA CO1-1-G-OO2.

References

Aljaro, M.E., and Montenegro, G. 1981. Growth of dominant Chilean shrubs in the Andean Cordillera. *Mount. Res. Dev.* 1:287–291.

Araya, S., and Ávila, G. 1981. Rebrote de arbustos afectados por fuego en el matorral Chileno. *An. Mus. Hist. Nat., Valparaíso* 14:107–113.

Armesto, J.J., and Gutiérrez, J.R. 1978. El efecto del fuego en la estructura de la vegetación de Chile central. *An. Mus. Hist. Nat., Valparaíso* 11:43–48.

Armesto, J.J., and Martínez, J.A. 1978. Relations between vegetation structure and slope aspect in the Mediterranean region of Chile. *J. Ecol.* 66:881–889.

Armesto, J.J., and Pickett, S.T.A. 1985. A mechanistic approach to the study of succession in the Chilean matorral. *Rev. Chil. Hist. Nat.* 58:9–17.

Arroyo, M.T.K., and Cavieres, L. 1997. The Mediterranean-type climate flora of central Chile. What do we know and how can we assure its protection? In *Taller Internacional sobre aspectos ambientales, ideológicos, éticos y políticos en el debate sobre bioprospección y uso de recursos genéticos en Chile*, eds. G. Montenegro and B.T. Timmermann, pp. 48–56. Santiago: Noticiero de Biología, Sociedad de Biología de Chile.

Aschmann, H. 1991. Human impact on the biota of Mediterranean-climate regions of Chile and California. In *Biogeography of Mediterranean Invasions*, eds. R.H. Groves, and F. di Castri, pp. 33–42. New York: Cambridge University Press.

Aschmann, H., and Bahre, C. 1977. Man's impact on the wild landscape. In *Convergent Evolution of Chile and California Mediterranean Climate Ecosystems*, ed. H.A. Mooney, pp. 73–84. Stroudsburg, PA: Dowden, Hutchinson and Ross.

Ávila, G., Aljaro, M.E., and Silva, B. 1981. Observaciones en el estrato herbáceo después del fuego. *An. Mus. Hist. Nat., Valparaíso* 14:99–105.

Ávila, G., Montenegro, G., and Aljaro, M.E. 1988. Incendios en la vegetación Mediterránea. In *Ecología del paisaje en Chile central: Estudios sobre sus espacios montañosos*, eds. E.R. Fuentes, and S. Prenafeta, pp. 81–88. Santiago: Ediciones Universidad Católica de Chile.

Bahre, C.J. 1979. Destruction of the natural vegetation of north-central Chile. *Univ. Cal. Pub. Geog.* 23:1–117.

Brown, N.A.C. 1993. Promotion of germination of fynbos seeds by plant-derived smoke. *New Phytol.* 123:575–584.

Carter, S. 1973. *A comparison of pattern of herb and shrub growth in comparable sites in Chile and California.* M.S. thesis. California State University, San Diego.

Cincotta, R.P., Wisnewski, J., and Engelman, R. 2000. Human population in the biodiversity hotspots. *Nature* 404:990–992.

Cody, M.L., and Mooney, H.A. 1978. Convergence versus nonconvergence in Mediterranean-type ecosystems. *Ann. Rev. Ecol. Syst.* 9:265–321.

Corporación Nacional Forestal (CONAF). 1998. Impacto del fuego sobre el medio ambiente. Unidad de Gestión Manejo del Fuego, pp. 1–16. Santiago

Cowling, R.M., Rundel P.W., Lamont, B.B., Arroyo, M.K., and Arianoutsou, M. 1996. Plant diversity in Mediterranean-climate regions. *Trends Ecol. Evol. (TREE)* 11: 362–368.

Cruz, A., and Moreno, J. 1997a. Fire intensity effects on plants of *Erica australis* with modified lignotuber TNC content. In *Eighth Conference on Mediterranean-Type Ecosystems in a Changing World, Proceedings*, 20. University of California: San Diego.

Cruz, A., and Moreno, J. 1997b. Seasonal course of TNC in *Erica australis*, a lignotuber plant from western Spain. In *Eighth Conference on Mediterranean-Type Ecosystems in a Changing World, Proceedings*, 21. San Diego, CA.

Cruz, A., and Moreno, J. 1997c. Resprouting of *Erica australis* along a resource availability gradient: relationship to plant TNC reserves. In *Eighth Conference on Mediterranean-Type Ecosystems in a Changing World, Proceedings*, 21. University of California: San Diego.

Dafni, A., Cohen, D., and Noy-Meir, I. 1981. Life-cycle variation in geophytes. *Ann. Missouri Bot. Garden* 68:652–660.

Dallman, P.R. 1998. *Plant Life in the World's Mediterranean Climates.* Berkeley: California Native Plant Society and University of California Press.

Davis, S.D., Heywood, V.H., Herrera, O., MacBryde, J., Villalobos, J., and Hamilton, A.C. 1997. *Centers of Plant Diversity: A Guide and Strategy for Their Conservation.* Cambridge, U.K.: IUCN Publications Unit.

de Lange, J.H., and Boucher, C. 1990. Autecological studies on *Audouinia capitata* (Bruniaceae). I. Plant-derived smoke as a seed germination cue. *S. Afri. J. Bot.* 56:700–703.

Dixon, K.W., Roche, S., and Pate, J.S. 1995. The promotive effect of smoke derived from burnt native vegetation on seed germination of Western Australian plants. *Oecologia* 101:185–192.

Fuentes, E.R., and Espinoza, G. 1986. Resilience of shrublands in central Chile: A vulcanism-related hypothesis. *Interciencia* 11:164–165.

Fuentes, E.R., and Muñoz, M.R. 1995. The human role in changing landscape in central Chile: implications for intercontinental comparison. In *Ecology and biogeography of mediterranean ecosystems in Chile, California, and Australia*, eds. M.T.K. Arroyo, P.H. Zedler, and M.D. Fox, pp. 401–417. New York: Springer-Verlag.

Fuentes, E.R., Montenegro, G., Rundel, P., Arroyo, M.T.K., Ginocchio, R., and Jaksic, F.M. 1995. Functional approaches to biodiversity in the Mediterranean-type ecosystem of central Chile. In *Mediterranean-Type Ecosystems: The Function of Biodiversity*, eds. G.W. Davis, and D.M. Richardson, pp. 185–228. Berlin: Springer-Verlag.

Ginocchio, R., and Montenegro, G. 1992. Interpretation of metameric architecture in dominant shrubs of the Chilean matorral. *Oecologia* 90:451–456.

Ginocchio, R., Holmgren, M., and Montenegro, G. 1994. Effect of fire on plant architecture in Chilean shrubs. *Rev. Chil. Hist. Nat.* 67:177–182.

Gómez, M. 2000. *Respuestas morfofisiológicas de rebrotes producidos después del fuego a partir de lignotuber, en Cryptocarya alba (Mol.). Looser, en el matorral de Chile central.* MS, thesis. Universidad de Chile, Santiago.

Hanes, T.L. 1971. Succession after fire in the chaparral of southern California. *Ecol. Monogr.* 41:27–52.

Hoffman, A.E. 1989. Chilean geophyte monocotyledons: Taxonomic synopsis and conservation status. In *Red Book of Chilean Terrestrial Conservation Status*, ed. I. Benoit, pp. 140–151. Santiago: Corporación Nacional Forestal (CONAF).

Hoffmann, A.J., and Hoffmann, A.E. 1978. Comportamiento fenológico de plantas de la cordillera de los Andes. *Arch. Biol. Exp.* 11:188.
Hoffmann, A.J., and Hoffmann, A.E. 1982. Altitudinal ranges of phanerophytes and chamaephytes in central Chile. *Vegetatio* 48:151–163.
Hoffmann, A.J., Liberona, F., and Hoffmann, A.E. 1998. Distribution and ecology of geophytes in Chile: Conservation threats to geophytes in Mediterranean-type region. In *Landscape Degradation and Biodiversity in Mediterranean-Type Ecosystems*, ed. P. Rundel, pp. 231–253. Berlin: Springer-Verlag.
Instituto Nacional de Estadísticas (INE). 1995. *Chile. Ciudades, pueblos y aldeas.* Santiago: INE.
Keeley, J.E. 1977. Seed production, seed population in soil and seedling production after fire for two congeneric pairs of sprouting and nonsprouting chaparral shrubs. *Ecology* 58:820–829.
Keeley, J.E. 1981. Reproductive cycles and fire regimes. In *Proceedings of Conference on Fire Regimes and Ecosystem Properties*, eds. H.A. Mooney, T.M. Bonnicksen, N.L. Christensen, J.E. Lotan, and W.A. Reiners, pp. 231–277. USDA Forest Service, Gen. Tech. Rep. WO-26.
Keeley, J.E. 1984. Factors affecting germination of chaparral seeds. *Bull. S. Cal. Acad. Sci.* 83:113–120.
Keeley, J.E. 1986a. Resilience of Mediterranean shrub communities to fire. In *Resilience in Mediterranean-Type Ecosystems*, eds. B. Dell, A.J.M. Hopkins, and B.B. Lamont, pp. 95–112. Dordrecht: Junk.
Keeley, J.E. 1986b. Seed germination patterns of *Salvia mellifera* in fire-prone environments. *Oecologia* 71:1–5.
Keeley, J.E. 1987. Role of fire in seed germination of woody taxa in California chaparral. *Ecology* 68:434–443.
Keeley, J.E. 1993. Smoke-induced flowering in the fire-lily *Cyrtanthus ventricosus*. *S. Afri. J. Bot.* 59:638.
Keeley, J.E. 1994. Seed germination patterns in fire-prone Mediterranean-climate regions. In *Ecology and Biogeography of Mediterranean Ecosystems in Chile, California and Australia*, eds. M.T.K. Arroyo, P.H. Zedler, and M.D. Fox, pp. 239–273. New York: Springer-Verlag.
Keeley, J.E., and Bond, W.J. 1997. Convergent seed germination in South African fynbos and Californian chaparral. *Plant Ecol.* 133:153–168.
Keeley, J.E., and Fotheringham, C.J. 1997. Trace gas emissions in smoke-induced seed germination. *Science* 276:1248–1250.
Keeley, S.C., and Johnson, A.W. 1977. A comparison of the pattern of herb and shrub growth in comparable sites in Chile and California. *Am. Midl. Nat.* 97:120–132.
Keeley, J.E., and Keeley, S.C. 1984. Post fire recovery of California coastal sage scrub. *Am. Midl. Nat.* 111:105–117.
Keeley, J.E., and Swift, C.C. 1995. Biodiversity and ecosystem functioning in Mediterranean-climate California. In *Mediterranean-Type Ecosystems: The Function of Biodiversity*, eds. G.W. Davies, and D.M. Richardson, pp. 121–183. Berlin: Springer-Verlag.
Keeley, J.E., Fotheringham, C.J., and Morais, M. 1999. Reexamining fire suppression impacts on brushland fire regimes. *Science* 284:1829–1832.
Keeley, J.E., Brooks, A., Bird, T., Cory, S., Parker, H., and Usinge, E. 1986. Demographic structure of Chaparral under extended fire-free conditions. In *Proceedings of the Chaparral Ecosystem Research Conference*, eds. F.J. Kruger, D.T. Mitchell, and J.U.M. Jarvis, pp. 133–137. Davis: California Water Source Center, University of California.
Keeley, J.E., Zedler, P.H., Zammit, C.A., and Stohlgren, T.J. 1989. Fire and demography. In *The California Chaparral: Paradigms Reexamined*, ed. S.C. Keeley, pp. 151–153. Los Angeles: Natural History Museum of Los Angeles County.

Kruger, F.J. 1983. Plant community diversity and dynamics in relation to fire. In *Mediterranean-Type Ecosystems: The Role of Nutrients*, eds. F.J. Kruger, D.T. Mitchel, and J.U.M. Jarvis, pp. 446–472. New York: Springer-Verlag.

Le Maitre, D.C., and Brown, P.J. 1992. Life cycles and fire-stimulated flowering in geophytes. In *Fire in South African Mountain Fynbos*, eds. B.W. van Wilgen, D.M. Richardson, F.J. Kruger, and H.J. van Hensbergen, pp. 145–160. Berlin: Springer-Verlag.

Matthei, O. 1995. *Manual de las malezas que crecen en Chile*. Santiago: Alfabeta Impresores.

Minnich, R.A. 1989. Chaparral fire history in San Diego County and adjacent northern Baja California: An evaluation of natural fire regimes and the effects of suppression management. In *The California Chaparral: Paradigms Reexamined*, ed. S.C. Keeley, pp. 37–47. Los Angeles: Natural History Museum of Los Angeles County.

Montenegro, G., and Ginocchio, R. 1993. Modular interpretation of architecture in shrub species. *An. Acad. Bras. Ci.* 65:189–202.

Montenegro, G., and Ginocchio, R. 1995. Ecomorphological characters as a resource for illustrating growth-form convergence in matorral, chaparral and mallee. In *Ecology and Biogeography of Mediterranean Ecosystems in Chile, California and Australia*, eds. M.T.K. Arroyo, P.H. Zedler, and M.D. Fox, pp. 160–176. New York: Springer-Verlag.

Montenegro, G., and Teillier, S. 1988. Species richness. In *Mediterranean-Type Ecosystems: A Data Source Book*, ed. R.L. Specht, pp. 811–893. Dordrecht: Kluwer Academic.

Montenegro, G., Aljaro, M.E., and Kummerow, J. 1979a. Growth dynamics of Chilean matorral shrubs. *Bot. Gaz.* 14:114–119.

Montenegro, G., Aljaro, M.E., and Arrieta, A. 1979b. Dinámica de crecimiento y productividad de especies dominantes en un transecto altitudinal de la Cordillera de los Andes. In *Proceedings Seminario sobre el Programa de Investigación Integrada "Impacto del hombre en los ecosistemas de montaña"*, MAB 6, UNESCO.

Montenegro, G., Ávila, G., and Schatte, P. 1983. Presence and development of lignotubers in shrubs of the Chilean matorral. *Can. J. Bot.* 61:1804–1808.

Montenegro, G., Ginocchio, R., and Segura, A. 1996. Effect of global change on natural ecosystems: matorral of central Chile as a case of study of global change through effect of fire and UV-B radiation. IAI Fire workshop. Oregon, September 9–13.

Montenegro, G., Aljaro, M.E., Walkowiak, A., and Saenger, R. 1981. Seasonality, growth and net productivity of herbs and shrubs of the Chilean matorral. In *Dynamics and Management of Mediterranean-Type Ecosystems*, eds. C.C. Conrad, and W.C. Oechel, pp. 135–141. USDA Forest Service, *Gen. Tech. Rep.* PSW 58.

Montenegro, G., Ávila, G., Osorio, R., and Gómez, M. 1989. Chile. In *Plant Phenomorphological Studies in Mediterranean-Type Ecosystems*, ed. G. Orshan, pp. 347–387. Dordrecht: Kluwer Academic.

Montenegro, G., Patrick, G., Echenique, P., Gómez, M. and Timmermann, B. Mechanisms toward a sustainable use of *Chorizanthe vaginata* Benth, var. *maritima* Remy: A medicinal plant from Chile. *Phyton, Int. J. Exp. Bot.* 68:91–106.

Mooney, H.A. 1977. *Convergent Evolution of Chile and California Mediterranean Climate Ecosystems*. Stroudsburg, PA: Dowden, Hutchinson and Ross.

Mooney, H.A., and Parsons, D.J. 1973. Structure and function of the California chaparral and example from San Dimas. In *Mediterranean Ecosystems: origin and Structure*, eds. F. di Castri and H.A. Mooney, pp. 83–112. New York: Springer-Verlag.

Mooney, H.A., Hamburg, S.P., and Drake, J.A. 1986. The invasions of plants and animals into California. In *Ecology of Biological Invasions of North America and Hawaii*, eds. H.A. Mooney and J.A. Drake, pp. 250–272. New York: Springer-Verlag.

Mooney, H.A., Dunn, E.L., Shropshire, F., and Song, L. 1970. Vegetation comparisons between the mediterranean climatic areas of California and Chile. *Flora* 159:480–496.

Muñoz, M.R., and Fuentes, E.R. 1989. Does fire induce shrub germination in the Chilean matorral? *Oikos* 56:177–181.

Nierig, W.A., and Lowe, C.H. 1984. Vegetation of the Santa Catalina mountains: Community types and dynamics. *Vegetatio* 58:3–28.

Palmer, T. 1993. *California's Threatened Environment.* Washington, DC: Island Press.

Parsons, D.J. 1976. Vegetation structure in the Mediterranean climate scrub communities of California and Chile. *J. Ecol.* 64:435–447.

Parsons, D.J., and Moldenke, A.R. 1975. Convergence in vegetation structure along analogous climatic gradients in California and Chile. *Ecology* 56:950–957.

Pickett, S.T.A., and White, P.S. 1985. *The Ecology of Natural Disturbance and Patch Dynamics.* San Diego: Acadanic Press.

Rundel, P.W. 1975. Trichocereus in the mediterranean zone of central Chile. *Cactus Succ. J.* 46:86–88.

Rundel, P.W. 1977. Population variability in the genus *Trichocereus* (Cactaceae) in central Chile. *Plant Syst. Evol.* 127:1–9.

Rundel, P.W. 1981a. The matorral zone of central Chile. In *Mediterranean-Type Shrublands*, eds. F. di Castri, D. Goodall, and R.L. Specht, pp. 175–201. The Hague: Elsevier.

Rundel, P.W. 1981b. Structural and chemical components of flammability. In *Fire regimes and ecosystem properties*, eds. H.A. Mooney, T.M. Bonnicksen, N.L. Christianson, J.E. Lotan, and W.A. Reiners, pp. 183–207. Washington DC: USDA Forest Service Gen. Tech. Rep. WO-26.

Rundel, P.W. 1981c. Fire as an ecological factor. In *Physiological Plant Ecology. I.*, eds. O.L. Lange, P.S. Nobel, C.B. Osmond, and H. Ziegler, pp. 501–538. New York: Springer-Verlag.

Rundel, P.W. 1993. Adaptative significance of some morphological and physiological characteristics in mediterranean plants: Facts and fallacies. In *Time-Scales of Water Stress Response of Mediterranean Biota*, eds. F. di Castri, and J. Roy, pp. 119–140. Berlin: Springer-Verlag.

Rundel, P.W. 1996. Monocotyledoneous geophytes in the California flora. *Madrono* 43: 355–368.

Rundel, P.W., and Weisser, P.J. 1975. La Campana, a new national park in central Chile. *Biol. Conserv.* 8:35–46.

Rundel, P.W., Baker, G.A., Parsons, D.J., and Stohlgren, T.J. 1987. Postfire demography of resprouting and seedling establishment by *Adenostoma fasciculatum* in the California chaparral. In *Plant Response to Stress: Functional Analysis in Mediterranean Ecosystems*, eds. J.D. Tenhunen, F.M. Catarino, O.L. Lange, and W.C. Oechel, pp. 575–596. Berlin: Springer-Verlag.

Segura, A.M., Holmgren, M., Anabalón J.J.,and Fuentes E.R. 1998. The significance of fire intensity in creating local patchiness in the Chilean matorral. *Plant Ecol.* 139: 259–264.

Specht, R.L. 1988. *Mediterranean-Type Ecosystems: A Data Source Book.* Dordrecht: Kluwer Academic.

Stone, E.C. 1951. The stimulative effect of fire on the flowering of the golden brodiaea (*Brodiaea ixiodes* Wats. var. *lugens* Jeps.). *Ecology* 32:534–537.

Thrower, N.J.W., and Bradbury, D.E. 1977. *Chile-California Mediterranean Scrub Atlas: A Comparative Analysis.* Stroudberg, PA: Dowden, Hutchinson and Ross.

Trabaud, L.V. 1987. Dynamic after fire of sclerophyllous communities in the Mediterranean Basin. *Ecol. Med.* 13:25–37.

Wells, P.V. 1969. The relation between mode of reproduction and extent of speciation in woody genera of the California chaparral. *Evolution* 23:264–267.

Zunino, S., and Riveros, G. 1990. Cartografia de los incendios forestrales en la 5 region. *An. Mus. Hist. Nat., Valparaíso* 21:89–94.

Appendix

Species list of coastal and mid-elevation matorral in central Chile (designed by G. Montenegro for medicinal plants regeneration in central Chile, Grant NIH-NSF 2UO1 TW 00316-06). Growth forms are phanerophyte (P), geophyte (G), hemicryptophyte (H), therophyte (T) and chameophyte (Ch).

Scientific name	Growth form	Organ	Family
I. COASTAL MATORRAL			
Adesmia angustifolia H. et A.	P	Lignotuber	Papilionaceae
Aextoxicon punctatum R. et P.	P	Lignotuber	Aextoxicaceae
Alstroemeria haemantha R. et P.	G	Bulb	Amaryllidaceae
Alstroemeria pelegrina L.	G	Bulb	Amaryllidaceae
Anemone decapetala Ard.	H	Root crown	Ranunculaceae
Apium sellowianum Wolff.	H	Root crown	Umbelliferae
Aristolochia chilensis Bridges ex. Lindl.	H	Seed	Aristolochiaceae
Astragalus amatus Clos.	T	Seed	Papilionaceae
Baccharis concava (R. et P.) Pers.	P	Lignotuber	Compositae
Baccahris linearis (R. et P.) Pers.	P	Lignotuber	Compositae
Bahia ambrosioides Lag.	P	Lignotuber	Compositae
Bipinnula fimbriata (Poepp.) Johnst.	G	Seed	Orchidaceae
Brodiaea porrifolia (Poepp.) Meigen	G	Bulb	Liliaceae
Calandrinia arenaria Cham	T	Seed	Portulacaceae
Calandrinia sericea H. et A.	T	Seed	Portulacaceae
Calandrinia grandiflora Lindl.	H	Root crown	Portulacaceae
Calystegia soldanella (L.) Roem. et Schult.	T	Seed	Convolvulaceae
Camassia biflora (R. et P.) Coc.	G	Bulb	Liliaceae
Carpobrotus aequilaterus (Haw.) N. E. Br.	H	Root crown	Aizoaceae
Chloraea bletioides Lindl.	G	Rhizome	Orchidaceae
Chloraea chrysantha Poepp.	G	Rhizome	Orchidaceae
Chloraea galeata Lindl.	G	Rhizome	Orchidaceae
Chloraea disoides Lindl.	G	Rhizome	Orchidaceae
Chorizanthe vaginata Benth.	H	Root crown	Polygonaceae
Cissus striata R. et P.	P	Lignotuber	Vitaceae
Citronella mucronata (R. et P.) D. Don	P	Lignotuber	Icacinaceae
Colletia ulicina Gill. et Hook.	P	Lignotuber	Rhamnaceae
Cristaria glaucophylla Cav.	T	Seed	Malvaceae
Erigeron fasciculatus Colla	H	Root crown	Compositae
Euphorbia portulacoides L.	T	Seed	Euphorbiaceae
Fluorensia thurifera (Mol.) DC.	Ch	Root crown	Compositae
Frankenia chilensis K. Presl. ex Roem. et Schult.	Ch	Root crown	Frankeniaceae
Fuchsia lycioides Andr.	P	Lignotuber	Onagraceae
Glandularia laciniata (L.) Schnack et Covas	T	Seed	Verbenaceae

Scientific name	Growth form	Organ	Family
Glandularia sulphurea (D. Don) Schnack et Covas	H	Root crown	Verbenaceae
Gnaphalium viravira Mol.	T	Seed	Compositae
Haplopappus foliosus DC.	P	Lignotuber	Compositae
Hippeastrum advenum (Ker-Gawl.) Herb.	G	Bulb	Amaryllidaceae
Hippeastrum rhodolirion Baker	G	Bulb	Amaryllidaceae
Leucheria cerberoana Remy	T	Seed	Compositae
Leucocoryne ixioides (Hook.) Lindl.	G	Bulb	Liliaceae
Linum chamissons Schiede	T	Seed	Linaceae
Llaqunoa glandulosa (H. et A.) D. Don	P	Seed	Sapindaceae
Lobelia Tupa L.	Ch	Root crown	Campanulaceae
Lupinus microcarpus Sims.	T	Seed	Papilionaceae
Lycium chilense Miers. ex A. DC.	P	Seed	Solanaceae
Malesherbia fasciculata D. Don	T	Seed	Malesherbiaceae
Margyricarpus pinnatus (Lamb.) O.K.	Ch	Root crown	Rosaceae
Monnina angustifolia DC.	T	Seed	Polygalaceae
Myrceugenia exsucca (DC.) Berg.	P	Lignotuber	Myrtaceae
Nicotiana acuminata (Graham) Hook.	T	Seed	Solanaceae
Nolana crassulifolia Poepp.	Ch	Root crown	Nolanaceae
Nolana sedifolia Poepp.	Ch	Root crown	Nolanaceae
Ochagavia carnea (Beer) L. B. Sm. et Looser	H	Root crown	Bromeliaceae
Oenothera acaulis Cav.	H	Root crown	Onagraceae
Oenothera affinis Cambess.	H	Root crown	Onagraceae
Oxalis carnosa Mol.	H	Root crown	Oxalidaceae
Oxalis laxa H. et A.	H	Root crown	Oxalidaceae
Peumus boldus Mol.	P	Lignotuber	Monimiaceae
Phycella ignea Lindl.	G	Bulb	Amaryllidaceae
Podanthus mitiqui Lindl.	P	Seed	Compositae
Pouteria splendens (A. DC.) O.K.	P	Seed	Sapotaceae
Proustia cuneifolia D. Don	P	Lignotuber	Compositae
Puya chilensis Mol.	H	Rhizome	Bromeliaceae
Ribes punctatum R. et P.	P	Lignotuber	Saxifragaceae
Schizantus litoralis Phil.	T	Seed	Solanaceae
Schizantus pinnatus R. et P.	T	Seed	Solanaceae
Scyphanthus elegans D. Don	H	Root crown	Loasaceae
Senecio cerberoanus Remy	Ch	Root crown	Compositae
Sisyrinchium junceum E. Mey. ex K. Presl.	G	Rhizome	Iridaceae
Solanum maritimun Meyen ex Nees	Ch	Root crown	Solanaceae
Solenomelus pedunculatus (Gill. ex Hook.) Hochr.	G	Rhizome	Iridaceae
Sphacele salviae (Lindl.) Briq.	Ch	Root crown	Labiatae
Sphaeralcea obtusiloba (Hook.) D. Don	Ch	Root crown	Malvaceae
Stachys albicaulis Lindl.	H	Root crown	Labiatae
Trichocereus litoralis (Johow) Looser	P	Seed	Cactaceae
Trichopetalum plumosum (R. et P.) Macbr.	G	Bulb	Liliaceae

Scientific name	Growth form	Organ	Family
Tweedia confertiflora (Dcne.) Malme	H	Seed	Asclepiadaceae
Verbena litoralis H. B. K.	H	Seed	Verbenaceae
Verbena porrigens Phil.	H	Seed	Verbenaceae
Vicia vicina Clos.	T	Annual	Papilionaceae
II. MID ELEVATION MATORRAL			
Acacia caven (Mol.) Mol.	P	Lignotuber	Mimosaceae
Adenopeltis serrata (W. Aiton) Johnst.	P	Lignotuber	Euphorbiaceae
Adesmia arborea Bert.	P	Lignotuber	Papilionaceae
Adesmia phylloidea Clos	T	Seed	Papilionaceae
Adesmia radicifolia Clos	T	Seed	Papilionaceae
Adesmia viscosa Gill. ex H. et A.	T	Seed	Papilionaceae
Alonsoa meridionalis (L. F.) O.K.	T	Seed	Scrophulariaceae
Amsinckia calycina (Moris) Chater	T	Seed	Boraginaceae
Anemone decapetala Ard.	H	Root crown	Ranunculaceae
Azara celastrina D. Don	P	Lignotuber	Flacourtiaceae
Azara dentata R. et P.	P	Lignotuber	Flacourtiaceae
Azara petiolaris (D. Don) Johnst.	P	Lignotuber	Flacourtiaceae
Azara serrata R. et P.	P	Lignotuber	Flacourtiaceae
Baccharis linearis (R. et P.) Pers.	P	Lignotuber	Compositae
Baccharis marginalis DC.	P	Lignotuber	Compositae
Baccharis racemosa (R. et P.) DC.	P	Lignotuber	Compositae
Beilschmiedia mersii (Gay) Kosterm	P	Lignotuber	Lauraceae
Berberis actinacantha Mart.	P	Lignotuber	Berberidaceae
Berberis chilensis Gill. ex Hook.	P	Lignotuber	Berberidaceae
Berberis grevilleana Gill. ex H. et A.	P	Lignotuber	Berberidaceae
Berberis montana Gay.	Ch	Lignotuber	Berberidaceae
Buddleja globosa Hope.	P	Seed	Buddlejaceae
Caesalpinia spinosa (Mol.) O.K.	P	Lignotuber	Caesalpiniaceae
Calceolaria ascendens Lindl.	G	Tuber	Scrophulariaceae
Calceolaria corymbosa R. et P.	G	Tuber	Scrophulariaceae
Calceolaria hypericina Poepp. ex DC.	G	Tuber	Scrophulariaceae
Calceolaria petiolaris Cav.	G	Tuber	Scrophulariaceae
Calceolaria polyfolia Hook.	G	Tuber	Scrophulariaceae
Cassia closiana Phil.	P	Lignotuber	Caesalpiniaceae
Centaurea chilensis H. et A.	T	Seed	Compositae
Cestrum parqui L'Herit.	P	Lignotuber	Solanaceae
Chusquea quila Kunth	G	Rhizome	Gramineae
Citronella mucronata (R. et P.) D. Don	P	Lignotuber	Icacinaceae
Clarkia tenella (Cav.) Lews et Lewis	T	Seed	Onagraceae
Colletia spinosa Lam.	P	Lignotuber	Rhamnaceae
Colliguaja odorifera Mol.	P	Lignotuber	Euphorbiaceae
Colliguaja salicifolia Gill. et Hook.	P	Lignotuber	Euphorbiaceae
Collomia biflora (R. et P.) Brand	T	Seed	Polemoniaceae
Conanthera bifolia R. et P.	G	Bulb	Tecophilaeaceae
Conanthera campanulata (D. Don.) Lindl.	G	Bulb	Tecophilaeaceae
Conanthera trimaculata (D. Don.) Meigen	G	Bulb	Tecophilaeaceae
Convolvulus chilensis Pers.	G	Rhizome	Convolvulaceae

Scientific name	Growth form	Organ	Family
Corynabutilon ceratocarpum (H. et A.) Kearney	Ch	Root crown	Malvaceae
Crinodendron patagua Mol.	P	Lignotuber	Elaeocarpaceae
Cryptocarya alba (Mol.) Looser	P	Lignotuber	Lauraceae
Diplolepis menziesii Schult	Ch	Seed	Asclepiadaceae
Discaria trinervis (Gill. ex H. et A.) Reiche	P	Lignotuber	Rhamnaceae
Drimys winteri J.R. et G. Forster	P	Lignotuber	Winteraceae
Eccremocarpus scaber R. et P.	P	Seed	Bignoniaceae
Ephedra andina Poepp. ex C.A. Mey	P	Lignotuber	Ephedraceae
Escallonia revoluta (R. et P.) Pers.	P	Lignotuber	Saxifragaceae
Escallonia illinita K. Presl.	P	Lignotuber	Saxifragaceae
Escallonia pulverulenta (R. et P.) Pers.	P	Lignotuber	Saxifragaceae
Escallonia rosea Griseb	P	Lignotuber	Saxifragaceae
Escallonia rubra (R. et P.) Pers.	P	Lignotuber	Saxifragaceae
Eupatorium glechonophyllum Less.	P	Lignotuber	Compositae
Eupatorium salvia Colla	P	Lignotuber	Compositae
Fabiana imbricata R. et P.	Ch	Seed	Solanaceae
Fuchsia magellanica Lam.	P	Lignotuber	Onagraceae
Geranium berterianum Colla ex Savi.	H	Seed	Geraniaceae
Gethyum atropurpureum Phil.	G	Bulb	Liliaceae
Gochnatia foliolosa (D. Don) D. Don ex H. et A.	P	Lignotuber	Compositae
Gymnophyton isatidicarpum (K. Presel. ex DC.) Math et Const.	Ch	Root crown	Umbelliferae
Haplopappus canescens (Phil.) Reiche	Ch	Root crown	Compositae
Haplopappus integerrimus (H. et A.) Hall.	Ch	Root crown	Compositae
Haplopappus multifolius Phil. ex Reiche	Ch	Root crown	Compositae
Haplopappus paucidentatus Phil.	Ch	Root crown	Compositae
Homalocarpus dichotomus (Poepp. ex DC.) Math. et Const.	T	Seed	Umbelliferae
Jubaea chilensis (Mol.) Baillon	P	Seed	Palmae
Kageneckia oblonga R. et P.	P	Lignotuber	Rosaceae
Larrea nitida Cav.	P	Seed	Zygophyllaceae
Lathyrus subandinus Phil.	T	Seed	Papilionaceae
Leucheria cerberoana Remy	T	Seed	Compositae
Lithrea caustica (Mol.) H. et A.	P	Lignotuber	Anacardiaceae
Llagunoa glandulosa (H. et A.) G. Don	P	Seed	Sapindaceae
Loasa pallida Gill. ex Arn.	T	Seed	Loasaceae
Loasa sigmoidea Urban et Gilg.	T	Seed	Loasaceae
Loasa tricolor Ker-Gawl.	T	Seed	Loasaceae
Loasa triloba Domb. ex A.L. Juss.	T	Seed	Loasaceae
Lobelia excelsa Bonpl.	Ch	Root crown	Campanulaceae
Lobelia polyphylla H. et A.	Ch	Root crown	Campanulaceae
Luma chequen (Mol.) A. Gray	P	Lignotuber	Myrtaceae
Madia sativa Mol.	T	Seed	Compositae

Scientific name	Growth form	Organ	Family
Malesherbia fasciculata D. Don.	T	Seed	Malesherbiaceae
Malesherbia lirana Gay.	T	Seed	Malesherbiaceae
Maytenus boaria Mol.	P	Lignotuber	Celastraceae
Moscharia pinnatifida R. et P.	T	Seed	Compositae
Muehlenbekia hastulata (J. E. Sm.) Johnst.	Ch	Lignotuber	Polygonaceae
Mutisia decurrens Cav.	P	Seed	Compositae
Mutisia acerosa Poepp. ex Less.	P	Seed	Compositae
Mutisia spinosa R. et P.	P	Seed	Compositae
Mutisia subulata R. et P.	P	Seed	Compositae
Myoschilos oblonga R. et P.	P	Lignotuber	Santalaceae
Myrceugenia rufa (Colla) Skottsb. ex Causel	P	Lignotuber	Myrtaceae
Notanthera heterophylla (R. et P.) D. Don	P	Seed	Loranthaceae
Nothofagus obliqua (Mirb.) Oerst.	P	Lignotuber	Fagaceae
Oxalis articulata Savigni	G	Bulb	Oxalidaceae
Pasithea coerulea (R. et P.) D. Don	G	Rhizoma	Liliaceae
Passiflora pinnatistipula Cav.	P	Seed	Passifloraceae
Persea lingue (R. et P.) Ness ex Kopp.	P	Lignotuber	Lauraceae
Peumus boldus Mol.	P	Lignotuber	Monimiaceae
Phacelia magellanica (Lam.) Coville	H	Seed	Hydrophyllaceae
Phycella ignea Lindl.	T	Bulb	Amaryllidaceae
Podanthus mitiqui Lindl.	P	Lignotuber	Compositae
Porlieria chilensis Johnst.	P	Lignotuber	Zygophyllaceae
Pouteria splendens (A. DC.) O.K.	P	Seed	Sapotaceae
Prosopis chilensis (Mol.) Stuntz	P	Lignotuber	Mimosaceae
Proustia pyrifolia DC.	P	Lignotuber	Compositae
Psoralea glandulosa L.	P	Lignotuber	Papilionaceae
Puya chilensis Mol.	H	Rhizoma	Bromeliaceae
Puya coerulea Lindl.	H	Rhizoma	Bromeliaceae
Puya venusta Phil.	H	Rhizoma	Bromeliaceae
Puya berteroniana Mez.	H	Rhizoma	Bromeliaceae
Quillaja saponaria Mol.	P	Lignotuber	Rosaceae
Retanilla ephedra (Vent.) Brongn.	P	Lignotuber	Rhamnaceae
Rhaphithamnus spinosum (A. L. Juss.) Mold.	P	Lignotuber	Verbenaceae
Rhodophiala rhodolirion (Baker) Traub.	G	Bulb	Amaryllidaceae
Ribes polyanthes Phil.	Ch	Lignotuber	Saxifragaceae
Salix humboldtiana Wild.	P	Lignotuber	Salicaceae
Salpiglossis sinnuata R. et P.	Ch	Root crown	Solanaceae
Satureja gilliesii (Graham.) Briq.	Ch	Lignotuber	Labiatae
Schinus latifolius (Gill. ex Lindl.) Engles	P	Lignotuber	Anacardiaceae
Schinus montanus (Phil.) Engles	P	Lignotuber	Anacardiaceae
Schinus polygamus (Cav.) Cabr.	P	Lignotuber	Anacardiaceae
Schizanthus pinnatus R et P.	T	Seed	Solanaceae
Senecio cerberoanus Remy	Ch	Root crown	Compositae

Scientific name	Growth form	Organ	Family
Senecio eruciformis Remy	Ch	Root crown	Compositae
Senecio fistulosus Poepp. ex Less.	Ch	Root crown	Compositae
Senecio yegua (Colla) Cabr.	P	Root crown	Compositae
Senna arnottiana (Gill. ex H. et A.) Irw. et Barneby	P	Lignotuber	Caesalpiniaceae
Sisyrinchium junceum E. Mey. ex K. Presl.	G	Rhizome	Iridaceae
Solanum ligustrinum Lodd.	P	Lignotuber	Solanacea
Solenomelus sisyrinchium (Griseb.) Pax. ex Diels.	G	Rhizome	Iridaceae
Sphacele salviae (Lindl.) Briq.	Ch	Root crown	Labiatae
Sphaeralcea obtusiloba (Hook.) G. Don.	Ch	Root crown	Malvaceae
Sophora macrocarpa J.E. Sm.	P	Lignotuber	Papilionaceae
Talguenea quinquenervia (Gill. et Hook.) Johnst.	P	Lignotuber	Rhamnaceae
Tessaria absinthioides (H. et A.) DC.	Ch	Root crown	Compositae
Teucrium bicolor J.E. Sm.	Ch	Root crown	Labiatae
Trevoa trinervis Miers.	P	Lignotuber	Rhamnaceae
Trichocereus chiloensis (Colla) Briton et Rose.	P	Seed	Cactaceae
Trichocline aurea (D. Don.) Reiche	T	Seed	Compositae
Trichopetalum plumosum (R. et P.) Macbr.	G	Bulb	Liliaceae
Triptilion spinosum R. et P.	T	Seed	Compositae
Triptilon gibbosum Remy	T	Seed	Compositae
Tristerix aphyllus (Miers ex DC.) Van Tiegh. ex B. et W.	P	Seed	Loranthaceae
Tristerix tetrandus (R. et P.) Mart.	P	Seed	Loranthaceae
Tristerix verticillatus (R. et P.) Barlow et Wiens	P	Seed	Loranthaceae
Tropaeolum tricolor Sweet.	G	Tuber	Tropaeolaceae
Verbena cinerascens Schauer	Ch	Root crown	Verbenaceae
Vicia magnifolia Clos.	T	Seed	Papilionaceae
Vicia vicina Clos.	T	Seed	Papilionaceae
Viviania crenata (Hook) G. Don	Ch	Seed	Vivianaceae
Viviania marifolia Cav.	Ch	Seed	Vivianaceae

4. Practical Implications

15. Management Implications of Fire and Climate Changes in the Western Americas

Penelope Morgan, Guillermo E. Defossé, and Norberto F. Rodríguez

Fires have shaped the structure, composition, and function of temperate ecosystems worldwide. In many forest, shrubland, and grassland ecosystems of temperate and boreal zones, biomass production exceeds decomposition. When lightning or people ignite fires, and when the weather and climatic conditions are conducive, this accumulated dead and live biomass burns. Particularly when these fires burn in extremely hot, dry, windy conditions, they threaten people and their property. That fire has played an important ecological role in these ecosystems makes fire management challenging, for ecological integrity and sustainability depend on fires and other disturbances (Pickett and White 1985).

In the temperate and boreal zones of western North and South America, fire regimes have changed in response to *both* climate and human action, though the relative influence varies. Fire regimes (occurrence, frequency, severity, intensity, and extent; Pickett and White 1985) reflect both the physical and sociopolitical environment, and they influence the type and abundance of fuel and therefore fire behavior and effects through time. Forest fire occurrence and effects are intimately linked with climate (Weber and Flannigan 1997; Flannigan, Stocks, and Weber, Chapter 4, this volume). Climate influences lightning occurrence (Price and Rind 1994) as well as fire behavior and effects. Fire, climate, and landscape-scale heterogeneity interact (Miller and Urban 1999; Swetnam and Betancourt 1998). There is growing evidence that the temperature increases associated with global climate change may be most pronounced at higher latitudes (IPCC 2001; Flannigan, Stocks, and Weber, Chapter 4, this volume) where temperate and boreal forests are found. The effects of climate change on vegetation will be mediated through fire and other disturbances (Swetnam and Betancourt 1998; Flannigan, Stocks, and Weber, Chapter 4, this volume). Changes in the global climate will alter the fire disturbance patterns that so strongly influence the structure, composition, and function of forest and other wildland ecosystems. Those altered fire regimes will be important determinants of rates and directions of ecosystem change, and they have powerful feedback to global climate change through their influence on carbon, nitrogen and water cycles (Flannigan, Stocks, and Weber, Chapter 4, this volume).

Scientists have begun to explain the complex interactions among fire, climate, vegetation, and land use across time and space. Such research is sorely needed (Schmoldt et al. 1999), for the human population is growing and the climate is changing. An understanding of fire regimes, how they have changed through time, and their interaction with climate is critical to fire management decisions today and for a future that will be shaped by a different climate and increasingly intensive and extensive land use by people.

Humans have long sought to control fire ignition, spread and effects. Fire management goals reflect the social, cultural, political, legal, and biophysical conditions, as well as the broader goals for natural resource management (Pyne, Andrews, and Laven 1996; Chandler et al. 1983). People both suppress and use fire to achieve a variety of goals. Fires may be suppressed and fuels are often managed to protect human life and property, prescribed burns are sometimes purposefully ignited to burn debris, enhance habitat for plants or animals or to restore ecological conditions, and in some areas lightning-ignited fires are managed to allow fire to play a natural or seminatural role (Pyne, Andrews, and Laven 1996). In 1996 the Argentine federal government started a National Fire Management Plan to support provincial states with human and material resources to fight wildland fires (Dentoni and Cerne 1999). The United States has recently adopted a national fire plan (*http://www.fireplan.gov/*) focused on suppressing severe wildland fires, reducing hazardous fuels, rehabilitating fire damage and restoring ecosystems, and assisting people in local communities. Both the Canadian (Canadian Forest Service 2001) and U.S. plans state that fire should assume a more natural role. Fire managers are only beginning to understand and plan for the synergy between fire and climate change. For instance, only the Canadian fire management plan (Canadian Forest Service 2001) reflects concerns over the effects of global climate change.

Because forests can either emit or absorb carbon, depending on their use, forest and fire management will be important in efforts to mitigate human-induced climate change. Increasingly, land managers will be pressured to manage land to absorb carbon. This will be a new challenge faced by land managers, particularly in fire-prone environments.

This chapter focuses on the practical, management implications of the fire and climate change research that is reported in the earlier chapters of this volume. We start with an overview of fire management goals and strategies, and then draw some parallels among vegetation, climate and land use history in the temperate and boreal zones of North and South America. We then contrast the role of land use and climate in influencing change in three major fire regimes. We conclude with the implications for the future and challenges for fire managers as they use the information from this book. Although our comments are primarily focused on temperate and boreal forests, the management implications extend to the woodlands, shrublands, and grasslands. Our comments are directed to scientists, as well as to land managers and wildland fire management specialists involved in planning and implementing fire management programs at regional and national levels.

Fire Management

Fire management goals reflect the social, cultural, political, legal, and biophysical conditions, as well as the broader goals for natural resource management (Pyne, Andrews, and Laven 1996; Chandler et al. 1983). Natural resource management goals vary greatly across the spectrum of landownership (e.g., private nonindustrial, private industrial, and public, including municipal, county, state or provincial, tribal and federal). Even within the same land ownership, land management objectives can be very different. For instance, while all federal lands managed by the U.S. Forest Service are considered for multiple uses, some are managed primarily for timber and other commodities, while others are managed primarily to provide wildlife habitat or recreation, and still others are protected as wilderness areas. Units of the extensive national park systems in both North and South America are managed for a combination of visitor recreation, protection of natural features, and maintenance of natural or historical conditions or processes. Some national parks (e.g., Grand Teton National Park in the United States, and national reserves in Argentina) are open to livestock grazing. The many other natural areas, including reserves, wildlife refuges, and state or provincial parks are typically smaller, have a heavier emphasis on visitor recreation, and often have ecological objectives more narrowly focused on individual species or groups of species. For these reasons fire management is more consistently focused there on suppression alone, with some notable exceptions. State and provincial lands are often logged or grazed for economic returns, with attendant fire management that largely focuses on protection from fire and other disturbances that will impact those commercial uses. Private industrial lands are typically managed intensively for timber production with commercial tree plantations and harvesting on relatively short rotations for timber and fiber production. For instance, many forests in the Valdivian region of Chile have been commercially logged since 1912 and intensively harvested since the 1980s (Veblen and Alaback 1999). Private, nonindustrial lands vary greatly in ownership and owner objectives, but protection from fire is an almost universal goal of managers and landowners.

The goals for fire management include (1) reducing fire hazard to protect human life and property or ecological values, (2) altering vegetation composition and structure to enhance habitats for plants or animals, (3) restoring ecological conditions and integrity, and (4) managing for natural or seminatural conditions (i.e., with minimal human impact) in parks, wilderness areas, or natural areas (Pyne, Andrews, and Laven 1996). Of these, reducing fire hazard to protect human life and property is the most widely applied, particularly near towns or in municipal watersheds. Fuels management to reduce fire hazard is often accomplished mechanically, although prescribed fires are commonly used following timber harvest. Mechanical, burning, or other treatments to reduce fire hazard are often legal requirements following logging on private, state, and federal lands, and prescribed burns are sometimes used to burn debris and prepare sites for tree regeneration following logging. Similarly prescribed burns are used for favoring habitat for particular plant and wildlife species. Restoration is much less common than the first two objectives. Ecological restoration is often accomplished with cutting, burning, or a combination of treatments designed to alter vegetation composition and structure and to restore past conditions and ecological integrity (Arno and Hardy 1996). Managing for fire as a natural process is largely limited to a few of the larger wilderness areas, national parks, and nature preserves. Although U.S. and Canadian policy allows for lightning and human-ignited fires to burn under carefully prescribed conditions, fire suppression is the most common management decision in parks and wilderness areas. In the United States, policy changes since 1988 sharply limit the conditions when lightning-ignited fires are managed to accomplish resource benefits (Parsons and Landres 1998). As a result many of the most ecologically significant fires (those that are large and intense) are being suppressed in all wilderness areas in the United States (Parsons and Landres 1998) and in most national parks in both North and South America.

Fire management goals are typically accomplished through some combination of suppression, planned and natural ignitions, and fire surrogates (grazing, mowing, logging, etc.) (Christensen 1995), as well as through education (Pyne, Andrews, and Laven 1996). Natural ignitions are lightning fires managed to burn within prescribed limits of time, place, fuels, threats to people and their property, and so on. Surrogates, including logging and prescribed fires with planned ignitions, can approximate some aspects of fire disturbance (e.g., some changes in structure, fuel reduction) but are less likely to simulate many of the functional effects of fires. Fuels management can include reducing debris by burning or chopping, converting to less combustible types, and isolating fuels through systems of fuel breaks or areas of limited access (Pyne, Andrews, and Laven 1996). Fires are suppressed when the risks to people or their property from fires and smoke is unacceptable, or when resources could be damaged. In the United States, fire suppression strategies include controlling fire by extinguishing it, containing a fire within firelines along its actively burning perimeter, or confining fire to an area defined by topographic and other boundaries beyond which the fire will not be allowed to spread (Pyne, Andrews, and Laven 1996). Confine and contain strategies may result in additional area being burned if the lines are far from the flames.

Fire management is expensive. Wildland fire management represents 25% of the cost of forest management in Canada (Canadian Forest Service 2001). In the United States, federal agencies spent an average of $629,905,720 in each of the last five years on fire management (*http://www.nifc.gov/stats/wildlandfirestats.html#costs*). Large, severe fire events account for a majority of the total area burned over time (Strauss, Bednar, and Mees 1989) and resource losses, as well as threats to people and their property (Maciliwain 1994; Defossé et al. 2001). For instance, 2% to 3% of all fires that exceeded 200 ha in size accounted for 98% of the area burned from 1950 to 1995 in Canada (*http://nofc.forestry.ca/fire/frn/English/frames.htm*; Amiro et al. 2001). In Canada's 417 million hectares of forest, about 10,000 fires occur each year, burning an average of 2.5 million ha/yr (*http://nofc.forestry.ca/fire/frn/English/frames.htm*). In the United States, between 1919 and 1999, on average, more than 13,000 fires burned more than 500,000 ha each year, but the area burned was highly variable from year to year. In Chile, more than 5000 wildfires burned approximately 50,000 ha of land each year between 1989 and 1994 in 29 million hectares of native forests, shrublands, and grasslands (*http://www.2.ruf.uni-freiburg.de/fireglobe/iffn/country/cl/cl_2.htm*). In Argentina, fires increased in number and size from 1997 (281 thousand hectares burned in 4774 fires) to 2000 (2.8 million hectares burned in 10,596 fires) (SRNyDS 1997, 1998, 1999; SDSyPA 2000).

Historical range of variability (HRV) is widely used by forest managers in the United States and Canada in planning for sustainability and conservation of biological diversity (Landres, Morgan, and Swanson 1999; Swetnam, Allen, and Betancourt 1999), as well as in ecological restoration (White and Walker 1997). Similar concepts have long provided management direction for many parks and wilderness areas in the United States and Canada (Christensen 1995; Parsons and Landres 1998; *http://www.parcscanada.pch.gc.Canada/library/fire/fire_e.htm*). Cissel, Swanson, and Weisberg (1999) used historical disturbance regimes (fire and landslides) to guide management. In the United States, departures from historical fire frequencies have been used to target restoration (Caprio and Graber 2000), to estimate areas at risk to catastrophic fires (GAO 1999; Hardy et al. 2001) and as a baseline for national regional, and local fire planning (Hann and Bunnell 2001). Using natural variation in management is grounded on ecological premises (Landres, Morgan, and Swanson 1999). However, recent reviews suggest that HRV has greater value in understanding and evaluating ecosystem change, and in communicating about the type and degree of change to be expected in ecosystems, than it does in determining management goals (Landres, Morgan, and Swanson 1999; Swetnam, Allen, and Betancourt 1999; Holling and Meffe 1996). Thus management should be informed by past variation, but even those management goals focused on restoring natural processes and conditions will more appropriately focus on ecological integrity, sustainability, and resilience for current and future conditions (Pavlik 1996; White and Walker 1997).

Fire management is central to ecosystem management, a framework that has been widely adopted for management in the United States and Canada.

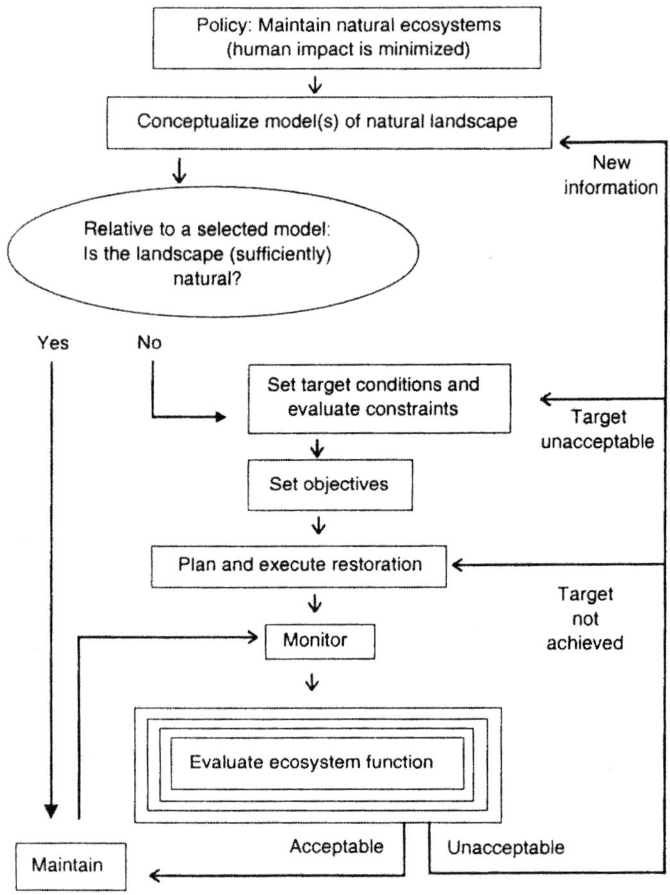

Figure 15.1. In the Sequoia and Kings Canyon National Parks in California, prescribed fire is used as part of a management framework to restore natural fire regimes (from Keeley and Stephenson 2000).

Christensen et al. (1996) summarize ecosystem management as including intergenerational sustainability—with goals built on sound ecological models and understanding of ecological complexity, ecosystem dynamics, context and scale—as well as the role of humans in ecosystems and their accountability. Further we must be humble and include enough future flexibility to accommodate uncertainty, surprise, and limits to our knowledge (Christensen et al. 1996; Landres, Morgan, and Swanson 1999). One model program for restoring fire while incorporating the best available knowledge about long-term fire history and climate change is in the Sierras (Fig. 15.1). This effort uses models based on fire history, ecosystem processes, and climate (Swetnam 1993; Miller and Urban 1999, 2000; Millar and Woolfenden 1999).

Similarities in Environment, History, and Fire Management Policy

The following discussion focuses on commonalities in environment, human history, and fire management policy in temperate and boreal forest zones of North and South America. We use three broad fire regime classes to frame our discussion. These are associated with wet forests (Agee 1993; Veblen and Alaback 1996), subalpine and boreal forests (Agee 1993; Flannigan et al. 1998), warm and dry forests (Swetnam and Betancourt 1990, 1998; Veblen et al., Chapter 9, this volume), chaparral and matorral (Armesto, Vidiella, and Jimenez 1995; Fuentes and Muñoz 1995), shrub-steppe, and grasslands. There are, of course, distinct differences in climate, land use and evolutionary history (Veblen and Alaback 1996; Armesto, Vidiella, and Jimenez 1995; Fuentes and Muñoz 1995), as well as in legal mandates and the sociopolitical environment. We focus on the implications of changing land use and climate for fire management.

Fires have shaped the structure, composition, and function of forest, woodland, shrubland, and grassland ecosystems, as well as the human response to them. These ecosystems are shaped as well by episodic droughts associated with ocean conditions, including ENSO (Swetnam and Betancourt 1990, 1998; Flannigan, Stocks, and Weber, Chapter 4, this volume; Kitzberger and Veblen, Chapter 10, this volume) and others (Baker, Chapter 5, this volume), as well as by a legacy of past climate change and disturbance.

In both North and South America, indigenous people used fires to manipulate the vegetation around them (Pyne 1982; Claraz 1988; Veblen et al., Chapter 9, this volume), although the extent and degree of influence clearly varied through time and from place to place. Following the first contacts with Europeans, the populations of indigenous peoples declined sharply, first through introduced diseases and then through war and other means of displacement in both North America (Pyne 1982) and southern South America (Roux 1987). That and the very intensive livestock grazing that followed, along with fire suppression, roads, and settlement of Euro-Americans in valleys, reduced the fire frequency dramatically in many ecosystem types early in the 1900s in western North America and in southern South America (Tortorelli 1947; Pyne 1982; Veblen et al., Chapter 9, this volume).

Large wildfires caused by lightning or by indigenous people covered large areas, and were the dominant fire events a century ago in the forests, Monte, and steppe zones of the Patagonian region of Argentina and the matorral region of Chile (Musters 1871; Claraz 1988; Veblen and Lorenz 1988). Similarly fires were extensive in western North America prior to 1935. Forests were burned to facilitate mining, logging, and agriculture in both North and South America in the late 1800s and early 1900s. In dry forests, woodlands and shrublands, fewer surface fires occurred following the introduction of domestic livestock. Intensive grazing dramatically reduced the abundance of fine fuels that affected the spread of surface fires. In South America, domestic livestock have been grazed intensively for more than 60 years in Chile and Argentina, and introduced deer species, European hares, and wild boars have had major impacts on vegetation dynamics

(Armesto, Vidiella, and Jimenez 1995; Veblen and Alaback 1996). Ranchers, soldiers, and Euro-American settlers also suppressed fires (Pyne 1982, 1995).

Logging commenced with colonization by Europeans. Readily accessible areas were logged early. Logging became more intensive and extensive as population increased, technology became available, and railroads and road networks facilitated transportation. Early logging was selective; clear-cutting became much more prevalent in the 1960s in both North and South America, particularly in the most productive wet forests (Agee 1993; Veblen and Alaback 1996). Rapid, extensive clearing of valleys in the Pacific Northwest and California occurred in the mid-1800s to 1940s (Veblen and Alaback 1996; Pyne, Andrews, and Laven 1996).

Based on the assumptions that fires were destructive, the fire policy imposed by Euro-Americans was one of suppression. Efforts to detect and attack fires became increasingly effective as funding, trained manpower, and technology became widely available in the 1930s and 1940s in both North and South America. This was triggered by public awareness and concern after large wildfires occurred (e.g., the fires of 1910 and 1933 in the United States), and after World War II when airplanes and smoke jumpers were increasingly used to fight fires (Pyne 1982, 1995; Pyne, Andrews, and Laven 1996).

Today increasingly effective fire suppression and diverse policies for land use have attempted to exclude fires from many wildland ecosystems. Especially in the dry temperate forests, shrublands, and grasslands, fuels have accumulated. Thus, fire suppression and other land uses have increased the potential for future fires to be intense and severe. When fires are intense and large, they can exceed our capacity to suppress them. Furthermore values at risk have increased as more people have moved to and built homes in areas that once burned often, particularly in the rural areas adjacent to towns and cities (Davis 1989; Haltenthoff 1994; Hirsch 2000; Rodríguez 1999).

Localized, but extensive invasions by exotic plants (e.g., *Bromus tectorum* in the shrub-steppe of the Great Basin (Knick and Rotenberry 1997), *Rosa eglanteria* and *Spartium junceum* in the forest-steppe ecotone of northern Patagonia, and others in the chaparral of California (Keeley and Fotheringham, Chapter 8, this volume) often fuel fires that are so frequent and extensive that the structure, function, and pattern of vegetation is greatly altered. People have established extensive plantations of introduced pine and other tree species as part of government-sponsored afforestation efforts. Many of these plantations are at risk from and fuel fires (Rodríguez 1997).

There are more than 15 million hectares at risk to stand-replacing fires in the conterminous United States, mostly in the warm, dry forests and in the shrublands and grasslands of the western United States (GAO 1999). With 8 of the 10 fastest growing states in the United States in the west, the risk to people and property continues to increase. In recent years enormous, intense fires have defied fire-fighting efforts and burned until fuels or weather limited them. In southern Argentina and Chile large fires occurred in 1986–87, 1993–94, and 1997–98 (*http://www2.ruf.uni-freiburg.de/fireglobe/iffn/country/cl/cl_3.htm*, *http://www2.*

ruf.uni-freiburg.de/fireglobe/iffn/country/ra/ra_8.htm), and also in the summer of 2000–2001. In that season and as an example, the city of Puerto Madryn in Argentina was surrounded by a wildfire that burned 30 thousand hectares of shrub-grassland, threatening people and properties for about a week until it was completely extinguished. 25 people died fighting the fire (Dentoni et al. 2001). Just as in the United States and Canada, the risk to people and their property is enhanced as the areas prone to large, intense fires are now increasingly populated, especially near towns and cities (Fig. 15.2). Large, severe fire events may become more common in the future in the western Americas through the influence of human-induced changes in climate and vegetation. Climate change drives ecological change through its effects on the rates of fire ignition (e.g., lightning) and spatial patterns of wildfires (Price and Rind 1994; Baker, Chapter 5, this volume). Humans have altered the climate, which affects the probability of ignition by lightning, fire occurrence, fire behavior, length of the fire season, and the effects of fire on vegetation, animals, soil, and air.

Fire management policies and science are evolving from fire suppression to fire management in response to scientific understanding of the ecological role of fire in ecosystems. Fire suppression is recognized as one of the leading threats to the integrity of wilderness and natural areas (Christensen 1995). While policy reviews following the large fires of 1988, 1994, and 2000 in the western United States and 1996 and 2000–2001 fires in Argentina have emphasized fire suppression capabilities, they have also broadened fire management to encompass active prescribed burning and restoration to enhance ecological integrity and natural resource sustainability (Mann and Plummer 1999; Hann and Bunnell 2001; Hardy et al. 2001). Policies are often reviewed and modified following large fire events in which the public felt threatened, or when firefighters die.

Scientists, managers, and the general public are more aware of the complex ecological roles played by wildfire. Global efforts to address climate change (United Nations Framework Convention on Climate Change, known as the Kyoto protocol), conservation of biological diversity (International Convention on Biological Diversity), and natural resource sustainability (Working Group on Criteria and Indicators for the Conservation of and Sustainable Management of Temperate and Boreal Forests, known as the Montreal Process) reflect commitments from many nations to reduce anthropogenic carbon emissions, conserve biological diversity, and practice sustainable forest management. Although these commitments are not yet mandatory, many government and nongovernment organizations are working to implement them. Accomplishing these goals in fire-prone environments will, by necessity, require progressive fire management.

Fire Regimes

Fire regimes have changed within the last century, but the degree and type of change varies with fire regime and geographic location in both North and South America. Although fire regimes have always changed in response to variations

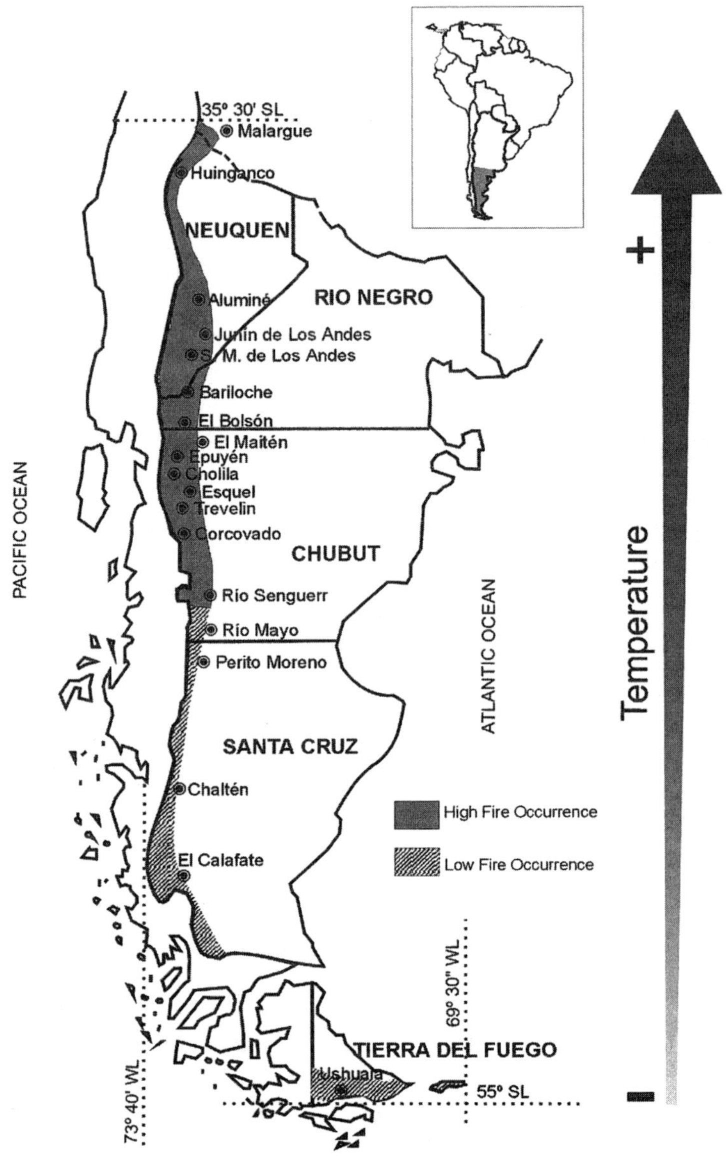

Figure 15.2. In the Patagonian region of Argentina, many of the population centers occur in high fire occurrence along the base of the Andes. As in the lake region of Chile, and in the western United States, much of the rural population growth is occurring in scenic areas which are also prone to fires.

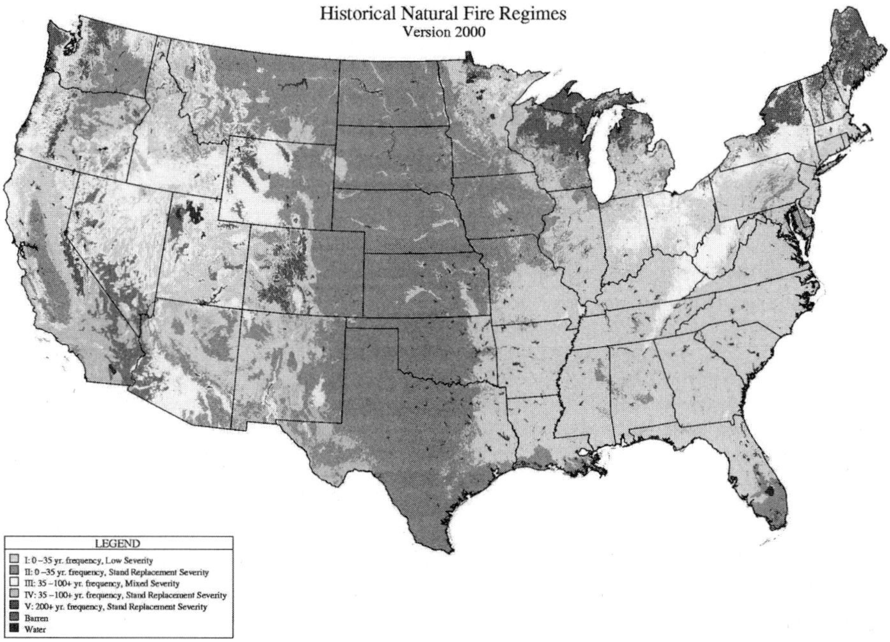

Figure 15.3. Mapped historical fire regimes classes for the conterminous United States (from *http://www.fs.fed.us/fire/fuelman*; Hardy et al. 2001).

in climate (Clark 1990; Swetnam 1993), these recent changes are commonly attributed to land use.

We use three broad fire regime classes to frame our discussion of the degree to which fire regimes have changed and why in the temperate and boreal forest zones of North and South America. Hardy et al. (2001; *http://www.fs.fed.us/fire/fuelman*) grouped historical fire regimes into three broad classes based upon fire frequency (<35 yr, 35 to 100 yr, and 200+ yr between fires) prior to intensive Euro-American settlement and then mapped them for the conterminous United States (Fig. 15.3). Similar maps have been developed for Canada (Canadian Forest Service 2001). These were mapped using rules based on expert opinion informed by fire history, fuels, and succession research; by necessity some judgments were made with little supporting information. Few data were available for some geographic areas and biophysical settings, and they were largely lacking for others, so the accuracy, which was not assessed, likely varies greatly. Clearly, historical fire frequency varied through time (e.g., Swetnam 1993; Swetnam and Betancourt 1998). Also we do not have very comprehensive information on the extent of stand-replacing fires historically, and the data for judging this has not been collected systematically across the landscape, potentially biasing our interpretation of historical fire regimes (Morgan et al. 2001; Baker and Ehle 2001). Unfortunately, such limitations are unavoidable at this scale. Nonetheless, such maps are useful for strategic planning, particularly when current fire regimes are contrasted with historical fire regimes (Morgan et al. 2001; Hardy et al. 2001).

Fire Regimes with Very Frequent Fires

Fire historically occurred very frequently (at intervals of less than 35 yr) on 61% of the land area in the conterminous United States (Hardy et al. 2001; http://www.fs.fed.us/fire/fuelman) (Fig. 3). Currently fires are both less frequent and more likely to be severe on 43% (grasslands) to 59% (forests) of the area on which these fire regimes occurred prior to the 1850s—making these fire regimes the most changed by land use. This fire regime encompasses a variety of ecosystems, including woodlands, dry forests, such as the Monte region in northern Patagonia in Argentina (Kitzberger and Veblen 1997; Kitzberger and Veblen 1999; Veblen et al. 1999, Chapter 9, this volume; Dentoni et al. 2001) and ponderosa pine and Douglas-fir forests of North America (Agee 1993; Covington and Moore 1994; Swetnam and Betancourt 1990, 1998; but see Baker and Ehle 2001 and Shinneman and Baker 1997 for contrast), as well as shrub-steppe (Wright and Bailey 1982) and most grasslands (Wright and Bailey 1982). As a result of fire exclusion, woody debris has accumulated, the size and frequency of regeneration gaps has declined (Stephenson 1999; Veblen et al., Chapter 9, this volume), and composition and age structure have changed (Veblen et al., Chapter 9, this volume; Covington and Moore 1994; Swetnam and Betancourt 1998).

Chaparral and other ecosystems found in Mediterranean climates in both North and South America also historically experienced fires at very frequent intervals (less than 35 yr) (Hardy et al. 2001; *http://www.fs.fed.us/fire/fuelman*). Evergreen sclerophyll shrublands and low forest dominate in the Mediterranean-type climates on the west coasts of Chile and California. Although the vegetation shares broad physiognomic similarities, there are distinct differences (Armesto, Vidiella, and Jimenez 1995; Fuentes and Muñoz 1995). For instance, lightning fires have always been more common in the chaparral (Armesto, Vidiella, and Jimenez 1995; Fuentes and Muñoz 1995). The structure, composition, and dynamics of both chaparral (California) and matorral (Chile) vegetation are greatly influenced by people through fire, grazing, agriculture, urban and rural development, and other land uses (Minnich 1983; Armesto, Vidiella, and Jimenez 1995; Fuentes and Muñoz 1995; Keeley, Fotheringham, and Morais 1999). In contrast to other parts of this fire regime, suppression has not diminished fire on the landscape in chaparral systems (Keeley, Fotheringham, and Morais 1999). Most of the area that burns each year does so under high winds; most ignitions are by people. Significantly increased number of fires and area burned per decade from 1910 to 1990 were correlated with increasing population density rather than with fire suppression (Keeley, Fotheringham, and Morais 1999). In fact fire frequency has increased and often resulted in conversion to grasslands dominated by nonnative species (Keeley, Fotheringham, and Morais 1999). Introduced annual grasses have similarly transformed parts of the shrub-steppe and other ecosystems in North America (Knick and Rotenberry 1997).

Fire Regimes with Fires of Intermediate Frequency

Where fires historically occurred less frequently (i.e., every 35 to 100 yr), they had mixed or stand-replacing effects on the dominant overstory (Hardy et al. 2001; http://www.fs.fed.us/fire/fuelman). These fire regimes historically occurred on approximately 34% of the conterminous United States. They too have changed with land use, so that fires are now more likely to be severe on more than half (56%) of the area that historically burned at moderate frequency. Examples of ecosystems typical of these fire regimes are the mixed conifer forests of middle elevations in the mountains and foothills of both North America (Agee 1993; Arno 1980) and South America (Veblen et al., Chapter 9, this volume).

In semiarid forests and adjacent shrub-steppe of northern Patagonia, the largest fire years occur when fires burn following periods that are sufficiently wet to support the growth of grasses as abundant and continuous fine fuels (Kitzberger and Veblen, Chapter 10, this volume). This also holds true for the dry forests of the southwestern United States (Swetnam and Betancourt 1990, 1998) where grasses grow in abundance during wet years.

Fire Regimes with Infrequent Fires

The third group of fire regimes identified by Hardy et al. (2001) were those with relatively infrequent fires (more than 200-yr intervals, between typically stand-replacing fires). Where these fire regimes occurred historically, fires now typically occur less frequently, and they are more severe on about one-quarter (27%) of the 5% of the conterminous United States where these fire regimes occurred historically (Hardy et al. 2001; http://www.fs.fed.us/fire/fuelman). The ecosystems associated with these fire regimes are moist forests of subalpine zones (Agee 1993; Arno 1980) and near coastal regions (Veblen and Alaback 1996) and interior warm, mesic forests (Veblen and Alaback 1996; Agee 1993; Arno 1980). Extensive fires are associated with extended droughts in these fire regimes (Agee 1993; Swetnam and Betancourt 1990, 1998; Rollins et al. 2000a,b, 2001). Similar trends hold for Patagonia in Argentina (Veblen et al., Chapter 9, this volume).

Wet forests are characterized by high precipitation (>1400 mm), cool summers, and mild winters (Veblen and Alaback 1996). Such forests are more extensive in Chile than in Argentina due to the rain-shadow effect of the Andes. In South America the temperate rainforests are a mix of evergreen conifer and broadleaf tree species (Veblen and Alaback 1996), with gradual changes in species composition and decreasing richness with increasing latitude. The pattern is similar but more gradual in North America (Veblen and Alaback 1996). Similar forests are also found to a limited extent in the Rocky Mountains (Agee 1993). Disturbances are prevalent. Wet forests historically experienced fire every 100 to 300 years overall, with more frequent fires at lower latitudes and further east or wherever seasonal drying was more pronounced (Agee 1993; Veblen and Alaback 1996; Amiro et al. 2001).

The Relative Importance of Land Use and Climate

These fire regimes vary along environmental gradients. Fire is a major controlling disturbance all along these gradients, but its ecological role, degree of influence by land use, and current departure from past conditions vary from site to site. Forest fires were historically more frequent at low elevations and less frequent at high elevations (Swetnam 1993; Baker, Chapter 5, this volume; Veblen et al., Chapter 9, this volume), and more frequent on drier aspects and watersheds than mesic ones (Heyerdahl, Brubaker, and Agee 2001). Baker (Chapter 5, this volume) hypothesized that the relative importance of fuels decreases and the importance of fire weather increases as one moves along the environmental gradient described above from warm, dry sites at low elevations to relatively cold and moist sites at high elevations. Rollins et al. (2000a,b, 2001) demonstrated this with empirical data for two contrasting Rocky Mountain wilderness areas in the United States.

Regional fire events, when many extensive fires occur, are typically associated with widespread droughts (Swetnam 1993; Swetnam and Betancourt 1990, 1998), account for the majority of the area burned (Strauss, Bednar, and Mees 1989; Kitzberger and Veblen, Chapter 10, this volume). During such years the threats to people and their property are highest (Maciliwain 1994; Defossé et al. 2001). Thus weather is a major driver of severe fire events in particular, and climate of fire regimes in general (Clark 1990; Swetnam and Betancourt 1998; Miller and Urban 1999).

Land use and human activity is also critical in determining fire patterns. While land use has influenced all three fire regimes, humans have most directly influenced fires where fires once occurred most frequently (Pyne 1982; Veblen et al., Chapter 9, this volume). Resulting trends in dry forests include increased tree density and invasion of trees into shrublands, woodlands, and grasslands. Together these trends have led to an increasingly continuous fuel, both from the ground to the tree crowns (i.e., fuel ladders), and from tree crown to tree crown across landscapes (Covington and Moore 1994; Kitzberger and Veblen 1997, 1999; Swetnam and Betancourt 1998; Kitzberger and Veblen, Chapter 10, this volume).

In the Andean-Patagonian region of Argentina, people currently ignite more fires than lightning does (Kitzberger and Veblen, Chapter 10, this volume). For example, Rodríguez (1999) reported that people ignited more than 60% of all fires that occurred during the 1990s in the northern area of that region. The statistics for Canada are similar (Canadian Forest Service 2001). Similarly most of the fires that occur in the Mediterranean climate regions of California and Chile are caused by people, as there is little lightning (Haltenhoff 1994; Armesto, Vidiella, and Jimenez 1995; Fuentes and Muñoz 1995; Keeley, Fotheringham, and Morais 1999). However, for fire spread, weather is a more important controlling factor than ignition source (Kitzberger and Veblen, Chapter 10, this volume; Keeley, Fotheringham, and Morais 1999).

The relative importance of land use and climate is important to fire managers, for it determines to some extent the potential for changes in land management policy to affect fire regimes. We can readily manipulate vegetation and fuels, and through education, patrols, and other fire suppression measures, we can alter the probability of ignition by people. We can do less to alter climate and weather.

Changes in fire regimes through time are only partially explained by climate and climate variability (Weber and Flannigan 1997; Flannigan, Stocks, and Weber, Chapter 4, this volume; Veblen et al., Chapter 9, this volume). This is because land use is also important. While climate is a driver of fire patterns in all fire regimes, land use and fire suppression have most altered fire regimes in dry forests and elsewhere where historically frequent surface fires were fueled by grasses and other fine fuels. Thus fire frequency and severity have changed dramatically with the initiation of intensive grazing in dry forests (Swetnam and Baisan 1996; Swetnam and Betancourt 1990, 1998), with the onset of increasingly effective fire suppression (Rollins et al. 2000a,b, 2001), and other land uses. Where stand-replacing fires were the norm, these changes are less pronounced. For instance, Keeley, Fotheringham, and Morais (1999) found that fire size had not changed in the twentieth century in chaparral ecosystems in southern California. Large fires occurred during droughts and were little influenced by fuel management. Climate has an overriding importance at both broad and fine scales (Swetnam and Betancourt 1998; Heyerdahl, Brubaker, and Agee 2001), particularly for extreme fire events. Human impacts are ubiquitous as well, but more pronounced in altering fire regimes where fires were historically frequent (Hardy et al. 2001) and where human population density is high and land use is intense (e.g., chaparral in California; Keeley, Fotheringham, and Morais 1999). Fire regimes have changed more in locations where human influence is greatest (Hardy et al. 2001).

Implications for Future Fire Management

We must better understand the complex interrelationships among fire, climate, vegetation, topography, and land use (Fig. 15.4) if we are to effectively manage fire as climate changes (Overpeck, Rind, and Goldberg 1990). Understanding the linkages among fire, climate, land use, and vegetation is useful as a reference or baseline for understanding and evaluating ecosystem change (Morgan et al. 1994; Landres, Morgan, and Swanson 1999; Swetnam, Allen, and Betancourt 1999). Historical range of variability in fire frequency and vegetation composition is widely used by natural resource managers in North America as a reference in determining goals for restoration and sustainability (Landres, Morgan, and Swanson 1999; Mann and Plummer 1999). The degree of departure of current from historical fire regimes has been included in broad ecological assessments and national strategic planning for fire management because managers find it useful for identifying areas of low ecological integrity, accumulating fuels and associated fire risk, and to prioritize restoration or other active management (Brown et al. 1994; Landres, Morgan, and Swanson 1999; Caprio and Graber 2000; Hardy et al. 2001; Hann and Bunnell 2001). Characterizing past fire

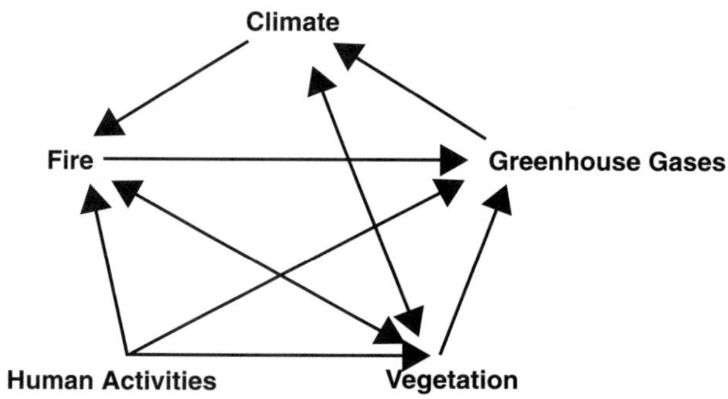

Figure 15.4. The linkages among fire, climate vegetation, the atmosphere, and land use are complex (from Canadian Forest Service 2001).

regimes is also useful for parameterizing and validating ecosystem models, and for extrapolating point and other local information to a continuous map (Morgan et al. 2001). Mechanistic models can be parameterized using empirically defined fire regimes and fire–climate–landscape relationships (Keane and Long 1998; Keane and Finney, Chapter 2, this volume).

However, taking full advantage of the lessons of history depends on identifying the drivers of change, such as land use and ocean temperatures (Swetnam, Allen, and Betancourt 1999) and their interactions. To do so, we need models that link fire behavior and fire effects (including hydrologic processes) to vegetation, land use, climate change, weather, and topography. In particular, spatially explicit models that use remotely sensed data and our best understanding of ecosystem processes would be most helpful (Keane, Burgan, and van Wagtendonk 2001; Keane and Finney, Chapter 2, this volume).

We are only just beginning to do strategic fire management at the landscape scale. To some degree our ability to do so is hampered by our relatively poor understanding of the spatial fire effects at landscape scales, which integrate the regional forcing by climate with the effects of local vegetation and topography. Also we know more about changing fire frequency at points than we do about other, especially spatial, aspects of fire regimes, such as fire severity, rotation, and spatial pattern (Morgan et al. 2001). The vegetation mosaics that develop with mixed fire regimes at middle elevations are complex and little understood, yet they occur widely in mesic forests (Agee 1993, 1998). Fires create and are influenced by spatial pattern (Agee 1998). Mixed severity fire regimes, in particular, create complex mosaics of vegetation. Our fire history data are limited in geographic extent, primarily to the dry forests that historically burned in nonlethal fires (Morgan et al. 2001). In grasslands we are often limited to archival records of actual fire events, which are typically limited to the years since the late 1800s at best, or to relatively coarse temporal resolution and geographically localized records provided by paleoecological records. In boreal forests and elsewhere where stand-replacing fires occur, we can reconstruct the spatial pattern of the last fire, but we don't know much about previous fires. Many landscapes have combinations of all of these vegetation types and associated fire regimes that are themselves changing in response to past climate, making strategic fire management more challenging.

Human-induced climate change will have dramatic impacts on fires. The number of lightning fires may increase by 30% (Price and Rind 1994) with a doubling of the carbon dioxide content of the atmosphere. Flannigan et al. (1998) suggest that the fire weather index, a measure of variables influencing fire intensity, may increase by two to five times under the same scenario. Future vegetation patterns may be very different than today (Bartlein, Whitlock, and Shafer 1997). Changing climate would have both direct and indirect effects on vegetation. Past climate changes have altered species distributions, influenced biological diversity, and altered tree mortality and disturbance rates. Coupled with effects on fire occurrence and severity, climate change could result in ecological, economic, and social consequences (Crutzen and Goldammer 1993). Further a positive feedback is possible, with higher carbon dioxide content in the atmosphere leading to more fires and more fires reducing forest cover and its potential to act as a carbon sink (Amiro et al. 2001). If extensive plantations are established to offset carbon emissions elsewhere, and they burn, those plantations could be sources rather than sinks for atmospheric carbon.

The area burned by fires in the western United States declined as fire suppression efforts became increasingly effective until the 1950s (Fig. 15.5). After that date more area burned despite increasing efforts and expenses in fire man-

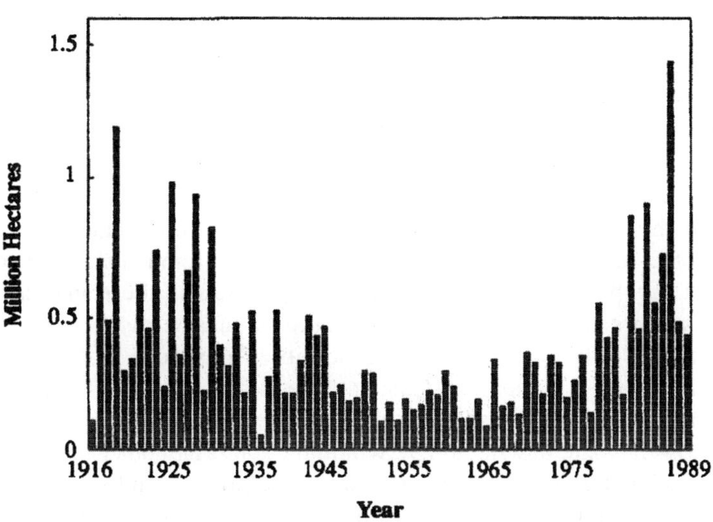

Figure 15.5. Area burned by wildfire in the western United States, 1915 to 1990 (from Agee 1993) reflect the effect of increasingly successful fire suppression efforts early in the century, and then an increase in area burned despite continued and increasing expenditures for fire suppression and fire management programs.

agement programs (Agee 1993; Hann and Bunnell 2001). A similar trend in Canada has been attributed to the combined effects of fire suppression and climate change (Amiro et al. 2001). Recent national assessments reflect increased risk to human life and property, as well as ecosystem health, streams, and native species (Hann and Bunnell 2001). The focus on fuels management as part of national fire management programs in the United States and Canada are motivated by concerns over public safety and the hope that fire suppression costs ($U.S. 350 million each year in Canada; Amiro et al. 2001) will be reduced. Hann and Bunnell (2001) suggest that with restoration and maintenance on 2% of the land base each year, many of the trends for the twentieth century could be reversed. However, their projections did not include climate change. We will have to rely heavily on prescribed fire and fire surrogates to subsidize lightning ignitions, even in very large wilderness and natural areas, but especially in small ones, and certainly in the majority of other lands. Lightning fires alone cannot recreate natural landscapes fragmented by roads, invaded by introduced species and heavily used by people. Fires will be suppressed whenever they threaten commercial timber and houses.

Future fires are likely to be severe and intense in response to a climate that is both more variable and changing in response to human action (Weber and Flannigan 1997; Swetnam and Betancourt 1998; Flannigan, Stocks, and Weber, Chapter 4, this volume). In Canada the annual area burned is projected to increase by 50% (Flannigan et al. 1998; Amiro et al. 2001).

Furthermore socioeconomic trends will augment this trend. With human population increases, the number of houses in the wildland–urban interface is growing, facilitated by improved communication systems (cellular phones, Internet, etc.) and a greatly improved transportation infrastructure. These trends are most pronounced in rural counties near wilderness areas, parks, or other amenities. For instance, such counties grew 13% in the 1970s and 34% from 1980 to 1987, compared to an average of 6.9% for other rural counties in the United States (Pyne 1995); trends are similar elsewhere. People moving to the interface from urban areas do not expect fire to occur around their dwellings, and often do not understand the propensity of wildland ecosystems to burn frequently. Another problem is that the lay public, the media, and some environmentalists do not fully understand the threats that increased amounts of fuels in those interface areas pose to managers and firefighters attempting to suppress fires in extreme conditions (Davis 1989; Hirsch 2000). The resort city of Bariloche in Argentina is an example. Surrounded by lakes and ancient forests of the Nahuel Huapi National Park, fire suppression was (and still is) the prevalent fire policy. Today more and more city dwellers are moving to live within the forest and other interface areas, where vegetation has been allowed to accumulate, no trees and shrubs were permitted to be cut, and very little fuels management has been done for years. This situation poses a tremendous danger for people and property, especially if dry and windy conditions prevail during the summer.

Managing fire effectively depends on understanding how fire, climate, vegetation, land use, and topography interact. Over extensive areas, fires now occur less frequently, but with potentially more severe effects on plants, animals, soils, and water. While fire patterns have not changed to the same degree everywhere, the changes often threaten people and their property, as well as long-term ecological integrity and sustainability. Based on the case studies presented in previous chapters, climate clearly influences fire frequency and severity. Further, simulation models of climate change and fire suggest that disturbances, including fire, insects, wind, and weather, will accelerate the rates of forest change due to climate shifts (Weber and Flannigan 1997; Flannigan, Stocks, and Weber, Chapter 4, this volume; Kitzberger and Veblen, Chapter 10, this volume). Many of the disturbances, including fire, are directly and indirectly influence by climate. This has important management implications. Unfortunately, separating the effects of human-induced climate change from the jointly contributing and interacting factors of land use, climate variability, fire, and other disturbances is challenging. Simulation models enable us to study the interactions among fire vegetation, climate, topography, fuels, and land use (Keane and Finney, Chapter 2, this volume). It is critical to do so in a way that is cognizant of the biophysical and social context, for the characteristics of the surrounding landscape and the legacy of the past will influence the response of fire and vegetation to climate change (Baker, Chapter 5, this volume; Veblen et al., Chapter 9, this volume).

Fires are often synchronous across widely separate areas with distinctly different forest types and land use, and those synchronous events are correlated with ENSO or other interannual global climate variations (Swetnam and Betancourt 1990, 1998; Swetnam and Baisan, Chapter 6, this volume; Kitzberger and Veblen, Chapter 10, this volume). This suggests an opportunity to fire managers. Because ENSO can be forecast in advance, fire managers should strategically plan accordingly, targeting prescribed burning for those years in which fire events are less likely to be synchronous, and devoting most of the fire personnel to fires suppression in years, like 2000 in western north America, when the climate forecasts suggest extensive and severe wildfires are likely over large geographic areas (Swetnam and Betancourt 1998). Synchronous events have tremendous implications for the combined threats to people and property, for they can quickly overwhelm our ability to suppress them. They are also likely to "reset" succession over large areas, potentially contributing to a positive feedback with an increasingly homogeneous landscape (Veblen et al., Chapter 9, this volume).

Addressing Fire Management Goals and Challenges in a Changing Future

How will and should fire managers and the stakeholders in their decisions respond to our growing understanding of the interactions between fire regimes and climate, as represented by the material in this book? First, we must accept that fires will occur, and their timing and intensity will be greatly influenced by climate change. In fact fires will mediate vegetation response to climate change (Swetnam and Betancourt 1998; Flannigan, Stocks, and Weber, Chapter 4, this volume). Second, it is clear that the risks to people and their property will continue to grow as more people settle in and otherwise use fire-prone environments. Keeley et al. (1999) recommend focusing fuels and fire management efforts in the strategic locations near towns to address the risk of ignition and fire risk. Third, continued attempts at fire exclusion may result in accumulating fuels wherever biomass production exceeds decomposition and removal. In such cases advance forecasting of ENSO and similar global circulation patterns affecting fire patterns will assist fire managers in strategically planning resource allocation to fire suppression (in those years where regional fire events are most likely) or prescribed burning (in other years and places where fire can be used effectively to accomplish resource management objectives) (Swetnam and Betancourt 1993, 1998). Fourth, we must analyze alternatives, including the implications of continued efforts at fire exclusion. The increasing availability of remote sensing, spatial analysis tools and models linking fire behavior and effects to climate change will assist scientists and managers in understanding the effects of alternative management and climate change scenarios (Miller and Urban 1999; Keane, Burgan, and van Wagtendonk 2001; Keane and Finney, Chapter 2, this volume). Thus we are poised to move from fire suppression to fire management as the dominant paradigm.

Managers find it challenging to incorporate our rapidly developing knowledge about fire, climate, and ecosystem dynamics with social values and fiscal and legal constraints. The public support and infrastructure for fire suppression are still far more developed than infrastructure for fire ecology research, and fire suppression paradigms still dominate most fire management programs. It will also depend on the social, political, and economic environment as well as biophysical factors. Nonetheless, we organize this part of our discussion around the fire management goals introduced in the beginning of this chapter. To this we add what we view will be an increasingly important goal, managing emissions.

Protecting People and Property

Protecting human life and property from both direct and indirect effects of fires will continue to be a major focus of fire management no matter what the land management goals are. Residential use of large areas adjacent to forests and parks means that fire hazard mitigation will be a major driver of fire management. Fire risk has and will continue to increase because there are more people in exurban areas, less grazing by domestic livestock contributes to more fuels on the landscape, more introduced grasses, shrubs, and trees fuel fires, and fuels are accumulating through our relatively effective fire exclusion. Fuels management programs are designed to reduce the likelihood that fires will grow out of control. In addition fire management programs focus on education to reduce accidental ignition of fires and to ensure that the landscaping around homes does not add to the risk.

Fuels management programs are generally more likely to be effective where fires burn less intensely, and where horizontal and vertical continuity of fuels influences fire spread. Fuels management is less effective for fires burning under very hot, dry, windy conditions. Unfortunately, there has been little assessment about whether fuels management can effectively mitigate fire hazard in a climate that is changing. This is one of the primary focuses of the Joint Fire Sciences Program of the US federal government agencies (*http://www.nifc.gov/joint_fire_sci/jointfiresci.html*) and of concerted research efforts elsewhere.

Clearly, climatically induced changes in fire regimes will greatly influence the structure, composition, and function of the forest ecosystems in North and South America (Veblen and Alaback 1996). However, fires in grasslands, shrublands, and woodlands often pose management challenges that are greater than in forest fires. More people die in grassland fires, and grassland fires more frequently threaten people and their property. For example, in 1994, 25 people died fighting a fire that burned close to Puerto Madryn, a small coastal city in the Patagonian region of Argentina (Dentoni et al. 2001). Indeed, most of the cases where firefighters or others have died in fires have been in nonforested or open woodland areas, including Mann Gulch in Montana (13 people died) and South Canyon/Storm King (14 people died there in 1994). More effort is focused on fire-related ecological and suppression research in forests than in these other ecosystems.

Enhancing Habitat

Prescribed fires are conducted to improve habitat for individual species or communities of plants and animals. It is possible that such efforts will increase as there is a growing body of research about the importance of fire in maintaining the landscape diversity on which many different birds, animals, and plants depend. Further the viability of many different plant and animal species is threatened by human action, including fragmentation and conversion of habitat, introduction of exotics, altered disturbance patterns, and exurban development. Conservation strategies increasingly focus at the landscape scale (Franklin 1993). Where species are endangered, land management activities, including fire management, are legally constrained in the United States.

Ecological Restoration

Efforts to restore ecological conditions, functions and integrity are increasingly common in ecosystems from prairies to forests. Many of these efforts have focused on restoring some "presettlement" (usually defined as prior to intensive Euro-American use of the land) structure or other condition, although some efforts focus on restoring native five regimes. The management mandate of many national parks in the United States, Canada, Chile, Argentina, and other countries of South America is often interpreted as requiring natural conditions.

Clearly, our deeper understanding of the ecological implications of climate change and the dynamics of ecosystems reinforces the need for a broader focus on restoring ecological integrity, resilience, and sustainability rather than on restoring some "vignette" or past condition (Pavlik 1996, White and Walker 1997, Landres et al. 1999). This is a focus on restoration of processes rather than structure (Stephenson 1999), though the most viable programs will integrate considerations of both process and structure.

Natural Process

Although the fire management in some wilderness areas, national parks, and nature preserves is focused on fire as a natural process, fires are often suppressed. Even in the areas where fires were historically infrequent, the fire rotation has changed in the twentith century (Rollins et al. 2000a,b, 2001; Baker 1992). Although it is possible that natural fire regimes will be restored in these and surrounding areas, it is most likely that such restoration will have to rely upon lightning ignitions. It is more likely that many of the ecologically significant fires will be suppressed (Parsons and Landres 1998), and it will be very challenging to establish prescribed fire programs approximating even a fraction of the frequency and extent of historical fires in most natural areas.

Managing Emissions

Smoke and atmospheric emissions from fire may well determine the future of fire management. Smoke emissions are of increasing concern because of the hazards to people who breath in particulates, reduced visibility especially in scenic areas but also along roads, and impacts on ozone (Riebau and Fox 2001). Concerns about smoke emissions have greatly altered fire management programs in many areas, and more changes will come. As efforts to mitigate the impacts of climate change grow, so will the impetus to sequester carbon in forests and grasslands. Such efforts must consider the impossibility of controlling all fires as well as the ecological consequences of not burning.

Land management agencies will get increasing pressure to sequester carbon, for instance, by planting trees. Historical ecological studies and projections suggest that this might be very challenging because extreme droughts and other weather events will make it difficult to prevent fires and smoke with the related emissions of carbon. In Argentina large afforestation programs have promoted planting trees in the shrub-steppe, and recent plantations have been justified as sequestering carbon. A central question is whether those trees can be harvested before they burn.

Conclusion

One of the clearest lessons from history is that fires have always occurred, and that they will continue to occur despite our efforts to detect and suppress them. The long history of fire in the temperate and boreal forests of North and South America emphasizes the prevalence and inevitability of fires. People have long feared, used, and sought to control fire (Pyne 1982, 1995), yet fires resist our efforts to control them completely. Thus fire managers should be cognizant not only of the complex interplay of fire, climate, vegetation, and land use but also of the need for managing for landscapes that are resilient to fire effects, and adapting our land use and housing patterns to the inevitability of fire occurrence.

National fire plans must address the implications of climate change for fire patterns. Flannigan et al. (1998) estimated that the annual area burned in Canada might increase by as much as 50%, especially in the West. The Canadian Forest Service (2001) is adapting fire management accordingly. In the United States, however, the national fire plan (*http://www.fireplan.gov*) recently adopted and currently being revised does not mention climate change or its implications. Similarly the Argentine National Fire Management Plan does not yet include climate change. Fire management in temperate and boreal forests of North and South America continue to focus primarily on suppression, although the natural role of fire is recognized as being important. Fire management must be broader than suppression, for even the most effective fire suppression cannot prevent all fires and that is not desirable ecologically or socially—because then the next fire that occurs may burn through accumulated fuels with greater intensity. Fire scientists and managers must work together to learn from one another about the complex interactions and synergies among fire, climate, vegetation, the atmosphere, and land use (including exurban development) (Fig. 4), and then to teach politicians, journalists, nongovernmental organizations, media communicators, teachers at all levels, and concerned citizens about the important role fire plays in nutrient cycling and other processes critical to ecosystem sustainability and the long-term implications of global climate change. To be successful, these national fire management programs must approach fire suppression as only one part of a more complex fire management strategy that includes fuels management, education and risk assessment, changes in land use, and land-use regulation and development, and they must recognize the reality of a changing climate. In short, we must change the dominant paradigm from one of fire suppression to fire management.

The global climate is changing under human influence (IPCC 2001). Humans have also greatly altered fire regimes through land use and introduction of exotic species. The synergies among climate, vegetation, land use, and fire has tremendous and challenging implications for the future. Particularly in dry forests, shrublands, and grasslands heavily used by people, there are powerful feedbacks between climate, fire, and vegetation (Flannigan, Stocks, and Weber, Chapter 4, this volume) that threaten long-term sustainability of the ecosystems on which we depend. Many have proposed sequestering carbon in forest ecosystems to mitigate the influence of fossil fuel emissions on global climate. Any such strategy to accumulate carbon in biomass must consider the likelihood that accumulated biomass will eventually fuel fires.

Climate change offers some great challenges to researchers. One is predicting the impact of climate change. A second is understanding the synergies among fire, vegetation, land use, atmosphere, and climate. A third is communicating those lessons clearly enough that managers, policy makers, and others can decide how the associated challenges in fire management should be addressed.

This volume offers much of use to managers, just as it raises further questions for scientists. The great challenge for the future is for scientists and managers to work together to anticipate how climate change, land use, and vegetation will interact with fire in the future.

References

Agee, J.K. 1993. *Fire Ecology of Pacific Northwest Forests.* New York: Island Press.

Agee, J.K. 1998. The landscape ecology of western fire regimes. *Northwest Sci.* 72(4): 24–34.

Amiro, B.D., Stocks, B.J., Alexander, M.E., Flannigan, M.D., and Wotton, B.M. 2001. Fire, climate change, carbon and fuel management in the Canadian boreal forest. *Int. J. Wildl. Fire* 10:405–413.

Armesto, J.J., Vidiella, P.E., and Jimenez, H.E. 1995. Evaluating causes and mechanisms of succession in the Mediterranean regions in Chile and California. In *Ecology and Biogeography of Mediterranean Ecosystems in Chile, California and Australia*, eds. M.T.K. Arroyo, P.H. Zedler, and M.D. Fox, pp. 418–434. New York: Springer-Verlag.

Arno, S.F. 1980. Forest fire history in the northern Rockies. *J. For.* 78(8):460–465.

Arno, S.F., and Hardy, C, eds. 1996. *The Use of Fire in Forest Restoration.* Gen. Tech. Rep. INT-GTR-341. Ogden, UT: USDA Forest Service, Intermountain Research Station.

Baker, W.L., and Ehle, D. 2001. Uncertainty in surface-fire history: The case of ponderosa pine forests in the Western United States. *Can. J. For. Res.* 31:1205–1226.

Bartlein, P.J, Whitlock, C., and Shafer, S.L. 1997. Future climate in the Yellowstone National Park region and its potential impact on vegetation. *Conserv. Biol.* 11(3): 782–792.

Brown, J.K., Arno, S.F., Barrett, S.W., and Menakis, J.P. 1994. Comparing the prescribed natural fire program with presettlement fires in the Selway-Bitterroot Wilderness. *Int. J. Wildl. Fire* 4:157–168.

Canadian Forest Service. 2001. Forest fire: Context for the Canadian Forest Service's science program. 2001. <*http://www.nrcan-rncan. gc. ca/cfs-scf/science/context_fire/ index_e.html*>. Accessed November 2001.

Caprio, A.C., and Graber, D.M. 2000. Returning fire to the mountains: Can we successfully restore the ecological role of pre-European fire regimes to the Sierra Nevada? In *Wilderness Science in a Time of Change Conference: Wilderness Ecosystems Threats and Management*, vol. 5, comps. D.N. Cole, S.F. McCool, W.T. Borrie, and J. O'Loughlin, pp. 233–241. USDA Forest Service Proceedings RMRS-P-15.

Chandler, C., Cheney, P, Thomas, P., Trabaud, L., and Williams, D. 1983. *Fire in Forestry.* New York: Wiley.

Christensen, N.L. 1995. Fire and wilderness. *J. Wilderness* 1:30–33.

Christensen, N.L., Baruska, A.M., Brown, J.H., and 10 colleagues. 1996. The report of the Ecological Society of America committee on the scientific basis for ecosystem management. *Ecol. Appl.* 6(3):665–691.

Cissel, J.H., Swanson, F.J., and Weisberg, P.J. 1999. Landscape management using historical fire regimes: Blue River, Oregon. *Ecol. Appl.* 9(4):1217–1231.

Claraz, J.E. 1988. *Viaje de exploración al Chubut 1865–1866.* Buenos Aires: Marymar.

Clark, J.S. 1990. Patterns, causes, and theory of fire occurrence during the last 750 yr in northwestern Minnesota. *Ecol. Monogr.* 60:135–169.

Covington, W.W., and Moore, M.M. 1994. Southwestern ponderosa forest structure changes since Euro-American settlement. *J. For.* 92:39–47.

Crutzen, P.J., and Goldammer, J.G. 1993. *Fire in the Environment: The Ecological, Atmospheric, and Climatic Importance of Vegetation Fires.* New York: Wiley.

Davis, J.B. 1989. The wildland-urban interface: What it is, where it is, and its fire management problems. *Fire Manag. Notes* 50(2):22–33.

Defossé, G.E., Rodríguez, N.F., Dentoni, M.C., Muñoz, M., and Colomb, H. 2001. Condiciones ambientales y bióticas asociadas al incendio "San Ramón" en Bariloche, Río Negro, Argentina, en el verano de 1999. In 1° *Reunión Binacional de Ecologia, XX Reunión Argentina de Ecologia, X Reunión de la Sociedad de Ecologia de Chile*, p. 91. Bariloche, Argentina.

Dentoni, M.C., and Cerne, V. 1999. *La atmósfera y los incendios.* Plan Nacional de Manejo del Fuego: Secretaría de Recursos Naturales y Desarrollo Sutentable. Buenos Aires. Argentina.

Dentoni, M.C., Defossé, G.E., Labraga, J.C., and del Valle, H.F. 2001. Atmospheric and fuel conditions related to the Puerto Madryn fire of 21 January, 1994. *Meteorol. Appl.* 8(3):361–370.

Flannigan M.D., Bergeron, Y., Engelmark, O., and Wotton, B.M. 1998. Future wildfire in circumboreal forests in relation to global warming. *J. Veg. Sci.* 9:469–476.

Franklin, J. 1993. Preserving biodiversity: Species, ecosystems, or landscapes? *Ecol. Appl.* 3:202–205

Fuentes, E.R., and Munoz, M.R. 1995. The human role in changing landscapes in central Chile: Implications for intercontinental comparisons. In *Ecology and Biogeography of Mediterranean Ecosystems in Chile, California and Australia*, eds. M.T.K. Arroyo, P.H. Zedler, and M.D. Fox, pp. 401–417. New York: Springer-Verlag.

Government Accounting Office (GAO). 1999. *Western national forests: a cohesive strategy is needed to address catastrophic wildfire threats.* GAO/RCED-99-65. Washington, DC. <*http:///www.gao.gov*> [10 September 2001].

Haltenhoff, H. 1994. Forest fires in Chile. IFFN No. 11. <*http://www.2.ruf.uni-freiburg.de/fireglobe/iffn/country/cl/cl_2.htm*> [10 October 2001].

Hann, W.J., and Bunnell, D.L. 2001. Fire and land management planning and implementation across multiple scales. *Int. J. Wildl. Fire* 10:389–403.

Hardy, C.C., Schmidt, K.M., Menakis, J.P., and Sampson, R.N. 2001. Spatial data for national fire planning and fuel management. *Int. J. Wildl. Fire* 10:353–372.

Heyerdahl, E.K., Brubaker, L.B., and Agee, J.K. 2001. Spatial controls of historical fire regimes: A multiscale example from the Interior West, USA. *Ecology* 82(3):660–678.

Hirsch, K. 2000. Canada's wildland urban interface: Challenges and solutions. Canadian Forest Service. <*http://www.nofc.forestry.ca/fire/frn/English/wui/UrbanInterface_e.htm*> [20 September 2001].

Holling, C.S., and Meffe, G.K. 1996. Command and control and the pathology of natural resource management. *Conserv. Biol.* 10(2):328–337.

Intergovernmental Panel on Climate Change. 2001. Climate change 2001: Impacts, adaptation and vulnerability. <*http://www.ipcc.ch/*> [1 May 2001].

Keane, R.E., and Long, D.G. 1998. A comparison of coarse-scale fire effects simulation strategies. *Northwest Sci.* 72:76–90.

Keane, R.E., Burgan, R., and van Wagtendonk, J. 2001. Mapping wildland fuels for fire management across multiple scales: Integrating remote sensing, GIS, and biophysical modeling. *Int. J. Wildl. Fire* 10:301–319.

Keeley, J.E., and Stephenson, N.L. 2000. Restoring natural fire regimes to the Sierra Nevada in an era of global change. In *Wilderness Science in a Time of Change Conference*: *Wilderness Ecosystems Threats and Management*, vol. 5. comps. D.N. Cole, S.F. McCool, W.T. Borrie, and J. O'Loughlin, pp. 255–265. USDA Forest Service Proceedings RMRS-P-15.

Keeley, J.E., Fotheringham, C.J., and Morais, M. 1999. Reexamining fire suppression impacts on brushland fire regimes. *Science* 284:1829–1832.

Kitzberger, T., and Veblen, T.T. 1997. Influences of humans and ENSO on fire history of Austrocedrus chilensis woodlands in northern Patagonia, Argentina. *Ecoscience* 4:508–520.

Kitzberger, T., and Veblen, T.T. 1999. Fire-induced changes in northern Patagonian landscapes. *Landscape Ecol.* 14:1–15.

Knick, S.T., and Rotenberry, J.T. 1997. Landscape characteristics of disturbed shrubsteppe habitats in southwestern Idaho (U.S.A.) *Landscape Ecol.* 12:287–297

Landres, P.B., Morgan, P., and Swanson, F.J. 1999. Overview of the use of natural variability concepts in managing ecological systems. *Ecol. Appl.* 9(4):1179–1188.

Maciliwain, C. 1994. Western inferno provokes a lot of finger-pointing, but little action. *Science* 370:585.

Mann, C.C., and Plummer, M.L. 1999. Call for "sustainability" in forests spark a fire. *Science* 283:1996–1998.

Millar, C.I., and Woolfenden, W.B. 1999. The role of climate change in interpreting historical variability. *Ecol. Appl.* 9:1207–1216.

Miller, C., and Urban, D.L. 1999. Forest pattern, fire and climate change in the Sierra Nevada. *Ecosystems* 2:76–87.

Miller, C., and Urban, D.L. 2000. Modeling the effects of fire management alternatives on Sierra Nevada mixed-conifer forests. *Ecol. Appl.* 10:85–94.

Minnich, R.A. 1983. Fire mosaics in southern California and northern Baja California. *Science* 219:1287–1294.

Morgan, P., Aplet, G.H., Haufler, J.B., Humphries, H.C., Moore, M.M., and Wilson, W.D. 1994. Historical range of variability: A useful tool for evaluating ecosystem change. *J. Sust. For.* 2(1/2):87–111.

Morgan, P., Hardy, C., Swetnam, T.W., Rollins, M.G., and Long, D.G. 2001. Mapping fire regimes across time and space: understanding coarse and fine-scale patterns. *Int. J. Wildl. Fire* 10:1–14.

Musters, G.C. 1871. *Vida entre los Patagones.* Buenos Aires: Solar/Hachete.

Overpeck, J.T., Rind, D., and Goldberg, R. 1990. Climate-induced changes in forest disturbance and vegetation. *Nature* 343:51–53.

Parsons, D.J., and Landres, P.B. 1998. Restoring natural fire to wilderness: how are we doing? In *Fire in Ecosystem Management: Shifting the Paradigm from Suppression to Prescription: Tall Timbers Fire Ecology Conference Proceedings*, eds. T.L. Prudent and L.A. Brennan, pp. 366–373. Tallahassee, FL: Tall Timbers Research Station.

Pavlik, B.B. 1996. Defining and measuring success. In *Restoring Diversity*, eds. D.A. Falk, C.I. Millar, and M. Olwell, pp. 127–155. Washington, DC: Island Press.

Pickett, S.T.A., and White, P.S. 1985. *Patch Dynamics and Natural Disturbance Regimes.* New York: Academic Press.

Price, C., and Rind, D. 1994. The impact of a 2 × CO_2 climate on lightning-caused fires. *J. Clim.* 7:1484–1494.

Pyne, S. 1982. *Fire in America.* Princeton: Princeton University Press.

Pyne, S. 1995. *World Fire: The Culture of Fire on Earth.* New York: Holt.

Pyne, S., Andrews, P.L., and Laven, R.D. 1996. *Introduction to Wildland Fire.* New York: Wiley.

Riebau, A.R., and Fox, D. 2001. The new smoke management. *Int. J. Wildl. Fire* 10: 415–427.

Rodríguez, N.F. 1997. Risk of mortality from wildfire in ponderosa pine (*Pinus ponderosa*) plantations in the Andean region of Patagonia, Argentina M.S. thesis. University of Idaho, Moscow.

Rodríguez, N.F. 1999. Incendios forestales, estadísticas, causas, combustibles (descomposición), medidas de prevención y control. In *Curso-taller de actualización en silvicultura de los bosques de ciprés de la Cordillera*, eds. G.A. Loguercio, M. Rajchenberg, N.F. Rodríguez, and P. Pantaenius, Río Negro, Argentina: El Bolsón.

Rollins, M., Swetnam, T.W., and Morgan, P. 2000a. Twentieth-century fire patterns in the Gila/Aldo Leopold Wilderness Complex in New Mexico and the Selway-Bitterroot wilderness area Idaho/Montana. In *Crossing the Millennium: Integrating Spatial Technologies and Ecological Principles for a New Age in Fire Management: Proceedings from the Joint Fire Science Conference and Workshop*, pp. 161–169. Moscow: University of Idaho.

Rollins, M., Swetnam, T.W., and Morgan, P. 2000b. Twentieth-century fire patterns in the Selway-Bitterroot Wilderness Area, Idaho/Montana and the Gila/Aldo Leopold Wilderness complex. In *Wilderness Science in a Time of Change Conference: Wilderness Ecosystems Threats and Management*, vol. 5, comps. D.N. Cole, S.F. McCool, W.T. Borrie, and J. O'Loughlin, pp. 283–287. USDA Forest Service Proceedings RMRS-P-15.

Rollins, M.G., Swetnam, T.W., and Morgan, P. 2001. Evaluating a Century of Fire Patterns in Two Rocky Mountain Wilderness Areas Using Digital Fire Atlases. *Can. J. For. Res.* 31:2107–2123.

Roux, C. 1987. *Las matanzas del Neuquén.* Buenos Aires: Plus Ultra Impresiones Sud Americana Ed.

Schmoldt, D.C., Peterson, D.C., Keane, R.E., Lenihan, J.M., McKenzie, D., Weise, D.R., and Sandberg, D.V. 1999. Assessing the effects of fire disturbance on ecosystems: A scientific agenda for research and management. Gen. Tech. Rep. PNW-455. Portland, OR: USDA Forest Service, Pacific Northwest Research Station.

Secretaría de Recursos Naturales y Desarrollo Sustentable (SRNyDS). 1997, 1998, 1999. *Estadísticas de incendios forestales.* Buenos Aires, Argentina: Dirección Nacional de Desarrollo Sustentable, Dirección de Recursos Forestales Nativos.

Secretaría de Desarrollo Sustentable y Política Ambiental (SDSyPA). 2000. *Estadísticas de incendios forestales.* Buenos Aires, Argentina: Dirección Nacional de Desarrollo Sustentable, Dirección de Recursos Forestales Nativos.

Shinneman, D.J. and Baker, W.L. 1997. Nonequilibrium dynamics between catastrophic disturbances and old-growth forests in ponderosa pine landscapes of the Black Hills. *Conserv. Biol.* 11:1276–1288.

Stephenson, N.L. 1999. Reference conditions for giant sequoia forest restoration: Structure, process and precision. *Ecol. Appl.* 9(4):1253–1265.

Strauss, D., Bednar, L., and Mees, R. 1989. Do one percent of forest fires cause ninety-nine percent of the damage? *For. Sci.* 35:319–328.

Swetnam, T.W. 1993. Fire history and climate change in giant sequoia groves. *Science* 262:885–889.

Swetnam, T.W., and Baisan, C.H. 1996. Historical fire regime patterns in the southwestern United States since A.D. 1700. In *Proceedings of the 2nd La Mesa Fire Symposium*, ed. C.D. Allen, pp. 11–32. Gen. Tech. Rep. RM-GTR-286. Fort Collins, CO: USDA Forest Service, Rocky Mountain Research Station.

Swetnam, T.W., and Betancourt, J.L. 1990. Fire–Southern Oscillation relations in the southwestern United States. *Science* 249:1017–1020.

Swetnam, T.W., and Betancourt, J.L. 1998. Mesoscale disturbance and ecological response to decadal climatic variability in the American Southwest. *J. Clim.* 11:3128–3147.

Swetnam, T.W., Allen, C.D., and Betancourt, J.L. 1999. Applied historical ecology: Using the past to manage for the future. *Ecol. Appl.* 9(4):1189–1206.

Tortorelli, L. 1947. *Los incendios de Bosques en la Argentina.* Buenos Aires: Ministerio de Agricultura de la Nación, Dirección Forestal.

Veblen, T.T., and Alaback, P.B. 1996. A comparative review of forest dynamics and disturbance in the temperate rainforests of North and South America. In *High-Latitude Rainforests and Associated Ecosystems of the West Coast of the Americas: Climate Hydrology, Ecology and Conservation*, eds. R.G. Lawford, P.B. Alaback, and E. Fuentes, pp. 173–213. New York: Springer-Verlag.

Veblen, T.T., and Lorenz, D.C. 1988. Recent vegetation changes along the forest/steppe ecotone in northern Patagonia. *Ann. Assoc. Am. Geog.* 78:93–111.

Veblen T.T., Kitzberger T., Villalba, R., and Donnegan, J. 1999. Fire history in northern Patagonia: The roles of humans and climatic variation. *Ecol. Monogr.* 69:47–67.

Weber, M.G., and Flannigan, M.D. 1997. Canadian boreal forest ecosystem structure and function in a changing climate: Impact on fire regimes. *Environ. Rev.* 5:145–166.

Weber, M.G., and Stocks, B.J. 1998. Forest fires and sustainability in the boreal forests of Canada. *Ambio* 27(7):545–550.

White, P.S., and Walker, J.L. 1997. Approximating nature's variation: selecting and using reference information in restoration ecology. *Conserv. Biol.* 5(4):338–349.

Wright, H.A., and Bailey, A.W. 1982. *Fire Ecology, United States and Southern Canada.* New York: Wiley.

Index

Ecological Studies

Volumes published since 1996

Volume 119
Freshwaters of Alaska: Ecological Synthesis (1997)
A.M. Milner and M.W. Oswood (Eds.)

Volume 120
Landscape Function and Disturbance in Arctic Tundra (1996)
J.F. Reynolds and J.D. Tenhunen (Eds.)

Volume 121
Biodiversity and Savanna Ecosystem Processes: A Global Perspective (1996)
O.T. Solbrig, E. Medina, and J.F. Silva (Eds.)

Volume 122
Biodiversity and Ecosystem Processes in Tropical Forests (1996)
G.H. Orians, R. Dirzo, and J.H. Cushman (Eds.)

Volume 123
Marine Benthic Vegetation: Recent Changes and the Effects of Eutrophication (1996)
W. Schramm and P.H. Nienhuis (Eds.)

Volume 124
Global Change and Arctic Terrestrial Ecosystems (1996)
W.C. Oechel (Ed.)

Volume 125
Ecology and Conservation of Great Plains Vertebrates (1997)
F.L. Knopf and F.B. Samson (Eds.)

Volume 126
The Central Amazon Floodplain: Ecology of a Pulsing System (1997)
W.J. Junk (Ed.)

Volume 127
Forest Design and Ozone: A Comparison of Controlled Chamber and Field Experiments (1997)
H. Sanderman, A.R. Wellburn, and R.L. Heath (Eds.)

Volume 128
The Productivity and Sustainability of Southern Forest Ecosystems in a Changing Environment (1998)
R.A. Mickler and S. Fox (Eds.)

Volume 129
Pelagic Nutrient Cycles (1997)
T. Andersen

Volume 130
Vertical Food Web Interactions (1997)
K. Dettner, G. Bauer, and W. Völkl (Eds.)

Volume 131
The Structuring Role of Submerged Macrophytes in Lakes (1998)
E. Jeppesen, M. Søndergaard, M. Søndergaard, and K. Christoffersen (Eds.)

Volume 132
Vegetation of the Tropical Pacific Islands (1998)
D. Mueller-Dombois and F.R. Fosberg

Volume 133
Aquatic Humic Substances (1998)
D.O. Hessen and L.J. Tranvik

Volume 134
Oxidant Air Pollution Impacts in the Montane Forests of Southern California: A Case Study of the San Bernardino Mountains (1999)
P.R. Miller and J.R. McBride (Eds.)

Volume 135
Predation in Vertebrate Communities: The Bialowieza Primeval Forest as a Case Study (1998)
B. Jedrzejewska and W. Jedrzejewski

Volume 136
Landscape Disturbance and Biodiversity in Mediterranean-Type Ecosystems (1998)
P.W. Rundel, G. Montenegro, and F.M. Jaksic (Eds.)

Volume 137
Ecology of Mediterranean Evergreen Oak Forests (1999)
F. Roda, J. Retana, C.A. Gracia, and J. Bellot (Eds.)

Volume 138
Fire, Climate Change, and Carbon Cycling in the Boreal Forest (2000)
E.S. Kasischke and B.J. Stocks (Eds.)

Volume 139
Responses of Northern U.S. Forests to Environmental Change (2000)
R.A. Mickler, R.A. Birdsey, and J. Hom (Eds.)

Volume 140
Rainforest Ecosystems of East Kalimantan: El Niño, Drought, Fire and Human Impacts (2000)
E. Guhardja et al. (Eds.)

Volume 141
Activity Patterns in Small Mammals: An Ecological Approach (2000)
S. Halle and N.-C. Stenseth (Eds.)

Volume 142
Carbon and Nitrogen Cycling in European Forest Ecosystems (2000)
E.-D. Schulze (Ed.)

Volume 143
Global Climate Change and Human Impacts on Forest Ecosystems: Postglacial Development, Present Situation and Future Trends in Central Europe (2001)
P. Joachim and U. Bernhard

Volume 144
Coastal Marine Ecosystems of Latin America (2001)
U. Seeliger and B. Kjerfve (Eds.)

Volume 145
Ecology and Evolution of the Freshwater Mussels Unionoida (2001)
B. Gerhard and W. Klaus (Eds.)

Volume 146
Inselbergs: Biotic Diversity of Isolated Rock Outcrops in Tropical and Temperate Regions (2001)
S. Porembski and W. Barthlott (Eds.)

Volume 147
Ecosystem Approaches to Landscape Management in Central Europe (2001)
J.D. Tenhunen, R. Lenz, and R. Hantschel (Eds.)

Volume 148
A Systems Analysis of the Baltic Sea (2001)
F.V. Wulff, L.A. Rahm, and P. Larsson (Eds.)

Volume 149
Banded Vegetation Patterning in Arid and Semiarid Environments: Ecological Processes and Consequences for Management (2001)
D.J. Tongway, C. Valentin, and J. Seghieri (Eds.)

Volume 150
Biological Soil Crusts: Structure, Function and Management (2001)
J. Belnap and O.L. Lange (Eds.)

Volume 151
Ecological Comparisons of Sedimentary Shores (2001)
K. Reise (Ed.)

Volume 152
Global Biodiversity in a Changing Environment: Scenarios for the 21st Century (2001)
F.S. Chapin III, O.E. Sala, and E. Huber-Sannwald (Eds.)

Volume 153
UV Radiation and Arctic Ecosystems (2002)
D.O. Hessen (Ed.)

Volume 154
Geoecology of Antarctic Ice-Free Coastal Landscapes (2002)
L. Beyer and M. Bölter (Eds.)

Volume 155
Conserving Biodiversity in East African Forests: A Study of the Eastern Arc Mountains (2002)
W.D. Newmark (Ed.)

Volume 156
Urban Air Pollution and Forests: Resources at Risk in the Mexico City Air Basin (2002)
M.E. Fenn, L.I. de Bauer, and T. Hernández-Tejeda (Eds.)

Volume 157
Mycorrhizal Ecology (2002)
M.G.A. van der Heijden and I.R. Sanders

Volume 158
Diversity and Interaction in a Temperate Forest Community: Ogawa Forest Reserve of Japan (2002)
T. Nakashizuka and Y. Matsumoto

Volume 159
Big-Leaf Mahogany: Genetics, Ecology, and Management (2003)
A.E. Lugo, J.C. Figueroa Colón, and M. Alayón (Eds.)

Volume 160
Fire and Climatic Change in Temperate Ecosystems of the Western Americas (2003)
T.T. Veblen, W.L. Baker, G. Montenegro, and T.W. Swetnam (Eds.)